NanoScience and Technology

The series NanoScience and Technology is focused on the fascinating nano-world, mesoscopic physics, analysis with atomic resolution, nano and quantum-effect devices, nanomechanics and atomic-scale processes. All the basic aspects and technology-oriented developments in this emerging discipline are covered by comprehensive and timely books. The series constitutes a survey of the relevant special topics, which are presented by leading experts in the field. These books will appeal to researchers, engineers, and advanced students.

Liangbing Hu · Feng Jiang · Chaoji Chen
Editors

Emerging Nanotechnologies in Nanocellulose

Springer

Editors
Liangbing Hu
Department of Materials Science
and Engineering
University of Maryland
College Park, MD, USA

Feng Jiang
Department of Wood Science
University of British Columbia
Vancouver, BC, Canada

Chaoji Chen
School of Resource and Environmental
Science
Wuhan University
Wuhan, Hebei, China

ISSN 1434-4904 ISSN 2197-7127 (electronic)
NanoScience and Technology
ISBN 978-3-031-14042-6 ISBN 978-3-031-14043-3 (eBook)
https://doi.org/10.1007/978-3-031-14043-3

This Springer imprint is published by the registered company Springer Nature Switzerland AG
The registered company address is: Gewerbestrasse 11, 6330 Cham, Switzerland

Preface

As the most abundant biopolymer on earth, cellulose has been serving as an essential raw material for thousands of years in supporting human well-being. Although the use of cellulose has experienced a declining trend with the introduction of high-performance synthetic polymers, we are now witnessing the revival of this old-school material to combat the negative ecological and environmental effects caused by the synthesis of polymers. On the other hand, the revival of cellulose is fundamentally supported by the advances in nanotechnologies, which unveils the potential of using cellulose to build high-performance materials. With such technical advancement, cellulose has been employed in more emerging applications covering optics, electronic, energy, environment, and health. Therefore, we find it is timely to compile the emerging nanotechnologies in nanocellulose.

In this book, the state-of-art nanotechnologies used in nanocellulose are summarized, with topics covering nanocellulose isolation, advanced structural design, mechanical engineering, and the emerging applications in light management, energy storage and harvesting, electronics, environmental remediation, water treatment, and biomedical applications. It is aimed to serve as a reference book for researchers and graduate students working in the field of nanocellulose. Each chapter was contributed by prominent researchers in their respective field, and the chapter was structured in a way not only to summarize the up-to-date research in the field but also to provide design principles and fundamental knowledge of the topic.

Chapter 1 discusses current understanding of plant cell wall biosynthesis and cellulose ultrastructure prerequisites for understanding nanocellulose isolation and also briefly summarizes common characterization of nanocellulose; Chap. 2 discusses the top-down processing of nanocellulose materials and their potential applications in various fields; Chap. 3 focuses on the recent developments in the design, synthesis, and application of novel bacterial nanocellulose-based multifunctional nanocomposites; Chap. 4 discusses the development of aerogels using nanocellulose as a building block, focusing on fabrication, structural control, and advanced applications; Chap. 5 discusses the design principle for making superstrong materials using nanocellulose as well as modeling approaches to explain the multiscale mechanics; Chap. 6 discusses using nanocellulose film for light management and the potential applications in photonic fields; Chap. 7 discusses using nanocellulose paper for the fabrication of nanocellulose-based electronics; Chap. 8 discusses the function of electrochemical energy storage devices and how nanocellulose can be used in the manufacturing of such devices; Chap. 9 describes recent advances in lignocellulose as a building block for advanced batteries materials; Chap. 10 provides a comprehensive overview of valorizing nanocellulose for water treatment through membrane filtration, absorption/adsorption, and solar-assisted water generation; Chap. 11 reviews the recent research on the functionalization, fabrication, and assembly of CNFs aiming at triboelectric nanogenerator (TENG) development for mechanical energy harvesting; Chap. 12 overviews the synthesis of different types of nanocellulose for the development of nanocellulose-based functional biomaterials for various biomedical applications.

College Park, USA Liangbing Hu
Vancouver, Canada Feng Jiang
Wuhan, China Chaoji Chen

Contents

Contributors

Adhikari Manjila Biofuels Institute, School of the Environment and Safety Engineering, Jiangsu University, Zhenjiang, China

Agarwal Umesh P. U.S. Department of Agriculture Forest Service, Forest Products Laboratory, Madison, WI, USA

Atta Omar Mohammad Biofuels Institute, School of the Environment and Safety Engineering, Jiangsu University, Zhenjiang, China

Cao Sisi Intel Corporation, Hillsboro, OR, USA

Chen Chaoji Hubei Key Laboratory of Biomass Resource Chemistry and Environmental Biotechnology, School of Resource and Environmental Science, Wuhan University, Wuhan, China

Chen Li-Feng CAS Key Laboratory of Mechanical Behavior and Design of Materials (LMBD), Department of Thermal Science and Energy Engineering, Hefei National Laboratory for Physical Sciences at the Microscale, University of Science and Technology of China, Hefei, Anhui, P. R. China

Chen Wenshuai Key Laboratory of Bio-Based Material Science and Technology, Ministry of Education, Northeast Forestry University, Harbin, P. R. China

Dai Ming Key Laboratory of Bio-Based Material Science and Technology, Ministry of Education, Northeast Forestry University, Harbin, P. R. China

Fang Zhiqiang State Key Laboratory of Pulp and Paper Engineering, South China University of Technology, Guangzhou, China

Farooq Umer Advanced Biomaterials & Tissue Engineering Centre, Huazhong University of Science and Technology, Wuhan, PR China

Gan Wentao Key Laboratory of Bio-Based Material Science and Technology of Ministry of Education, Northeast Forestry University, Harbin, China

Gu Long Department of Materials Science and Engineering, University of Wisconsin-Madison, Madison, WI, USA

Hou Gaoyuan State Key Laboratory of Pulp and Paper Engineering, South China University of Technology, Guangzhou, China

Jiang Feng Sustainable Functional Biomaterial Lab, Department of Wood Science, the University of British Columbia, Vancouver, BC, Canada

Jiang Qisheng Auragent Bioscience LLC, St. Louis, MO, USA

Kim Jung-Hwan School of Energy and Chemical Engineering, Ulsan National Institute of Science and Technology (UNIST), Ulsan, Republic of Korea

Kuang Yudi School of Biomedical Sciences and Engineering, South China University of Technology, Guangzhou International Campus, Guangzhou, P. R. China;
National Engineering Research Center for Tissue Restoration and Reconstruction, South China University of Technology, Guangzhou, P. R. China

Lee Donggue School of Energy and Chemical Engineering, Ulsan National Institute of Science and Technology (UNIST), Ulsan, Republic of Korea

Lee Sang-Young Department of Chemical and Biomolecular Engineering, Yonsei University, Seoul, Republic of Korea

Li Guanhui State Key Laboratory of Pulp and Paper Engineering, South China University of Technology, Guangzhou, China

Li Qing Key Laboratory of Bio-Based Material Science and Technology, Ministry of Education, Northeast Forestry University, Harbin, P. R. China

Li Teng Department of Mechanical Engineering, University of Maryland, College Park, MD, USA

Li Zhaoyang Department of Electromechanical Engineering and Centre for Artificial Intelligence and Robotics, University of Macau, Macau SAR, China

Ling Shengjie School of Physical Science and Technology, ShanghaiTech University, Shanghai, P. R. China

Manan Sehrish Biofuels Institute, School of the Environment and Safety Engineering, Jiangsu University, Zhenjiang, China;
Department of Biomedical Engineering, College of Life Science and Technology, Huazhong University of Science and Technology, Wuhan, PR China

Nyholm Leif Department of Chemistry-Ångström Laboratory, Uppsala University, Uppsala, Sweden

Pang Zhenqian Department of Mechanical Engineering, University of Maryland, College Park, MD, USA

Qiu Xueqing School of Chemical Engineering and Light Industry, Guangdong Provincial Key Laboratory of Plant Resources Biorefinery, Guangdong University of Technology, Guangzhou, China

Ray Upamanyu Department of Mechanical Engineering, University of Maryland, College Park, MD, USA

Shahzad Ajmal Biofuels Institute, School of the Environment and Safety Engineering, Jiangsu University, Zhenjiang, China

Singamaneni Srikanth Department of Mechanical Engineering and Materials Science, Institute of Materials Science and Engineering, Washington University in St. Louis, St. Louis, MO, USA

Ul-Islam Mazhar Department of Chemical Engineering, College of Engineering, Dhofar University, Salalah, Oman

Ullah Muhammad Wajid Biofuels Institute, School of the Environment and Safety Engineering, Jiangsu University, Zhenjiang, China;
Department of Biomedical Engineering, College of Life Science and Technology, Huazhong University of Science and Technology, Wuhan, PR China

Wang Xudong Department of Materials Science and Engineering, University of Wisconsin-Madison, Madison, WI, USA

Wang Zhaohui College of Materials Science and Engineering, Hunan University, Changsha, China

Xia Qinqin Key Laboratory of Bio-Based Material Science and Technology of Ministry of Education, Northeast Forestry University, Harbin, China

Yang Guang Biofuels Institute, School of the Environment and Safety Engineering, Jiangsu University, Zhenjiang, China

Yao Chunhua Department of Materials Science and Engineering, University of Wisconsin-Madison, Madison, WI, USA

Yu Haipeng Key Laboratory of Bio-Based Material Science and Technology, Ministry of Education, Northeast Forestry University, Harbin, P. R. China

Zhong Junwen Department of Electromechanical Engineering and Centre for Artificial Intelligence and Robotics, University of Macau, Macau SAR, 999078 China;
Wuhan National Laboratory for Optoelectronics, Huazhong University of Science and Technology, Wuhan, 430074 China

Zhou Jun Wuhan National Laboratory for Optoelectronics, Huazhong University of Science and Technology, Wuhan, China

Zhu J. Y. U.S. Department of Agriculture Forest Service, Forest Products Laboratory, Madison, WI, USA

Zhu Penghui Sustainable Functional Biomaterial Lab, Department of Wood Science, the University of British Columbia, Vancouver, BC, Canada

Zhu Shuze Department of Engineering Mechanics, Zhejiang University, Hangzhou, China

Zhu Yeling Sustainable Functional Biomaterial Lab, Department of Wood Science, the University of British Columbia, Vancouver, BC, Canada

Chapter 1
Nanocellulose: Native State, Production, and Characterization

J. Y. Zhu and Umesh P. Agarwal

Abstract This chapter first briefly discusses the current understanding of plant cell wall biosynthesis and cellulose ultra-structure prerequisites for understanding nanocellulose. From this fundamental understanding, the chapter is devoted to discussing conventional methods for producing two predominant types of cellulose nanomaterials (CNMs), namely cellulose nanocrystals (CNCs) and cellulose nanofibrils (CNFs). Although many types of chemical treatments can deconstruct lignocellulose into nanoscale materials, mechanistic analyses based on the understanding of cell wall structure indicated that oxidation and acid hydrolysis are the most efficient approaches for producing CNMs from delignified wood pulp fibers. Some degree of delignification is necessary for producing CNMs from raw lignocellulose. Recent advances in achieving sustainable production of CNMs are also briefly mentioned. This chapter also briefly covers methods for characterizing various properties of CNMs.

Nomenclature

BEP	Bleached kraft eucalyptus pulp
CDF	Combined delignification factor, (1.3)
CHF_X	Combined hydrolysis factor for xylan (X), (1.5)
CHF_G	Combined hydrolysis factor for cellulose (G, glucan), (1.1)
CNCs	Cellulose nanocrystals, refers to a class of individually separated short crystalline cellulosic nanoparticles with aspect ratios of approximately 25 or smaller and good crystallinity. CNCs are often produced by concentrated acid hydrolysis unless high cellulose crystallinity is demonstrated
LCNCs	CNCs containing lignin

J. Y. Zhu (✉) · U. P. Agarwal
U.S. Department of Agriculture Forest Service, Forest Products Laboratory, One Gifford Pinchot Dr., Madison, WI 53726, USA
e-mail: junyong.zhu@usda.gov

L. Hu et al. (eds.), *Emerging Nanotechnologies in Nanocellulose*, NanoScience and Technology, https://doi.org/10.1007/978-3-031-14043-3_1

S-CNCs	CNCs produced by hydrolysis using concentrated sulfuric acid (S)
CNFs	Cellulose nanofibrils refers to nanoscale cellulosic fibrils that can be either individually separated or physically entangled or interconnected as fibril networks
LCNFs; T-CNFs	CNFs containing lignin; TEMPO-mediated oxidation (T)
CNMs	Cellulose-based nanomaterials
LCNMs	CNMs containing lignin
CNWs	Cellulose nanowhiskers (short nanofibrils) with morphology similar to CNCs
LCNWs	Lignin containing CNWs.
CSR	Cellulosic solid residues from (acid or enzymatic) hydrolysis of cellulosic materials
DP	Cellulose degree of polymerization
L_R	Fraction of lignin retained on fractionated lignocellulosic solids, (1.4).
X_R	Fraction of xylan retained on fractionated lignocellulosic solids, (1.6).
LNPs	Lignin nanoparticles
NPMs	Nanoscale (lignocellulosic) polymeric materials

1.1 Introduction on Cellulose Nanomaterials

Cellulose is a renewable polymer and abundant in many regions of the world. Except in cotton, with cellulose content of 95%, cellulose coexists with hemicelluloses, lignin, and other minor components such as extractives in lignocellulosic biomass. This facilitates the utilization of cellulose for interesting and diverse applications, but the process of separating cellulose from the other components may prove difficult (as discussed later in this section). Typical cellulose content in wood is approximately 40–45% [95, 144] and varies with wood species.

In plant cell walls, cellulose is synthesized by plasma-membrane-localized cellulose synthase complexes [80]. A synthesized elementary cellulose fibril consists of many (16–36) cellulose chains [27, 35, 40, 103]. Each cellulose chain is made of thousands of glucan units connected by β-1–4 linkage. For wood, cellulose chain length or degree of polymerization is on the order of 10,000 [110]. Cellulose chains in a fibril are well organized and contain intrachain and interchain molecular hydrogen bonds (H-bonds), i.e., intrachain H-bonds between O(6) and O(2)H and between O(3)H and ring O(5) (Fig. 1.1a) and interchain H-bonds between O(3) and O(6)H (Fig. 1.1a). The cellulose chain conformation is stabilized by these two types of H-bonds. The inter-cellulose chain hydrogen bonding aggregates elementary cellulose fibrils to form large cellulose fibrils at various scales (or sizes) (Fig. 1.2), from a few nanometers (elementary fibrils) to tens of nanometers (nanofibrils) and submicron

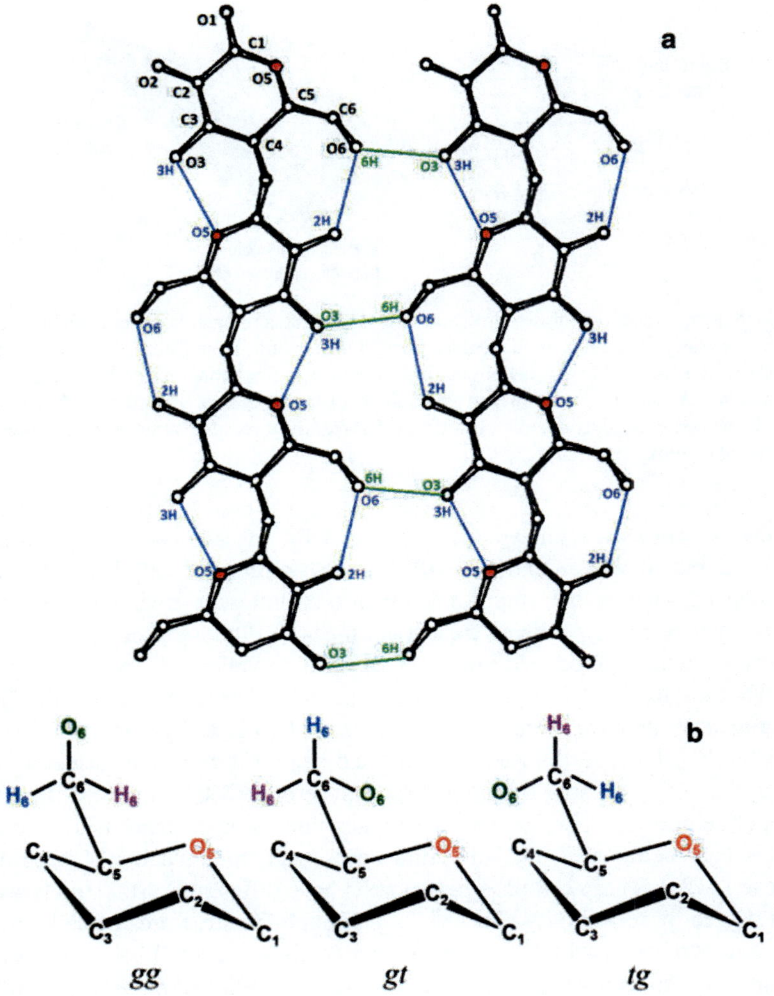

Fig. 1.1 Schematics of cellulose molecular structure. **a** intra- and interchain hydrogen bonding, adapted from [48] with permission from Elsevier©, **b** local conformations of the cellulose $C(6)H_2OH$ group [139] with permission from American Chemical Society©

fibrils (microfibrils). Note that the traditional terminology of "microfibrils" often referred to fibrils in nanometer scale [40, 44]. This traditional terminology became confusing when the recent terminology of "cellulose nanofibrils" came along. In this chapter, the term "microfibrils" means microfibrillated cellulose with dimensions from submicrometer to a few micrometers.

The extensive intra- and inter-cellulose chain hydrogen bonding makes cellulose very stable and insoluble in water and in many solvents. The hydroxyl groups of neighboring cellulose molecules form chains of hydrogen bonds and are the cause of a fiber's tensile strength. Cellulose exists in several forms and is polymorphic.

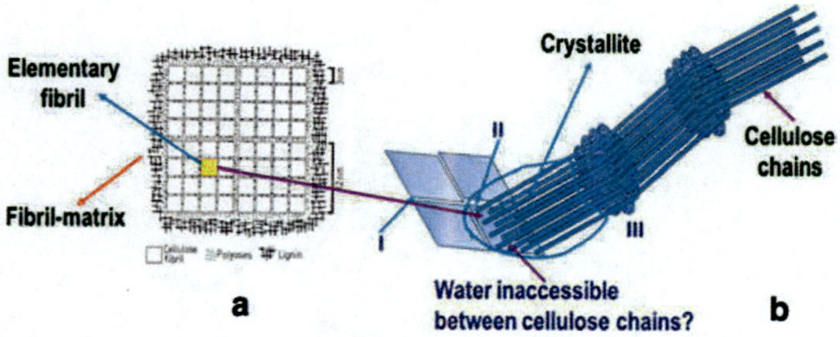

Fig. 1.2 Schematics of cellulose ultra-structure: **a** Radial aggregation in cell walls (through hydrogen bonding) of cellulose elementary fibrils into nanofibrils and fibril matrices from [40] with permission from TAPPI©, **b** Longitudinal cellulose aggregation with crystalline and disordered regions (drawn based on Rowland and Roberts [103]), I: coalesced surface with high order, II: readily accessible slightly disordered surfaces, III: readily accessible surfaces of strain-disorder tilt and twist regions

Four distinct forms are known—I, II, III, and IV. In cellulose I two crystalline forms—Iα and Iβ were identified [10]. The crystallinity of cellulose I materials varies depending upon their source, and for wood cellulose, there is a question about its native state being crystalline or noncrystalline [4]. Cellulose II, or regenerated and mercerized cellulose, produced from solubilized and swelled cellulose, respectively, is another most used polymorph. Between these two polymorphic phases, the degree of hydrogen bonding is different and the chain and $C(6)H_2OH$ group conformations also differ (Fig. 1.1b). Cellulose III (produced by liquid ammonia or various amines) and cellulose IV (prepared by treatment in glycerol at 260 °C) are the additional phases of cellulose, although some researchers think that the latter is a mixture of cellulose I and cellulose II (Dr. R.H. Atalla, personal communication) or disordered cellulose I [122]. These cellulose polymorphs have different crystal structures and varied degree of crystallinity (Table 1.1). Although the structures of celluloses Iβ and II consist of two-chain $P2_1$ monoclinic unit cells, cellulose II has an antiparallel structure with chains stacked in opposite polarity [65]. Cellulose polymorphs differ in several properties, including morphology, density, crystallinity, moisture absorption, and mechanical properties, which influence the applications of these materials.

The preceding discussion indicates that cellulosic materials in plants have a hierarchical structure from nanoscale fibrils, to micrometer fibers, to tens of centimeters (diameter) wood logs. The latter are used in the production of cellulose nanomaterials (CNMs). This cellulose structure in cell walls shows that nanoscale cellulosic materials exist in plant biomass. Separating these nanoscale cellulose fibrils results in a natural polymeric CNM with greatly increased interfacial surface area and therefore with very strong hydrogen bonding. Given the facts that plant biomass is renewable and biodegradable and cellulose has great mechanical strength [79, 113], CNMs as a class of nanomaterials have substantial advantages for a variety of applications (e.g., hydrogels, aerogels, emulsions, nanocomposites, polymer grafting, electronics, drug delivery, cements) compared with traditional nonrenewable polymeric materials [70],

Table 1.1 Cellulose polymorphs, crystal structures

Polymorph	Space group	Number of chains	References
Cellulose Iα	P1	1	[89]
Cellulose Iβ	$P2_1$	2	[88]
Cellulose II	$P2_1$	2	[65]
Cellulose III_I*	$P2_1$	1	[123]
Cellulose IV_I**	P1	2	[47]
Cellulose IV_{II}**	P1	2	[47]

*Cellulose III_I is made from cellulose I; ** Celluloses IV_I and IV_{II} are made, respectively, from cellulose I and cellulose II

when it can be produced using sustainable separation methods [141]. This potential has attracted great interest in CNMs in the broad scientific community.

The vast amount of strong hydrogen bonding in the molecular structure of cellulose and the matrix barriers from hemicelluloses and lignin make separating nanoscale cellulosic fibrils from plant biomass difficult. Historically, extensive mechanical fibrillation with significant energy input was first applied to successfully separate micro- to nanoscale cellulose fibrils, or cellulose nanofibrils (CNFs), from cellulose fibers [119]. With the application of enzymes or chemicals, the extent of fibrillation and therefore the energy input for fibrillation was substantially decreased [53, 96, 107]. Cellulose nanocrystals (CNCs), a special class of CNMs, were discovered in cellulose hydrolysis experiments using concentrated mineral acid solutions [11, 86, 105]. Acid hydrolysis can isolate crystallized cellulose, and by solubilizing less ordered cellulose, it can yield only highly crystalline cellulose particles. CNCs are generally a type of CNM with a typical length of 50–500 nm, degree of polymerization (DP) between 100 and 300 [12, 87, 105], and diameter of 3–10 nm, with a high degree of crystallinity. CNCs as crystalline and individually separated rod-like particles also have liquid crystal properties [37, 73, 100] that may be well suited for optical applications [8, 38] in addition to conventional applications in rheology [90, 92] and composites [39, 135].

Typically, both CNCs and CNFs do not contain lignin but may very well contain some levels of hemicelluloses because they are mostly produced from commercial bleached chemical pulps. Recently, lignin-containing cellulose nanomaterials (LCNMs) have shown some interesting properties resulting from the presence of lignin. For example, LCNMs generally have some hydrophobicity, which is favorable for most composite processing. LCNMs also can absorb UV light. Producing LCNMs, such as lignin-containing cellulose nanocrystals (LCNCs) [5, 17] or lignin-containing cellulose nanofibrils (LCNFs) [17, 111], also has the advantage of using less-processed fibers, i.e., without bleaching, or even using untreated lignocelluloses directly [18, 24, 76].

Cellulose as a natural polymer in plant biomass is renewable and biodegradable. However, developing environmentally sustainable and cost-effective processes for separating CNMs and/or LCNMs from plant biomass is the key to achieving true sustainability and reaping the full benefits of cellulose for a variety of applications. In the context of CNMs, the objectives of this chapter are to (1) present a detailed description of several existing CNM and LCNM production technologies (giving readers a general understanding of nanomaterials from lignocelluloses), along with a brief discussion of developed processes that have the potential of achieving sustainability (giving readers a better understanding of barriers for commercial utilization of cellulose nanomaterials), and (2) present some currently used state-of-the-art methods in the characterization of CNM.

1.2 Producing Cellulose Nanomaterials

CNCs are supposed to be highly crystalline. The term CNC is not scientifically defined. In other words, the qualified range of crystallinity is not specified. In some literature, the rod-like CNMs produced by enzymatic hydrolysis [7] or oxidation [54, 68, 81, 82] or through sonic fibrillation [116] were termed as CNCs. The limited accuracy in measuring of CNC crystallinity by the commonly used x-ray diffraction technique [93, 108] causing more difficulties in defining CNCs. Furthermore, all existing crystallinity measurements using X-ray diffraction [108], NMR [66], and Raman [6] are spatially averaged and cannot determine the crystallinity of individual cellulose nanofibril/nanocrystal. Based on the crystallite model of natural cellulose (cellulose I) shown in Fig. 1.2b [44, 103], CNCs are isolated crystallites with the same diameter of elementary fibrils. Because most reported CNCs were from cellulosic materials derived from chemical pulping which promoted cellulose chain consolidation, hornification, and crystallization [3].

1.2.1 Concentrated Mineral Acid Hydrolysis for Producing Cellulose Nanocrystals (CNCs)

Early studies on cellulose hydrolysis associated with CNC production used aqueous sulfuric acid solution at 2.5 N (~22 wt%) and boiling temperature with a long hydrolysis time [86, 105]. This approach achieved a CNC yield of approximately 30–40%. A later study found that cellulose can be hydrolyzed into dispersible colloidal nanoparticles when sulfuric acid concentration was increased to 65 wt% [84]. The improved dispersion, which resulted in a stable colloidal suspension (Fig. 1.3a, b), is due to the formation of a half-ester sulfate group (HSO_3) at the cellulose C6 position during cellulose hydrolysis by sulfuric acid at high concentrations (schematically shown in Fig. 1.3c). The sulfate group provides CNCs with electrostatic repulsion to disperse, an advantage for aqueous processing. Since then, sulfuric acid concentration at approximately 64 wt% has been used as the standard for CNC production. Reaction time was decreased to approximately 60 min at approximately 50 °C [15, 21, 29, 32,

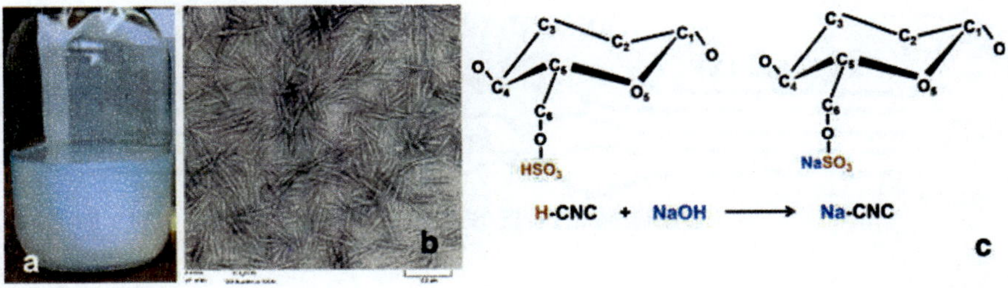

Fig. 1.3 **a** A photo of CNC colloidal suspension from 64% sulfuric acid hydrolysis of a commercial bleached pulp from [4]; **b** A TEM image showing well-dispersed CNCs (scale bar = 0.2 μm) from [29]; **c** scheme showing the conversion of acid form CNC (H-CNC) to sodium form of CNC (Na-CNC) through neutralization using NaOH from [139] with permission from American Chemical Society©

52, 83, 99, 130]. Through neutralization using sodium hydroxide, the HSO_3 group at the C6 position on CNCs can be converted to a sodium form ($NaSO_3$). For convenience, CNCs with the original form of half-ester sulfate group (HSO_3) are called acid form CNCs (H-CNC) and the hydroxide neutralized CNCs are called sodium form CNCs (Na-CNC) (Fig. 1.3c). The neutralized sodium form CNCs also have greater thermal stability than the acid form [121]. Although hydrochloric acid [9, 11, 86, 136], hydrochloric acid vapor [64], and phosphoric acid [26, 120] were also used for CNC production, the low cost of sulfuric acid together with the formation of the half-ester sulfate group (concentrated phosphoric acid can also form half-ester group) made sulfuric acid the preferred choice for CNC production. That is the case not only at laboratory scales but also at pilot and commercial scales [34, 98]. Despite maturity of the concentrated sulfuric acid hydrolysis process for producing CNCs, two key technical issues were not addressed for nearly 70 years: low CNC yield of only approximately 50% and sulfuric acid recovery. Both are detrimental to achieving sustainability in terms of cost and environmental impact. Here, we present some recent progress on CNC yield improvement and tailoring CNC properties.

As early as in the 1950s, Battista [11] reported that drastic hydrolysis conditions resulted in significant cellulose loss to sugars, whereas mild acid hydrolysis conditions can induce cellulose crystallization to result in decreased or no loss of cellulose. He proposed a cellulose crystallization model and a degradation model as shown in Fig. 1.4 for mild and drastic hydrolysis, respectively. Furthermore, Battista found that when cellulosic materials were first pretreated under mild hydrolysis conditions and then by drastic hydrolysis conditions, it resulted in significantly lower cellulose loss, which indirectly substantiated the cellulose crystallization model under mild conditions. Unfortunately, this earlier work was not successfully applied to improve CNC yield, despite several CNC yield optimization studies that were conducted through the factorial design of experiments in a range of conditions. These optimization studies failed to find the so-called "mild conditions" to improve CNC yield [21, 52]. Partly because the "mild condition" that Battista referred to was merely for decreasing cellulose loss, it was not the same condition for improving CNC yield

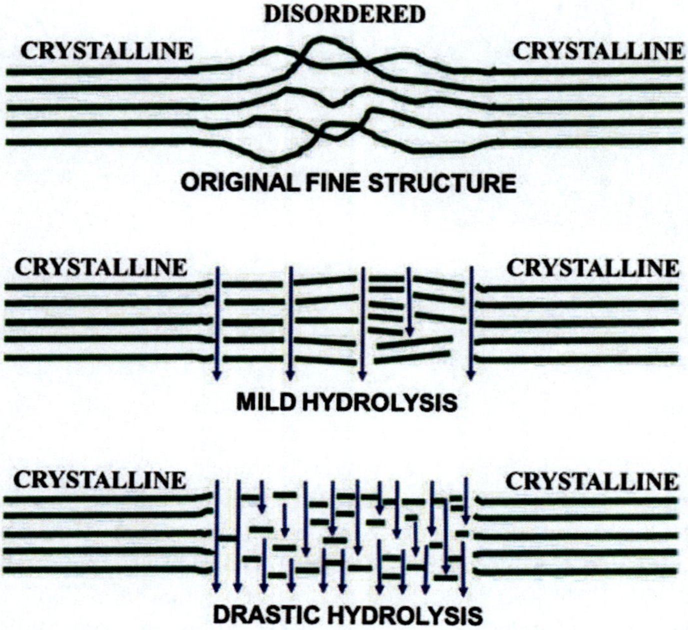

Fig. 1.4 Cellulose hydrolysis models using concentrated mineral acid, redrawn from [11] with permission from American Chemical Society©

because the insolubilized cellulose (from the Battista "mild condition") still needed to be sufficiently depolymerized for it to become CNCs. Upon H_2SO_4 hydrolysis, cellulose depolymerization/degradation, cellulosic solids sulfation, and production of CNCs took place abruptly and simultaneously [52]. This suggests that the relationship of CNC yield with respect to the independent hydrolysis variables of acid concentration, reaction temperature, and reaction time may be difficult to resolve by conventional statistical experimental design and linear regression data analysis. Under CNC production conditions, the rapid hydrolysis reaction at high acid concentrations makes CNC optimization extremely difficult if not impossible, especially when a very large acid step was used according to the conventional statistical experimental design, such as approximately 10 and 24 wt% in the work of Bondeson et al. [21] and [52], respectively.

Until recently, an optimization by Zhu and others at the USDA Forest Products Laboratory was focused on small ranges of acid concentrations between 55 and 66 wt% and temperatures between 40 and 60 °C after the first round of optimization [130]. The study discovered a very narrow operating window, e.g., acid between 58 and 60 wt% at approximately 54–58 °C, which not only significantly decreased cellulose loss to less than 10% but also produced a good CNC yield of approximately 55–60%, with approximately 10–20% of original cellulose in the form of cellulosic solid residues (CSR) rather than soluble sugars. This illustrates that a slightly lower acid concentration of 55–60 wt%, compared with the standard 64 wt%, may crystallize cellulose and preserve cellulose from degradation into sugars, as suggested by Battista (Fig. 1.4). The crystallized cellulose has a significantly lower hydrolysis rate than disordered cellulose [137], which results in CSR. CNC yield might be further improved by further hydrolyzing the CSR either in a separate step or simply in the same step by slightly tuning the hydrolysis condition. The CSR can also be fibrillated into CNFs with low energy input [128, 130] as they have already been substantially hydrolyzed. Therefore, this study by Zhu's group verified the co-existence of CNC and non-CNC cellulosic solids (CSR) and provided opportunity to achieve optimization of CNC yield.

A first-order reaction-kinetics-based cellulose hydrolysis model for predicting CNC and CSR yield and other major degradation products such as sugars and furans were developed [126]. As shown in Fig. 1.5, three routes can be used to describe acid hydrolysis of cellulose for CNC production: (1) The predominant route, represented by reaction rate constant k_1, is crystalline cellulose hydrolysis into CNCs. The CNCs can be further hydrolyzed into glucose at a rate constant k_2. (2) The second route is the direct hydrolysis of disordered cellulose into glucose at rate constant k_3; this fraction of cellulose has a very fast hydrolysis rate. (3) The hydrolysis of hemicelluloses to sugars such as xylan to xylose at rate constat k_4. The glucose from routes 1 and 2 and the hemicellulosic sugars from route 3 can be further degraded into furan such as hydroxymethylfurfural (HMF) and furfural at rate constants k_4 and k_6, respectively. With the assumption of first-order kinetics, the CNC and CSR yields can be predicted as

$$Y_{CNC} = \frac{C_{Cel0} k_1 \gamma}{C_{CSR0}(k_1 + k_3 - k_2)} \{\exp(-k_2 t) - exp[-(k_1 + k_3)t]\} \times 100\% \quad (1.1)$$

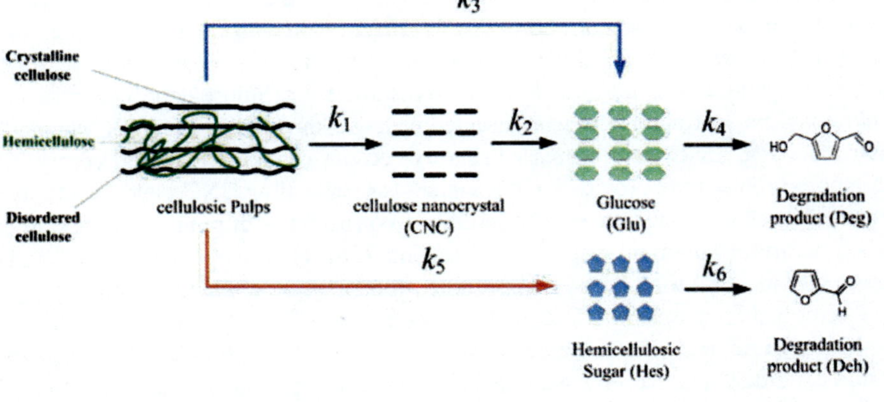

Fig. 1.5 Reaction kinetic model for concentrated acid hydrolysis of bleached fibers only containing cellulose and hemicelluloses from [126] with permission from American Chemical Society©

$$Y_{CSR} = \frac{C_{Cel0}}{C_{CSR0}}\{1 - \gamma[1 - \exp(-(k_1 + k_3)t]]\} + \frac{X_{n0}}{C_{CSR0}}[1 - \delta_x(1 - \exp(-k_5 t)] \times 100\% \quad (1.2)$$

where C_{Cel0}, X_{n0}, and C_{SR0} are the initial cellulose, xylan, and fiber mass concentration in the pseudo-homogeneous system (g/L). γ and δ_x are degrees of hydrolysis for cellulose and xylan, respectively [137]. Reaction coefficients k_1, k_3, and k_5 were found to strongly depend on sulfuric acid concentration (C_{SA}), as shown below for bleached eucalyptus pulp fibers, while k_2 can be neglected for conditions without significant CNC degradation to sugars, such as acid concentration below 62 wt% for reaction time less than 1 h below 45 °C

$$k_1 = 4.806 \times exp\left[-\frac{36350}{RT}\right] {C_{SA}}^{3.816} \quad (1.3a)$$

$$k_3 = 24.12 \times exp\left[-\frac{40220}{RT}\right] {C_{SA}}^{3.257} \quad (1.3b)$$

$$k_5 = 0.0010 \times exp\left[-\frac{31440}{RT}\right] {C_{SA}}^{7.512} \quad (1.3c)$$

$$\gamma = 1 - \frac{60.826}{{C_{SA}}^{1.975} {R_0}^{0.131}} \quad (1.4a)$$

$$\delta_x = 1 - \frac{43.312}{{C_{SA}}^{2.185} {R_0}^{0.097}} \quad (1.4b)$$

$$R_0 = exp\left[\frac{T - 25}{14.75}\right] \quad (1.4c)$$

An experimental study by Zhu's group [29] discovered that CNC yield decreased suddenly beyond acid concentration 62 wt%, in agreement with kinetic model predictions [126]. As shown in Fig. 1.6, acid concentration is the determinate factor controlling CNC yield. CNC yield was low due to insufficient cellulose depolymerization below acid concentration of 58 wt%, in agreement with the literature [21, 52], whereas above 62 wt%, CNC yield decreased sharply because of rapid cellulose hydrolysis to glucose. This study provided a mechanism for controlling CNC yield by simply controlling acid concentration. The finding is consistent with a more recent study using a careful factorial design of experiments [36]. Hydrolysis temperature and especially reaction time in the range studied (35 °C $\le T \le$ 80 °C; 15 min $\le t \le$ 240 min) had no dominant effects on CNC yield, as predicted [126].

The sulfuric acid concentration range of 58–62 wt% can be considered as the "mild condition" that has long been searched for. This reported mild acid concentration range was from measurements in the initial prepared acid solution and differs from the actual concentration in the reacting solution; however, its significance should

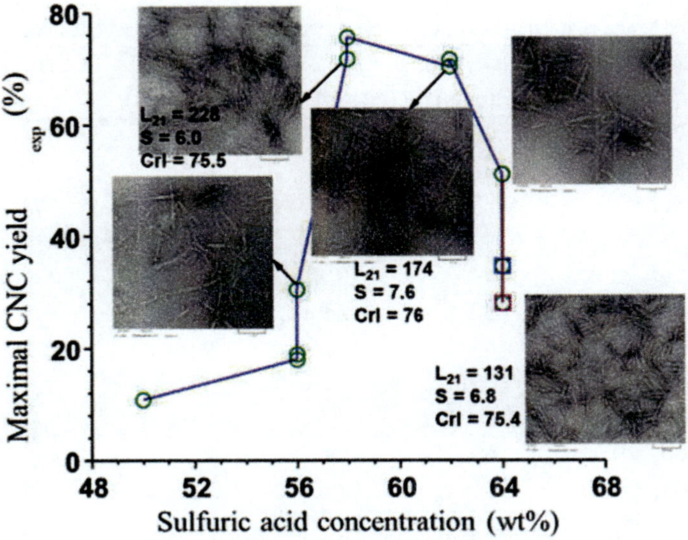

Fig. 1.6 Effects of acid concentration on maximal CNC yield and morphology. Here maximal CNC yield represents the highest yield obtainable in ranges of reaction temperatures and durations at a given acid concentration. Modified from [29]. All scales in TEM image are 200 nm. L_{21} is the length–weight averaged CNC length in nm; S is CNC sulfur content in mg/g, and CrI is XRD measured CNC crystallinity index. All scale bars = 0.2 μm

not be underestimated, because the standard acid concentration of 64 wt% used for comparison was also from measurements in the initial acid solution. Water loss during reaction and variations in fiber sources both affect the optimal acid concentration in a reacting solution. The cellulose crystallization model under mild conditions proposed by Battista (Fig. 1.4) explains the underlying mechanism for the observed CNC yield improvement. Laboratory bench-scale experiments by Zhu's group verified that high CNC yield of more than 75 wt% [29] using bleached eucalyptus kraft pulp fibers with cellulose content of only approximately 85 wt% (CNC yield of 90% based on cellulose content) was obtained without decreasing the crystallinity of the resultant CNCs (Table 1.2). This almost doubled CNC yield and therefore, substantially improved the economics of CNC production for commercial applications. The kinetic model presented above also predicted the yield of CSR under hydrolysis conditions below the optimal CNC yield condition, which is in agreement with experiments [126, 130]. Table 1.2 lists the length weighted mean length (L_{21}), sulfur content (S), and crystallinity (CrI) of CNCs produced from bleached kraft eucalyptus fibers at different sulfuric acid concentrations [29].

CNCs have interesting optical properties. Formation of the chiral nematic phase is affected by surface charge and CNC morphology [49, 62, 73]. For example, the iridescent colors and the order of the nematic layers of a CNC film can be well-controlled by the polydispersity of CNC size distribution [38]. Moreover, the length of the CNCs is directly related to chiral nematic pitch [15]. Therefore, it is desirable

Table 1.2 Yield and characteristics of CNCs produced from bleached eucalyptus kraft pulp fibers

Hydrolysis condition [a]	CNC yield (% fibers) [b]	L_{21} (nm)	S (mg/g CNCs)	CrI (%)
(56, 70, 90)	30	281	4.7	74
(58, 56, 60)	39	227	4.6	74
(58, 56, 120)	63	217	5.2	73
(58, 56, 180)	72	228	6.0	76
(58, 56, 210)	76	205	6.0	73
(62, 40, 75)	58	183	8.7	79
(62, 40, 105)	65	177	9.9	77
(62, 50, 75)	70	143	8.0	76
(62, 60, 30)	72	189	8.0	73
(62, 60, 60)	69	149	9.5	76
(64, 45, 15)	51	148	7.3	69
(64, 45, 45)	35	117	9.1	72
(64, 45, 75)	30	131	8.0	74
(64, 45, 105)	28	131	6.8	75

[a]Hydrolysis condition (xx, yy, zzz) represent: sulfuric acid concentration = xx wt%, T temperature = yy °C, and time = zzz min. [b]CNC yields were based on mass of feed fibers with cellulose content of 85% (not based on mass of cellulose in the fibers)

to be capable of tailoring CNC morphology for different applications. CNCs with a large aspect ratio may be more suitable as a polymer reinforcement component for composites. CNCs of varied lengths were produced by hydrolyzing microcrystalline cellulose using 6 N HCl under a range of conditions [136]. Under the standard sulfuric acid concentration of 64 wt% and 45 °C, properties CNCs from wood fibers were affected by reaction time and fiber loading [15]. A longer reaction time and lower fiber loading resulted in decreased CNC mean length and a smaller decrease in diameter, or an overall decrease in mean CNC aspect ratio. However, CNC yield was not reported. Decreased CNC length is also expected to be associated with CNC yield loss. A kinetic study found that although reaction time and temperature affect CNC yield, acid concentration is the dominant factor that dictates CNC yield, as discussed above [126]. Zhu's group clearly demonstrated (Fig. 1.6) that controlling sulfuric acid concentration can not only substantially increase CNC yield [126] but also tailor CNC morphology [29]. The average CNC length was more than doubled to more than 200 nm simply by using a slightly lower acid concentration than 64 wt% (Table 1.2). However, the change in CNC morphology was negligible when other reaction conditions (i.e., temperature and reaction time) were varied. For example, when reaction time was increased by 7 times from 15 to 105 min at acid 64 wt% and 45 °C, the average CNC length was only decreased by 11% from 148 mm to 131 nm, in agreement with an earlier study [15].

Zhu's group also showed that CNC crystallinity and sulfate group content can be tailored by properly controlling acid hydrolysis conditions [29]. Acid concentration of 62 wt%, slightly below the standard condition of 64 wt%, resulted in the greatest crystallinity and sulfation under an optimized temperature and reaction time. Furthermore, it appears that in general, a maximal CNC yield is in the vicinity of a maximal crystallinity and sulfation at a given acid concentration (Table 1.2). This is because a too mild condition causes insufficient cellulose depolymerization, insufficient hydrolysis of disordered cellulose, and insufficient sulfation, whereas a too severe condition can destroy crystallinity and hydrolyze CNCs into sugars to result in decreasing sulfation. Using acid concentration for controlling CNC yield while tailoring CNC morphology and surface properties is probably one of the most important discoveries in the past 70 yearson CNC production by concentrated sulfuric acid hydrolysis. This provides a practically viable approach to optimize CNC production based on application needs.

1.2.2 Mechanical Fibrillation for Producing Cellulose Nanofibrils (CNFs)

Cellulose nanofibrils (CNFs) refer to cellulosic fibrils with diameters of a couple of hundred nanometers or less. Fibrils greater than this size are often referred to as cellulose microfibrils (CMF) or microfibrillated cellulose. As discussed in Sect. 1, plant or wood cell walls have a hierarchical structure. CNFs are imbedded in the cell wall and can therefore be separated through simple mechanical fibrillation [58, 119, 129], as clearly shown by SEM and TEM imaging (Fig. 1.7a, b). Typically, pure mechanical fibrillation produces entangled fibril networks (Fig. 1.7b). Very different fibril morphologies were obtained depending on the extent of fibrillation as revealed by microscopy under different magnifications. Tree-like structures with distinct kinks as well as "soft looking" structures were reported [129] from mechanical fibrillated bleached hardwood fibers (Fig. 1.8a, b). Interconnected fibril bundles as large as 250 nm in diameter (Fig. 1.8c, d) and twisted cellulose fibrils were also present (Fig. 1.8a–d). Fibril whiskerization occurred under extended fibrillation (Fig. 1.8e) to result in CNC-like whiskers of approximately 10 nm in diameter and 100 nm in length (Fig. 1.8f). This diversity in CNFs can be beneficial for diversified applications. Mechanical fibrillated CNFs with entangled network structure are desirable for polymer reinforcement in composites through direct mixing [135] but may not be suitable for certain applications require alignment or electro-spinning.

From the pulp and paper industry, it is well known that mechanical pulping for producing fibers, tens of micrometers in diameter, from wood chips is energy intensive. This energy consumption is directly related to the amount of surface area produced through refining. A much greater increase in surface area is required in producing CNFs compared with producing fibers. Therefore, pure mechanical fibrillation of fibers to nanofibril scale is very energy intensive. A laboratory study by Zhu's

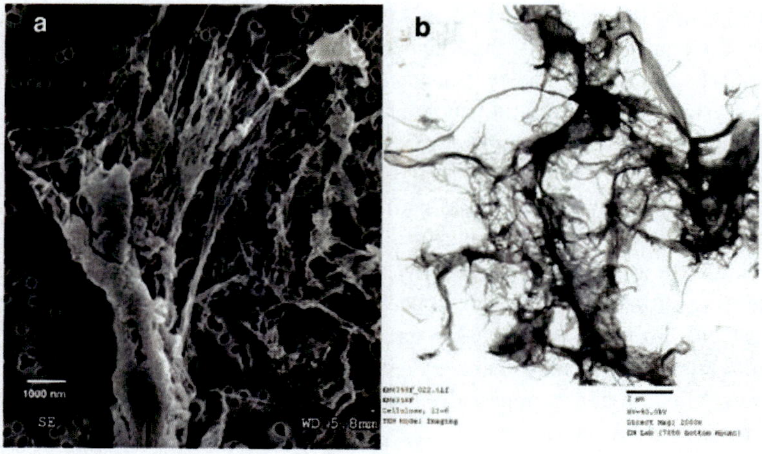

Fig. 1.7 **a** FE-SEM image showing separation of micro- and nanofibrils of a bleached wood pulp microfiber, scale = 1000 nm, and **b** TEM image showing structures of the mechanically fibrillated fibril, scale = 2 μm [129]

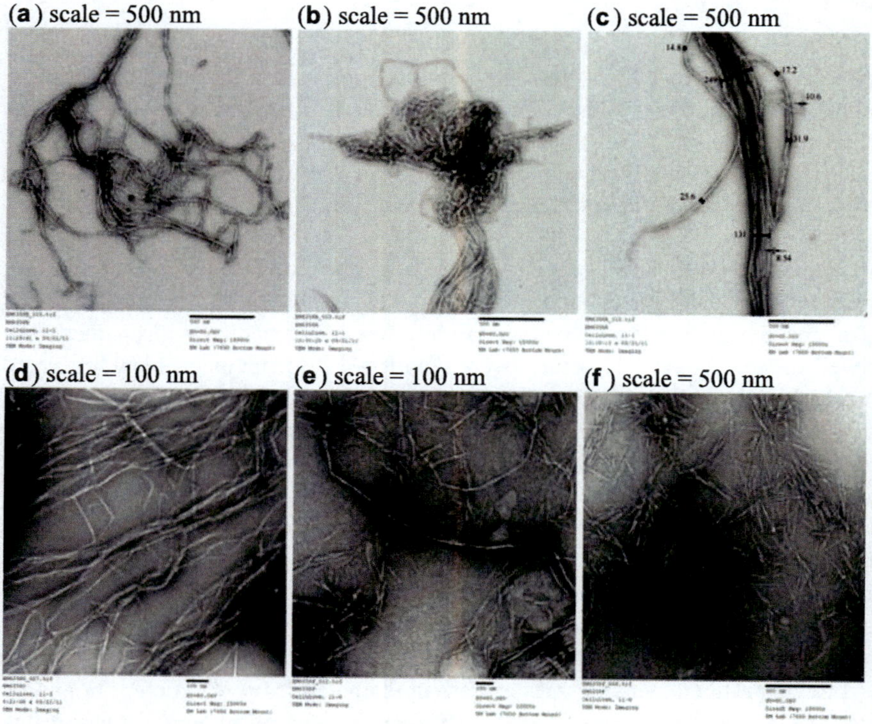

Fig. 1.8 TEM images show different CNF morphologies from pure mechanical fibrillation of bleached eucalyptus pulp from [129]

group indicated that energy consumption for fibrillating bleached eucalyptus pulp fibers to the nanoscale level can be on the order of 10 MJ/kg [55], although the energy requirement is expected to decrease with scale-up of the fibrillation process. It has been reported that applying chemical or enzymatic pretreatments can decrease energy consumption in refining, for example, for producing chemical-thermo-mechanical pulps in the pulp and paper industry or for producing fine substrates to improve cell wall accessibility to cellulase for sugar production via enzymatic saccharification [146]. Similarly, to produce CNFs, chemical and enzymatic treatments have shown substantial energy savings in mechanical fibrillation [97, 132]. TEMPO-mediated oxidation treatment substantially reduces mechanical fibrillation energy requirement [56].

1.2.3 Understanding Key Processes to Decreasing Mechanical Fibrillation Energy

Almost all chemical and enzymatic treatments can decrease energy consumption in mechanical fibrillation for producing CNFs. This is simply because all treatments loosen the structure of lignocellulosic fibers by removing or dissolving certain components. Indeed, alkali [125], acid [97, 128], organic solvent [31, 85], oxidation [42, 72, 107], ionic liquid [16], deep eutectic solvents [69, 109], acid hydrotropes [18, 25, 114], endoglucanase [53, 96, 132], xylanase [118, 138], and complex enzymes [143] have all been used with varied degrees of success. Nevertheless, these treatments introduced additional processing steps that added cost to production and environment. The key is to find an easy treatment with lowest cost and environmental impact. Mechanistic analyses can help identify which treatment may be most effective for CNF production. Here, we can analyze this from the following considerations:

I. Dissolution of lignin and hemicelluloses:

From the current understanding of cell wall structure (Fig. 1.2), a certain degree of delignification and hemicellulose removal is helpful to loosen up the cell wall structure for mechanical nanofibrillation [18, 24, 114], which explains why all the treatments can decrease energy requirements for mechanical fibrillation. For example, when medium density fibers and raw wood fibers (from birch) were treated by acid hydrotropes *p*-Toluenesulfonic acid (*p*-TsOH) [18] and maleic acid [24], the effect of lignin and hemicellulose removal on nanofibrillation was clearly seen. As listed in Table 1.3, based on AFM topographical measurements, greater extent of delignification and dissolution of hemicelluloses resulted in finer CNFs. Pure mechanical fibrillation of undelignified wood fibers to nanofibril size is extremely difficult if not impossible [55]. It has been recognized that xylan exists in two domains in certain plants [23, 61]: the major domain is the one that is coated on the surface of cellulose, and the minor domain is the part that links cellulose elemental fibrils. Therefore, the removal of hemicelluloses by acid treatment effectively cuts these linkages between cellulose chains, which facilitates mechanical fibrillation. Furthermore, delignification is always associated with some dissolution of hemicelluloses due to the cross-linking of lignin and xylan [50, 61]. However, a typical xylanase treatment alone is effective only in microfibrillation but not in nanofibrillation because of limited accessibility to xylan between nanofibrils [138].

Complete delignification and hemicellulose dissolution provide increased opportunity for hydrogen bonding among cellulose fibrils [111], which can hinder the separation of the fibrils. It has been shown that residual lignin facilitated both fibrillation (due to its lubrication effect) and separation of cellulose fibrils [17, 25, 101]. For instance, under the identical conditions of maleic acid treatment with a similar level of hemicellulose dissolution of approximately 30% and equal extent of mechanical fibrillation, commercial unbleached mixed hardwood pulp fibers with a higher lignin content of 17% were easier to fibrillate than the fibers with a lower lignin content of 4% from the same wood mix and the same pulp mill [17]. Specifically, the AFM-measured mean CNF height was 7.5 nm from the pulp with lignin content of 17% compared with 9.5 nm from the pulp with lignin content of 4%.

II. Depolymerization of cellulose:

This has not been well recognized as important to nanofibrillation of cellulose, partly because of the lack of understanding of cellulose fibril aggregation in the longitudinal direction. A recent study using dilute acid hydrolysis treatment of commercial bleached pulp fibers by Zhu's group proposed a rational model on the longitudinal cellulose ultra-structure [97]. As shown in Fig. 1.9, cellulose fibrils aggregate in the longitudinal direction to form long fibers. For example, typical commercial bleached softwood fibers are on the order of 1–2 mm long with a DP of 2000. Using a molecular size glucan unit of 5A, one can estimate the typical fiber length of only 1 μm, or approximately three orders of magnitude shorter than the measured fiber length of 1 mm. This huge difference cannot be simply attributed to issues in DP measurements. It is fibril longitudinal aggregation that resulted in millimeter-long fibers. Micrometer lengths rather than millimeter lengths have been observed for CNFs produced from all existing methods. The model shown in Fig. 1.9a suggests that typical nanoscale fibrils are much shorter than millimeter-long fibers. To separate these nanoscale fibrils

Table 1.3 Effect of delignification and dissolution of hemicelluloses on AFM-measured average height of mechanically fibrillated CNFs

Sample	Lignin	Xylan	Passes in fibrillation	CNF mean height (nm)
Birch MDF	**23.5**	**20.5**		
CNF-P50T80t20	16.0	15.0	5	51.1
CNF-P65T80t20	11.6	14.0	5	29.4
CNF-P80T80t20	7.2	12.2	5	15.3
Birch Fibers	**19.5**	**23.0**		
CNF-M60T90t30	15.8	14.0	5	12.7
CNF-M50T100t60	15.4	11.0	5	9.1

from the fiber matrix, one must cut or disintegrate fibers. Consequently, one cannot obtain millimeter-long nanofibrils. Experimental data on the mean length and height of CNFs produced under different acid treatment conditions with varied degrees of fibrillation (Fig. 1.9b) indicated an inverse correlation between their height and length (Fig. 1.9b) [97]. A similar inverse correlation was observed from CNFs produced from commercial pulp fibers dried differently (Fig. 1.9b) [19]. This indirectly verified the importance of cellulose depolymerization for nanoscale fibrillation. This also explains why pure endoglucanase treatment [53, 96, 132] that simply depolymerizes cellulose can facilitate nanofibrillation. Therefore, cutting fibril length or depolymerization of cellulose is a necessary condition to obtain thin fibrils through mechanical fibrillation.

III. Cellulose and lignin surface modification:

This approach includes cellulose or lignin sulfation, such as the formation of half-ester groups on the cellulose surface in concentrated sulfuric acid hydrolysis of cellulose in CNC production [128, 130] and esterification (carboxylation) of lignin and cellulose in concentrated di-carboxylic acid hydrolysis [24, 25, 30]. The negative surface charge groups, such as HSO_3^{-1} and COO^{-1}, result in a repulsion force between the fibrils, which facilitates fibrillation. The repulsion enhanced the lignin's lubrication effect previously discussed [101]. The effect of this electrostatic repulsive interaction can be best seen by comparing materials with similar levels of lignin

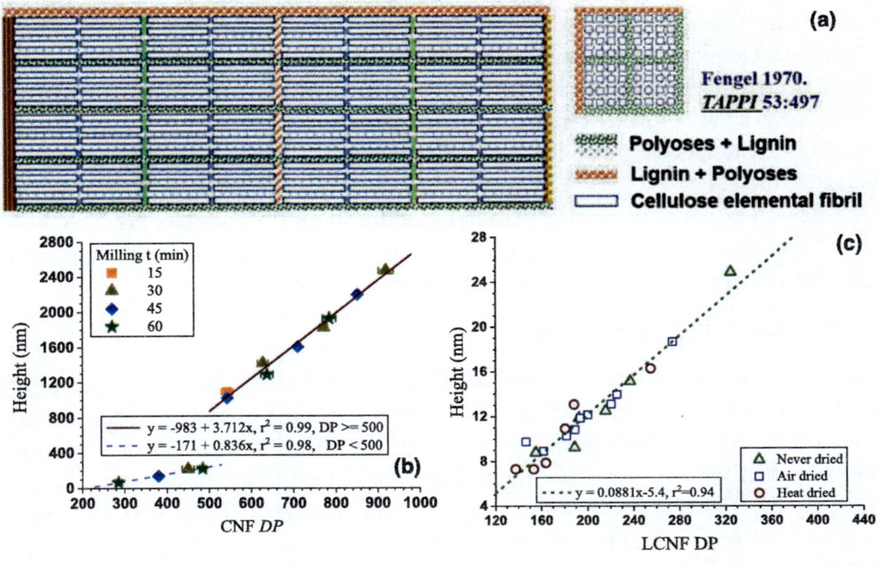

Fig. 1.9 **a** A schematic diagram of a wood fiber longitudinal hierarchical structure model that illustrates "natural" fibrils with certain length or aspect ratios that vary with fibril lateral dimension. Reproduced from [97]. **b** and **c** Correlations of CNF height with length: under different acid treatment conditions and extent of fibrillation [97], (**b**) using commercial fibers dried differently (**c**) [19]

Table 1.4 Effect of lignin carboxylation on mechanical fibrillation of birch wood lignocellulosic fibers to produce lignin-containing CNFs (LCNFs)

	LCNF-MA	LCNF-*p*-TsOH
Klason lignin (%)	12.9	12.8
Xylan (%)	13.8	15.7
Cellulose (%)	61.3	62.9
LCNF carboxyl group content (mml/g)	0.22	0.06
LCNF surface charge (mV)	−42	−32
LCNF mean diameter (nm)	12	23
AFM image[a]	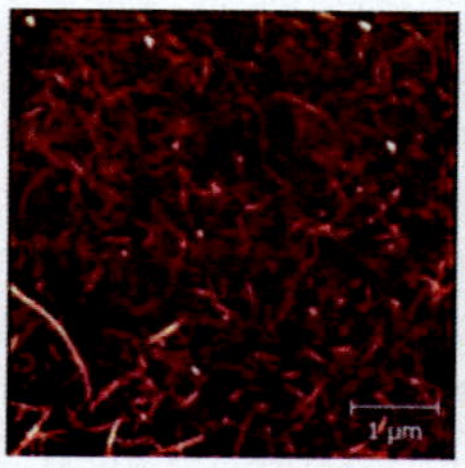	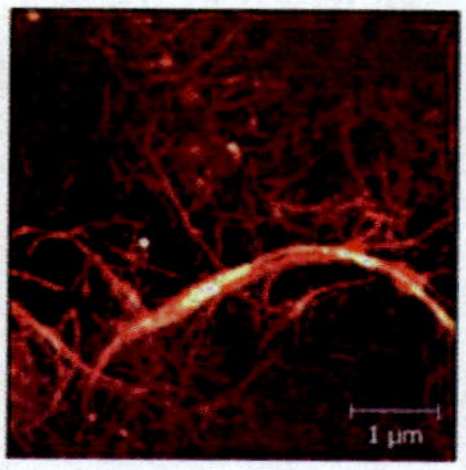

[a]One pass through microfluidization [25] with permission from John Wiley and Sons©

and hemicellulose dissolution but with and without charges. When birch wood was fractionated by two very different acid hydrotropes, *p*-TsOH and maleic acid (MA), under a similar level of delignification and hemicellulose dissolution and the same extent of mechanical fibrillation [25], the uncarboxylated lignocellulosic solids from *p*-TsOH fractionation resulted in CNFs with a much greater mean fibril diameter of 23 nm due to a lower surface charge compared with the 12-nm diameter for the CNFs from MA-fractionated lignocellulosic solids with carboxylation and a higher surface charge (Table 1.4).

On the other hand, the carboxylation of cellulose on fibril surface by TEMPO-mediated oxidation of native cellulose also facilitated fibrillation [56] in addition to oxidation of hydroxyl groups, as discussed in Sect. 1.2.4.

IV. Breaking hydrogen bonding:

Hydrogen bonding plays a major role in cellulose chain consolidation to form a rigid structure. In the context of isolating CNFs, breaking of the hydrogen bonds should be very effective for fibrillation. This explains why TEMPO-mediated oxidation is so robust in producing very fine and uniform fibrils with low energy input. Energy cost after the TEMPO oxidation is very low. This further explains why leaving a small amount of lignin in fibers can facilitate nanofibrillation [17, 101] because of lignin's role in preventing hydrogen bonding, as discussed earlier in this section.

Based on these analyses, the following can be concluded: (1) Oxidation is an effective treatment for nanofibrillation of delignified pulp fibers, as evidenced by the substantially decreased energy consumption in mechanical fibrillation of TEMPO-oxidized pulp fibers. (2) Unless using undelignified pulp fibers, alkali treatment is not preferred to acidic treatment, simply because acid is much more effective in depolymerizing cellulose and dissolving hemicelluloses. In addition, there is a cost associated with alkali recovery. (3) Lignin solvent-based techniques, including ionic liquids or gamma-Valerolactone (GVL) [31], though suitable to deconstruct plant cell walls, may over-deconstruct the cell wall for nanofibrillation because lignin's positive effect can be negated by its removal. Similar to the alkali recovery situation, there is a cost associated with solvent recovery. (4) Acid hydrotropes [142], with their ability to achieve rapid delignification, depolymerization of cellulose, and dissolution of hemicelluloses, all at atmospheric pressure, are suitable for treating undelignified fibers for producing LCNFs [141]. Especially, lignin and cellulose carboxylation achieved by maleic acid as a di-carboxylic acid hydrotrope—an FDA-approved indirect food additive (21CFR175-177) with low environmental impact—can facilitate mechanical fibrillation for producing LCNFs. We discuss some of these processes in the following subsections.

1.2.4 Chemical Oxidation of Cellulose for Decreasing Mechanical Fibrillation Energy

Among the oxidation pretreatment methods, TEMPO-mediated oxidation has received the most attention for its robust performance in breaking hydrogen bonds by oxidation on the exocyclic cellulose C6 carbon for effective separation of CNFs. This oxidation is often carried out using a mixture of TEMPO, NaBr, and NaClO at pH 10–11 and at room temperature for a few hours, depending on the chemical loading. Typical chemical loadings on cellulose mass basis are TEMPO, 0.25–1.6 wt%; NaBr, 2.5–15 wt%; and NaClO, 10–40% [56, 106, 107]. As depicted in Fig. 1.10, the C6 primary hydroxyls on the surfaces of cellulose fibrils can be selectively converted to sodium carboxyl groups through TEMPO-mediated oxidation. Furthermore, this conversion takes place in one of every two glucosyl units at the C6 position on the cellulose fibril surface [56]. Both oxidized TEMPO and NaBrO and/or NaClO reacted with cellulose to carboxylate the C6 carbon on the cellulose fibril surface [57]. The extensive breakage of hydrogen bonding and carboxylation on the fibril surface substantially facilitated the mechanical disintegration or separation of fibrils [107]. The carboxylation substantially improved the dispersion of the TEMPO-oxidized CNFs (T-CNFs) [56]. A carboxyl group content of 1.0 mmol/g resulted in a well-dispersed T-CNF suspension. The increase in carboxylation facilitated fibrillation and resulted in finer T-CNFs with improved optical transmittance [56]. TEMPO-mediated oxidation is quite effective on a variety of cellulose materials (Fig. 1.11). Furthermore, T-CNFs were also effective for producing LCNF from undelignified thermomechanical pulp fibers.

Periodate oxidation of cellulose [77] has also attracted interest recently for being able to produce carboxylate cellulose and facilitate fibrillation to produce CNFs. Periodate can oxidize the hydroxyl group at the cellulose C2 and C3 position into

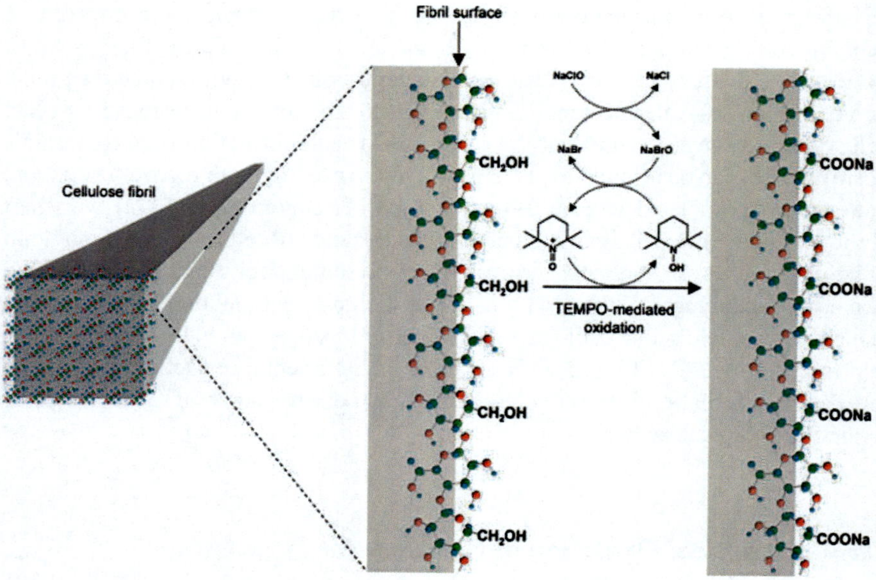

Fig. 1.10 A schematic model for the oxidation of C6 primary hydroxyls on native cellulose microfibril surfaces to C6 carboxylate groups using the TEMPO/NaBr/NaClO system [91] with permission from American Chemical Society©

aldehyde groups, which are subsequently oxidized into carboxyl group by sodium chlorite [72, 117]. Carboxyl group content as high as 1.8 mmol/g was achieved, although cellulose crystallinity was substantially decreased.

Oxidation using ammonium persulfate was also attempted for producing carboxylated CNCs [68]. More recently, the method has been used for producing CNFs [42].

Periodate, ammonium persulfate, and chlorite all have varied levels of toxicities, especially when they react with lignin. This issue needs to be handled appropriately while using these oxidation methods for producing CNFs.

1.2.5 Acid Hydrolysis to Facilitate Mechanical Fibrillation

From the point of view of cellulose depolymerization and hemicellulose dissolution, acid treatment should be effective. As discussed previously, using a mild sulfuric acid hydrolysis condition for producing CNCs from bleached kraft eucalyptus fibers resulted in CNCs as well as greatly hydrolyzed short fibers as CSR [130]. As shown in Fig. 1.12, the two CSR samples had a fiber length of approximately 50–200 μm compared with approximately 1.0 mm for typical hardwood fibers. The fiber shortening was due to significant cellulose depolymerization. Concentrated acid hydrolysis also dissolved a significant amount of hemicelluloses. The estimated xylan

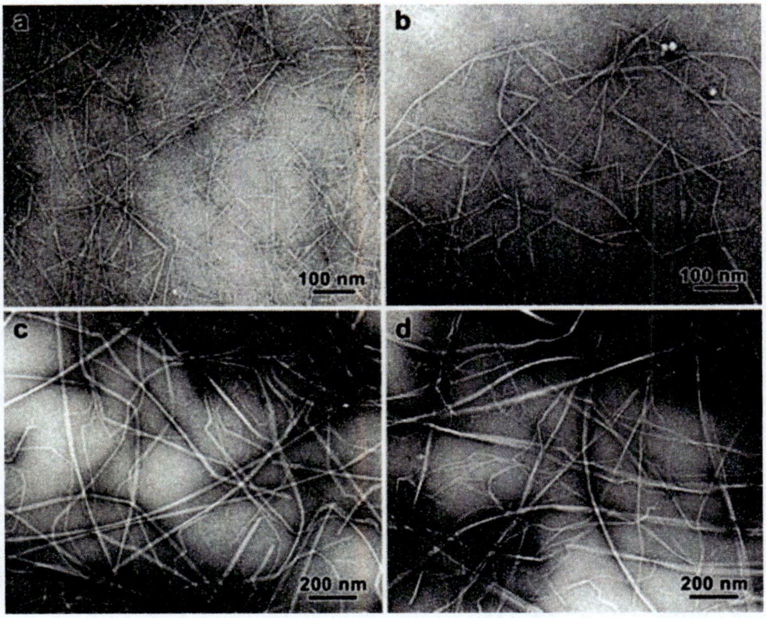

Fig. 1.11 Effect of feed fiber source on the morphology of resultant TEMPO-CNFs, from [107]: **a** bleached sulfite wood pulp, **b** cotton, **c** tunicin, **d** bacterial cellulose, with permission from American Chemical Society©

removal was 97% and 85%, respectively, for the two CSR samples shown (S62T40t70 and S58T60t40) based on measured xylose concentrations in the hydrolysates [126]. These CSRs were found to be readily fibrillated into individual separated nanofibrils (or whiskers) (Fig. 1.12). This also achieved integrated production of CNCs with CNFs.

To further illustrate this, the following equations present the modeling of CNF production using dilute acid treatment with subsequent disk milling. Kinetics of xylan dissolution under acidic conditions has been well developed [112]. By using a bi-phasic model [63, 78, 137], the fraction of xylan, X_R, retained on solids after acid hydrolysis of a lignocellulosic material can be predicted by a severity factor—combined hydrolysis factor (CHF) [97, 145].

$$X_R = (1 - \theta)\exp(-CHF_X) + \theta\exp(-f \cdot CHF_X) \quad (1.5)$$

$$CHF_X = \exp(\alpha - E_a/RT + \beta C) \cdot C \cdot t \quad (1.6)$$

where θ and $(1-\theta)$ in 1.5 were the fractions of the slow and fast reaction (hydrolysis) xylan, respectively, f is the ratio of the rate constants between the slow and fast xylan hydrolysis reactions, subscript X of *CHF* represents xylan, C is the initial molar concentration of acid; α and β are adjustable parameters, E_a is the apparent

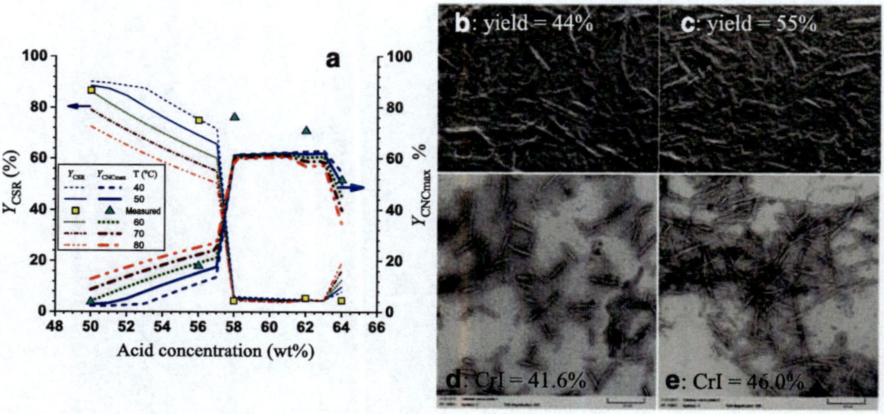

Fig. 1.12 Integrated production of CNCs and CNFs from bleached eucalyptus kraft pulp fibers using mild concentrated sulfuric acid hydrolysis. **a** effect of acid concentration on yields of CNCs and CSRs [126] with permission from American Chemical Society©, **b** and **c** CSR from sulfuric acid hydrolysis S62T40t70 (acid 62 wt% at 40 °C for 70 min) and S58T60t40, respectively [128] with permission from American Chemical Society©, **d** and **e** CNFs from the CSRs shown in **b** and **c**, respectively, after 5 passes through an 87-μm chamber of a microfluidizer, from [130]. Scale bar = 2 μm for b and c, = 0.2 μm for d **and e**

activation energy, R is the universal gas constant of 8.314 J/mole/K, and T is the temperature in kelvins.

Similarly, cellulose also exhibits bi-phase behavior with slow and fast reaction fractions during acid hydrolysis just like xylan [126, 134, 137]. Because the DP of the hydrolyzed lignocellulosic fibers is a good indication of the extent of cellulose degradation (depolymerization) by an acid, DP can also be predicted by a combined hydrolysis factor.

$$\frac{DP}{DP_{BEP}} = \left(1-\theta'\right)\exp(-CHF_X) + \theta' \exp\left(-f' \cdot CHF_X\right) \tag{1.7}$$

where θ' and $(1-\theta')$ in 1.7 were the fractions of the slow and fast reaction (hydrolysis) cellulose, respectively, and f' is the ratio of the rate constants between the slow and fast cellulose hydrolysis reactions.

Mechanical fibrillation of lignocellulosic materials can physically break up cellulose fibers to result in fine and short fibrils with decreased DP. When using a stone disk mill SuperMassColloider (SMC) to fibrillate dilute oxalic acid treated eucalyptus bleached kraft pulp (BEP) fibers, the DP of the fibrillated fibrils (nano to micro sizes) can be predicted by the following equation [97]:

$$\begin{aligned}\frac{DP}{DP_{BEP}} &= (1-\theta')\exp(-CHF_X) + \theta' \exp(-f' \cdot CHF_X) \\ &\quad -b(1+CHF_X^{\varepsilon})\left(\frac{t^m}{t_T^m}\right)\end{aligned} \tag{1.8}$$

where t^m is SMC milling time in min and $t_T^m = 60$ min is the maximal milling time as a normalizing parameter for all experiments. The parameters in (1.5–1.7) can be obtained by fitting experimentally measured xylan dissolution and cellulose DP data through acid hydrolysis in a range of acid concentrations. The parameters of b and ε in (1.8) can be obtained by fitting the measured DPs of CNFs from disk milling of acid-hydrolyzed CSRs. Excellent fittings were obtained (Fig. 1.13a) [97], indicating that (1.8) is a good model for predicting a CNF length-related parameter DP. All fitting parameters of dilute oxalic acid treatment of BEP fibers are listed in Table 1.5.

It is interesting to find that the CNF or CNC diameters or heights measured by AFM correlate to the CNF DPs very well for CNF or CNC samples obtained from different feedstocks, varied acid hydrolysis conditions, and degrees of fibrillation [19, 97] (Fig. 1.9b, c), suggesting that cellulose depolymerization is a necessary condition for producing cellulose nanomaterials. Furthermore, DP can be a good measure of CNF morphology for CNF samples produced using the same feedstock and same process but under different chemical or physical treatments, including

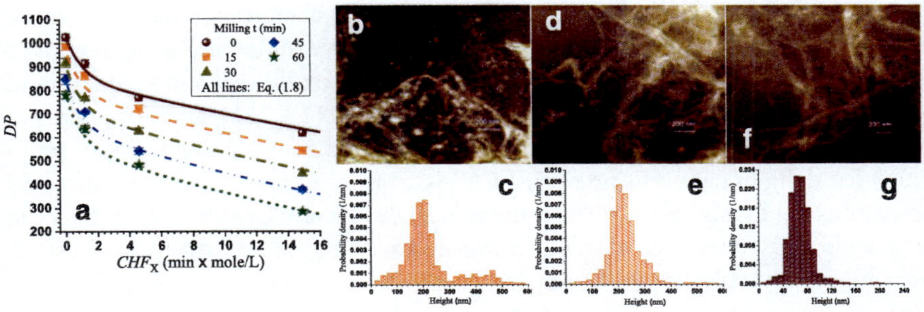

Fig. 1.13 Illustration of using DP to control CNF morphology from [97]. **a** Effects of acid hydrolysis severity CHFX on decrease in fibril DP through nanofibrillation using disk milling at different times, **b**, **d**, **f** AFM images of CNFs obtained from bleached eucalyptus pulp fibers oxalic acid treated at different severities and fibrillation times, with (**b**): DP = 450 ± 14, CHFx = 14.9, milling tm = 30 min, **d** DP = 485 ± 8, CHFx = 4.6, milling tm = 60 min, **f** DP = 286 ± 8, CHFx = 14.9, milling tm = 60 min; (**c**, **e**, **g**) are AFM-measured height (diameter) probability density functions of CNFs shown in B, D, F, respectively. Scale bar = 200 nm for B and D, = 100 nm for F

Table 1.5 Fitting parameters for (1.5–1.8)

Equation (1.5)	Equation (1.6)	Equation (1.7)	Equation (1.8)	Unit
	$\alpha = 8.21$			None
	$\beta = 20.7$			L/mol
	$E_a = 25{,}000$			J/mol
$\theta = 0.623$		$\theta' = 0.838$		None
$f = 0.049$		$f' = 0.020$		None
			$b = 0.137$	None
			$\varepsilon = 0.137$	None

varied extents of fibrillation. This is clearly illustrated in Fig. 1.13b–g [97]. The two CNF samples shown in Fig. 1.13b, d had similar DP and similar height distributions (comparing Fig. 1.13c–e), i.e., similar morphologies, despite that they were produced under different acid treatment severities and mechanical fibrillation times. The CNF shown in Fig. 1.13f had a lower DP than the CNFs shown in Fig. 1.13b, d. The CNFs in Fig. 1.13b, d also had lower heights than the CNF shown in Fig. 1.13f (Fig. 1.13g).

This discussion indicates that (1.8) can not only control CNF length (DP), but also control CNF diameter. The process presented in this subsection provides an effective way to control morphology in CNF production using acid hydrolysis with subsequent mechanical fibrillation.

1.2.6 Recent Advances and Production of Lignin-Containing Cellulose Nanomaterials

Recent advances in producing CNMs include using chemicals, other than those previously presented, that are easily recyclable and have low environmental impact. Specifically, these advances use low solubility di-carboxylic acids to facilitate acid recovery [25, 30, 59, 73], especially using maleic acid [17, 131], an FDA-approved indirect food additive (21CFR175-177). Maleic acid also has hydrotropic properties for delignification to produce LCNMs from raw lignocellulose directly, as discussed later in this Section. The carboxylated CNMs produced from di-carboxylic acid hydrolysis were advantageous compared with sulfated S-CNCs because of their higher crystallinity, greater thermal stability, and better surface modification [30]. Di-carboxylic-acid-produced CNCs also show chiral nematic properties with potential for optical and sensing applications [73]. One can also produce cellulose nanowhiskers (CNWs) with morphology similar to CNCs (Fig. 1.14) by directly fibrillating MA-treated cellulosic solids without separating CNCs through dialysis to reduce water usage. Interested readers can consult recent literature on the subject [124, 141].

LCNMs have attracted great interest recently for their hydrophobic and UV protection properties. LCNMs can be produced using commercial unbleached pulp fibers simply by mechanically fibrillating the unbleached pulp fibers [41, 55]. As previously discussed, application of chemical treatments can decrease energy input in fibrillation. Concentrated maleic acid hydrolysis was applied to produce LCNFs and LCNCs from commercial unbleached kraft pulp fibers with lignin content up to 17% [17].

Concentrated sulfuric acid hydrolysis was directly applied to hydrothermally treated Wiley-milled wood to produce LCNCs with a very high lignin content of approximately 30% [5]. Organosolv fractionation was also applied to wood directly to produce LCNFs and LCNCs [85]. As we stated previously, most chemical fractionation treatments are effective for producing cellulose nanomaterials.

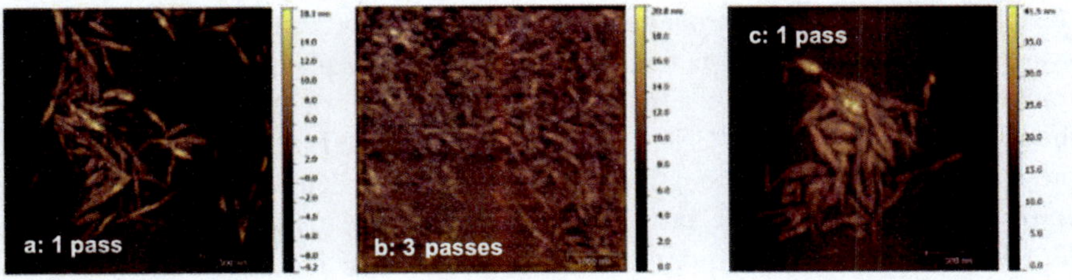

Fig. 1.14 AFM images of cellulose nanowhiskers (CNWs) produced by concentrated acid hydrolysis of lignocellulose followed by mechanical fibrillation of the hydrolyzed cellulosic solids residue (CSR). **a** and **b** CNWs from maleic acid hydrolysis of bleached eucalyptus kraft pulp fibers at M70T120t120 (70 wt% concentration at 120 °C for 120 min) with subsequent homogenization for 1 pass (**a**) and 3 passes (**b**), respectively. **c** LCNWs (lignin content of 7%) from *p*-TsOH hydrolysis of hammer milled birch wood at P60T80t60 (60 wt% concentration at 80 °C for 60 min) with subsequent homogenization for 1 pass. All scale bars = 500 nm. Reproduced from [124] and [141]

From a mechanistic point of view, as discussed in Sect. 2.3, recently developed low-temperature acid hydrotropic fractionation (AHF) [24, 28, 142] can solubilize lignin, depolymerize cellulose, and dissolve hemicelluloses, and hence, it is ideally suited for producing LCNMs directly from raw plant biomass, such as wood. This has been demonstrated recently using *p*-TsOH and maleic acid as hydrotropes on birch wood [18, 24, 25] and herbaceous biomass [76, 114, 115]. The degree of delignification can be controlled by a delignification factor [76, 114, 140], and the amount of hemicellulose dissolution can be controlled using (1.5) and a similar CHF_X as shown in (1.6):

$$L_R = \left(1 - \theta' - \theta'_R\right)e^{-\mathrm{CDF}} + \theta' \cdot e^{-f' \cdot \mathrm{CDF}} + \theta'_R \tag{1.9}$$

$$\mathrm{CDF} = \exp\left(\alpha' - \frac{E'}{RT} + \beta' C\right) \cdot C \cdot t \tag{1.10}$$

Therefore, the lignin content of the resultant LCNMs can be controlled by controlling AHF severity. Uniform CNFs were obtained even with only one pass through microfluidization under a range of delignification levels [18, 24, 25]. It is important to point out that maleic acid as a hydrotrope can carboxylate the resultant LCNFs through esterification with lignin and cellulose, which facilitated both fibrillation to further decrease mechanical energy consumption and dispersion of LCNFs.

1.3 Characterization of Cellulose Nanomaterials

For applications of CNMs, characterization is important because the properties of the nanomaterials are determined by their structural characteristics. For example, analysis of morphology (shape, size, and structure) is important because it determines the physical and chemical properties of CNMs (e.g., how many hydroxyl groups are available for a reaction). Also, the morphology of CNMs plays a role in the distribution of the particles, e.g., in a nanocomposite, which in turn affects the properties of the composite. Similarly, in CNCs, crystallinity and crystal size impact their physical properties, such as tensile strength, dimensional stability, density, chemical reactivity, and swelling. For those CNC properties that are sensitive to moisture, e.g., swelling, film thickness, and water vapor permeability, measurement of accessibility to water is important. One aspect of the CNMs is that when they are produced, they are in an aqueous suspension state. Nevertheless, the characterization of these materials rarely takes place in the suspension state. For analysis by many techniques, CNMs need to be dried. The way the nanomaterials consolidate upon drying can therefore have an influence on a particular characterization. Here, some of the methods and techniques commonly used to characterize CNMs are covered briefly, and for further follow-up, references to the literature are provided. From the literature on this topic, it is clear that significant challenges remain with regard to reproducible and reliable characterization of the CNMs, which must be overcome. A detailed review on this matter was recently published [43]. Table 1.6 provides a list of the techniques most often used to characterize nanocelluloses. These property measurements are briefly described in the following sections.

1.3.1 Morphology

As previously discussed, CNMs exist in several shapes and sizes depending upon the isolation/production procedures. Although in the suspension state, which is the native state, dynamic light scattering (DLS) has been used to obtain hydrodynamic "apparent particle size" [20], their morphology (shape, size, aspect ratio, and size distribution) has mostly been determined by TEM and AFM, although high-resolution SEM (e.g., FE-SEM) has occasionally been used [127]. The DLS results cannot be directly linked to the nanoparticle length or cross section because DLS provides information only on the average hydrodynamic diameter of the particles in colloidal suspensions and not on their internal aspects. Different shapes of CNMs include nanospheres, nanorods, nanocrystals, nanowhiskers, nanofibrils, and nanoclusters. Based on the morphology of the nanomaterials, their properties can be finely tuned. Therefore, it is essential to carry out a detailed study of the morphology of the CNMs and its effect on the properties.

All measurements of CNMs based on TEM images are limited by the potential biases due to aggregation of the nanoparticles. Therefore, identifying edges, clumping, imaging artifacts, and sample selection for statistical measurements becomes difficult [5]. Tendency of CNCs to aggregate is known [5, 22]. For example, from AFM and TEM imaging, it was reported that CNCs width by TEM was nearly twice as great as AFM height [22]. This was brought about by the CNCs that were laterally aggregated in the TEM images. Moreover, by TEM, laterally aggregated CNCs are also likely to result in longer length dimensions.

Table 1.6 Commonly used techniques to estimate various properties/characteristics of nanocelluloses

Technique	Property/characteristic						
	Crystallinity	Supramolecular structure	Morphology	Elemental analysis	Thermal stability	Surface character	Rheology measure
	–	–	–	–	–	–	–
XRD	√	√	–	–	–	–	–
Raman	√	√	–	–	–	√	–
NMR	√	√	–	–	–	–	–
IR	–	–	–	–	–	√	–
TEM	–	–	√	–	–	–	–
AFM	–	–	√	–	–	–	–
SEM	–	–	√	–	–	–	–
SIMS	–	–	√	√	–	–	–
DLS	–	–	√	√	–	–	–
CHNS EA	–	–	–	√	–	–	–
XPS	–	–	–	√	–	–	–
AES	–	–	–	√	–	–	–
ICP	–	–	–	√	–	–	–
EDX	–	–	–	√	–	–	–
TGA	–	–	–	–	√	–	–
Rheometer	–	–	–	–	–	–	√
ζ-potential	–	–	–	–	–	√	–
Conducto. titration	–	–	–	–	–	√	–

From a production process control point of view, rapid and quantitative measure of CNM morphology is important. In Sect. 2, we mentioned that DP can be a good measure of CNM morphology. Although DP measurement is quantitative, it is not straightforward. However, water retention value (WRV) has long been used in the pulp and paper industry to characterize the extent of fiber fibrillation because, for a fiber sample, it measures the water-holding capacity. Moreover, WRV was found to correlate well with a fibrous sample's accessibility to cellulase enzymes, cellulose substrate's pore surface area, and a substrate's cellulose enzymatic saccharification [75, 127]. Recently, a detailed CNM characterization was carried out to evaluate WRV as a measure of CNM morphology [51]. It was found that WRV performed better than DP, enzyme adsorption, and enzymatic cellulose digestibility in terms sensitivity to morphological changes of a CNF sample produced from bleached eucalyptus kraft pulp through different mechanical fibrillations. Because WRV measurement is simple and rapid, it can be a very useful tool for practical applications in industry as well as in laboratories.

1.3.2 Surface Charge (Zeta Potential) and Surface Characterization

The ζ-potential of a CNM sample is due to the charged groups at the surface of the particle and is a key indicator of the stability of its colloidal dispersions. Low value of the ζ-potential ($< \pm 20$ mV) of a colloidal CNM dispersion causes flocculation of the colloids, and the suspension becomes unstable. Stabilization of the colloidal CNM suspension is obtained via electrostatic repulsion. The zeta potential is often reported in literature and is traditionally used to evaluate relative changes in a CNM's colloidal stability. For CNCs, common values are −20 to −50 mV (not including HCl produced CNCs, which are not charged), and depending upon the extent of oxidation, CNFs produced by TEMPO-mediated oxidation can have values near −60 mV. To measure zeta potential, an electrophoretic mobility analyzer is used. For accurate measurements, some salts are added to ensure a finite double-layer thickness and a visually stable suspension (such as hydrophobically modified CNCs in water), so that the measured zeta potential is meaningful [98].

Another method to determine surface charge of CNMs is conductometric titrations [98] and references therein). The titration is carried out as an acid–base titration by adding aliquots of standardized NaOH solution to the suspension of the CNM, the conductivity and pH are continuously measured, and the equivalence point is determined. Because conductometric titrations must be performed on CNMs in acid form, in case the materials are in salt form (e.g., CNCs in sodium form, $-OSO_3Na$), they can be treated with ion-exchange resin [13, 14]. However, significant insight into particle aggregation and/or colloidal stability is not provided by charge measurements, particularly in media of varying pH and/or ionic strength [43], such measurements play a major role in the rheological [133], colloidal [33], interfacial [60], and reinforcing properties [104]. For instance, CNCs are effective as Pickering emulsion stabilizers, and the ones with lower charge produce more stable emulsions [60].

Surface characterization of CNMs is needed either when a functional group is pre-existent and is generated during the process of its production (e.g., sulfate groups in CNCs or carboxyl groups in TEMPO-CNFs) or in situations where CNMs are modified/derivatized for the purpose of dispersing them in hydrophobic polymer matrices. Although usually CNM surfaces are altered by covalent bonding, other kinds of chemical surface modification (e.g., surface adsorption) can also occur. For information on the latter type of modification, the reader is referred to a recent review [43].

The characterization of covalently bound groups can be carried out in situ using the traditional analytical methods (e.g., IR, NMR, Raman, XPS). Alternatively, the grafted compound can be isolated (either chemically or by the first removal of cellulose enzymatically) and then analyzed by techniques that have a high sensitivity for detection (e.g., liquid chromatography, mass spectrometry, liquid state NMR, IR). Once again, further details on the specific characterization methods and their use in surface characterization can be obtained from [43] where the original work was reviewed.

1.3.3 Crystallinity (CrI)

Cellulose CrI of CNMs has received a lot of attention because it is a very important material characteristic that plays an important role in their physical, mechanical, and chemical properties. For example, when crystallinity increases so does tensile strength, dimensional stability, and density, whereas properties such as chemical reactivity, sensitivity to moisture, and swelling decline. Crystallinity of CNMs can be measured using the same methods that are used for other cellulose materials. However, although numerous techniques/methods are available, e.g., XRD Segal [93, 108], NMR [43, 66], Raman [3, 6], to measure cellulose CrI, there is some uncertainty when it comes to defining what exactly each method is measuring. This is partly because the methods fail to properly deal with cellulose that is neither amorphous nor crystalline but is organized in some way [3]. Moreover, in the case of XRD, which is the most widely used technique, numerous methods exist to calculate the CrI and depending upon how a diffractogram is processed, different values of CrI are obtained [93]. In the case of Raman spectroscopy, two different methods, namely, 380-Raman and 93-Raman, have been used for crystallinity measurement of CNMs, and differing CrI values were obtained [3, 43]. In this context, it was recently reported that wood cellulose is not crystalline in its native state [4], and it is only when heated (thermally or hydrothermally) that it becomes partly crystalline [5]. Consequently, bleached pulp cellulose, a substrate for most CNMs, can be considered as partly crystalline. How the crystallinity of this cellulose is further modified upon production of CNMs depends largely upon the steps used in the processing of this pulp. Moreover, in the case of CNCs, it remains unclear why the crystallinity is not significantly higher compared with that of the pulp from which it was produced. Nevertheless, an explanation that accounts for this observation and that is supported by the recent findings [1, 3] is that both pulp cellulose (a composite of crystalline and aligned/organized cellulose) and crystalline cellulose (in CNCs) contribute in the same regions in various techniques (e.g., XRD, NMR, and Raman). Once the contribution from the aligned/organized cellulose is removed by acid hydrolysis, the crystallinity of the remaining crystalline portion does not increase significantly because the two contributions were similar to begin with. The similarity of the water accessibility data between pulp and pulp CNCs [4] further supported this explanation indicating that the water accessibility similar to the crystalline CNCs can be obtained by the aligned/organized cellulose in the pulp.

1.3.4 Supramolecular Structure

Cellulose crystallinity is one measure of the supramolecular structure of a CNM, but there are other ways to characterize it that provide additional information on the aggregated state. Such structural parameters are useful because the CNMs not only can be better defined but also can be produced according to the requirements of specific applications. In XRD, the crystallite size of CNMs can be calculated from the diffractogram [2] although there are different ways of accomplishing this (e.g., with or without fitting the diffractogram first to various contributing components). Similarly, using ^{13}C solid-state NMR, the number of hydroxyl groups on the surface of a CNM particle as well as its phase structure [67] can be investigated. Regarding CNCs, the researchers reported that instead of well-defined crystalline and noncrystalline regions, the particles comprised of a more finely varied distribution of surroundings. Yet another property, accessibility of the CNMs to water, was studied by Raman spectroscopy [4, 43]. This is an important property to measure because many properties of the nanomaterials are affected by CNM–water interactions [45]. As described in [4] and [43], the method for calculating the accessibility of a CNM to water is based upon calculating the Raman band intensity increase at 1380 cm^{-1} after, in a CNM sample, complete OH-to-OD exchange has occurred.

1.3.5 Elemental Analysis

The elemental composition of a CNM can be measured by several techniques. Some of the most useful methods for carbon, hydrogen, nitrogen, and sulfur elemental analysis (CHNS EA) are X-ray photoelectron spectroscopy (XPS), auger electron spectroscopy (AES), secondary ion mass spectrometry (SIMS), energy dispersive X-ray spectroscopy (EDS or EDX), and atomic emission spectroscopy techniques including inductively coupled plasma (ICP). A recent review [43] nicely summarizes applications of various methods for CMN elemental analysis and is a good reference for interested readers.

1.3.6 Thermal Stability

Thermal stability is an important property for functional CNM-based materials. For instance, in preparing thermoplastics, the nanomaterials must withstand the melt processing temperature. Thermal degradation of CNMs has been investigated using thermogravimetric analysis (TGA). A TGA instrument continuously measures mass of a sample while the temperature of the sample is increased over time. TGA is used to find the maximum degradation temperature of CNMs because beyond this temperature, they will degrade. Under an inert atmosphere, depending upon the chemical and physical characteristics of the CNCs and CNFs, their degradation temperature lies in the range of 220–350 °C [5, 46, 94]. Moreover, in the context of CNCs, it is well known that sulfated crystals have a significantly lower degradation temperature compared with nonsulfated crystals [74, 102]. Additionally, compared with the acid form, the sodium form of CNCs was found to be more stable [120]. The degradation temperature also depends upon the heating rate and the nature of the environment.

1.3.7 Rheology Measurement

Rheology measurements, which are carried out with a rheometer, have to do with the flow and deformation of the CNMs under applied forces. This measurement is important, for example, in understanding challenges in processing of CNM composites. The rheological response of CNMs depends upon several factors, such as the microstructure of nanomaterials, degree of dispersion, and interactions between the CNM and the solvent or matrix material (in a composite system). The viscoelastic response of the CNMs is shear dependent and is used to analyze the contributions of various nanomaterial characteristics—for example, concentration, particle size, morphology, and surface properties. Further details on this topic can be obtained from [43,71].

1.4 Comments on Technical and Economic Problems and Future Trends

Economic and sustainable production of CNM remains one of the main barriers to the commercialization of CNM and CNM-based products. Many research efforts have used CNM from the early production techniques, such as pure mechanical fibrillation, concentrated sulfuric acid hydrolysis, and TEMPO-mediated oxidation. Recent research trends and funding are heavily focused on new product development using CNM obtained from these early production techniques. This is partly because several pilot facilities have been developed that made these CNMs readily available to the broad scientific community. However, addressing the environmental concerns, the economics associated with energy input, and chemical recovery using the CNM production processes will be a great challenge. The research and development activities need to be further funded to develop novel CNM production processes that can address the issues of lowering costs and making production more sustainable. In this chapter, we described a few processes, such as the concentrated di-carboxylic acid hydrolysis, for integrated production of CNCs with CNFs and using CNC-like CNFs for most CNC applications.

Dewatering will be another challenge for transportation, drying, and application of CNMs in hydrophobic media. On-site production of CMF instead of CNM may be suitable for applications in paper products. Novel drying approaches such as foam drying and novel de-aggregation techniques for CNM dispersion need to be developed.

Finally, product development needs to focus on (1) cost for product substitution with marginal performance improvement; (2) novel usages with unique performance properties with orders of magnitude performance improvement; and (3) large-volume applications from a forest management perspective.

References

1. U.P. Agarwal, Analysis of cellulose and lignocellulose materials by raman spectroscopy: a review of the current status. Molecules **24**(9), 1659 (2019). https://doi.org/10.3390/molecules24091659
2. U.P Agarwal, Raman spectroscopy in the analysis of cellulose nanomaterials, in *Nanocelluloses: Their Preparation, Properties, and Applications*, ed. by U.P. Agarwal, R.H. Atalla, A. Isogai (American Chemical Society, Washington DC, 2017)
3. U.P. Agarwal, S.A. Ralph, R.S. Reiner, C. Baez, New cellulose crystallinity estimation method that differentiates between organized and crystalline phases. Carbohyd. Polym. **190**, 262–270 (2018)
4. U.P. Agarwal, S.A. Ralph, R.S. Reiner, C. Baze, Probing crystallinity of never-dried wood cellulose with Raman spectroscopy. Cellulose **23**, 125–144 (2016)
5. U.P. Agarwal, S.A. Ralph, R.S. Reiner, C.G. Hunt, C. Baez, R. Ibach, K.C. Hirth, Production of high lignin-containing and lignin-free cellulose nanocrystals from wood. Cellulose **25**(10), 5791–5805 (2018)
6. U.P. Agarwal, R.S. Reiner, S.A. Ralph, Celulose I crystallinity determination using FT-Raman spectroscopy: univariate and multivariate methods. Cellulose **17**(4), 721–733 (2010)
7. S.R. Anderson, D. Esposito, W. Gillette, J.Y. Zhu, U. Baxa, S.E. McNeil, Enzymatic preparation of nanocrystalline and microcrystalline cellulose. TAPPI J. **13**(5), 35–41 (2014)
8. J. Araki, M. Wada, S. Kuga, T. Okano, Biréfringent glassy phase of a cellulose microcrystal suspension. Langmuir **16**(6), 2413–2415 (2000)
9. J. Araki, M. Wada, S. Kuga, T. Okano, Flow properties of microcrystalline cellulose suspension prepared by acid treatment of native cellulose. Colloids Surf. A **142**, 75–82 (1998)

10. R.H. Atalla, D.L. VanderHart, Native cellulose: a composite of two distinct crystalline forms. Science **223**(4633), 283–285 (1984)
11. O.A. Battista, Hydrolysis and crystallization of cellulose. Ind. Eng. Chem. **42**(3), 502–507 (1950)
12. O.A. Battista, S. Coppicic, J.A. Howsmon, F.F. Morehead, W.A. Sisson, Level-off degree of polymerization: relation to polyphase structure of cellulose fibers. Ind. Eng. Chem. **48**(2), 333–335 (1956)
13. S. Beck, M. Méthot, J. Bouchard, Erratum to General procedure for determining cellulose nanocrystal sulfate half-ester content by conductometric titration (Cellulose). Cellulose **22**(1), 117 (2015a). https://doi.org/10.1007/s10570-014-0513-y
14. S. Beck, M. Méthot, J. Bouchard, General procedure for determining cellulose nanocrystal sulfate half-ester content by conductometric titration. Cellulose **22**(1), 101–116 (2015)
15. S. Beck-Candanedo, M. Roman, D.G. Gray, Effect of reaction conditions on the properties and behavior of wood cellulose nanocrystal suspensions. Biomacromol **6**, 1048–1054 (2005)
16. A.H. Bhat, I. Khan, M.A. Usmani, R. Umapathi, S.M.Z. Al-Kindy, Cellulose an ageless renewable green nanomaterial for medical applications: an overview of ionic liquids in extraction, separation and dissolution of cellulose. Int. J. Biol. Macromol. **129**, 750–777 (2019)
17. H. Bian, L. Chen, H. Dai, J.Y. Zhu, Integrated production of lignin containing cellulose nanocrystals (LCNC) and nanofibrils (LCNF) using an easily recyclable di-carboxylic acid. Carbohyd. Polym. **167**, 167–176 (2017)
18. H. Bian, L. Chen, R. Gleisner, H. Dai, J.Y. Zhu, Producing wood-based nanomaterials by rapid fractionation of wood at 80 °C using a recyclable acid hydrotrope. Green Chem. **19**, 3370–3379 (2017)
19. H. Bian, L. Chen, H. Dai, J.Y. Zhu, Effect of fiber drying on properties of lignin containing cellulose nanocrystals and nanofibrils produced through maleic acid hydrolysis. Cellulose **24**(10), 4205–4216 (2017)
20. Y. Boluk, C. Danumah, Analysis of cellulose nanocrystal rod lengths by dynamic light scattering and electron microscopy. J. Nanoparticle Res. **16**(1) (2014)
21. D. Bondeson, A. Mathew, K. Oksman, Optimization of the isolation of manocrystals from microcrystalline cellulose by acid hydrolysis. Cellulose **13**, 171–180 (2006)
22. A. Brinkmann, M. Chen, M. Couillard, Z.J. Jakubek, T. Leng, L.J. Johnston, Correlating cellulose nanocrystal particle size and surface area. Langmuir **32**(24), 6105–6114 (2016)
23. M. Busse-Wicher, T.C.F. Gomes, T. Tryfona, N. Nikolovski, K. Stott, N.J. Grantham, D.N. Bolam, M.S. Skaf, P. Dupree, The pattern of xylan acetylation suggests xylan may interact with cellulose microfibrils as a twofold helical screw in the secondary plant cell wall of Arabidopsis thaliana. Plant J. **79**(3), 492–506 (2014)
24. C. Cai, K. Hirth, R. Gleisner, H. Lou, X. Qiu, J.Y. Zhu, Maleic acid as a dicarboxylic acid hydrotrope for sustainable fractionation of wood at atmospheric pressure and $\leq$ 100 °C: Mode and utility of lignin esterification. Green Chem. **22**(5), 1605–1617 (2020). https://doi.org/10.1039/C9GC04267A
25. C. Cai, J. Li, K. Hirth, G.W. Huber, H. Lou, J.Y. Zhu, Comparison of two acid hydrotropes for sustainabl fractionation of birch wood. Chemsuschem **13**, 4649–4659 (2020). https://doi.org/10.1002/cssc.202001120
26. S. Camarero Espinosa, T. Kuhnt, E.J. Foster, C. Weder, Isolation of thermally stable cellulose nanocrystals by phosphoric acid hydrolysis. Biomacromol **14**(4), 1223–1230 (2013)
27. N.C. Carpita, Update on mechanisms of plant cell wall biosynthesis: How plants make cellulose and other (1→4)-β-D-Glycans. Plant Physiol. **155**(1), 171–184 (2011)
28. L. Chen, J. Dou, Q. Ma, N. Li, R. Wu, H. Bian, D.J. Yelle, T. Vuorinen, S. Fu, X. Pan, J.Y. Zhu, Rapid and near-complete dissolution of wood lignin at $\leq$ 80 °C by a recyclable acid hydrotrope. Sci. Adv. **3**(9), e1701735 (2017)
29. L. Chen, Q. Wang, K. Hirth, C. Baez, U.P. Agarwal, J.Y. Zhu, Tailoring the yield and characteristics of wood cellulose nanocrystals (CNC) using concentrated acid hydrolysis. Cellulose **22**, 1753–1762 (2015)

30. L. Chen, J.Y. Zhu, C. Baez, P. Kitin, T. Elder, Highly thermal-stable and functional cellulose nanocrystals and nanofibrils produced using fully recyclable organic acids. Green Chem. **18**, 3835–3843 (2016)
31. M. Chen, Q. Ma, J.Y. Zhu, D. Martin Alonso, T. Runge, GVL pulping facilitates nanocellulose production from woody biomass. Green Chem. **21**(19), 5316–5325 (2019)
32. Y. Chen, C. Liu, P.R. Chang, X. Cao, D.P. Anderson, Bionanocomposites based on pea starch and cellulose nanowhiskers hydrolyzed from pea hull fibre: effect of hydrolysis time. Carbohyd. Polym. **76**(4), 607–615 (2009)
33. F. Cherhal, F. Cousin, I. Capron, Influence of charge density and ionic strength on the aggregation process of cellulose nanocrystals in aqueous suspension, as revealed by small-angle neutron scattering. Langmuir **31**(20), 5596–5602 (2015)
34. G. Delepierre, O.M. Vanderfleet, E. Niinivaara, B. Zakani, E.D. Cranston, Benchmarking cellulose nanocrystals Part II: new industrially produced materials. Langmuir **37**(28), 8393–8409 (2021)
35. S.Y. Ding, M.E. Himmel, The maize primary cell wall microfibril: a new model derived from direct visualization. J. Agric. Food Chem. **54**(3), 597–606 (2006)
36. S. Dong, M.J. Bortner, M. Roman, Analysis of the sulfuric acid hydrolysis of wood pulp for cellulose nanocrystal production: a central composite design study. Ind. Crops Prod. **93**, 76–87 (2016)
37. X.M. Dong, D.G. Gray, Induced circular dichroism of isotropic and magnetically-oriented chiral nematic suspensions of cellulose crystallites. Langmuir **13**(11), 3029–3034 (1997)
38. A.G. Dumanli, H.M. Van Der Kooij, G. Kamita, E. Reisner, J.J. Baumberg, U. Steiner, S. Vignolini, Digital color in cellulose nanocrystal films. ACS Appl. Mater. Interf. **6**(15), 12302–12306 (2014)
39. S.J. Eichhorn, A. Dufresne, M. Aranguren, N.E. Marcovich, J.R. Capadona, S.J. Rowan, C. Weder, W. Thielemans, M. Roman, S. Renneckar, W. Gindl, S. Veigel, J. Keckes, H. Yano, K. Abe, M. Nogi, A.N. Nakagaito, A. Mangalam, J. Simonsen, A.S. Benight, A. Bismarck, L.A. Berglund, T. Peijs, Review: Current international research into cellulose nanofibres and nanocomposites. J. Mater. Sci. **45**(1), 1–33 (2010)
40. D. Fengel, Ultrastructural behavior of cell wall polysaccharides. Tappi **53**(3), 497–503 (1970)
41. A. Ferrer, E. Quintana, I. Filpponen, I. Solala, T. Vidal, A. Rodríguez, J. Laine, O.J. Rojas, Effect of residual lignin and heteropolysaccharides in nanofibrillar cellulose and nanopaper from wood fibers. Cellulose **19**(6), 2179–2193 (2012)
42. I. Filipova, V. Fridrihsone, U. Cabulis, A. Berzins, Synthesis of nanofibrillated cellulose by combined ammonium persulphate treatment with ultrasound and mechanical processing. Nanomaterials **8**(9), 640 (2018). https://doi.org/10.3390/nano8090640
43. E.J. Foster, R.J. Moon, U.P. Agarwal, M.J. Bortner, J. Bras, S. Camarero-Espinosa, K.J. Chan, M.J.D. Clift, E.D. Cranston, S.J. Eichhorn, D.M. Fox, W.Y. Hamad, L. Heux, B. Jean, M. Korey, W. Nieh, K.J. Ong, M.S. Reid, S. Renneckar, R. Roberts, J.A. Shatkin, J. Simonsen, K. Stinson-Bagby, N. Wanasekara, J. Youngblood, Current characterization methods for cellulose nanomaterials. Chem. Soc. Rev. **47**(8), 2609–2679 (2018)
44. A. Frey-Wyssling, The fine structure of cellulose microfibrils. Science **119**(3081), 80–82 (1954)
45. H. Fukuzumi, T. Saito, A. Isogai, Influence of TEMPO-oxidized cellulose nanofibril length on film properties. Carbohyd. Polym. **93**(1), 172–177 (2013)
46. H. Fukuzumi, T. Saito, T. Wata, Y. Kumamoto, A. Isogai, Transparent and high gas barrier films of cellulose nanofibers prepared by TEMPO-mediated oxidation. Biomacromol **10**, 162–165 (2009)
47. E.S. Gardiner, A. Sarko, Packing analysis of carbohydrates and polysaccharides. 16. The crystal structures of celluloses IV_I and IVI_{II}. Can. J. Chem. **63**, 173–180 (1985)
48. K.H. Gardner, J. Blackwell, The hydrogen bonding in native cellulose. BBA – Gen. Subj. **343**(1), 232–237 (1974)
49. M. Giese, L.K. Blusch, M.K. Khan, M.J. MacLachlan, Functional materials from cellulose-derived liquid-crystal templates. Angew. Chem. Int. Edit. **54**, 2888–2910 (2015)

50. J.H. Grabber, J. Ralph, R.D. Hatfield, Model studies of ferulate - Coniferyl alcohol cross-product formation in primary maize walls: Implications for lignification in grasses. J. Agric. Food Chem. **50**(21), 6008–6016 (2002)
51. F. Gu, W. Wang, Z. Cai, F. Xue, Y. Jin, J.Y. Zhu, Water retention value for characterizing fibrillation degree of cellulosic fibers at micro and nanometer scales. Cellulose **25**(5), 2861–2871 (2018)
52. W.Y. Hamad, T.Q. Hu, Structure–process–yield interrelations in nanocrystalline cellulose extraction. Can. J. Chem. Eng. **88**, 392–402 (2010)
53. M. Henriksson, G. Henriksson, L.A. Berglund, T. Lindström, An environmentally friendly method for enzyme-assisted preparation of microfibrillated cellulose (MFC) nanofibers. Eur. Polymer J. **43**, 3434–3441 (2007)
54. M. Hirota, K. Furihata, T. Saito, T. Kawada, A. Isogai, Glucose/glucuronic acid alternating co-polysaccharides prepared from TEMPO-oxidized native celluloses by surface peeling. Angew. Chem. Int. Ed. **49**(42), 7670–7672 (2010)
55. I.C. Hoeger, S.S. Nair, A.J. Ragauskas, Y. Deng, O.J. Rojas, J.Y. Zhu, Mechanical deconstruction of lignocellulose cell walls and their enzymatic saccharification. Cellulose **20**(2), 807–818 (2013)
56. A. Isogai, T. Saito, H. Fukuzumi, TEMPO-oxidized cellulose nanofibers. Nanoscale **3**(1), 71–85 (2011)
57. T. Isogai, T. Saito, A. Isogai, TEMPO electromediated oxidation of some polysaccharides including regenerated cellulose fiber. Biomacromol **11**(6), 1593–1599 (2010)
58. S. Iwamoto, A.N. Nakagaito, H. Yano, Nano-fibrillation of pulp fibers for the processing of transparent nanocomposites. Appl. Phys. A Mater. Sci. Process. **89**, 461–466 (2007)
59. C. Jia, L. Chen, Z. Shao, U.P. Agarwal, L. Hu, J.Y. Zhu, Using a fully recyclable dicarboxylic acid for producing dispersible and thermally stable cellulose nanomaterials from different cellulosic sources. Cellulose, 1–16 (2017)
60. I. Kalashnikova, H. Bizot, P. Bertoncini, B. Cathala, I. Capron, Cellulosic nanorods of various aspect ratios for oil in water Pickering emulsions. Soft Matter **9**(3), 952–959 (2013)
61. X. Kang, A. Kirui, M.C. Dickwella Widanage, F. Mentink-Vigier, D.J. Cosgrove, T. Wang, Lignin-polysaccharide interactions in plant secondary cell walls revealed by solid-state NMR. Nat. Commun. **10**(1), 347 (2019). https://doi.org/10.1038/s41467-018-08252-0
62. J.A. Kelly, A.M. Shukaliak, C.C.Y. Cheung, K.E. Shopsowitz, W.Y. Hamad, M.J. MacLachlan, Responsive photonic hydrogels based on nanocrystalline cellulose. Angew. Chem. Int. Ed. **52**(34), 8912–8916 (2013)
63. T. Kobayashi, Y. Sakai, Hydrolysis rate of pentosan of hardwood in dilute sulfuric acid. Bull. Agric. Chem. Soc. Jpn. **20**, 1–7 (1956)
64. E. Kontturi, A. Meriluoto, P.A. Penttilä, N. Baccile, J.M. Malho, A. Potthast, T. Rosenau, J. Ruokolainen, R. Serimaa, J. Laine, H. Sixta, Degradation and crystallization of cellulose in hydrogen chloride vapor for high-yield isolation of cellulose nanocrystals. Angew. Chem. Int. Ed. **55**(46), 14455–14458 (2016)
65. P. Langan, Y. Nishiyama, H. Chanzy, A revised structure and hydrogen-bonding system in cellulose II from a neutron fiber diffraction analysis. J. Am. Chem. Soc. **121**(43), 9940–9946 (1999)
66. P.T. Larsson, K. Wickholm, T. Iversen, A CP/MAS 13 C NMR investigation of molecular ordering in celluloses. Carbohyd. Res. **302**(1–2), 19–25 (1997)
67. C.H. Lemke, R.Y. Dong, C.A. Michal, W.Y. Hamad, New insights into nano-crystalline cellulose structure and morphology based on solid-state NMR. Cellulose **19**(5), 1619–1629 (2012)
68. A.C.W. Leung, S. Hrapovic, E. Lam, Y. Liu, K.B. Male, K.A. Mahmoud, J.H.T. Luong, Characteristics and properties of carboxylated cellulose nanocrystals prepared from a novel one-step procedure. Small **7**(3), 302–305 (2011)
69. P. Li, J.A. Sirviö, A. Haapala, H. Liimatainen, Cellulose nanofibrils from nonderivatizing urea-based deep eutectic solvent pretreatments. ACS Appl. Mater. Interf. **9**(3), 2846–2855 (2017)

70. T. Li, C. Chen, A.H. Brozena, J.Y. Zhu, L. Xu, C. Driemeier, J. Dai, O.J. Rojas, A. Isogai, L. Wågberg, L. Hu, Developing fibrillated cellulose as a sustainable technological material. Nature **590**(7844), 47–56 (2021)
71. J. Liao, K.A. Pham, V. Breedveld, Rheological characterization and modeling of cellulose nanocrystal and TEMPO-oxidized cellulose nanofibril suspensions. Cellulose **27**(7), 3741–3757 (2020)
72. H. Liimatainen, M. Visanko, J.A. Sirviö, O.E.O. Hormi, J. Niinimaki, Enhancement of the nanofibrillation of wood cellulose through sequential periodate-chlorite oxidation. Biomacromol **13**(5), 1592–1597 (2012)
73. C. Lin, B. Chen, Y. Liu, Y. Chen, M. Liu, J.Y. Zhu, Carboxylated cellulose nanocrystals with chiral nematic property from cotton by dicarboxylic acid hydrolysis. Carbohyd. Polym. **264** (2021)
74. N. Lin, A. Dufresne, Surface chemistry, morphological analysis and properties of cellulose nanocrystals with gradiented sulfation degrees. Nanoscale **6**(10), 5384–5393 (2014)
75. X. Luo, J.Y. Zhu, Effects of drying-induced fiber hornification on enzymatic saccharification of lignocelluloses. Enzyme Microb. Technol. **48**(1), 92–99 (2011)
76. Q. Ma, J. Zhu, R. Gleisner, R. Yang, J.Y. Zhu, Valorization of wheat straw using a recyclable hydrotrope at low temperatures (≤90 °C). ACS Sustain. Chem. Eng. **6**, 14480–14489 (2018). https://doi.org/10.1021/acssuschemeng.8b03135
77. E. Maekawa, T. Koshijama, Properties of 2,3-dicarboxy cellulose combined with various metallic ions. J. App. Polym. Sci. **29**, 2289–2297 (1984)
78. M.T. Maloney, T.W. Chapman, A.J. Baker, Dilute acid hydrolysis of paper birch: Kinetics studies of xylan and acetyl-group hydrolysis. Biotechnol. Bioeng. **27**(3), 355–361 (1985)
79. R.E. Mark, *Cell wall mechanics and tracheids* (Yale University Press, 1967)
80. H.E. McFarlane, A. Döring, S. Persson, The cell biology of cellulose synthesis. Ann. Rev. Plant Biol. **65**, 69–94 (2014)
81. S. Montanari, M. Roumani, L. Heux, M.R. Vignon, Topochemistry of carboxylated cellulose nanocrystals resulting from TEMPO-mediated oxidation. Macromolecules **38**(5), 1665–1671 (2005)
82. R.J. Moon, A. Martini, J. Nairn, J. Simonsen, J. Youngblood, Cellulose nanomaterials review: structure, properties and nanocomposites. Chem. Soc. Rev. **40**, 3941–3994 (2011)
83. J.L. Moran, V.A. Alvarez, V.P. Cyras, A. Vazquez, Extraction of cellulose and preparation of nanocellulose from sisal fibers. Cellulose **15**, 149–159 (2008)
84. S.M. Mukherjee, H.J. Woods, X-ray and electron microscope studies of the degradation of cellulose by sulphuric acid. Biochim. Biophys. Acta **10**, 499–511 (1953)
85. K. Nelson, T. Retsina, V. Pylkkanen, R. O'Connor, Processes and apparatus for producing nanocellulose, and compositions and products produced therefrom. U.S. Patent No. 9,187,865 B2 (2015)
86. R.F. Nickerson, J.A. Habrle, Cellulose intercrystalline structure. Ind. Eng. Chem. **39**, 1507–1512 (1947)
87. Y. Nishiyama, U.J. Kim, D.Y. Kim, K.S. Katsumata, R.P. May, P. Langan, Periodic disorder along ramie cellulose microfibrils. Biomacromol **4**(4), 1013–1017 (2003)
88. Y. Nishiyama, P. Langan, H. Chanzy, Crystal structure and hydrogen-bonding system in cellulose Iβ from synchrotron X-ray and neutron fiber diffraction. J. Am. Chem. Soc. **124**(31), 9074–9082 (2002)
89. Y. Nishiyama, J. Sugiyama, H. Chanzy, P. Langan, Crystal structure and hydrogen bonding system in cellulose Iα from synchrotron X-ray and neutron fiber diffraction. J. Am. Chem. Soc. **125**(47), 14300–14306 (2003)
90. H. Oguzlu, C. Danumah, Y. Boluk, Colloidal behavior of aqueous cellulose nanocrystal suspensions. Curr. Opin. Colloid Interface Sci. **29**, 46–56 (2017)
91. Y. Okita, T. Saito, A. Isogai, Entire surface oxidation of various cellulose microfibrils by TEMPO-mediated oxidation. Biomacromol **11**(6), 1696–1700 (2010)
92. W.J. Orts, L. Godbout, R.H. Marchessault, J.F. Revol, Enhanced ordering of liquid crystalline suspensions of cellulose microfibrils: a small angle neutron scattering study. Macromolecules **31**(17), 5717–5725 (1998)

93. S. Park, J.O. Baker, M.E. Himmel, P.A. Parilla, D.K. Johnson, Cellulose crystallinity index: Measurement techniques and their impact on interpreting cellulase performance. Biotechnol. Biofuels **3**, 10 (2010). https://doi.org/10.1186/1754-6834-3-10
94. L. Petersson, I. Kvien, K. Oksman, Structure and thermal properties of poly(lactic acid)/cellulose whiskers nanocomposite materials. Compos. Sci. Technol. **67**(11–12), 2535–2544 (2007)
95. R.C. Pettersen, The chemical composition of wood, in *The Chemistry of Solid Wood. Advances in Chemistry Series 207*, ed. by R.M. Rowell (American Chemical Society. Washington D.C. 1984), pp. 115–116
96. M. Pääkko, M. Ankerfors, H. Kosonen, A. Nykänen, S. Ahola, M. Österberg, J. Ruokolainen, J. Laine, P.T. Larsson, O. Ikkala, T. Lindström, Enzymatic hydrolysis combined with mechanical shearing and high-pressure homogenization for nanoscale cellulose fibrils and strong gels. Biomacromolecules **8**, 1934–1941 (2007)
97. Y. Qin, X. Qiu, J.Y. Zhu, Understanding longitudinal wood fiber ultra-structure for producing cellulose nanofibrils using disk milling with dilute acid prehydrolysis. Sci. Rep. **6**, 35602 (2016)
98. M.S. Reid, M. Villalobos, E.D. Cranston, Benchmarking cellulose nanocrystals: From the laboratory to industrial production. Langmuir **33**(7), 1583–1598 (2017)
99. J.-F. Revol, H. Bradford, J. Giasson, R.H. Marchessault, D.G. Gray, Helicoidal self-ordering of cellulose microfibrils in aqueous suspension. Int. J. Biol. Macromol. **14**, 170–172 (1992)
100. J.-F. Revol, L. Godbout, X.-M. Dong, D.G. Gray, H. Chanzy, G. Maret, Chiral nematic suspensions of cellulose crystallites; phase separation and magnetic field orientation. Liq. Cryst. **16**, 127–134 (1994)
101. E. Rojo, M.S. Peresin, W.W. Sampson, I.C. Hoeger, J. Vartiainen, J. Laine, O.J. Rojas, Comprehensive elucidation of the effect of residual lignin on the physical, barrier, mechanical, and surface properties of nanocellulose films. Green Chem. **17**, 1853–1866 (2015)
102. M. Roman, W.T. Winter, Effect of sulfate groups from sulfuric acid hydrolysis on the thermal degradation behavior of bacterial cellulose. Biomacromol **5**(5), 1671–1677 (2004)
103. S.P. Rowland, E.J. Roberts, Nature of accessible surfaces in the microstructure of cotton cellulose. J. Polym. Sci. Part A-1 Polym. Chem. **10**(8), 2447–2461 (1972)
104. R. Rusli, K. Shanmuganathan, S.J. Rowan, C. Weder, S.J. Eichhorn, Stress transfer in cellulose nanowhisker composites—Influence of whisker aspect ratio and surface charge. Biomacromolecules **12**(4), 1363–1369 (2011)
105. B.G. Rånby, The colloidal properties of cellulose micelles. Discuss. Faraday Soc. **11**, 158–164 (1951)
106. T. Saito, A. Isogai, TEMPO-mediated oxidation of native cellulose. The effect of oxidation conditions on chemical and crystal structures of the water-insoluble fractions. Biomacromolecules **5**, 1983–1989 (2004)
107. T. Saito, Y. Nishiyama, J.L. Putaux, M. Vignon, A. Isogai, Homogeneous suspensions of individualized microfibrils from TEMPO-catalyzed oxidation of native cellulose. Biomacromolecules **7**(6), 1687–1691 (2006)
108. L. Segal, J.J. Creely, A.E. Martin, C.M. Conrad, An empirical method for estimating the degree of crystallinity of native cellulose using the X-ray diffractometer. Text. Res. J. **29**, 786–794 (1959)
109. J.A. Sirviö, M. Visanko, H. Liimatainen, Deep eutectic solvent system based on choline chloride-urea as a pre-treatment for nanofibrillation of wood cellulose. Green Chem. **17**(6), 3401–3406 (2015)
110. E. Sjostrom, *Wood Chemistry: Fundamentals and Application*, 2nd edn. (Academic Press Inc., San Diego, 1992)
111. K.L. Spence, R.A. Venditti, O.J. Rojas, Y. Habibi, J.J. Pawlak, The effect of chemical composition on microfibrillar cellulose films from wood pulps: Water interactions and physical properties for packaging applications. Cellulose **17**(4), 835–848 (2010)
112. E.L. Springer, Hydrolysis of Aspenwood xylan with aqueous solutions of hydrochloric acid. TAPPI **49**(3), 102–106 (1966)

113. A. Šturcová, G.R. Davies, S.J. Eichhorn, Elastic modulus and stress-transfer properties of tunicate cellulose whiskers. Biomacromol **6**(2), 1055–1061 (2005)
114. C. Su, K. Hirth, Z. Liu, Y. Cao, J.Y. Zhu, Acid hydrotropic fractionation of switchgrass at atmospheric pressure using maleic acid in comparison with p-TsOH: Advantages of lignin esterification. Ind. Crops Prod. **159**, 113017 (2021). https://doi.org/10.1016/j.indcrop.2020.113017
115. C. Su, K. Hirth, Z. Liu, Y. Cao, J.Y. Zhu, Maleic acid hydrotropic fractionation of wheat straw to facilitate value-added multi-product biorefnery at atmospheric pressure. GCB Bioenergy **13**2021bhttps://doi.org/10.1111/gcbb.12866
116. L. Tang, B. Huang, Q. Lu, S. Wang, W. Ou, W. Lin, X. Chen, Ultrasonication-assisted manufacture of cellulose nanocrystals esterified with acetic acid. Biores. Technol. **127**, 100–105 (2013)
117. A. Tejado, M.N. Alam, M. Antal, H. Yang, T.G.M. van de Ven, Energy requirements for the disintegration of cellulose fibers into cellulose nanofibers. Cellulose **19**(3), 831–842 (2012)
118. T.M. Tenhunen, M.S. Peresin, P.A. Penttilä, J. Pere, R. Serimaa, T. Tammelin, Significance of xylan on the stability and water interactions of cellulosic nanofibrils. React. Funct. Polym. **85**, 157–166 (2014)
119. A.F. Turbak, F.W. Snyder, K.R. Sandberg, Microfibrillated cellulose, a new cellulose product: properties, uses, and commercial potential. J. Appl. Polym. Sci. Appl. Polym. Symp. **37**, 815–827 (1983)
120. O.M. Vanderfleet, D.A. Osorio, E.D. Cranston, Optimization of cellulose nanocrystal length and surface charge density through phosphoric acid hydrolysis. Philos. Trans. Royal Soc. A: Math., Phys. Eng. Sci. **376**(2112) (2018)
121. O.M. Vanderfleet, M.S. Reid, J. Bras, L. Heux, J. Godoy-Vargas, M.K.R. Panga, E.D. Cranston, Insight into thermal stability of cellulose nanocrystals from new hydrolysis methods with acid blends. Cellulose **26**(1), 507–528 (2019)
122. M. Wada, L. Heux, J. Sugiyama, Polymorphism of cellulose I family: reinvestigation of cellulose IV_I. Biomacromolecules **5**(4), 1385–1391 (2004)
123. M. Wada, Y. Nishiyama, P. Langan, X-ray structure of ammonia-cellulose I: new insights into the conversion of cellulose I to cellulose III_I. Macromolecules **39**(8), 2947–2952 (2006)
124. H. Wang, J.J. Zhu, Q. Ma, U.P. Agarwal, R. Gleisner, R. Reiner, C. Baez, C. Liu, J.Y. Zhu, Pilot-scale production of cellulosic nano whiskers with similar morphology to cellulose nanocrystals. Front. Bioeng. Biotechnol. **8**, 565084 (2020). https://doi.org/10.3389/fbioe.2020.565084
125. H. Wang, M. Zuo, N. Ding, G. Yan, X. Zeng, X. Tang, Y. Sun, T. Lei, L. Lin, Preparation of nanocellulose with high-pressure homogenization from pretreated biomass with cooking with active oxygen and solid alkali. ACS Sustain. Chem. Eng. **7**(10), 9378–9386 (2019). https://doi.org/10.1021/acssuschemeng.9b00582
126. Q. Wang, X. Zhao, J.Y. Zhu, Kinetics of strong acid hydrolysis of a bleached kraft pulp for producing cellulose nanocrystals (CNCs). Ind. Eng. Chem. Res. **53**(27), 11007–11014 (2014)
127. Q.Q. Wang, Z. He, Z. Zhu, Y.-H.P. Zhang, Y. Ni, X.L. Luo, J.Y. Zhu, Evaluations of cellulose accessibilities of lignocelluloses by solute exclusion and protein adsorption techniques. Biotechnol. Bioeng. **109**(2), 381–389 (2012)
128. Q.Q. Wang, J.Y. Zhu, J.M. Considine, Strong and optically transparent films prepared using cellulosic solid residue (CSR) recovered from cellulose nanocrystals (CNC) production waste stream. ACS Appl. Mater. Interf. **5**(7), 2527–2534 (2013)
129. Q.Q. Wang, J.Y. Zhu, R. Gleisner, T.A. Kuster, U. Baxa, S.E. McNeil, Morphological development of cellulose fibrils of a bleached eucalyptus pulp by mechanical fibrillation. Cellulose **19**(5), 1631–1643 (2012)
130. Q.Q. Wang, J.Y. Zhu, R.S. Reiner, S.P. Verrill, U. Baxa, S.E. McNeil, Approaching zero cellulose loss in cellulose nanocrystal (CNC) production: recovery and characterization of cellulosic solid residues (CSR) and CNC. Cellulose **19**(6), 2033–2047 (2012)
131. R. Wang, L. Chen, J.Y. Zhu, R. Yang, Tailored and integrated production of carboxylated cellulose nanocrystals (CNC) with nanofibrils (CNF) through maleic acid hydrolysis. Chem. Nano Mater. **3**(5), 328–335 (2017)

132. W. Wang, M.D. Mozuch, R.C. Sabo, P. Kersten, J.Y. Zhu, Y. Jin, Production of cellulose nanofibrils from bleached eucalyptus fibers by hyperthermostable endoglucanase treatment and subsequent microfluidization. Cellulose **22**, 351–361 (2015)
133. Q. Wu, Y. Meng, S. Wang, Y. Li, S. Fu, L. Ma, Harper, D. 2014. Rheological behavior of cellulose nanocrystal suspension: Influence of concentration and aspect ratio. J. Appl. Polym. Sci. **131**(15)
134. Q. Xiang, Y.Y. Lee, R.W. Torget, Kinetics of glucose decomposition during dilute-acid hydrolysis of lignocellulosic biomass. Appl. Biochem. Biotechnol. **113–16**, 1127–1138 (2004)
135. X. Xu, F. Liu, L. Jiang, J.Y. Zhu, D. Haagenson, D.P. Wiesenborn, Cellulose nanocrystals vs. cellulose nanofibrils: a comparative study on their microstructures and effects as polymer reinforcing agents. ACS Appl. Mater. Interf. **5**(8), 2999–3009 (2013)
136. H. Yu, Z. Qin, B. Liang, N. Liu, Z. Zhou, L. Chen, Facile extraction of thermally stable cellulose nanocrystals with a high yield of 93% through hydrochloric acid hydrolysis under hydrothermal conditions. J. Mater. Chem. A **1**(12), 3938–3944 (2013)
137. X. Zhao, Y. Zhou, D. Liu, Kinetic model for glycan hydrolysis and formation of monosaccharides during dilute acid hydrolysis of sugarcane bagasse. Biores. Technol. **105**, 160–168 (2012)
138. H. Zhou, F. St. John, J.Y. Zhu, Xylanase pretreatment of wood fibers for producing cellulose nanofibrils: a comparison of different enzyme preparations. Cellulose **26**(1), 543–555 (2019)
139. H. Zhu, W. Luo, P.N. Ciesielski, Z. Fang, J.Y. Zhu, G. Henriksson, M.E. Himmel, L. Hu, Wood-derived materials for green electronics, biological devices, and energy applications. Chem. Rev. **116**(16), 9305–9374 (2016)
140. J. Zhu, L. Chen, R. Gleisner, J.Y. Zhu, Co-production of bioethanol and furfural from poplar wood via low temperature (≤90 °C) acid hydrotropic fractionation (AHF). Fuel **254**, 115572 (2019). https://doi.org/10.1016/j.fuel.2019.05.155
141. J.Y. Zhu, U.P. Agarwal, P.N. Ciesielski, M.E. Himmel, R. Gao, Y. Deng, M. Morits, M. Österberg, Towards sustainable production and utilization of plant-biomass-based nanomaterials: a review and analysis of recent developments. Biotechnol. Biofuels **14**, 114 (2021). https://doi.org/10.1186/s13068-021-01963-5
142. J.Y. Zhu, L.H. Chen, C. Cai, Acid hydrotropic fractionation of lignocelluloses for sustainable biorefinery: advantages, opportunities, and research needs. Chemsuschem **14**(15), 3031–3046 (2021). https://doi.org/10.1002/cssc.202100915
143. J.Y. Zhu, R. Sabo, X. Luo, Integrated production of nano-fibrillated cellulose and cellulosic biofuel (ethanol) by enzymatic fractionation of wood fibers. Green Chem. **13**(5), 1339–1344 (2011)
144. J.Y. Zhu, X.S. Zhuang, Conceptual net energy output for biofuel production from lignocellulosic biomass through biorefining. Prog. Energy Combust. Sci. **38**(4), 583–589 (2012)
145. W. Zhu, C.J. Houtman, J.Y. Zhu, R. Gleisner, K.F. Chen, Quantitative predictions of bioconversion of aspen by dilute acid and SPORL pretreatments using a unified combined hydrolysis factor (CHF). Process Biochem. **47**, 785–791 (2012)
146. W. Zhu, J.Y. Zhu, R. Gleisner, X.J. Pan, On energy consumption for size-reduction and yield from subsequent enzymatic sacchrification of pretreated lodgepole pine. Biores. Technol. **101**(8), 2782–2792 (2010)

Chapter 2
Top-Down Processing of Nanocellulose Materials

Chaoji Chen, Wentao Gan, and Qinqin Xia

Abstract This chapter discusses the top-down processing of nanocellulose materials and their potential applications in various fields. The hierarchically porous structure of biomass, the source material for producing top-down nanocellulose materials, is first introduced to provide basic understanding of its intrinsic structure. With such understanding, a variety of fabrication and modification strategies such as delignification, densification, patterning, surface functionalization and hybridization have been developed to tune the structure and properties of top-down nanocellulose materials. The rich tunability of structure and properties imparts these top-down nanocellulose materials with multiple functions for a wide range of applications in the fields of lightweight structural materials, thermal management, light management, water-energy technologies, electronics and so on. With advantageous features of inherited aligned structure, tunable structure and properties, improved performance and potentially low cost and environmental impacts, top-down nanocellulose materials have emerged as promising candidate of value-added sustainable materials for our society.

2.1 Introduction

Cellulose is the most abundant biopolymer on Earth and widely used in various applications such as pulp and paper industry, packaging, biomedical, personal care, electronics, energy storage, environmental remediation, constructions and so on [1]. Nanoscale cellulose materials are generally called nanocellulose, which can be fabricated by bottom-up or top-down process. In a bottom-up process, the source material is first broken down to nanoscale building blocks (e.g., cellulose nanofibrils, cellulose nanocrystals, cellulose nanofibers), which are further assembled into macroscale

C. Chen (✉)
Hubei Key Laboratory of Biomass Resource Chemistry and Environmental Biotechnology, School of Resource and Environmental Science, Wuhan University, Wuhan 430079, China
e-mail: chenchaojili@whu.edu.cn

W. Gan · Q. Xia
Key Laboratory of Bio-Based Material Science and Technology of Ministry of Education, Northeast Forestry University, Harbin 150040, China

L. Hu et al. (eds.), *Emerging Nanotechnologies in Nanocellulose*, NanoScience and Technology, https://doi.org/10.1007/978-3-031-14043-3_2

structure. This type of nanocellulose materials or structures can be called bottom-up nanocellulose materials or structures. Such bottom-up strategy has been widely explored in the past few decades, boosting the academical research and industrial practice of nanocellulose materials [2]. Recently, another strategy that is different from the bottom-up one, called top-down strategy, has emerged as a new approach for nanocellulose material processing and fabrication [3, 4]. In a top-down process, the source material as a whole is directly processed into the final product (called top-down nanocellulose material), without involving the breaking down of it into nanoscale building blocks and subsequent assembling of these building blocks. Compared with bottom-up strategy, top-down process demonstrates some unique advantageous features such as low cost, low energy consumption, easy for bulk processing at a large scale, inherited structure from the source biomass (e.g., aligned structure), unique properties. For these reasons, top-downTop down process has witnessed a rapid development in the past decade [4].

In this chapter, we discuss recent progress on the top-down processing of nanocellulose materials. Taking wood as an example, we first introduce the intrinsic structure and composition of the source material, as the basis for the subsequent discussion of fabrication and functionalization strategies. Different top-down fabrication and modification strategies are then discussed, including delignification, densification, twisting, patterning, surface functionalization and hybridization. These fabrication and modification strategies have led to the rich tuning of structures and functions of top-down nanocellulose materials, enabling them to be used in various application fields such as lightweight structural materials, thermal management, optical devices, water-energy devices and electronics. Finally, we summarize the concluding remarks and discuss the future challenges and opportunities of top-down nanocellulose materials development.

2.2 Intrinsic Structure and Fabrication Strategies of Top-Down Nanocellulose Materials

Cellulose is mainly biosynthesized by biomass. Understanding the intrinsic structure of biomass and cellulose is critical for further fabrication and utilization of nanocellulose materials. In this section, we will take wood as an example and discuss its intrinsic structure and top-down fabrication strategies of nanocellulose materials.

2.2.1 Intrinsic Structure and Compositions

Biomass, renewable resource energy, is organic material and contains stored energy from the sun, include wood, [5] bamboo, [6, 7] cotton, [8] grass, [9] straw [10] and so on. Wood is one of the most abundant and versatile natural materials, which has been widely used in furniture, buildings and tools and as bio-fuel for thousands of years since ancient time. In the past century, wood is expanding through new uses. Wood is considered as a cellular solid mainly composed of parallel hollow tubes, which can be classified into two major groups termed softwoods (gymnosperms) and hardwoods (angiosperms). Cell as the unit structure of wood can be roughly divided into tracheidal, parenchymal and libriformal cells in terms of their transportation, storage and mechanical strengthening functions, respectively [11]. Wood comprises basic units called cells and is a robust assembly of three primary biopolymer composite materials: cellulose (40–50 wt%) hemicelluloses (15–35 wt%) and lignin (25–35 wt%) (Fig. 2.1) [12–14]. Wood cell walls generally are composed of the middle lamella, the thin primary wall and the much thicker secondary wall. It does not have an obvious interface between primary and middle lamella, thus, they are often collectively termed the "compound middle lamella." The secondary wall, containing the majority of the mass of wood, is typically composed of the distinction of three layers (called S1, S2 and S3), in which S2 accounts for more than 80% of the cell wall volume [15]. The presence of microfibrils in the various cell wall layers oriented at different angles. The microfibrils within the cell wall have an axial direction in the S2 layer and a transverse direction in the outer layers. Cellulose is mostly concentrated in the secondary cell wall, while lignin in the middle lamella. In cell wall, cellulose is a linear polymer of anhydro-D-glucopyranose units linked by β- (1 → 4) glycosidic bonds. The cellulose molecules are linked laterally by hydrogen bonds into linear bundles. Cellulose present as semi-crystalline elementary fibrils to provide the skeleton of wood. The microfibrils can be further divided into cellulose elementary fibrils with diameters of 2–4 nm and hemicelluloses is polymers of anhydrosugar units and contains several different sugar units, specifically glucomannan between the elementary fibrils and xylan at the surface link the fibrils to the surrounding lignin [16]. Lignin, the second most abundant terrestrial polymer, is an amorphous polymer of phenylpropanoids to associate closely with hemicellulose. Hemicellulose and lignin as binder glue the cellulose fibrils in the cell wall together [17].

With a unique hierarchically porous cellular structure, wood exhibits remarkably high strength, stiffness and toughness. Many other lignocellulosic biomasses such as bamboo, grass, straw and bagasse show similar hierarchically porous structure and compositions to wood. The understanding of their unique hierarchical structure and lignocellulosic compositions is critical for the top-down fabrication and modification of nanocellulose materials from these sourced materials.

2.2.2 *Fabrication Strategies*

Several strategies have been used to fabricate nanocellulose materials via top-down process, such as delignification, densification, surface functionalization and hybridization. The different strategies with different treating processes lead to substantially different structures, properties and applications.

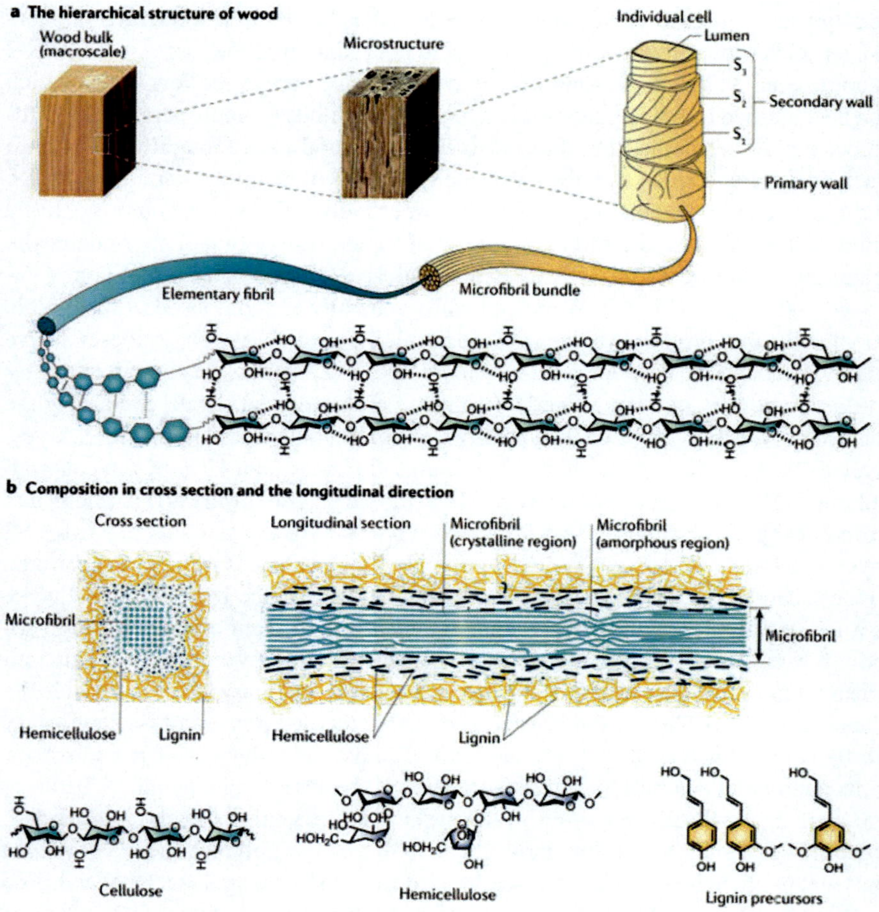

Fig. 2.1 **a** Wood has a hierarchical porous structure with hollow channels aligned along the longitudinal direction. The secondary walls of individual wood cells feature three layers (S1, S2 and S3) that surround the interior lumen (that is, the void space in the center of the cell). Cellulose, hemicelluloses and lignin are the three major components that form the rigid cell wall of wood. **b** The cross section and longitudinal views of the structure of the wood cell wall reveal that the cellulose microfibrils embedded in the lignin and hemicellulose matrix have both crystalline and amorphous microfibril regions. Copyright 2020 Springer Nature Limited [5]

2.2.2.1 Delignification

Delignification is one common top-down strategy to liberate the cellulose nanofibers, during which lignin was in-situ removed from the biomass matrix. Various delignification processes have been carried out under different conditions with respect to structure and concentration of the substrate (lignin), temperature, pH and type of the delignifying reagent. The general methods for lignin separation from biomass (e.g., wood, bamboo, straw and so on) involve different delignifying agents such as acid, alkali, organic solvent and ionic liquid. Acid and alkali are the main delignifying agents and used for a variety of delignification processes.

Acid method. The aromatic structure of lignin undergoes oxidative ring-opening reactions in acidic $NaClO_2$-based delignification processes (Fig. 2.2), during which acidic $NaClO_2$ solution will initiate a electrophilic attack on activated ortho- or para-positions by a radical species. Firstly, the phenolic hydroxyl groups in lignin were eliminated hydrogen to generate phenoxy and mesomeric cyclohexadienonyl radicals. Subsequently, these radicals continue to combine with the ClO_2^- to generate hydrolyzed or undergo elimination of the para-substituent to give quinonoid or muconic acid-type structures. Finally, the formation of acidic groups makes the lignin more soluble in water, leading to the separation of lignin from the cell wall of biomass [18, 19].

Alkali method. The two major alkaline processes for producing cellulose involve the degradation of lignin structure by hot alkali solution boiling process and the improvement of cellulosic whiteness by H_2O_2 boiling.

Lignin, the second most abundant biomass component, can be removed by sodium hydroxide and sodium sulfite. Alkali solvent system mainly cleaves the ether bonds between the structural units of lignin while maintaining the integrity of the carbon–carbon bond, as a result, the lignin with three-dimensional network structure can be degraded into some small segments and dissolved in the alkali solution. The phenolic α-phenyl ether, phenolic α-alkyl ether, phenolic β-aryl ether and methyl aryl ether in lignin will be activated and induced electron transfer by hot alkali solution. During this process, alkaline reagents do not directly engage in substitution reactions with lignin but rather react with quinone methides or other intermediates of the enone type [18, 20]. In an aqueous alkaline environment, OH^- in alkali solution reacts with phenolic hydroxyl group of lignin to form phenolic anion. The presence of a free phenolic hydroxyl group is easily converted into the quinone methide intermediates (Fig. 2.3a, Scheme 1) and subsequently the instable α-substituent can be eliminated (Fig. 2.3b, Scheme 2). It should be noted that β-aryl ether bond can be broken only after neighboring group participates in the reaction. OH^- reacts with α-carbon (C_α) to form epoxy compounds and β-carbon atom (C_β) of β-aryl ether bonds was attacked by sulfite ion to eliminate the β-aryl substituent. Thus, it will result in the cleavage

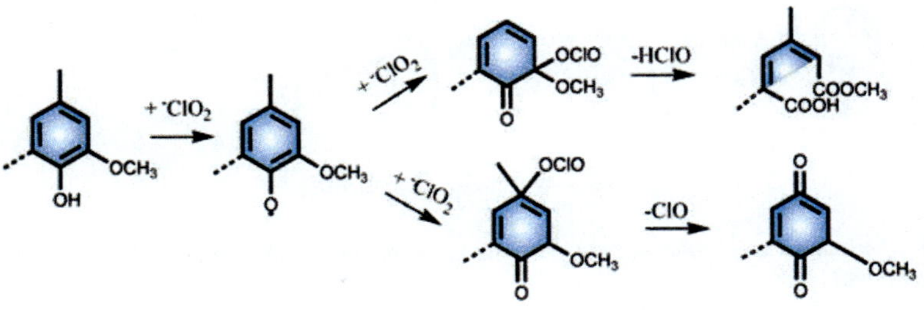

Fig. 2.2 The reaction of phenolic lignin units in acid $NaClO_2$ solution

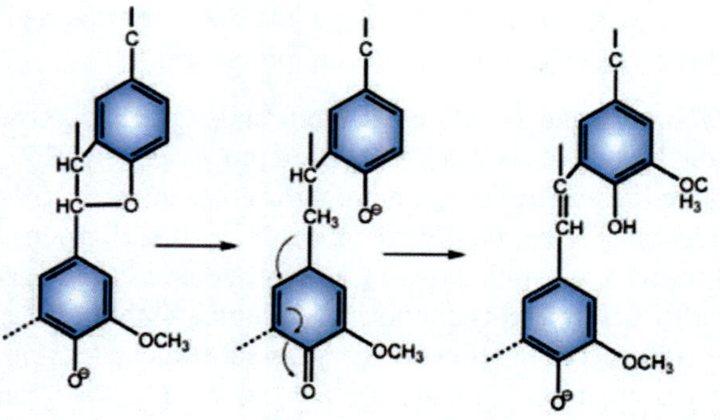

(a) **Scheme 1**. phenolic α-phenyl ether.

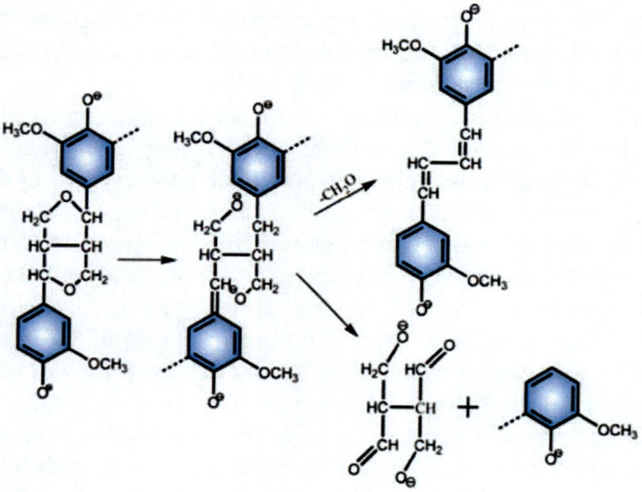

(b) **Scheme 2**. phenolic α-alkyl ether.

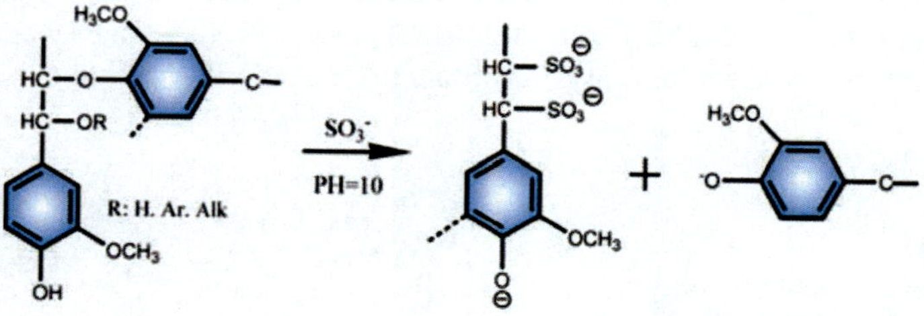

(c) **Scheme 3**. phenolic β-aryl ether.

Fig. 2.3 The reactions of **a** phenolic α-phenyl ether, **b** phenolic α-alkyl ether and **c** phenolic β-aryl ether in $NaOH/Na_2SO_3$ solution

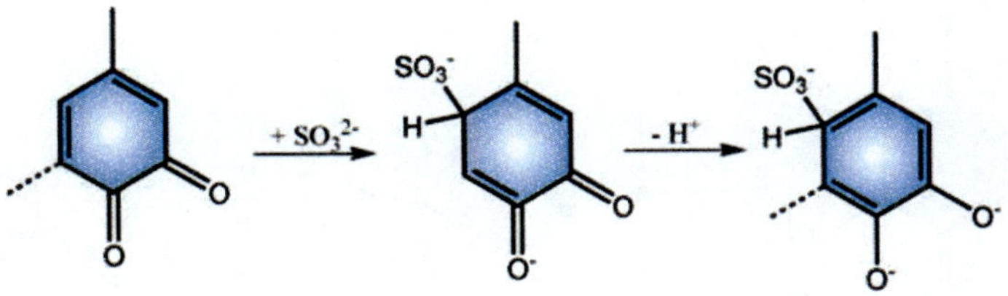

Fig. 2.4 Addition of sulfite ions to quinonoid structures to form aromatic sulfonic acid structures

of β-aryl ether bonds to form small fragments and dissolved in the alkali solution (Fig. 2.3c, Scheme 3).

Lignin was attacked to generate the intermediates of the quinone with brown color and the degradable fragments to dissolve in alkaline solution. Sodium sulfite can be used as a nucleophile to add to quinones to give aromatic sulfonic acid structures, as shown in Fig. 2.4. These structures may also be removed by reductive nucleophilic processes.

Sodium hydroxide applied in the alkaline delignification stage is not a bleaching agent as such H_2O_2. H_2O_2 can be used as bleaching agent and will decompose into H_2O and HOO^- in alkaline condition.

$$H_2O_2 + OH^- = HOO^- + H_2O$$

The perhydroxyl anion HOO reacts with quinonoid structures to produce oxirane, dioxetane and hydroxyquinone structures by molecular rearrangement. The latter are unstable and will further generate alkaline and/or oxidative degradation products, which convert the chromophoric quinonoid structure into the colorless structure (Fig. 2.5). Thus, H_2O_2 are capable of bleaching and delignifying the pulp to improve the brightness. The degradation products from the lignocellulosic matrix can dissolve in alkaline solution.

Organic solvent. Organosolv fractionation is a promising process for effective separation of lignin from biomass (Fig. 2.6). A wide range of organic solvents, including acetone, methanol, ethanol, ketones, ethylene glycol are used in biomass fractionation [21].

Ethanol. Lignin can be separated from biomass by ethanol pretreatment [22]. The β-O-4 bond in lignin was broken. In this process, the G-type lignin can be selectively degraded and transformed to S-type or p-hydroxyphenyl-type lignin. Lignin after ethanol pretreatment will degrade into small fragments with low molecular weight and dissolve in ethanol solution [22].

Valerolactone. γ-valerolactone (GVL), a natural chemical extracted from biomass, is an effective solvent for lignin separation from biomass. Particularly, the aryl-ether bonds will dramatically cleave under the condition of GVL/H_2O (v:v = 80:20) system, resulting in a significant reduction of amount of β-O-4′ linkages [23, 24].

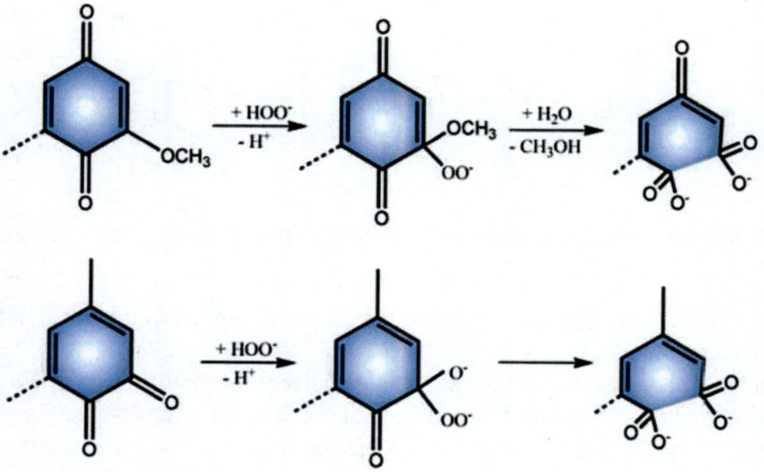

Fig. 2.5 Addition of hydroperoxide ions to side chain structures

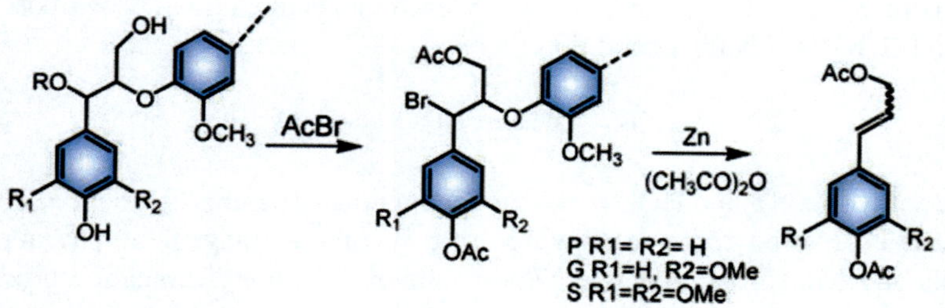

Fig. 2.6 The reaction of β-O-4′ linkages in organic solution

G–S type in the lignin component undergoes remarkable degradation. But lignin will generate repolymerization in this system, which leads to more condensed lignin.

Ionic liquid. Ionic liquids (ILs) have emerged as attractive solvents for biomass pretreatment to obtain nanocellulose and cellulose-based material. They have enhanced activity, high thermal stabilities and low vapor pressures. The ILs properties can be easily tailored by regulating the type of anions and cations, enabling ILs to suit various biomass and optimal processing parameters. Many studies have shown that ionic liquids could destruct the linkage and recalcitrance of lignocellulosic biomass by hydrogen bonding and strong ionic effects, resulting in effective separation of lignin. After ILs pretreatment, the extract efficiency of lignin can be up to 60.4% [25].

Recently an IL-mediated pretreatment process was developed to extracts lignin from maple wood powder. Up to 44% lignin extracted from biomass, but the anion as a hydrogen bond acceptor interacted with the hydroxyl group of cellulose, thus

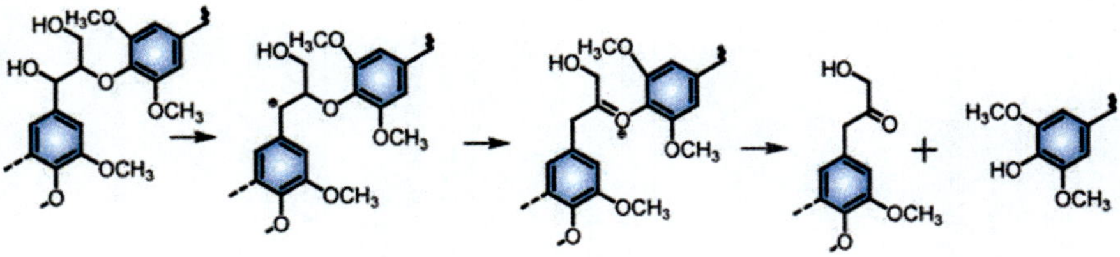

Fig. 2.7 The possible reaction pathway of the cleavage of the β-O-4 compound

disrupting the crystalline structure and hydrogen bond network structure of cellulose. It generally decreases the yield of cellulose [26, 27].

Deep eutectic solvent. Deep eutectic solvents (DESs) as analogous ionic liquids are easy to prepare, low cost, nontoxic, and environmentally friendly, which is promising for lignocellulosic biomass pretreatment and bio-chemicals production. A DES is generally a combination of two or more components which are able to associate with each other by hydrogen bond interaction [28].

DES displays effective lignin fractionation during lignocellulosic biomass pretreatment. Firstly, in acidic DES media, acidolysis occurs in lignin to form a positive proton at C_α. Then, a resonance-stabilized positive charge is formed after a deprotonation and protonation reaction (Fig. 2.7). The C_β of β-O-4 bond was hydrated and deprotonated, forming many lignin fragments containing ketone and phenol groups [29] (Table 2.1).

2.2.2.2 Densification

Compression. Density is the best index of the strength of biomass such as wood; higher density generally indicates greater strength. Densification technology is one of the most commonly used and oldest solid improvement methods. Radial compression to achieve permanent deformation of delignified wood cell wall with aligned cellulose nanofibrils, thereby, increased density and enhanced mechanical properties is expected to be a promising strategy for preparing functional cellulose based materials (Fig. 2.8). Recently, a top-down wood delignification approach combined with densification has been attracting increasing attention in processing high-performance structural materials [30]. A two-step method involves the totally removal of lignin and hemicellulose from the natural wood followed by hot-pressing, resulting in a tightly packed wood cell wall, which greatly enhance hydrogen bond formation among bare cellulose fibers after delignification. The main goal of densifying delignified wood is to increase its mechanical properties (e.g., tensile strength, toughness and hardness). As reported by Hu et al. [30] densifying the delignified wood leads to a completely collapsed wood cell wall with highly aligned cellulose nanofibrils, coining a new type of structural materials. More recently, several studies have delved into the mechanical enhancement mechanism of densified delignified wood [31, 32]. Besides the

Table 2.1 Various delignification approaches

Method	Chemicals	Temp. (°C)	Time	Environmental impact	Cellulose maintenance	Advantage	Disadvantage
Acid	$NaClO_2$ + Acetic acid	80	6–12 h	Toxic gas	High	Effective, economical, high brightness	Cause loss of pulp strength, toxic gas
Alkali	NaOH + Na_2SO_3 + H_2O_2	80–100	7–48 h	Black liquid	High	Effective and economical, low chemical cost, provide chloride-free effluent for recovery	Darkens
Organic	Ethanol + water	180	80 min	Environmentally friendly	High	Low chemical cost, provide chloride-free effluent for recovery	High temperature and high press, specific equipment
	Valerolactone	170	1 h	Environmentally friendly	High	Effective, provide chloride-free effluent for recovery	High temperature and high press, specific equipment
Ionic liquid/ Deep eutectic solvent	[Emim]Ac	125	1 h	Slight toxicity	Low	Effective and fast, provide chloride-free effluent for recovery	High chemical cost
	[Hpy]Cl + [Hmim]Cl	100	30 min	Slight toxicity	Low	Effective and fast, provide chloride-free effluent for recovery	High chemical cost

(continued)

Table 2.1 (continued)

Method	Chemicals	Temp. (°C)	Time	Environmental impact	Cellulose maintenance	Advantage	Disadvantage
	ChCl/lactic acid	120	4	Environmentally friendly	High/Medium	Effective and fast, low chemical cost, provide chloride-free effluent for recovery	The aggregation of lignin on cellulose surface

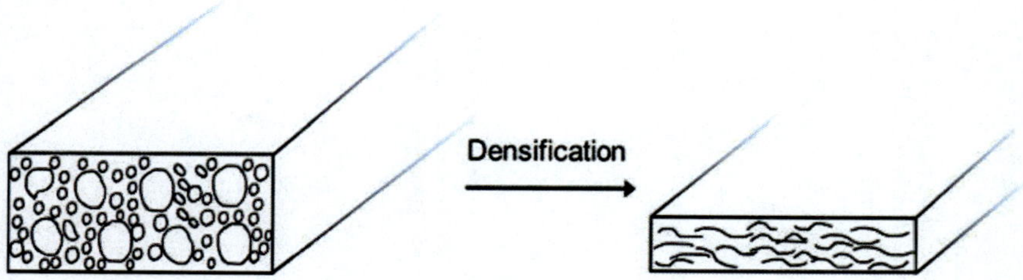

Fig. 2.8 Graphical illustration of the densification process

hydrogen bonding, the reversible interlocks and topographic stacking between the wood cell walls also contributed to the strong cellulose scaffold formation.

Twisting. Twisting is another effective approach to densify biomass structure (Fig. 2.9). Aligned cellulose nanofibers in a dense structure can greatly improve the mechanical performance of fiber-based materials, expanding their applications in photovoltaics, energy storage, thermal management and wearable electronic devices. For example, a facile top-down approach combining delignification and subsequent twisting was developed to fabricate macroscopic wood fibers comprising aligned cellulose nanofibers directly from natural wood [33]. Twisting leads to the collapse of the wood strip into a dense fiber with a circular cross-section and a diameter of a few hundred micrometers. Twisting process can be potentially used in structural engineering of biomass toward multiple functional applications.

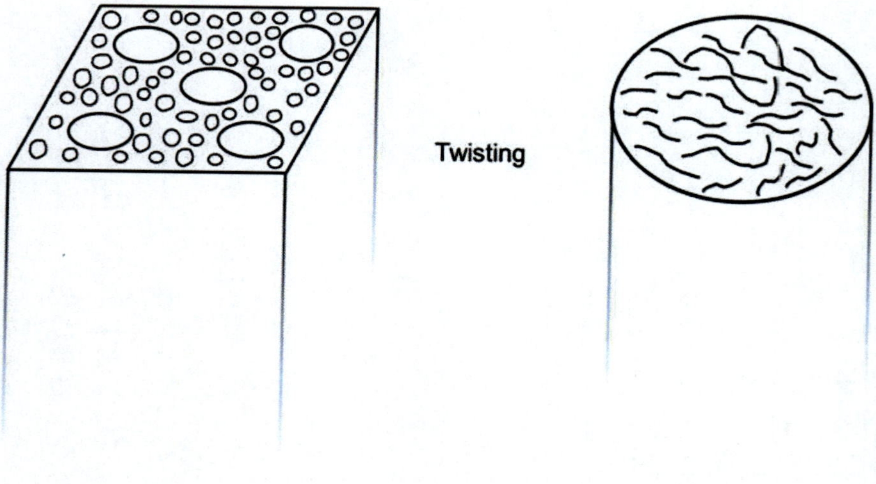

Fig. 2.9 Graphical illustration of the twisting process

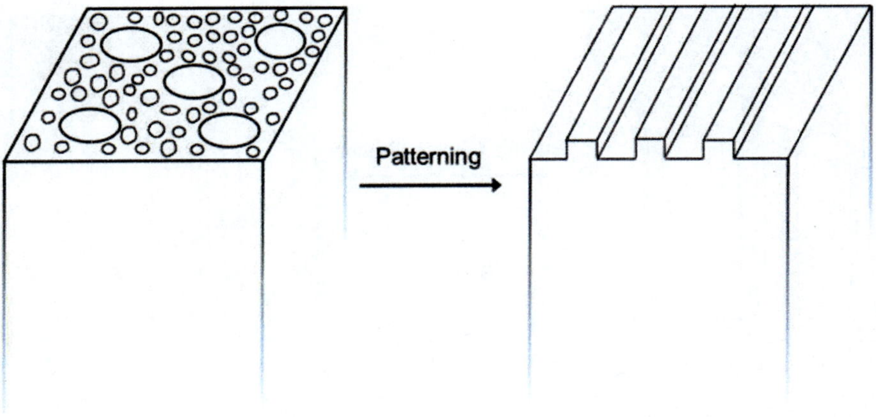

Fig. 2.10 Graphical illustration of the patterning process

2.2.2.3 Patterning

Patterning was developed recently to engineer the surface structure of biomass such as wood (Fig. 2.10). The hierarchical multi-fibrous structure of wood with nanofibril aggregates of ~30 nm in diameter enables the imprinting treatment on its surface. Delignification was applied to release cellulose fibril aggregates from the bondage of lignin prior to imprinting. Covered with H_2O molecules, the fibrils in the wet delignified wood are able to move freely and change their relative positions under loading pressure. When wood became dry, precision patterns were obtained by hydrogen bonds between the adjacent fibrils. Through such smart top-down patterning process, Hu et al. [34] achieved precision imprinted structures on a wood surface with various shapes and sizes, largely extending the potential use of wood to emerging fields such as optics, biology, electronics, etc.

2.2.2.4 Surface Functionalization

Hydroxyl groups in cellulose nanofibrils chains offer extensive possibilities for surface functionalization because of its high reactivity. Chemical modifications can obviously change the charges distribution on the cellulose surface, which is very useful in aiding cellulose functionalization by decreasing agglomeration and endowing new properties (Fig. 2.11). For example, Hu et al. [35] developed a highly conductive nanofluidic membrane directly from delignified wood followed by chemical oxidation. Via 2,2,6,6-Tetramethylpiperidine-1-oxyl radical (TEMPO) oxidation, the long-range ordered cellulose nanofibers with high surface charge density can be obtained due to the converting of hydroxyl groups to carboxyl groups, leading to a high ion transport efficiency. Fu et al. [21] obtained a hydrophobic/oleophilic

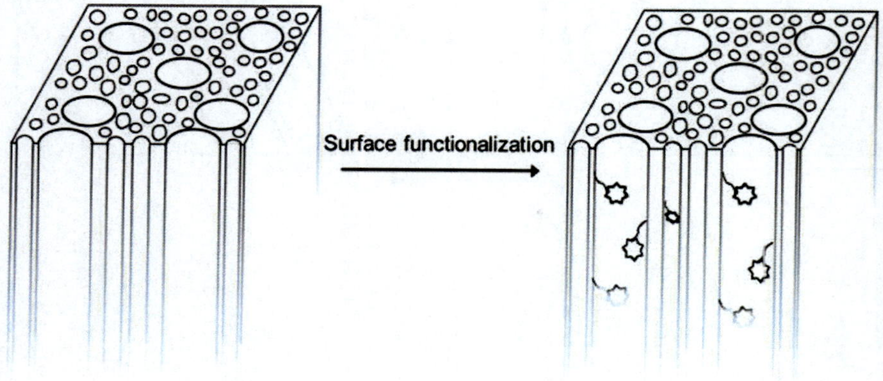

Fig. 2.11 Graphical illustration of the surface functionalization process

delignified wood foam via the delignification and epoxy-amine groups modification. After the graft of epoxy-amine groups on cellulose framework, a low surface energy surface of wood foam can be obtained, thus showing excellent oil–water separation capability.

2.2.2.5 Hybridization

Coating. Hybridization coating is another important treatment strategy to fabricate top-down functional nanocellulose composite (Fig. 2.12). Using top-down approach, hierarchical and porous biomass such as wood with directional cellulose nanofibers provides an ideal platform for nanoparticles hybridization, which can be used to implement new functionalities that go beyond their natural properties. For example, Chen and co-workers [36] successfully prepared plasmonic wood by decorating palladium nanoparticles (Pd NPs) both on the surface and inside wood's mesostructures. Pd NPs was reduced by lignin and capped and bound on the hydroxyls of cellulose and hemicellulose, forming a stable Pd NPs decorating wood membrane. In another study, Hu et al. [37] coated wood membrane with narrow bandgap semiconductor of $CuFeSe_2$ nanoparticles (NPs) for high-efficiency solar steam generation.

Infiltration. Infiltration is another hybridization strategy for biomass engineering (Fig. 2.13). Especially, the natural hierarchical structure that encompasses vertically aligned microchannels and multiscale pores of biomass facilitates the infiltration of external components. For example, Zhu et al. [38] prepared anisotropic transparent wood via top-down method to remove lignin subsequently hybridized index-matching polymer to fulfill the high transparency. The open lumina and well-distributed cellulose nanofibers in delignified wood allow fast penetration of

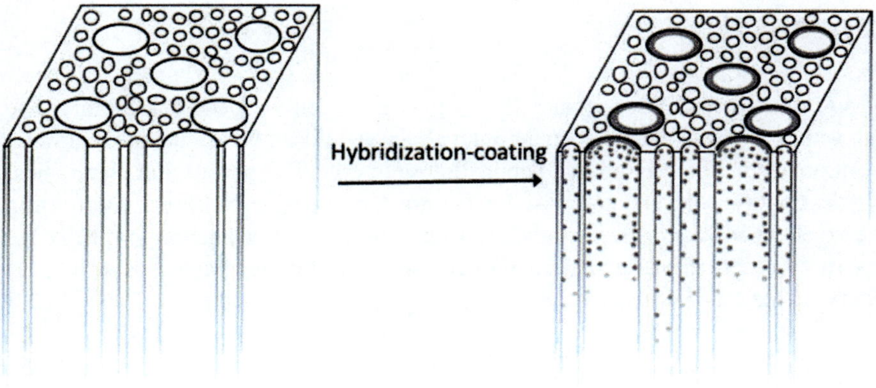

Fig. 2.12 Graphical illustration of the hybridization (coating) process

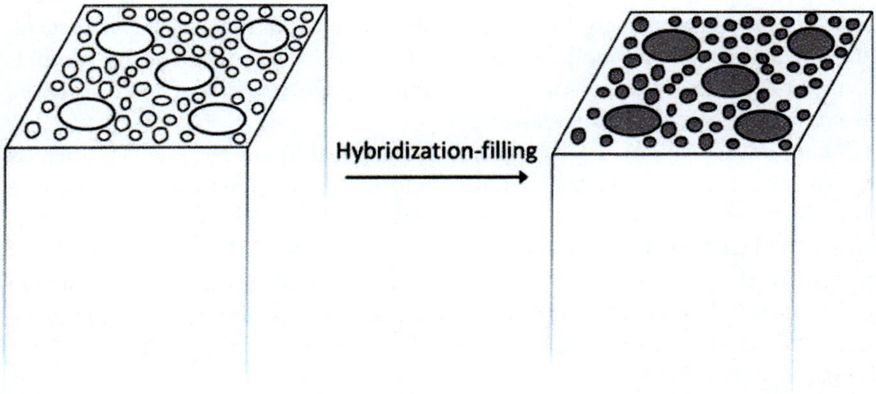

Fig. 2.13 Graphical illustration of the hybridization (infiltration) process

polymer, resulting in a counterintuitive transparent wood. In addition, Fu et al. [39] found the inorganic clay can be easily impregnated into nanocellulosic wood scaffold and formed a nanostructure wood hybrid with outstand fire-retardancy. The clay nanoplatelets are well distributed around the wood cell wall via the strong hydrogen bond between clay and cellulose nanofibers, resulting in an enhanced thermal stability. Hybridizing top-down nanocellulose materials with metals, ceramic and other 2D nanomaterials to form the organic/inorganic composites represents a promising direction for converting nanocellulose-based resources into functional materials and devices.

2.3 Structures

Lignocellulosic biomass is generally composed by cell wall and cell lumina. The cell wall is a typical composite material with the cellulose fibers being the reinforcement embedded in a lignin and hemicellulose matrix. The porous and hierarchical biomass with directional cellulose fibers attracted intensive interest in transferring the excellent characteristic into advanced nanocellulose-based materials with diverse macroscopic structures such as 3D porous foam, 2D films and papers, as well as 1D fibers via the top-down approach.

2.3.1 *3D Structure (Porous Foam Structure and Dense Bulk Structure)*

3D porous foams are widely used in a broad range of applications including bio-scaffolds, sensors, energy storage, water treatment and thermal management. However, the generation of cellulose-based 3D foams is usually based on the "bottom-up" method, involving energy- and time-extensive pretreatments such as disruption, decomposition, extraction and reassembly.

Taking advantages of the porous and hierarchical structure, biomass such as wood can be directly converted to 3D foam materials by lignin and hemicellulose removal from the cell wall through a "top-down" approach. The significant removal of lignin and hemicellulose exposes large amounts of wrapped cellulose nanofibrils and destroys the thin cell walls of natural wood (Fig. 2.14a). The 3D porous structure can be controlled through different drying processes. Using the partial densification, a densified wood structure with collapsed cell wall was obtained through the slight hot-pressing at 60 °C until its totally dried (Fig. 2.14b) [40]. Based on the freeze drying, natural wood after chemical treatment was converted to wood aerogel with lamellar structure and numerous stacked arched layers (Fig. 2.14c, d). Inheriting the cellulose arrangement in natural wood, the 3D porous wood foam with layer-by-layer structure, shows numerous promising characteristics of high porosity, low density, large specific surface area, good mechanical properties, outstanding elasticity, superb thermal and electrical insulation, as well as tunable surface chemistry [41]. Furthermore, the lamellar structure of wood foam can be well preserved even after high-temperature carbonization (Fig. 2.14e) [42]. The fact that wood foam can form a compressible carbon sponge with excellent electric conductivity opens up more possibility for intelligent electronic devices by using renewable wood-based materials [42, 43].

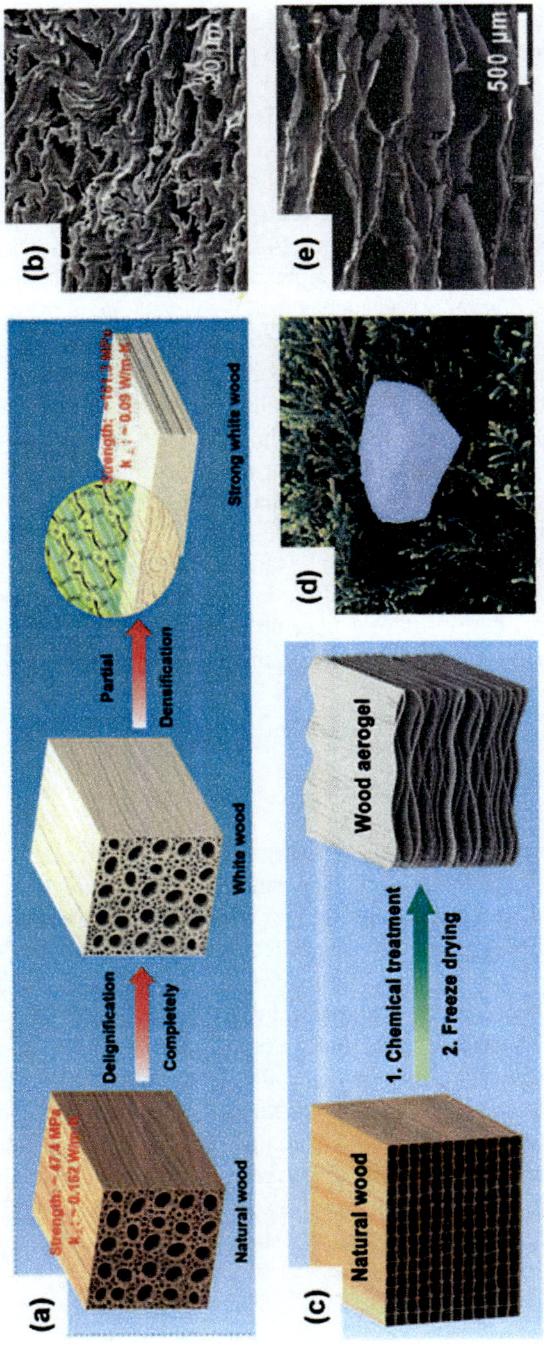

Fig. 2.14 3D top-down nanocellulose structure. **a** Schematic showing the two-step top-down fabrication process of strong white wood. **b** Microstructure of the strong white wood. **c** Schematic showing the two-step top-down fabrication process of wood aerogel. **d, e** SEM and optical images of wood aerogel. Copyright 2018 American Chemical Society, [43] 2019 WILEY–VCH [40]

2.3.2 2D Structure (Film)

Through a top-down process (e.g., delignification plus densification), biomass such as wood could be directly converted into 2D film or paper. The top-down method includes two steps: the delignification process for removing colorful lignin and a partial of hemicellulose, and a pressing process for densifying cellulose nanofibrils to remove the light-scattering air pocket. Subsequently treated by shear pressing, two kinds of transparent paper (e.g., anisotropic transparent paper and isotropic transparent paper) can be obtained (Fig. 2.15a–c) [44, 45]. The wood block is sliced perpendicular to longitudinal direction with a preset thickness. After delignification, a soft and easily processed white wood slice is well preserved due to the removal of lignin binder, which can be further transformed to a thin film during hot-pressing. Finally, the vertically aligned wood channels are completely crushed into an isotropic transparent paper with a randomly distributed cellulose fibers by longitudinal hot-pressing (Fig. 2.15d–f). Meanwhile, an anisotropic transparent paper with highly aligned cellulose fibers can be obtained by shear pressing (Fig. 2.15g), where the alignment of cellulose fibers can be clearly observed in the cross-sectional SEM image (Fig. 2.15h, i). In addition to adjusting the pressing direction, the structural anisotropy in transparent wood film can also be tailored by tuning the cutting direction. Different angle between the sawing direction and the longitudinal direction in natural wood corresponds to different fiber arrangement in treated wood film, resulting in different anisotropic properties (e.g., mechanical property) [44].

2.3.3 1D Structure (Fiber)

1D nanoscale cellulose fiber exhibits extraordinary mechanical performance owning to its defect-free molecular structure. Cellulose nanofibrils with theoretical Young's modulus up to 120–140 GPa and a tensile strength of 6–7 GPa has great potential in many areas such as textiles, construction and reinforced composites [46]. Recently, a novel top-down approach for fabricating macroscopic wood fibers with highly aligned cellulose nanofibers directly from natural wood through delignification and densification, following by twisting is demonstrated (Fig. 2.16a) [33]. The diameter of the twisted wood-textile fibers can be controlled by tuning the cross-sectional dimension of delignified wood strips. After twisting, the delignified wood strip is rubbing into a tightly intertwined fiber with a circular cross-section and less defects (Fig. 2.16b–d). The aligned microchannels with a certain twist angle and well oriented cellulose fibers are clearly presented in wood-textile fiber surface (Fig. 2.16e–g), which is conducted to the mechanical property enhancement.

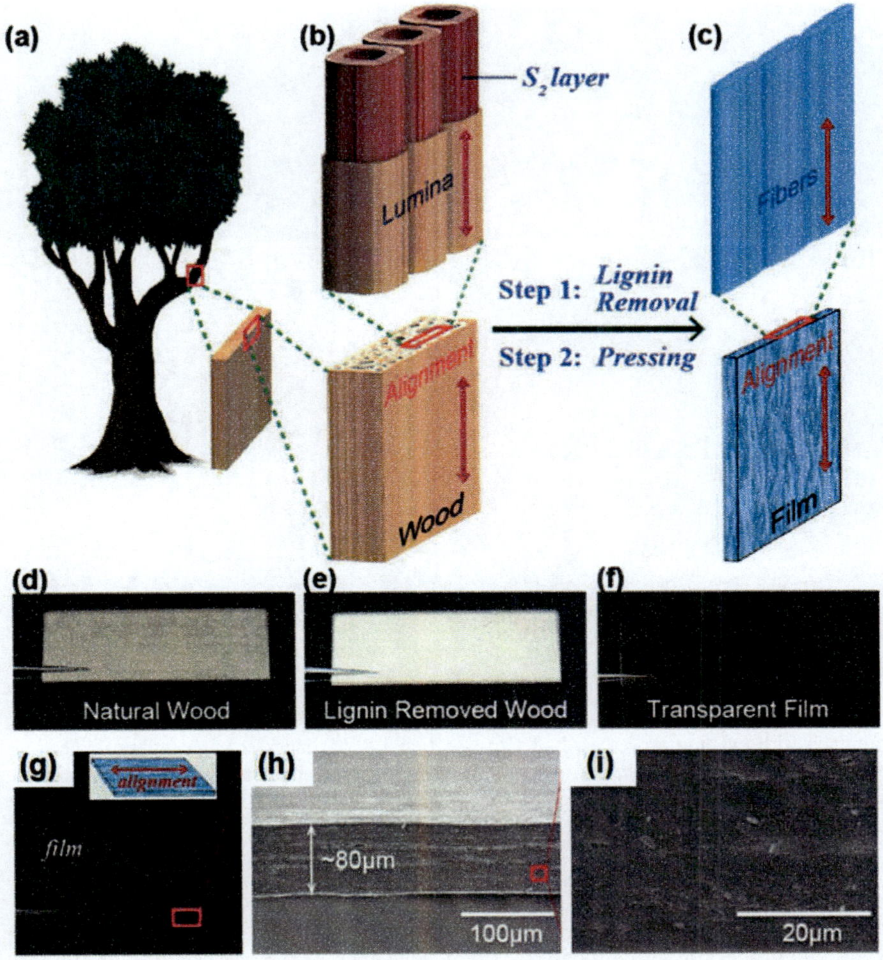

Fig. 2.15 Top-down fabrication of 2D nanocellulose material from wood. **a** Wood microstructures. **b** The two-step fabrication process. Step 1: lignin removal, which leads to the change of wood color from yellow to white while maintaining the cellulose framework in natural wood. Step 2: mechanical pressing, which leads to about a five-time thickness reduction. **c** The lumina are fully closed after mechanical pressing. During Step 1 and Step 2, both the microscale alignment for lumen and the nanoscale alignment for CNFs in natural wood are well preserved in the resulting transparent film. **d–f** Photos of the color changes from (**d**) yellow natural wood to (**e**) white wood and finally to a (**f**) transparent film. **g–i** Picture and SEM images of transparent film pressed from the white wood. It is dense and the anisotropic property is well preserved. Copyright 2017 WILEY–VCH [44]

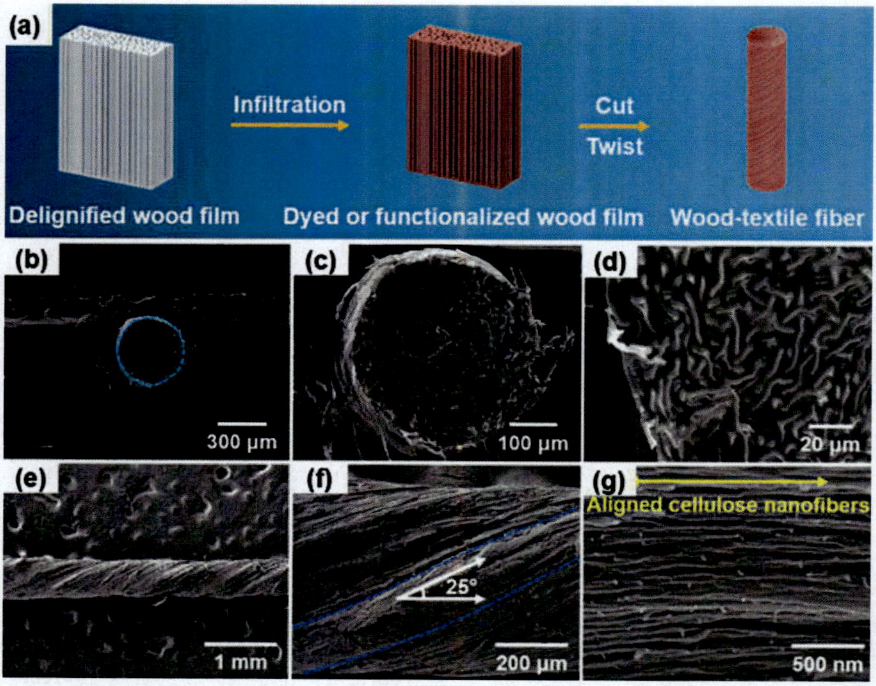

Fig. 2.16 1D wood fiber. **a** The schematic of wood fiber preparation via the delignification combined with twisting processes. **b–d** Cross-sectional SEM images of wood fiber, indicating tightly intertwined cellulose fibers. **e–g** SEM images of wood fiber, demonstrating the aligned cellulose nanofibers along the longitudinal direction. Copyright 2018 WILEY–VCH [33]

2.4 Applications of Top-Down Nanocellulose Materials

Top-down processing of nanocellulose materials enables the preservation of aligned structure of native biomass. Taking wood as an example, its derived top-down nanocellulose materials generally process a highly aligned structure, which is substantially different from nanocellulose materials fabricated from a bottom-up process. Moreover, the structure and properties of top-down nanocellulose materials could be facilely tuned via various physicochemical approaches. These features combined make top-down nanocellulose materials functional in a wide range of applications such as lightweight structural materials, thermal management, optical devices, water-energy nexus and electronics.

2.4.1 Lightweight Structural Materials

Lightweight structural materials have been identified as a critical demand in construction, automotive and aerospace since the reduction of structural mass, thereby saving energy and reducing the greenhouse gas emission. Although some biomass such as natural wood has been used as structural material for construction since ancient time, the insufficient mechanical properties (e.g., strength and toughness) is unsatisfactory for many advanced applications such as skyscraper, automobile and space exploration. Recently, a top-down wood modification approach has attracted great attentions in processing high-performance structural materials (Fig. 2.17a–c) [30]. Bulk natural wood was directly processed into a super-strong structural material with a more than tenfold enhancement in mechanical strength via the partial delignification combined with densification. Benefitting from the densely packed cell walls with aligned cellulose nanofibers, the densified wood demonstrated a high specific tensile strength of up to 422.2 MPa cm^3 g^{-1}, which is superior to natural wood, typical plastics, even of lightweight titanium alloy (Ti6Al4V) without additional additives [30]. More recently, Xiao et al. reported a cell wall engineering strategy to shape flat wood sheets into various 3D structures, such as honeycomb, zigzag and corrugated structures without ripping or tearing [47]. The resulting 3D-molded wood exhibits six-fold enhancement in tensile strength than that of natural wood, and even comparable to the widely used lightweight aluminum alloy but with only 1/3 of its density, demonstrating great potential in advanced engineering fields that require both high mechanical strength and formability (Fig. 2.17d–f).

In addition to the high mechanical strength and excellent formability, another benefit of these top-down nanocellulose materials is their potentially lower cost and less energy consumption compared with those bottom-up counterparts. More attractively, they generally demonstrate much lower environmental impacts than those petrochemical-based structural materials such as plastics and carbon fibers, as well as the widely used structural materials such as steels and alloys. Such intriguing environmental sustainability along with their high mechanical strength and good formability positions top-down nanocellulose materials as a promising candidate of next-generation sustainable lightweight structural material.

2.4.2 Thermal Management

Thermal management materials featured cost-effective energy efficiency are vital to substantially lower the energy consumption and decrease carbon footprint. Biomass such as wood naturally exhibits low thermal conductivity and anisotropic thermal transport owing to its semi-crystalline structure, specific orientation of cellulose fibrils and porous structure. Tuning the biopolymeric components and porous structure can be an effective strategy in manipulating the thermal properties of biomass.

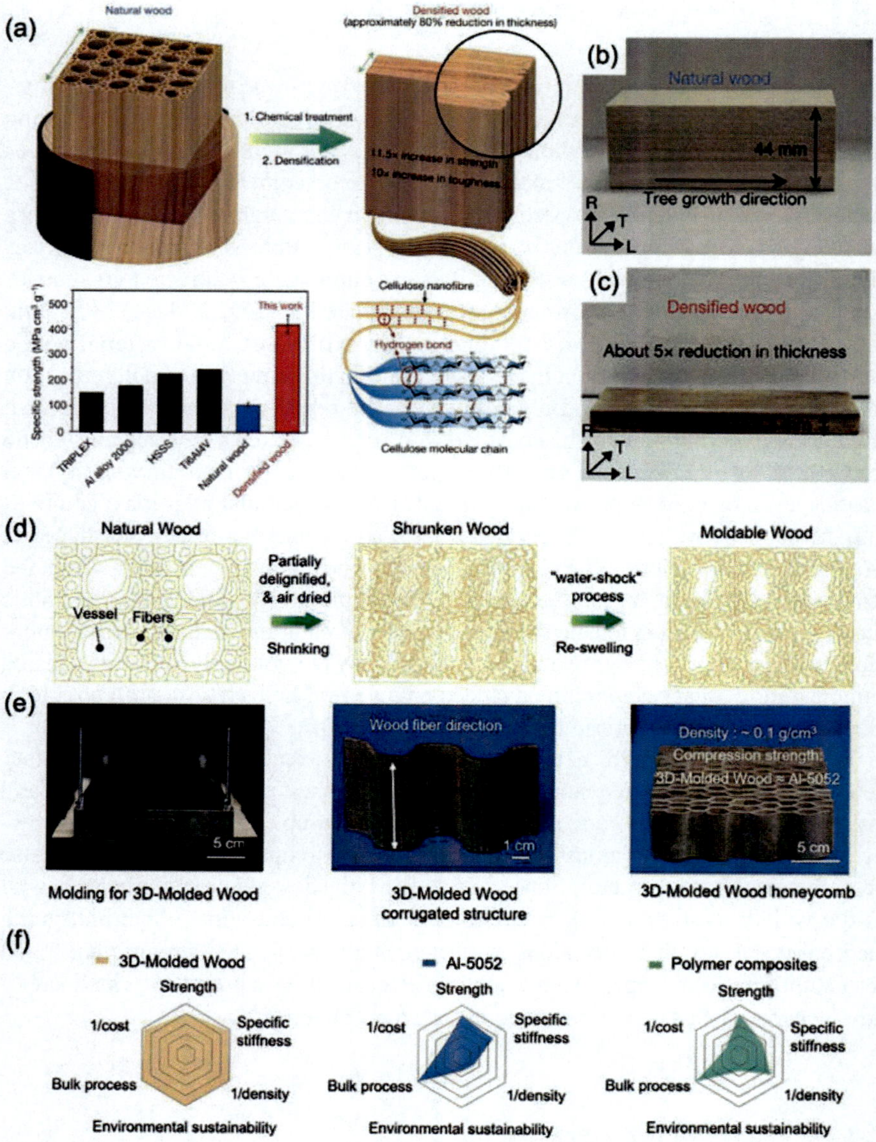

Fig. 2.17 Lightweight wood structural materials. **a** Processing approach and mechanical performance of densified wood via the delignification and densification. Photographs of (**a**) natural wood and (**b**) densified wood. **c** The schematic of wood cell wall structure during the fabrication of the moldable wood. **d** Photographs of 3D-molded wood structure made from wood veneers. **e** Radar plots comparing the performance of 3D-molded wood, Al-5052 alloy and typical polymer composites. Copyright 2018 Springer Nature, [30] 2021 AAAS [45]

The most common thermal management application of top-down nanocellulose materials is thermal insulation. Recently, Li et al. [41] reported the excellent anisotropy of thermal conductivity of a hierarchal alignment of cellulose nanofibrils (denoted as nanowood) (Fig. 2.18a). Therein, nanowood exhibits anisotropic thermal properties with an extremely low thermal conductivity in the transverse direction (0.03 W/m·K) and high thermal conductivity in the axial direction (0.06 W/m·K) owing to the hierarchically aligned nanofibrils within the highly porous cellulose framework. As a result, the thermal conductivity of nanowood along the axial direction can be more efficient thermal dissipation than that of the transverse direction, leading to preventing local overheating on the illuminated side (Fig. 2.18b). In another study, He et al. [40] developed a high-performance top-down nanocellulose material, called strong white wood, with improved mechanical properties and thermal insulation for energy efficient high-rise buildings by pore structure engineering. The fabrication approach of such thermal insulating strong white wood material involves a complete delignification treatment and partial densification process (pore structure control), which induces strong hydrogen bonds forming between the cellulose nanofibers and preserve cellulose nanofibers' alignment, leading to efficient decrease of the conductive heat flow (Fig. 2.18c, d). Song et al. [43] developed a top-down fabrication strategy (i.e., removing lignin and hemicellulose) to directly convert natural wood into wood aerogel with a layered structure. Such wood aerogel exhibited not only excellent compressibility but also an extremely low thermal conductivity of 0.028 W/m·K along the axial direction, which is 3.6-times lower than that of the natural wood material (Fig. 2.18e–g).

More recently, another attractive thermal management capability of top-down nanocellulose materials, called radiative cooling, was revealed by Hu's group [48]. A material with passive radiative cooling capability could emit more heat than the amount of solar irradiance it absorbs, which offers a perpetual path to dissipate heat from the structure through the atmospheric transparent window into the ultracold universe without energy consumption, thus being able to efficiently regulate the interior temperatures of buildings in an energy efficient way. Hu's group found that top-down nanocellulose material could be an idea candidate for passive radiative cooling through rational engineering of bulk wood material. The passive radiative cooling top-down nanocellulose material called cooling wood was fabricated by completely removing lignin and partial hemicellulose from wood followed by mechanical pressing. Such treatment endows wood with the low optical loss of the cellulose fibers and the disordered photonic structure, resulting in continuous subambient cooling during both day and night. As shown in Fig. 2.18 h, i, an average below-ambient temperature of >9 °C during the night and >4 °C between 11 am and 2 pm were archived, demonstrating excellent radiative cooling performance.

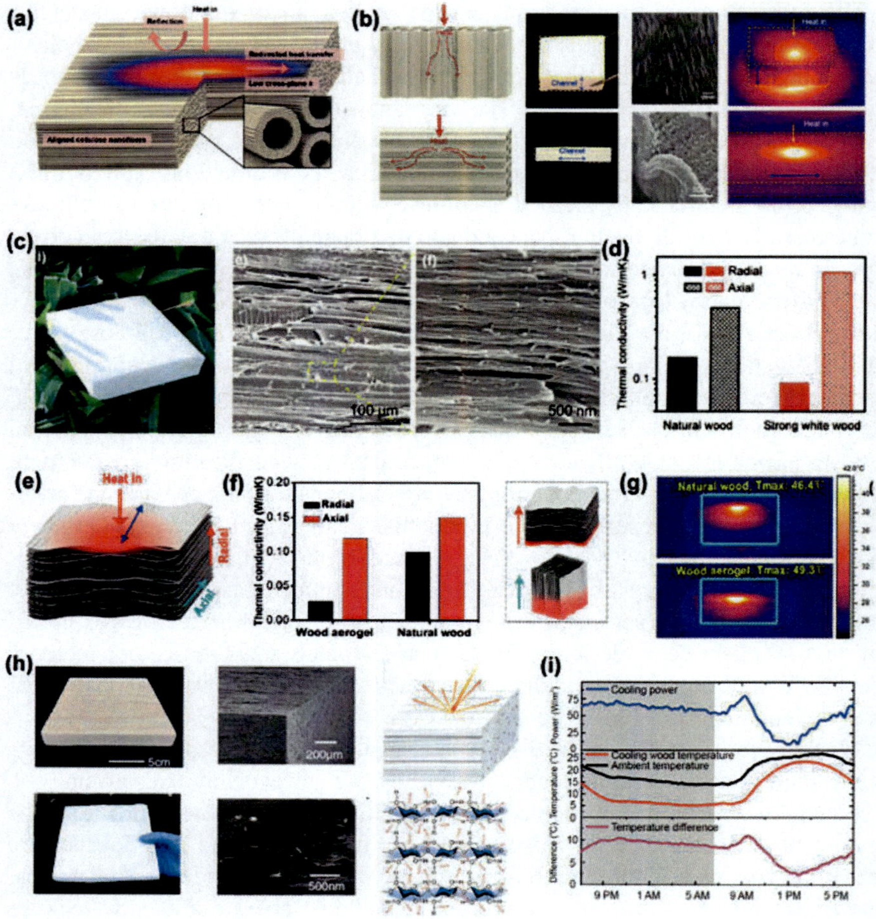

Fig. 2.18 Thermal management applications of top-down nanocellulose materials. **a** Schematics of the thermally insulating properties of the nanowood. **b** Schematic representation of heat conduction along the wood cell walls as axial heat transfer, whereas heat conduction across the cell walls and hollow channels (that is, the lumen and the nanosized pores inside the fibril walls) is referred to as transverse heat transfer. **c** Photo image and the SEM images in TL plane of the strong white wood. **d** Thermal conductivities of natural wood and strong white wood along two different directions, radial and axial directions denote perpendicular and parallel to tree growth direction, respectively. **e** Schematic of the anisotropic thermally insulating properties of the wood aerogel. **f** Thermal conductivities of natural wood and wood aerogel, perpendicular (radial) and parallel (axial) to the cellulose nanofiber alignment direction. **g** IR images of natural wood and wood aerogel, showing the temperature distribution along the in-plane surface. **h** Photos of a board and SEM of natural wood and cooling wood. SEM image of partially aligned cellulose nanofibers of the cooling wood. Schematic showing the wood structure strongly scattering solar irradiance. **i** Twenty-four–hour continuous measurement of the 200 mm–by–200 mm cooling wood. Copyright 2019, 2018 AAAS, [41, 48] 2019 WILEY–VCH [40]

2.4.3 *Optical Applications*

Recently, top-down nanocellulose materials have also been used for optical applications. One elegant example is the invention of transparent wood [49]. Native wood can be made transparent via the top-down approach involved in chemical delignification to remove light-absorbed lignin, followed by polymer immersion to eliminate the light-scattering air gaps between cellulose nanofibrils. Apart from the high transparency of transparent wood, the most outstanding advantage of transparent wood is the great light management capability, which not only can guide sunlight through the wood channels, but also exhibits a macroscopic light propagation effect with a large haziness (Fig. 2.19a) [38]. These unique light management capabilities of transparent wood ensure its potential use as transparent substrate for energy efficient buildings, [50] solar cells, [51] heat-shielding windows, [52] and luminescent devices [53]. Hu et al. [54] firstly introduced the transparent wood as the basic window or roof to replace glass for energy efficient buildings. They developed a transparent wood with the broadband optical transparency over the visible wavelength range larger than 85% and haze larger than 95% (Fig. 2.19b, c). The super high optical haze of transparent wood can create a uniform and consistent daylight distribution without glare effect. Furthermore, the thermal insulation with the low thermal conductivity of transparent wood can retard the flow of heat through the building roof and provide a structural barrier between house and outside environment, which make house warmer in winter and cooler in the summer (Fig. 2.19d). Zhu et al. [55] successfully prepared a highly transparent wood composite with low-tortuosity channels, which can be used as a broad range light management layer for overall improving the energy conversion efficiency up to 18% in a GaAs solar cell owning to the outstanding light scattering capability.

More recently, aesthetic patterns were integrated to the transparent wood composites without compromising their optical performance, imparting them with a new feature called "aesthetics." [56, 57] There are two top-down strategies to fabricate aesthetic, patternable transparent wood composites: (1) selective delignification combined with polymer infiltration to make wood transparent while preserving the intrinsic aesthetic patterns (i.e., growth rings) of the native wood; [56] and (2) patternable delignification via solar-assisted chemical brushing treatment combined with polymer infiltration such that the covered parts of wood remain unchanged while the uncovered parts turn transparent [57]. Such aesthetic designability largely extends the application fields of transparent wood composites and improves their added values, representing an exciting direction for value-added utilization of biomass in the optical fields and beyond.

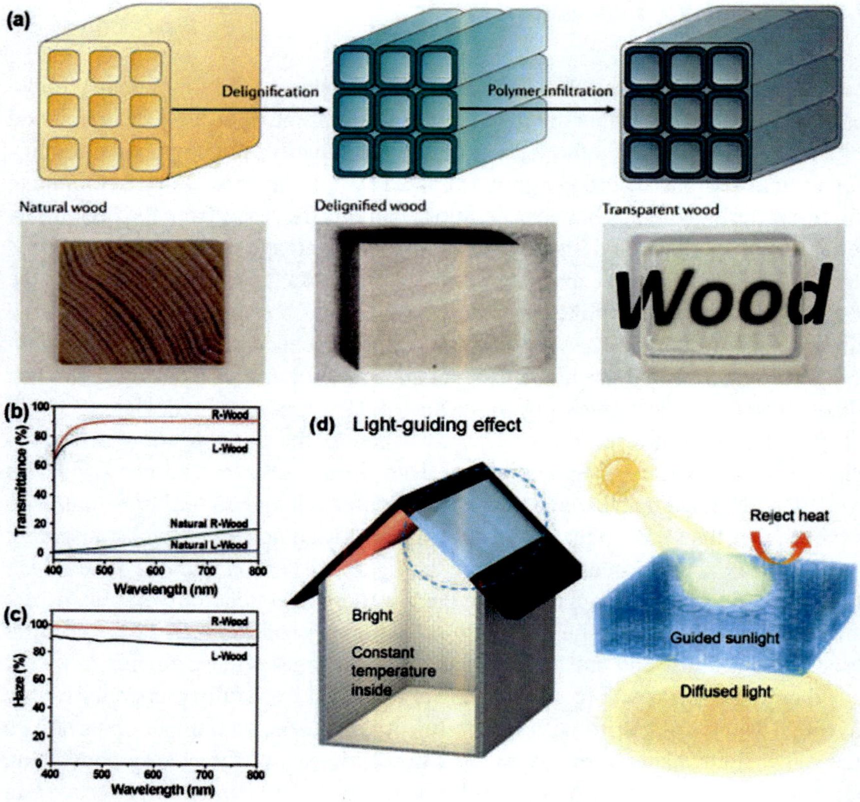

Fig. 2.19 **a** Wood can be made transparent via a two-step top-down process involving delignification and polymer infiltration. First, wood is chemically treated to remove lignin from the cell walls and it is then infiltrated with a polymer of a refractive index that matches the refractive index of the wood substrate. The photographs show the color change of wood from brown (natural wood) to white (delignified wood) to transparent. **b** Total transmittance for nature R-wood, nature L-wood, transparent R-wood and transparent L-wood. **c** Both R- and L-wood composites show high transmittance haze. **d** A schematic of the wave-guiding effect of transparent wood. Light can be guided along the wood fibers independent of the angle of the incident beam. Copyright 2020 Springer Nature, [5] Copyright 2016 WILEY–VCH [38, 54]

2.4.4 Water-Energy Nexus

Water scarcity and energy shortage are two most stressing challenges of modern society [58]. Developing sustainable water-energy technologies is highly demanded. Top-down nanocellulose materials generally preserve the aligned, hierarchically porous structure from the native biomass (e.g., wood), thus being attractive for selective multiscale mass transport, which is essential for water-energy applications [5]. For example, after modified (e.g., carbonization or delignification combined with carbon and nanoparticle coating), wood can be used as electrode scaffold for electrochemical energy storage (Fig. 2.20a) [59, 60]. The hierarchical pores inherited from the native wood provides separated pathways for ions, electrons and even gas transport, imparting the modified wood based electrode excellent electrochemical performance. When used as filter for water treatment, wood can be modified via metallic nanoparticle coating (Fig. 2.20b) [36, 61]. The pollutant could be easily removed when the contaminant water went through the hierarchical pores. Solar evaporator is another water treatment device, which utilizes solar energy to heat up the evaporative surface, as such the water evaporation will be sped up and vapor condensed and collected as clean water [58]. Wood is a naturally functional evaporator, with abundant pores for water transport, a low thermal conductivity for heat localization at the evaporative surface. With suitable treatment to improve its light absorption capability, e.g., surface carbonization, photothermal particle coating, the evaporation rate of top-down nanocellulose based evaporator improves substantially, demonstrating great potential for affordable clean water production (Fig. 2.20b) [62, 63].

Although excellent water treatment performance (both filter and solar evaporator) has been achieved, the water stability of these top-down nanocellulose based materials remains as a potential challenge. Long-term durability of the device needs further investigation before this type of products can finally go into the market as a reliable and competitive product.

2.4.5 Electronics

Consumer electronics, such as computer, smartphone, tablet, wearable healthcare devices and artificial intelligence have brought us many conveniences. However, they are typically made by some non-renewable and non-biodegradable substrates and conductive circuit, which is regarded as electronic waste that is frequently discarded at the end of its service life. In response to the challenges of sustainable development, the new generation of green electronic devices fabricated by renewable and biodegradable materials exhibit huge potentials [64].

As the most abundant, renewable and biodegradable biomass, wood is an attractive musical instrument material and can be used to construct many orchestral instruments such as violin, piano and flute, due to the lamellar structure with aligned wood fibers and good vibration properties, easy processability, as well as dielectricity. Through suitable top-down processing, wood can be made into super-strong and ultrathin top-down nanocellulose material for use in acoustic devices with much improved performance. Gan et al. firstly reported a wood vibrating diaphragm with a record-small thickness of less than 10 μm through a simple top-down approach involved in partial delignification followed by densification (Fig. 2.21a) [65]. The high tensile strength and Young's modulus of wood diaphragm caused by reduced mechanical defects of cellulose nanofibrils in wood cell wall enable excellent acoustic properties

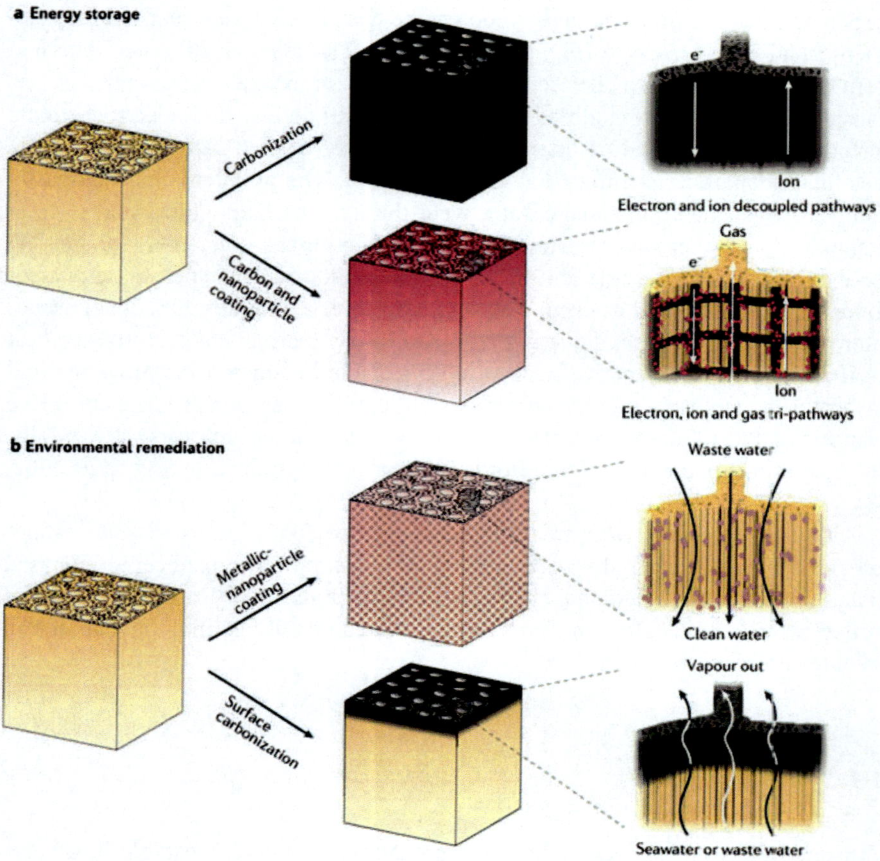

Fig. 2.20 **a** The construction of wood-based energy-storage devices with decoupled pathways for fast transport of ions and electrons. These pathways can be achieved by the carbonization of the wood or by partial delignification and coating with carbon nanotubes and/or metallic nanoparticles to form triple pathways for the fast transport of ions, electrons and gas. **b** For environmental-remediation applications, decorating metallic catalyst nanoparticles or functional groups on the surface of the cell walls can yield a wood-based filter with a high throughput of water and high efficiency. Alternatively, a surface-carbonized bilayer wood evaporator, which offers strong light absorption, fast water transport and evaporation, and good thermal insulation, can be used for solar-assisted production of clean water. Copyright 2020 Springer Nature [5]

with a high resonance frequency and large displacement amplitude, which are much better than these of polypropylene diaphragm in a commercial speaker.

In addition to structural compaction, converting wood cell walls into carbonaceous material with high electrical conductivity can also be implemented in various electronic devices and sensors. Chen et al. [42] demonstrated that the substantial removal of lignin and hemicellulose combined with high-temperature carbonization change the chemical components and structure of wood, leading to the formation of a superb elastic wood carbon sponge with a high electrical conductivity instead

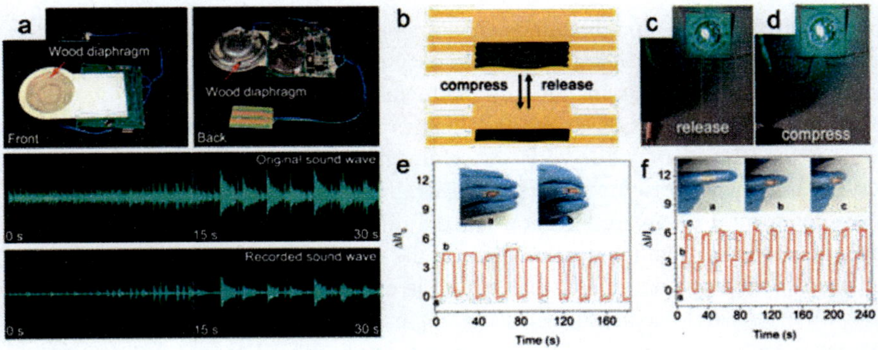

Fig. 2.21 **a** Photographs of the speaker with wood diaphragm and the sound wave of the original song and recorded sound wave, showing the good acoustic properties of the speaker prepared by wood diaphragm. **b** Schematic of the wood carbon sponge for strain sensor under compression and release condition. **c, d** Photographs of the LER light with wood carbon sponge sensor under release and compression. **e, f** Current changes under different progressive motions, demonstrating the promising application for monitoring human body movements. Copyright 2019 Springer Nature, [65] 2018 Elsevier [42]

of brittle wood carbon (Fig. 2.21b). The compressive wood carbon sponge with a spring-like lamellar structure exhibited a maximum strain as high as 80% under the compressive stress of 70 kPa, while the carbonized wood without lamellar structure was collapsed into small pieces under the same compressive strain. Based on the sensitive electrical responses upon compression, a flexible strain sensor was assembled by using the wood carbon sponge as the substrate and copper tape as the coating layers (Fig. 2.21b). After compression, the connected light-emitting diode (LED) becomes bright, demonstrating the increased electronic conductivity of wood carbon sponge sensor (Fig. 2.21c, d). Furthermore, the wood carbon sponge sensor was also chipped onto a human finger, the sensitive current changes can be detected by controlling the bending of the finger motion, suggesting its promising application for monitoring human body movements (Fig. 2.21b–f). However, as such wood-based sensor failed to detect the faint electronic pulses such as the blood pressure and cardiac impulse, further improving its fitness and sensitivity is necessary.

2.5 Concluding Remarks and Prospects

Top-down processing of nanocellulose materials has become one of the most powerful strategies for cellulose based functional materials development. Intensive progress has been made on the fabrication and modification techniques, as well as functionalization and applications. Moreover, the recent advances of nanotechnologies on the synthesis, characterizations and device science and engineering have accelerated the development of top-down nanocellulose materials. On the other hand, theoretical simulation methodology development has enabled the deeper understanding of the processing-structure–property-function relationships of top-down nanocellulose materials, which in turn facilitates their utilization.

Despite these progresses, several challenges remain. First, most progresses were made in the laboratory, while very few achievements had been made in the market, leaving huge gaps between academic research and commercialization. The major obstacles toward commercialization are the relatively small manufacturing scale, higher cost than the competitive products in the market, as well as the reliability and uniformity of the products. However, with the continuously reduced manufacturing cost, more matured manufacturing technologies and driving force from the government in pursuing sustainable development and carbon neutrality goal, more and more products made from top-down nanocellulose are expected in the near future. From a scientific point of view, the uniform chemical modification of the bulk raw material with a large size remains challenging due to the slow transport and diffusion of chemicals within these large-sized samples. Deeper understanding of the diffusion behavior and mechanism is needed to address this issue. Material stability and fire safety are also challenging to achieve for nanocellulose based materials due to their hydrophilicity and flammability. Strategies that can improve their water stability and fire safety yet without losing their performance and sustainability (e.g., biodegradability, low environmental impact) are needed. Finally, the true sustainability of top-down nanocellulose materials need to be evaluated comprehensively through life-cycle-assessment (LCA) and techno-economic assessment (TEA), including the entire cycle loop ranging from raw materials, processing, product serving, to the final disposal after its serving life. Maintaining the sustainability of top-down nanocellulose materials should be the most primary design principle in bio-based sustainable functional material development.

References

1. T. Li, C. Chen, A.H. Brozena, J.Y. Zhu, L. Xu, C. Driemeier et al., Developing fibrillated cellulose as a sustainable technological material. Nature **590**(7844), 47–56 (2021). https://doi.org/10.1038/s41586-020-03167-7
2. B. Thomas, M.C. Raj, A.K. B, R.M. H, J. Joy, A. Moores, et al., Nanocellulose, a versatile green platform: from biosources to materials and their applications. Chem. Rev. **118** (24), 11575–11625. (2018). https://doi.org/10.1021/acs.chemrev.7b00627
3. J. Gan, Y. Wu, F. Yang, X. Wu, Y. Wang, J. Wang, UV-filtering cellulose nanocrystal/carbon quantum dot composite films for light conversion in glass windows. ACS Appl. Nano Mater. (2021). https://doi.org/10.1021/acsanm.1c03082
4. C. Chen, L. Hu, Nanoscale ion regulation in wood-based structures and their device applications. Adv. Mater. **33**(28) (2020). https://doi.org/10.1002/adma.202002890
5. C. Chen, Y. Kuang, S. Zhu, I. Burgert, T. Keplinger, A. Gong et al., Structure–property–function relationships of natural and engineered wood. Nat. Rev. Mater. **5**(9), 642–666 (2020). https://doi.org/10.1038/s41578-020-0195-z
6. Z. Li, C. Chen, H. Xie, Y. Yao, X. Zhang, A. Brozena et al., Sustainable high-strength macrofibers extracted from natural bamboo. Nat. Sustain. (2021). https://doi.org/10.1038/s41893-021-00831-2

7. Z. Li, C. Chen, R. Mi, W. Gan, J. Dai, M. Jiao et al., A strong, tough, and scalable structural material from fast-growing bamboo. Adv. Mater. **32**(10), e1906308 (2020). https://doi.org/10.1002/adma.201906308
8. Y. Liu, B. Guo, Q. Xia, J. Meng, W. Chen, S. Liu et al., Efficient cleavage of strong hydrogen bonds in cotton by deep eutectic solvents and facile fabrication of cellulose nanocrystals in high yields. ACS Sustain. Chem. Eng. **5**(9), 7623–7631 (2017). https://doi.org/10.1021/acssuschemeng.7b00954
9. J. Li, C. Chen, W. Gan, Z. Li, H. Xie, M. Jiao et al., A bio-inspired, hierarchically porous structure with a decoupled fluidic transportation and evaporative pathway toward high-performance evaporation. J. Mater. Chem. A **9**(15), 9745–9752 (2021). https://doi.org/10.1039/d0ta11385a
10. X. Wang, Q. Xia, S. Jing, C. Li, Q. Chen, B. Chen et al., Strong, hydrostable, and degradable straws based on cellulose-lignin reinforced composites. Small **17**(18), e2008011 (2021). https://doi.org/10.1002/smll.202008011
11. A. Tampieri, S. Sprio, A. Ruffini, G. Celotti, I.G. Lesci, N. Roveri, From wood to bone: multistep process to convert wood hierarchical structures into biomimetic hydroxyapatite scaffolds for bone tissue engineering. J. Mater. Chem. **19** (28) (2009). https://doi.org/10.1039/b900333a
12. V. Eta, J.-P. Mikkola, Deconstruction of Nordic hardwood in switchable ionic liquids and acylation of the dissolved cellulose. Carbohyd. Polym. **136**, 459–465 (2016). https://doi.org/10.1016/j.carbpol.2015.09.058
13. T. Heinze, O.A. El Seoud, A. Koschella, Production and characteristics of cellulose from different sources, in *Cellulose Derivatives. Springer Series on Polymer and Composite Materials* (2018), pp. 1–38
14. E. Hosseini Koupaie, S. Dahadha, A.A. Bazyar Lakeh, A. Azizi, E. Elbeshbishy, Enzymatic pretreatment of lignocellulosic biomass for enhanced biomethane production-A review. J. Environ. Manag. **233**, 774–784 (2019). https://doi.org/10.1016/j.jenvman.2018.09.106
15. V. Hlavata, P. Kuklik, J. Celler, J. Vanerek, Microfiber angle and its effect on wood cell behavior. Adv. Mater. Res. **1144**, 88–93 (2017). https://doi.org/10.4028/www.scientific.net/AMR.1144.88
16. B. Ghislain, B. Clair, Diversity in the organisation and lignification of tension wood fibre walls—A review. IAWA J. **38**(2), 245–265 (2017). https://doi.org/10.1163/22941932-20170170
17. S.S. Hindi, R.A. Abohassan, Cellulosic microfibril and its embedding matrix within plant cell wall. Int. J. Innov. Res. Sci. Eng. Technol. **5**(3), 2727–2734 (2016)
18. J. Gierer, Chemistry of delignification. Wood Sci. Technol. **20**(1), 1–33 (1986)
19. G. Pipon, C. Chirat, D. Lachenal, Comparative effect of ozone, chlorine dioxide, and hydrogen peroxide on lignin: reactions affecting pulp colour in the final bleaching stage. Holzforschung **61**(6), 628–633 (2007). https://doi.org/10.1515/hf.2007.100
20. J. Gierer, The chemistry of delignification. A general concept (1982)
21. Q. Fu, F. Ansari, Q. Zhou, L.A. Berglund, Wood nanotechnology for strong, mesoporous, and hydrophobic biocomposites for selective separation of oil/water mixtures. ACS Nano **12**(3), 2222–2230 (2018). https://doi.org/10.1021/acsnano.8b00005
22. Y. Li, Y. Liu, W. Chen, Q. Wang, Y. Liu, J. Li et al., Facile extraction of cellulose nanocrystals from wood using ethanol and peroxide solvothermal pretreatment followed by ultrasonic nanofibrillation. Green Chem. **18**(4), 1010–1018 (2016). https://doi.org/10.1039/c5gc02576a
23. H.Q. Le, A. Zaitseva, J.P. Pokki, M. Stahl, V. Alopaeus, H. Sixta, Solubility of organosolv lignin in gamma-valerolactone/water binary mixtures. Chemsuschem **9**(20), 2939–2947 (2016). https://doi.org/10.1002/cssc.201600655
24. M. Wu, J.-K. Liu, Z.-Y. Yan, B. Wang, X.-M. Zhang, F. Xu et al., Efficient recovery and structural characterization of lignin from cotton stalk based on a biorefinery process using a γ-valerolactone/water system. RSC Adv. **6**(8), 6196–6204 (2016). https://doi.org/10.1039/c5ra23095k
25. F.-L. Wang, S. Li, Y.-X. Sun, H.-Y. Han, B.-X. Zhang, B.-Z. Hu et al., Ionic liquids as efficient pretreatment solvents for lignocellulosic biomass. RSC Adv. **7**(76), 47990–47998 (2017). https://doi.org/10.1039/c7ra08110c

26. J. Shi, S. Pattathil, R. Parthasarathi, N.A. Anderson, J. Im Kim, S. Venketachalam et al., Impact of engineered lignin composition on biomass recalcitrance and ionic liquid pretreatment efficiency. Green Chem. **18**(18), 4884–4895 (2016)
27. H. Wu, M. Mora-Pale, J. Miao, T.V. Doherty, R.J. Linhardt, J.S. Dordick, Facile pretreatment of lignocellulosic biomass at high loadings in room temperature ionic liquids. Biotechnol. Bioeng. **108**(12), 2865–2875 (2011). https://doi.org/10.1002/bit.23266
28. Y.T. Tan, A.S.M. Chua, G.C. Ngoh, Deep eutectic solvent for lignocellulosic biomass fractionation and the subsequent conversion to bio-based products—A review. Bioresour. Technol. **297**, 122522 (2020). https://doi.org/10.1016/j.biortech.2019.122522
29. Q. Xia, Y. Liu, J. Meng, W. Cheng, W. Chen, S. Liu et al., Multiple hydrogen bond coordination in three-constituent deep eutectic solvents enhances lignin fractionation from biomass. Green Chem. **20**(12), 2711–2721 (2018). https://doi.org/10.1039/c8gc00900g
30. J. Song, C. Chen, S. Zhu, M. Zhu, J. Dai, U. Ray et al., Processing bulk natural wood into a high-performance structural material. Nature **554**(7691), 224–228 (2018). https://doi.org/10.1038/nature25476
31. M. Frey, D. Widner, J.S. Segmehl, K. Casdorff, T. Keplinger, I. Burgert, Delignified and densified cellulose bulk materials with excellent tensile properties for sustainable engineering. ACS Appl. Mater. Interf. **10**(5), 5030–5037 (2018). https://doi.org/10.1021/acsami.7b18646
32. K. Li, S. Wang, H. Chen, X. Yang, L.A. Berglund, Q. Zhou, Self-Densification of highly mesoporous wood structure into a strong and transparent film. Adv. Mater. **32**(42), e2003653 (2020). https://doi.org/10.1002/adma.202003653
33. C. Jia, C. Chen, Y. Kuang, K. Fu, Y. Wang, Y. Yao et al., From wood to textiles: top-down assembly of aligned cellulose nanofibers. Adv. Mater. **30**(30), e1801347 (2018). https://doi.org/10.1002/adma.201801347
34. D. Huang, J. Wu, C. Chen, X. Fu, A.H. Brozena, Y. Zhang et al., Precision imprinted nanostructural wood. Adv. Mater. **31**(48), e1903270 (2019). https://doi.org/10.1002/adma.201903270
35. T. Li, S.X. Li, W. Kong, C. Chen, E. Hitz, C. Jia, et al., A nanofluidic ion regulation membrane with aligned cellulose nanofibers. Sci. Adv. **5**, eaau4238 (2019). https://doi.org/10.1126/sciadv.aau4238
36. F. Chen, A.S. Gong, M. Zhu, G. Chen, S.D. Lacey, F. Jiang et al., Mesoporous, three-dimensional wood membrane decorated with nanoparticles for highly efficient water treatment. ACS Nano **11**(4), 4275–4282 (2017). https://doi.org/10.1021/acsnano.7b01350
37. H. Liu, C. Chen, H. Wen, R. Guo, N.A. Williams, B. Wang et al., Narrow bandgap semiconductor decorated wood membrane for high-efficiency solar-assisted water purification. J. Mater. Chem. A **6**(39), 18839–18846 (2018). https://doi.org/10.1039/c8ta05924a
38. M. Zhu, J. Song, T. Li, A. Gong, Y. Wang, J. Dai et al., Highly anisotropic, highly transparent wood composites. Adv. Mater. **28**(26), 5181–5187 (2016). https://doi.org/10.1002/adma.201600427
39. Q. Fu, L. Medina, Y. Li, F. Carosio, A. Hajian, L.A. Berglund, Nanostructured wood hybrids for fire-retardancy prepared by clay impregnation into the cell wall. ACS Appl. Mater. Interf. **9**(41), 36154–36163 (2017). https://doi.org/10.1021/acsami.7b10008
40. S. He, C. Chen, T. Li, J. Song, X. Zhao, Y. Kuang, et al., An energy-efficient, wood-derived structural material enabled by pore structure engineering towards building efficiency. Small Methods **4**(1) (2019). https://doi.org/10.1002/smtd.201900747
41. T. Li, J. Song, X. Zhao, Z. Yang, G. Pastel, S. Xu, et al., Anisotropic, lightweight, strong, and super thermally insulating nanowood with naturally aligned nanocellulose. Sci. Adv. **4**(3), eaar3724 (2018)
42. C. Chen, J. Song, S. Zhu, Y. Li, Y. Kuang, J. Wan et al., Scalable and sustainable approach toward highly compressible, anisotropic Lamellar carbon sponge. Chem. **4**(3), 544–554 (2018). https://doi.org/10.1016/j.chempr.2017.12.028
43. J. Song, C. Chen, Z. Yang, Y. Kuang, T. Li, Y. Li et al., Highly compressible, anisotropic aerogel with aligned cellulose nanofibers. ACS Nano **12**(1), 140–147 (2017). https://doi.org/10.1021/acsnano.7b04246

44. M. Zhu, Y. Wang, S. Zhu, L. Xu, C. Jia, J. Dai, et al., Anisotropic, transparent films with aligned cellulose nanofibers. Adv. Mater. **29**(21) (2017). https://doi.org/10.1002/adma.201606284
45. M. Zhu, C. Jia, Y. Wang, Z. Fang, J. Dai, L. Xu et al., Isotropic paper directly from anisotropic wood: top-down green transparent substrate toward biodegradable electronics. ACS Appl. Mater. Interf. **10**(34), 28566–28571 (2018). https://doi.org/10.1021/acsami.8b08055
46. R.J. Moon, A. Martini, J. Nairn, J. Simonsen, J. Youngblood, Cellulose nanomaterials review: structure, properties and nanocomposites. Chem. Soc. Rev. **40**(7), 3941–3994 (2011). https://doi.org/10.1039/c0cs00108b
47. S. Xiao, C. Chen, Q. Xia, Y. Liu, Y. Yao, Q. Chen et al., Lightweight, strong, moldable wood via cell wall engineering as a sustainable structural material. Science **374**(6566), 465–471 (2021)
48. T. Li, Y. Zhai, S. He, W. Gan, Z. Wei, M. Heidarinejad et al., A radiative cooling structural material. Science **364**(6442), 760–763 (2019)
49. S. Fink, Transparent wood–a new approach in the functional study of wood structure. Holzforschung **46**, 403 (1992)
50. R. Mi, T. Li, D. Dalgo, C. Chen, Y. Kuang, S. He et al., A clear, strong, and thermally insulated transparent wood for energy efficient windows. Adv. Func. Mater. **30**(1), 1907511 (2020)
51. C. Jia, T. Li, C. Chen, J. Dai, I.M. Kierzewski, J. Song et al., Scalable, anisotropic transparent paper directly from wood for light management in solar cells. Nano Energy **36**, 366–373 (2017). https://doi.org/10.1016/j.nanoen.2017.04.059
52. Z. Yu, Y. Yao, J. Yao, L. Zhang, Z. Chen, Y. Gao et al., Transparent wood containing Cs x WO 3 nanoparticles for heat-shielding window applications. J. Mater. Chem. A **5**(13), 6019–6024 (2017)
53. W. Gan, S. Xiao, L. Gao, R. Gao, J. Li, X. Zhan, Luminescent and transparent wood composites fabricated by poly(methyl methacrylate) and γ-Fe2O3@YVO4:Eu3+ nanoparticle impregnation. ACS Sustain. Chem. Eng. **5**(5), 3855–3862 (2017). https://doi.org/10.1021/acssuschemeng.6b02985
54. T. Li, M. Zhu, Z. Yang, J. Song, J. Dai, Y. Yao, et al., Wood composite as an energy efficient building material: guided sunlight transmittance and effective thermal insulation. Adv. Energy Mater. **6**(22) (2016). https://doi.org/10.1002/aenm.201601122
55. M. Zhu, T. Li, C.S. Davis, Y. Yao, J. Dai, Y. Wang et al., Transparent and haze wood composites for highly efficient broadband light management in solar cells. Nano Energy **26**, 332–339 (2016). https://doi.org/10.1016/j.nanoen.2016.05.020
56. R. Mi, C. Chen, T. Keplinger, Y. Pei, S. He, D. Liu et al., Scalable aesthetic transparent wood for energy efficient buildings. Nat. Commun. **11**(1), 3836 (2020). https://doi.org/10.1038/s41467-020-17513-w
57. Q. Xia, C. Chen, T. Li, S. He, J. Gao, X. Wang, et al., Solar-assisted fabrication of large-scale, patternable transparent wood. Sci. Adv. **7**(5), abd7342 (2021). https://doi.org/10.1126/sciadv.abd7342
58. C. Chen, Y. Kuang, L. Hu, Challenges and opportunities for solar evaporation. Joule **3**(3), 683–718 (2019). https://doi.org/10.1016/j.joule.2018.12.023
59. C. Chen, Y. Zhang, Y. Li, Y. Kuang, J. Song, W. Luo, et al., Highly conductive, lightweight, low-tortuosity carbon frameworks as ultrathick 3d current collectors. Adv. Energy Mater. **7**(17) (2017). https://doi.org/10.1002/aenm.201700595
60. C. Chen, S. Xu, Y. Kuang, W. Gan, J. Song, G. Chen, et al., Nature-inspired tri-pathway design enabling high-performance flexible Li-O2 batteries. Adv. Energy Mater. **9** (9) (2019). https://doi.org/10.1002/aenm.201802964
61. S. He, C. Chen, G. Chen, F. Chen, J. Dai, J. Song et al., High-performance, scalable wood-based filtration device with a reversed-tree design. Chem. Mater. **32**(5), 1887–1895 (2020). https://doi.org/10.1021/acs.chemmater.9b04516
62. M. Zhu, Y. Li, G. Chen, F. Jiang, Z. Yang, X. Luo, et al., Tree-inspired design for high-efficiency water extraction. Adv. Mater. **29**(44) (2017). https://doi.org/10.1002/adma.201704107
63. C. Chen, Y. Li, J. Song, Z. Yang, Y. Kuang, E. Hitz, et al., Highly flexible and efficient solar steam generation device. Adv. Mater. **29**(30) (2017). https://doi.org/10.1002/adma.201701756

64. W. Li, Q. Liu, Y. Zhang, C. Li, Z. He, W.C.H. Choy, et al., Biodegradable materials and green processing for green electronics. Adv. Mater. **32**(33) (2020). https://doi.org/10.1002/adma.202001591
65. W. Gan, C. Chen, HT. Kim, et al., Single-digit-micrometer thickness wood speaker. Nat. Commun. **10**(5084) (2019). https://doi.org/10.1038/s41467-019-13053-0

Chapter 3
Recent Development of Multifunctional Nanocomposites Based on Bacterial Nanocellulose

Sisi Cao, Qisheng Jiang, and Srikanth Singamaneni

Abstract Bacterial nanocellulose (BNC) is an emerging nanomaterial that has drawn immense attention due to its unique physical and chemical properties. This chapter mainly focuses on the recent developments in the design, synthesis, and application of novel BNC-based multifunctional nanocomposites. The robust bio-mediated synthesis of nanocellulose network when combined with versatile in situ and post-modification strategies enables the incorporation of various functional nanomaterials into the nanocellulose network. These BNC-based multifunctional nanocomposites are highly promising for water treatment, energy harvesting, chemical sensing, catalysis, and life sciences. However, there are several outstanding challenges associated with BNC such as high production cost and durability that need to be addressed before these composites can be widely deployed in the real world. Despite the challenges, owing to their unique structure and properties, BNC-based nanocomposites will attract increased for various applications.

3.1 Introduction

Bacterial nanocellulose (BNC) is an emerging nanomaterial and has drawn particular interest because of 3D interconnected porous network, outstanding mechanical properties, high purity of cellulose, rich surface chemistry, and cost-efficient, scalable, and environment-friendly synthesis [1–5]. This chapter will focus on the recent development of novel BNC-based multifunctional nanocomposites, especially those realized via in situ bio-fabrication technique.

S. Cao
Intel Corporation, NE Century Blvd, Hillsboro, OR 97124, USA

Q. Jiang
Auragent Bioscience LLC, St. Louis, MO 63108, USA
e-mail: qjiang@auragentbio.com

S. Singamaneni (✉)
Department of Mechanical Engineering and Materials Science, Institute of Materials Science and Engineering, Washington University in St. Louis, St. Louis, MO 63130, USA
e-mail: singamaneni@wustl.edu

L. Hu et al. (eds.), *Emerging Nanotechnologies in Nanocellulose*, NanoScience and Technology, https://doi.org/10.1007/978-3-031-14043-3_3

From its discovery in 1819 and exploitation as food (*nata de coco*) in the Philippines, BNC has emerged as a commercial product in food, cosmetic, and medical industries, which produce large quantities of BNC each day and is also being widely investigated by researchers as a matrix for functional nanocomposites [6, 7]. BNC can be biosynthesized by the genera *Aerobacter*, *Acetobacter*, *Achromobacter*, *Agrobacterium*, *Alcaligenes*, *Azotobacter*, *Pseudomonas*, *Rhizobium*, *Sarcina*, and *Gluconacetobacter*. Among them, only *Gluconacetobacter* can produce large enough quantities of cellulose for commercial applications [8]. As a polysaccharide with a β-1,4-glycosidic linkage whose molecular formula is $(C_6H_{10}O_5)_n$, BNC can be produced through biochemical steps by bacteria with a continuous supply of oxygen and sugar [5, 9, 10]. Unlike the plant cellulose, which comprises of hemicelluloses, lignin, pectin, BNC composed of highly pure cellulose nanofibrils, offers the advantage of no extra processing to remove these unwanted impurities [11, 12]. The diameter of BNC nanofibrils ranges from 25 to 100 nm, while their length reaches several micrometers. As a biomaterial, the cellulose formation of BNC can be tailored in situ, such as the shape and supramolecular structure of the products [2]. Through biosynthesis, the BNC hydrogel with hierarchical structure possesses micropores with 10 um in diameter. After a thorough cleaning, the as-obtained BNC hydrogel exhibits high crystallinity (70–80%), high degree of polymerization (up to 8000), extremely high water content (~99% by weight), and moldability [13]. In terms of the above merits, BNC is recognized as an attractive platform for loading various functional materials via adsorption of pre-synthesized nanostructures, in situ formation of nanostructures on the nanoscale cellulose fibers or incorporation of nanomaterials during bio-fabrication [7, 14–23]. The resulted BNC-based materials with unique chemical, physical properties have been successfully applied to a broad range of areas such as catalysis, sensing, energy storage, textile industry, paper making, water filtration, and biomedical materials/devices [24] (Fig. 3.1).

3.2 Design Principles Underpinning BNC-Based Multifunctional Nanocomposite

So far, a few strategies have been demonstrated to realize functional BNC nanocomposites: (i) adsorption of nanomaterials on BNC after its formation; (ii) in situ formation of nanomaterials on BNC after its formation; (iii) pyrolyzing BNC to achieve carbon-based functional composite; (iv) incorporation of nanomaterials in BNC network during its bio-fabrication (Fig. 3.2).

BNC has abundant hydroxyl groups on the fibers constituting the inherent 3D porous network, enabling its facile adsorption of various functional nanomaterials. Its excellent mechanical strength allows BNC to be a promising support material. By simple immersion, drop-casting, vacuum filtration method, functional nanomaterials can be easily deposited on BNC to achieve functional composites [15, 26–29]. The

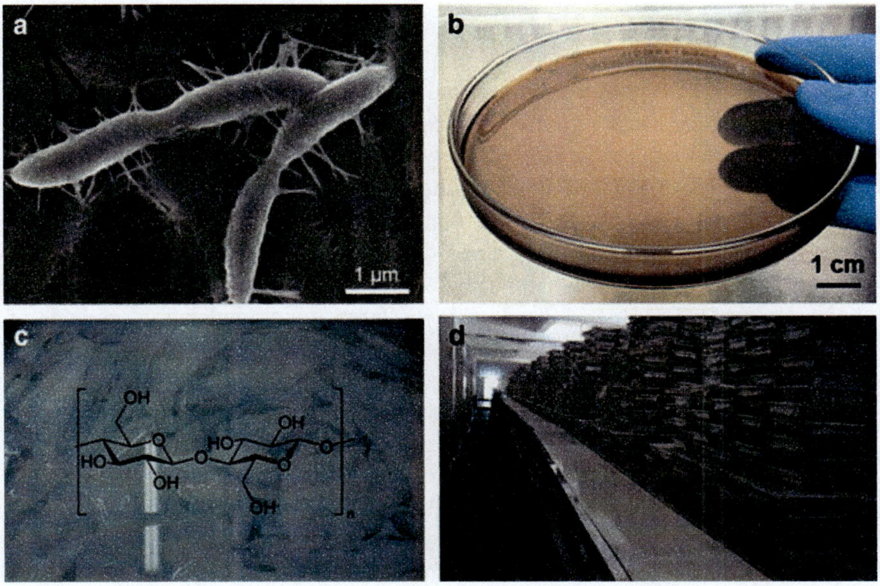

Fig. 3.1 **a** SEM image showing the cellulose-producing bacteria [25]. **b** Photograph of BNC broth in a glass Petri dish. Photographs of (**c**) large quantity of BNC produced by an industrial microbial fermentation and (**d**) a production line for BNC in a food factory [7]

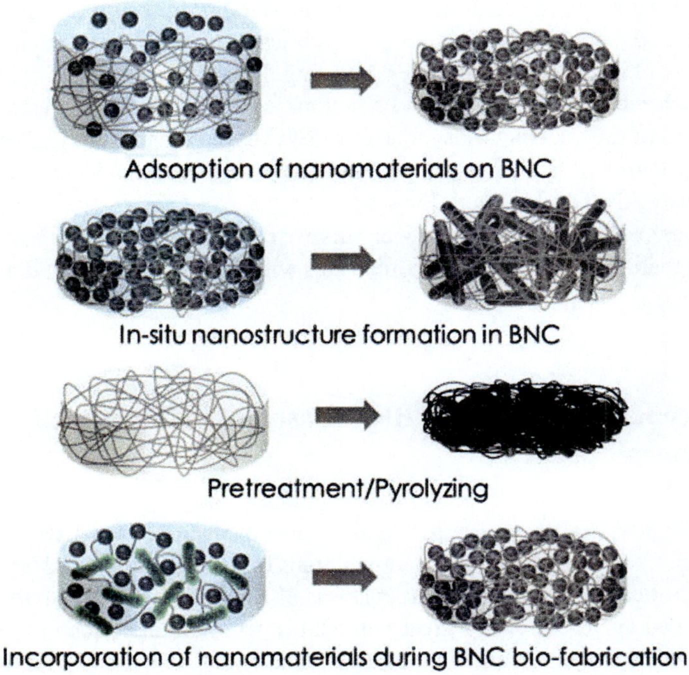

Fig. 3.2 Strategies for realization of BNC-based multifunctional nanocomposite

major advantages of this approach are the ease of fabrication and the wide choices of functional materials.

The abundance of hydroxyl groups and 3D porous network can also facilitate the penetration and adsorption of various metal ions (Au, Ag, Pt, Cu, Pd, Cd, etc.) onto the fibers, which can be subsequently reduced to form nanostructures in an in situ manner [13, 30–37]. Meanwhile, the ultrafine fibers and the space between them can prevent aggregation of nanostructures, ensuring the effective function of the as-synthesized nanostructures, leading to outstanding performance.

Pyrolysis of BNC is an efficient way to realize BNC-derived carbon-based functional nanocomposite [16, 19, 20, 38–44]. Different functionalization methods of the pyrolyzed BNC have been developed such as infiltrating polymers, doping with heteroatoms, or decorating with nanoparticles. In this approach, BNC is utilized as a promising raw material. However, pyrolyzed BNC lost its mechanical strength, hydrophilicity, abundant hydroxyl groups, and the process lacks scalability.

To fully exploit the unique bio-fabrication of BNC, additives can be added to the bacteria broth, which can alter the assembly and microstructure of BNC to achieve desired control of composition, morphology, and properties [45–47]. The presence of additives has been proven to influence the cellulose-producing bacterial cells or directly be incorporated in the BNC network during bio-fabrication [48, 49]. The introduction of nanomaterials as additives during the bio-fabrication can lead to a novel class of nanocomposites [2, 23, 50–55]. However, the choice of introduced materials has limitation since they should meet following requirements: (i) biocompatibility, not toxic to the bacteria; (ii) aqueous stability in a slight acidic aqueous environment (the optimal pH of BNC growth is around 3); (iii) large enough size to be trapped within the BNC network, preventing the leak during BNC washing. To counter the limitation of materials and functions added during bio-fabrication, post-modification of the one-spot biosynthesized BNC-based composites can be used to introduce additional nanomaterials or functions owing to the mechanical strength and stability of the BNC network [56–59].

Moreover, a combination of two or more strategies mentioned above can be adopted to realize BNC-based nanocomposites with desired structure and properties [60].

3.3 Recent Applications of BNC-Based Multifunctional Nanocomposite

Owing to abundant hydroxyl groups and high surface area, BNC is a highly attractive 3D platform for loading functional nanomaterials [59]. The as-prepared BNC-based nanocomposite inherits mechanical robustness of BNC network and the functionalities of loaded agents, showing great potential in various fields, such as advanced sensors, environmental remediation, catalysis, and energy storage [7]. For example, plasmonic nanoparticles can be immobilized on the BNC network via the hydrogen bonding between the plasmonic nanoparticles and cellulose nanofibers [15]. Owing to the 3D interconnected porous network, much higher loading of nanoparticles can be achieved on BNC compared to the 2D filter paper, thus plasmonic BNC aerogel achieved enhanced performance than 2D plasmonic paper. Apart from efficient loading, the fibrous network of BNC enables stable entrapment of functional materials, which ensures the long-term durable performance of BNC-based nanocomposites. For instance, PDA microspheres and GO can be physically trapped within the BNC network viabio-fabrication, and these BNC-based nanocomposites exhibit excellent robustness even after long-time exposure to sonication [52, 53]. On the other hand, after simple vacuum filtration on filter paper, the composites immediately disintegrate after subjecting to gentle mechanical agitation. Here, we have introduced various nanocomposite utilizing BNC as a versatile platform for diverse applications.

3.3.1 *Sensing*

3.3.1.1 BNC-Based Nanocomposites for Surface-Enhanced Raman Scattering (SERS) Substrate

After adsorbing on or in close to nanostructures of noble metal, the Raman scattering signal from molecules can be greatly enhanced, termed as surface-enhanced Raman scattering (SERS) (Fig. 3.3) [61–68]. Owing to the high sensitivity to detect trace level of chemicals, SERS is a highly promising transduction platform for chemical sensing spanning from forensics to homeland security [69–74].

Tian and co-workers have demonstrated that BNC loaded with highly dense plasmonic nanoparticles can serve as a highly promising candidate for SERS applications [15]. In this work, a BNC foam (aerogel) was simply immersed into a concentrated gold nanorods (AuNRs) solution to realize a SERS-active composite. Due to the

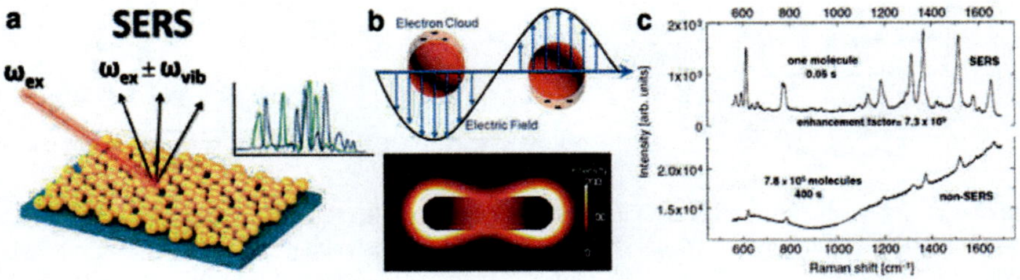

Fig. 3.3 **a** Surface-enhanced Raman spectroscopy (SERS) involves inelastic light scattering by molecules adsorbed onto corrugated metal surfaces such as silver or gold nanoparticles [79]. **b** Schematic drawing and Finite-difference time-domain (FDTD) modeling images showing the plasmonic hot-spot induced by the oscillation of electron cloud around gold nanorods [80]. **c** Standard Raman and SERS spectra of rhodamine 6G molecules with 633 nm laser excitation, indicating the powerful enhancement of SERS compared to classic Raman spectroscopy [81]

enormous available surface area and abundant hydroxyl functional groups within the 3D porous BNC network, an extremely high density of AuNRs can be loaded into the BNC foam. The plasmonic BNC foam presents significantly better SERS performance compared to conventional filter paper-based SERS substrate due to a much larger plasmonically active sensing volume under the laser footprint (Fig. 3.4). This plasmonic BNC foam can quantitatively detect a model analyte 2-naphthalenethiol (2-NT) down to 1 pM.

In terms of practical implementation, the SERS with simple and efficient sample collection is highly appealing. Among various sample collection methods, physical swabbing using a flexible substrate allows maximized sample collection efficiency on a solid surface and has been validated as an attractive method for real-world application [75–78].

A flexible BNC-based SERS swab has been developed using filtration method to locally deposit AuNRs on the top portion of a BNC substrate (Fig. 3.5) [26]. The naturally dried plasmonic BNC film with a high density of AuNRs within a few layers on the top surface within the light penetration depth (few tens of microns) and ultrasmooth surface demonstrates excellent SERS performance and can efficiently collect, recognize, and detect pathogenic bacteria.

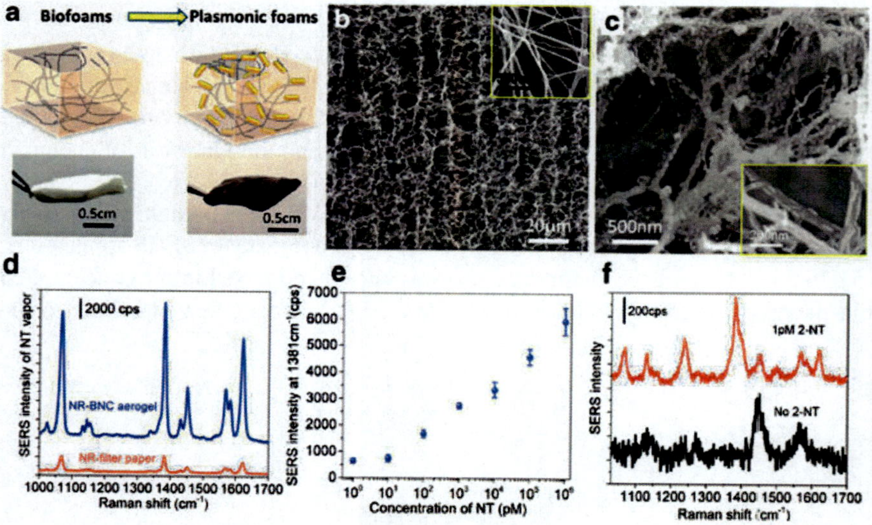

Fig. 3.4 **a** Schematic illustrating the fabrication of plasmonic BNC foam. **b** SEM images of cross section of BNC foam (inset: high-magnification SEM image showing individual fiber). **c** SEM images of cross section of plasmonic BNC foam with high AuNRs loading (inset: high-magnification SEM image showing individual fiber decorated by AuNRs). **d** SERS spectra obtained from plasmonic BNC foam and plasmonic filter paper. **e** Quantitative SERS detection of 2-NT by plasmonic BNC foam. **f** Zoom-in SERS spectra showing the ability of plasmonic BNC foam to detect 2-NT down to 1 pM [15]

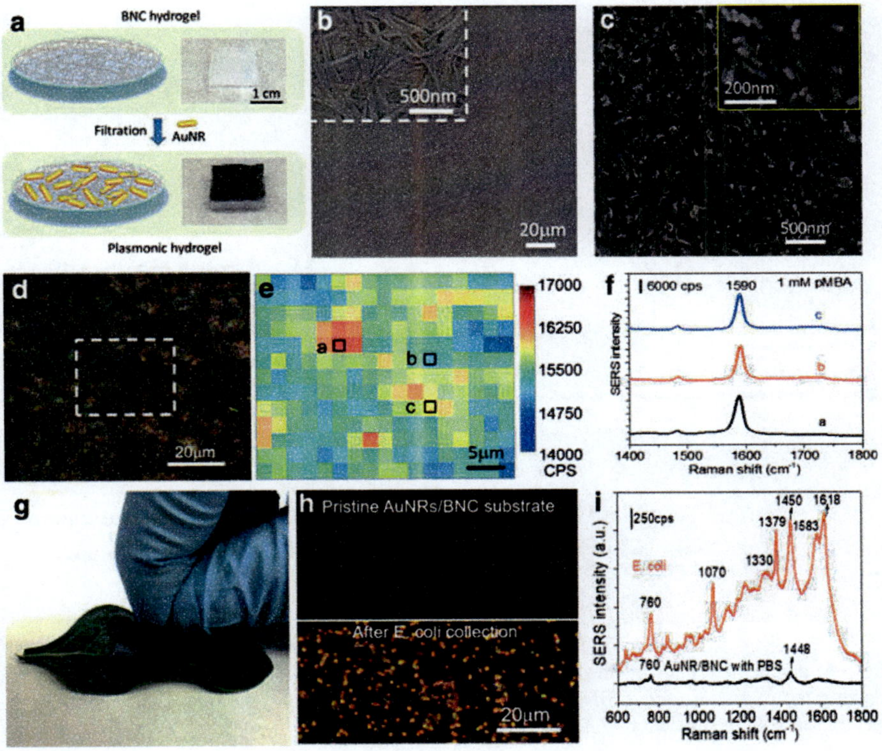

Fig. 3.5 **a** Schematic showing the fabrication of plasmonic BNC film. **b** SEM image showing the ultrasmooth surface of the BNC film. **c** SEM image showing the densely populated AuNRs embedded within the BNC film. **d** Optical image of a plasmonic BNC film. **e** SERS intensity map of 1590 cm^{-1} band from the area with a dashed line in (**d**). **f** Representative SERS spectra of pMBA collected from (**e**) suggesting the uniformity of SERS signal. **g** Optical image showing the easy swab can be performed using plasmonic BNC film on a spinach leaf. **h** Dark-field optical image of the plasmonic BNC film before and after *E. coli* collection. **i** Representative SERS spectra showing the ability to detect *E. coli* using plasmonic BNC film [26]

3.3.1.2 BNC-Based Nanocomposites for Optical Sensing of Biologically Relevant Analytes

BNC-based nanopaper adsorbed with plasmonic nanostructures and photoluminescent nanoparticles (PL-NPs) has recently developed as a powerful platform for optical sensing of multiple biologically relevant analytes (Fig. 3.6) [31]. In this work, silver nanoparticles (AgNPs), gold nanoparticles (AuNPs) were in situ formed on the BNC fibers and PL-NPs were modified using a two-step linkage method. The as-prepared BNC nanopaper demonstrates desired plasmonic and photoluminescent properties, enabling efficient optical sensing of small molecules, pathogens, and biomolecules. This platform holds great potential in medical, environmental, and food sensing.

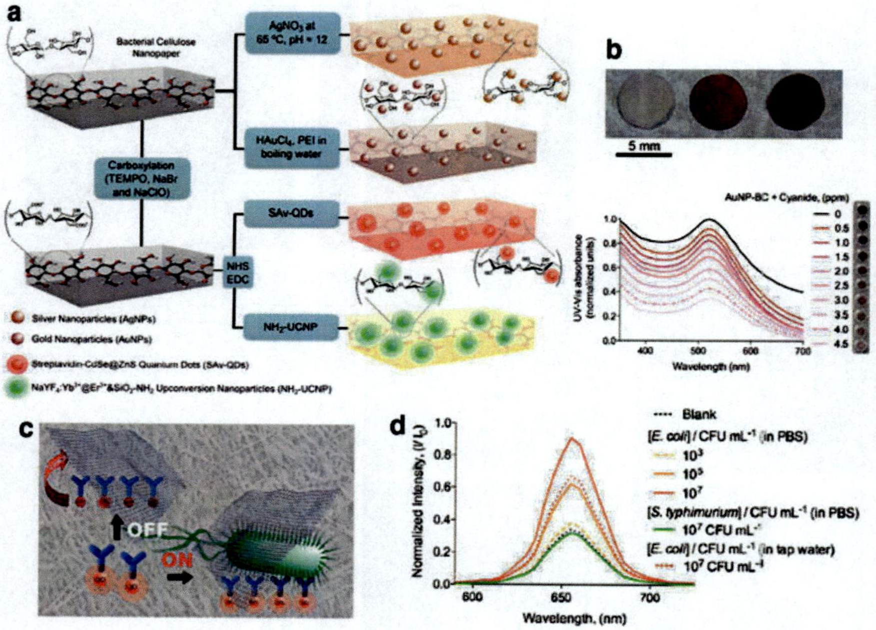

Fig. 3.6 **a** Schematic showing the fabrication of optically active BNC-composites. **c** Optical image showing the plasmonic-active BNC paper. **c** Schematic representing the biosensing platform. **d** Spectra showing the detection of *E.coli* [31]

3.3.2 Novel BNC-Based Functional Nanocomposites for Energy-Efficient Water Treatment

Owing to the limited freshwater source on the Earth, freshwater scarcity has been recognized as one of the biggest threats to global sustainable development in the twenty-first century. This crisis has been exacerbated by growing world's population and pollution caused by human activity, such as food production and manufacturing [82–84]. To alleviate the freshwater shortage, various water treatment technologies have been developed, but these technologies involve a large amount of consumption of fossil fuels or generation of greenhouse gases [85]. Thus, sustainable methods for clean water generation using renewable energy with minimized negative impacts on environment are highly attractive. Recent progress made in novel BNC-based composites reveals their great potential in tackling the above-mentioned challenge.

3.3.2.1 BNC-Based Composite Membranes for Energy-Efficient Water Treatment

As the abundant and green energy available on earth, solar energy has been utilized to alleviate the global freshwater shortage [86]. Among current solar-driven desalination technologies, solar steam generation converts sunlight to heat for freshwater generation, and it is highly appealing in resolving the freshwater shortage, especially in resource-limited regions [87, 88]. To maximize the evaporation efficiency, interfacial solar membranes enabling heat localization by floating the solar membranes at the water–air interface have been extensively validated within the past 4 years [60, 89]. Typically, most interfacial solar membranes are composed of a photothermal layer on the top and a supportive substrate on the bottom (Fig. 3.7). There are a few common features of these bilayered membranes: (i) the top evaporative layer enables efficient heat generation under sunlight irradiation and is comprised of photothermal materials, which exhibits high photothermal conversion efficiency and allows broadband light absorption; (ii) supportive substrates with low thermal conductivity reduce conductive heat transfer from the hot top layer to the bulk water, leading to heat localization on the evaporative region for more vapor generation; (iii) supportive substrates possess interconnected porous network with good hydrophilicity, which ensures adequate water supply from bulk water to the photothermal layer for vapor generation [60].

Recently, novel bilayered BNC nanocomposites has been realized through biofabrication, which efficiently integrates BNC and photothermal nanomaterials such

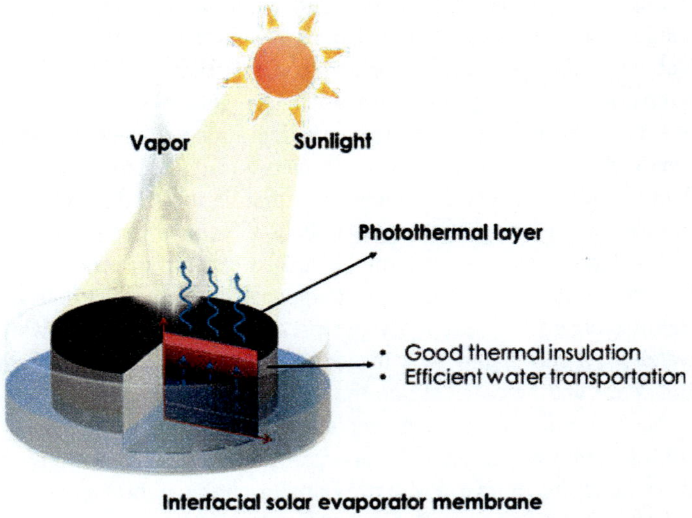

Fig. 3.7 Schematic illustration of a bilayered interfacial solar evaporator system and the design principles of substrates

as reduced graphene oxide, RGO, and Molybdenum disulfide, MoS_2) for highly efficient solar steam generation (Figs. 3.8, 3.9) [23, 52]. The RGO/BNC and MoS_2/BNC bilayered composites were achieved in a layered bio-fabrication manner: (i) static BNC growth to form a dense, thick bottom layer; (ii) introduction of fresh broth with nanomaterials to form the top layer. The cellulose-producing bacteria can secrete entangled ultrafine BNC fibers, sewing up these two layers to form a robust bilayered composite. This novel method opens up new possibilities for realizing multi-layered BNC composites. These bilayered BNC composites exhibited high light absorption (~96%), photothermal conversion, heat management, and water transportation, resulting in high solar thermal efficiency up to ~83% under simulated solar illumination at 10 kW/m^2. The facile fabrication of these novel bilayered BNC-based solar evaporator is highly scalable and cost-efficient.

Considering that solar evaporators are commonly used in aqueous environment, their excellent biodegradability is highly appealing, because the disposal or the leaching of non-degradable materials poses a significant concern on the environment and ecosystem [53]. Therefore, utilizing environment-benign and biodegradable materials to construct solar evaporators allows reduced negative effects on the environment and ensures long-term development.

In light of the above consideration, a solar evaporator comprised of BNC and polydopamine (PDA) particles has been developed (Fig. 3.9) [53]. The diameter of the PDA particles was tailored for maximum light absorption before the introduction into the BNC growth. This work demonstrated that besides 2D nanomaterials (e.g. RGO and MoS_2), spherical particles can also be successfully incorporated into the BNC network during its bio-fabrication. The PDA/BNC bilayered foam showed high solar steam generation performance under one sun (~78% efficiency). The novel, scalable processes for realizing completely biodegradable solar evaporator is a sustainable solution to alleviate worldwide water crisis.

Despite the high solar energy-to-steam efficiencies (60–90%), current solar steam generation technology lacks efficient vapor condensation and clean water collection systems, which decreases the overall solar energy-to-collected water efficiencies to only 30–60% [90]. Photothermal membrane distillation (PMD) has been developed and proven to be very promising to address the above issue [91]. Traditional membrane distillation (MD) is a thermally driven separation process based on a hydrophobic membrane and temperature gradient of feed and distillate water. The temperature difference (ΔT) leads to vapor pressure difference (ΔP) which drives the feed water to vaporize and pass through the membrane pores and the vapor will then condense and be collected on the distillate side. To generate the temperature difference, MD needs high consumption of electricity, which greatly hinders its application in remote regions and resource-limited settings where electricity is not readily available. In contrast, PMD uses a hydrophobic photothermal membrane to harvest sustainable solar energy to create interfacial heating, providing a high transmembrane temperature gradient for efficient vapor generation (Fig. 3.10) [85, 92].

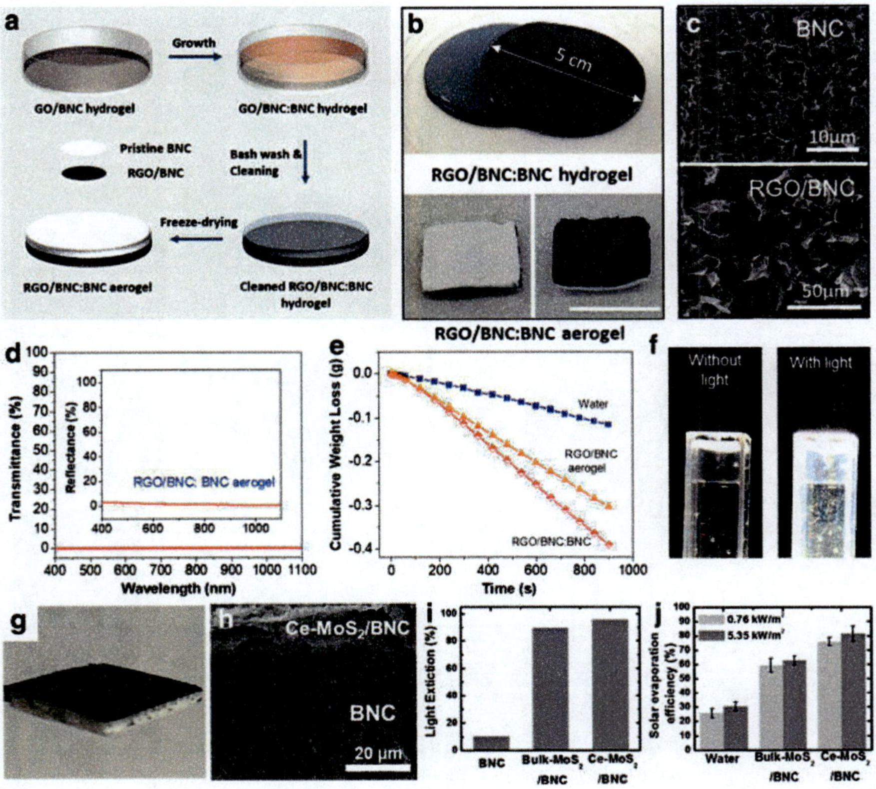

Fig. 3.8 **a** Schematic showing the bio-fabrication of bilayered RGO/BNC foam. **b** Optical image showing the wet, freeze-dried RGO/BNC foam. **c** SEM images showing the cross section of pristine BNC and RGO/BNC foams. **d** Transmittance and reflectance of RGO/BNC bilayered foam. **e** Plot showing the cumulative weight loss due to evaporation under solar illumination with RGO/BNC bilayered foam floating on water surface. **f** Optical images showing RGO/BNC bilayered foam floated on water surface and steam generation upon light illumination [52]. **g** Optical and cross-sectional. **h** SEM images of a ce-MoS_2/BNC foam. **i** Light extinction and **j** Solar evaporation efficiency comparison between BNC, Bulk-MoS_2/BNC and ce-MoS_2/BNC [23]

Compared to traditional MD membrane materials such as PVDF and PTFE, BNC offers several advantages to realize PMD membrane: (i) excellent biocompatibility; (ii) low toxicity; (iii) easy incorporation of photothermal materials.

Recently, a PMD membrane based on BNC and PDA particles, two environmentally friendly sustainable materials, has been developed [86]. In this work, a bilayered PMD membrane was realized by incorporating PDA particles during BNC bio-fabrication to form a thin photothermal layer and additional pristine BNC bio-fabrication to form a thin vapor transport layer (Fig. 3.11). The bilayered membrane was then freeze-dried and fluoro-silanized to achieve hydrophobicity. With the separation of photothermal layer and vapor transport layer, high porosity and vapor permeability can be maintained, without compromising either the vapor transport

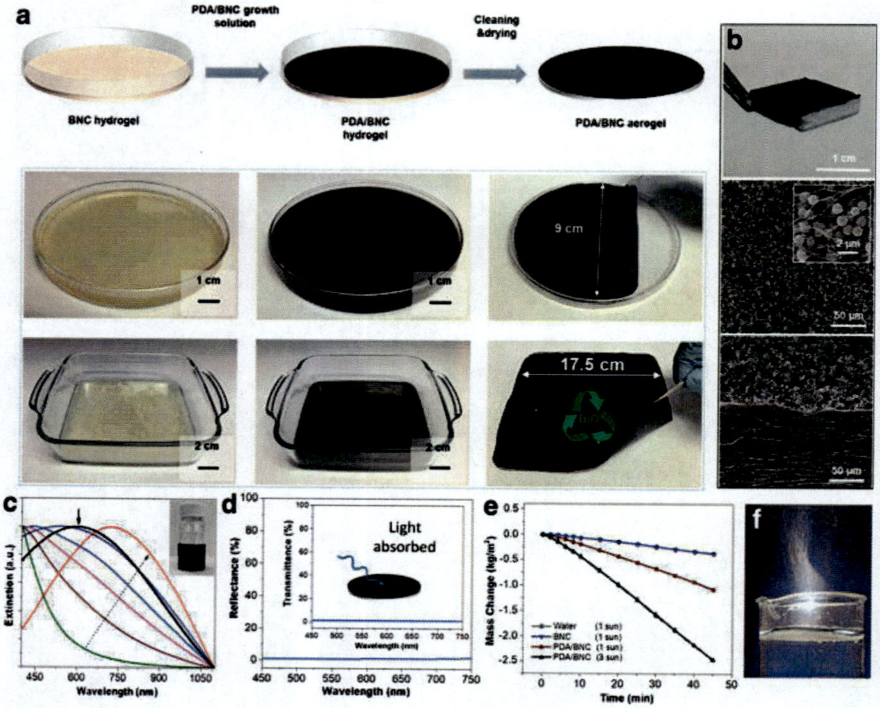

Fig. 3.9 **a** Schematic and optical images showing a completely biodegradable bilayered PDA/BNC solar evaporator achieved via bio-fabrication method. **b** Optical and SEM images showing the surface and cross section of a PDA/BNC foam. **c** Vis–NIR extinction spectra of PDA particles with different diameters. **d** Transmittance and reflectance spectra of PDA/BNC solar evaporator. **e** Plot and **f** optical pic showing the solar steam generation performance [53]

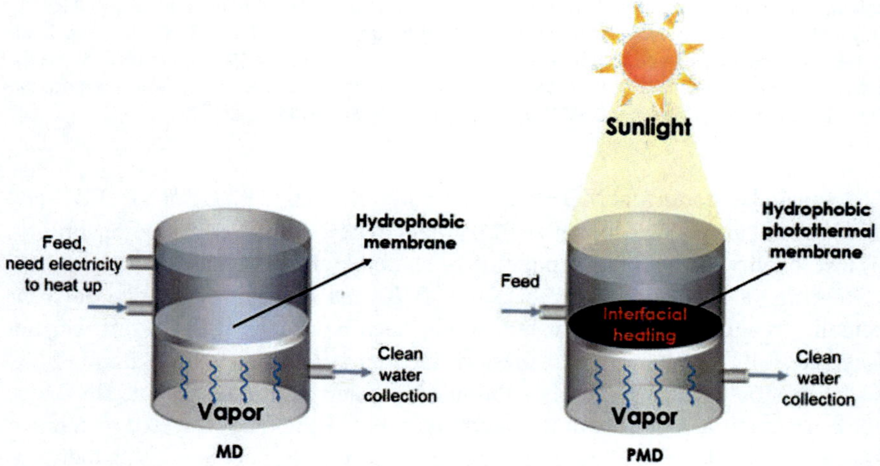

Fig. 3.10 Illustrated comparison between membrane distillation and photothermal membrane distillation systems

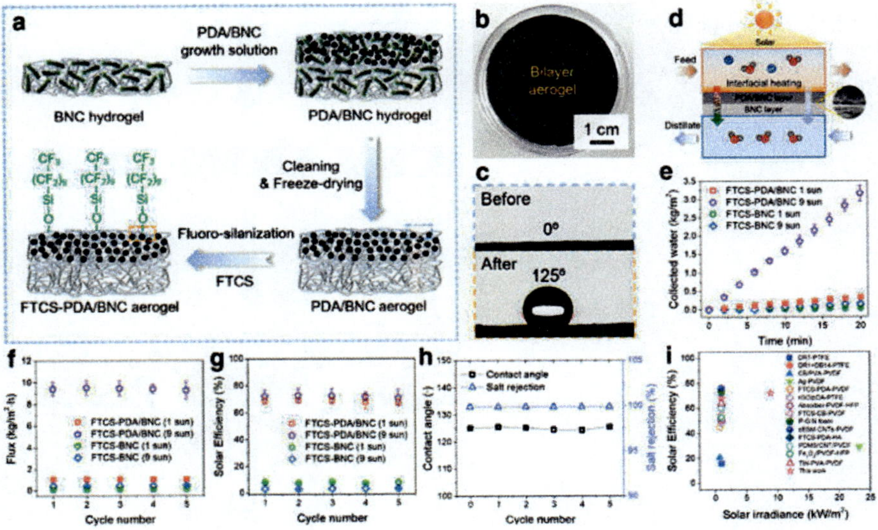

Fig. 3.11 **a** Schematic illustration showing the fabrication processes of the FTCS-PDA/BNC membrane. **b** Image of an FTCS-PDA/BNC membrane. **c** Water contact angle images of PDA/BNC membranes before and after FTCS treatment. **d** Schematic showing the PMD system using bilayered FTCS-PDA/BNC membrane. **e** The collected water (kg m^{-2}) from FTCS-BNC and FTCS-PDA/BNC membranes during PMD tests under 1 kW m^{-2} and 9 kW m^{-2}. **f** Water fluxes of FTCS-BNC and FTCS-PDA/BNC membranes during 5 cycles of PMD tests. Each cycle lasted 1 h. **g** Solar energy-to-collected water efficiencies of FTCS-BNC and FTCS-PDA/BNC membranes during 5 cycles of PMD tests. **h** Alterations of contact angles and salt rejections of FTCS-PDA/BNC membranes during 5 cycles of PMD tests. Error bars from triplicate tests were smaller than the symbols. **i** A comparison of solar energy-to-collected water efficiencies between our work and previous PMD works using ambient temperature feed water [86]

or photothermal ability. Thus, the PMD membrane demonstrated a high permeate flux of 1.0 kg m^{-2} h^{-1} under 1 sun (1 kW m^{-2}) and 68% solar energy-to-collected water efficiency during the treatment of saline water without any additional help from heat recovery systems. The PMD membrane demonstrated in this work shows great potential for the applications in decentralized resource-limited settings in remote, rural or disaster-affected areas. Additionally, due to its biocompatibility, the disposal of the PMD membrane does not pose threat to the environment and ecosystems.

3.3.2.2 BNC-Based Photothermal Membrane as a Solution for Tackling Membrane Biofouling

To alleviate freshwater shortage, various membrane technologies have been explored and validated for water treatment [93–95]. However, fouling (mineral fouling, organic fouling, and biofouling) and resultant compromised performance are common challenges for most of these membranes [96, 97]. Accounting for more than 45% of

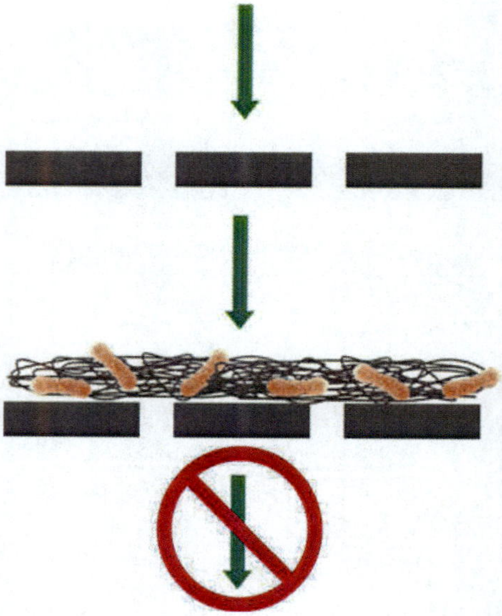

Fig. 3.12 Schematic illustration of membrane biofouling

membrane fouling, biofouling is the major shortcoming of various membrane technologies (e.g. reverse osmosis, ultrafiltration) due to the difficulty of the complete removal of biofilms (Fig. 3.12) [98–100].

However, most of the current techniques (adjusting pH, introducing disinfectants and biocides, utilizing antimicrobial nanomaterials) aiming to control biofouling are only effective for a short period of time as even if most biofilm is removed once, a small amount of residual cells can gradually adapt to imposed harsh environments and reform biofilms [101–108]. Therefore, seeking highly efficient and cost-effective methods or novel membrane materials to overcome membrane biofouling is significant for long-term stable performance in water treatment.

Converting incident light into heat to kill bacteria, photothermal effect has been demonstrated as an efficient option for tackling biofouling, obviating the need for harsh chemical treatments to achieve bacterial lysis [108].

A recently reported novel ultrafiltration membrane comprised of reduced graphene oxide (RGO) and BNC exhibits excellent anti-biofouling property via photothermally driven disinfection (Fig. 3.13) [54]. By introducing RGO during BNC bacteria-mediated growth, a membrane with inherent photothermal properties can be obtained. The layered packing of GO and BNC fibers achieved by this unique bio-fabrication approach yields a stacked structure, with a pore size (~2.2 nm in radius) which fits in ultrafiltration regime. Different from previously reported GO-based membranes, this new type of membrane with BNC as scaffold exhibited enhanced stability and durability: stable water flux under 100 psi pressure loading, aqueous stability over a wide pH range (4–9), and reusability after extensive vortex cleaning. Under renewable light sources or proposed low-energy LEDs, bacteria on the surface of RGO/BNC

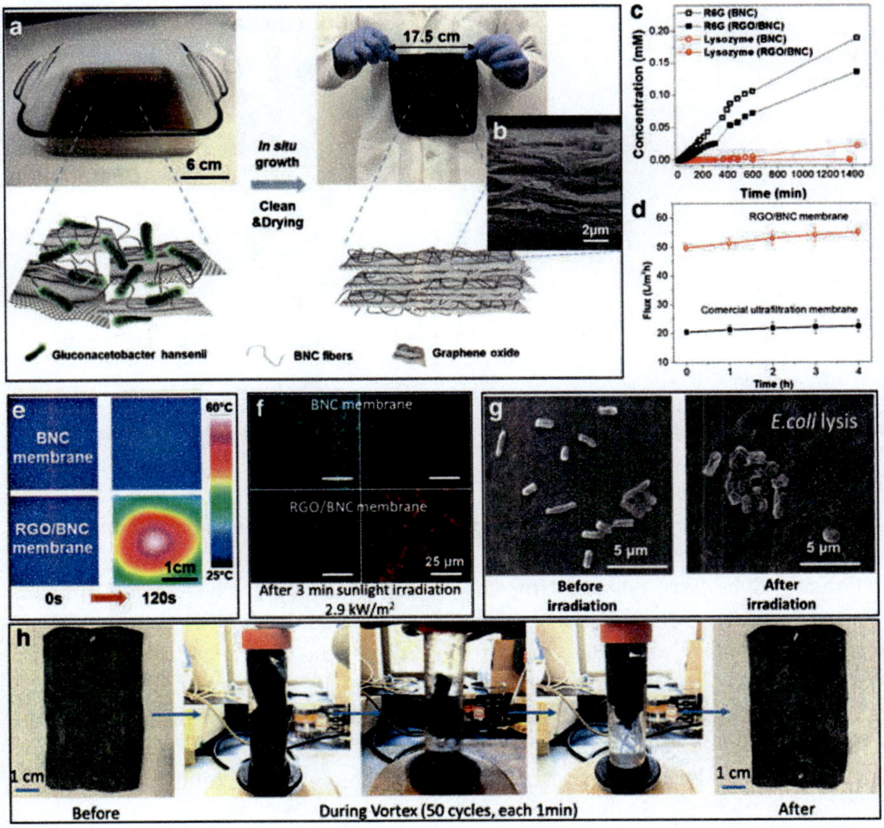

Fig. 3.13 **a** Schematic and optical images showing the fabrication of RGO/BNC membrane. **b** Cross-sectional SEM image of an RGO/BNC membrane. **c** Diffusion of model solutes through pure BNC and RGO/BNC membranes. **d** Water flux of RGO/BNC membranes (~8 μm) and commercial ultrafiltration membranes (YMGESP3001, GE) under 100 psi. **e** IR images showing photothermally enhanced surface temperature of membranes. **f** Bactericidal activity of RGO/BNC membrane on *E.coli* incubated on the surface revealed in fluorescence. **g** SEM images showing the bacterial lysis after the light irradiation. **h** Reusability of RGO/BNC membrane after multiple vortex washing [54]

membrane can be efficiently eliminated due to its excellent photothermal properties. This novel bio-fabrication of RGO/BNC membrane not only can provide an example of exploiting photothermal effect to tackle biofouling but also can advance the development of GO-based filtration membranes for water purification due to its unprecedented mechanical and chemical stability.

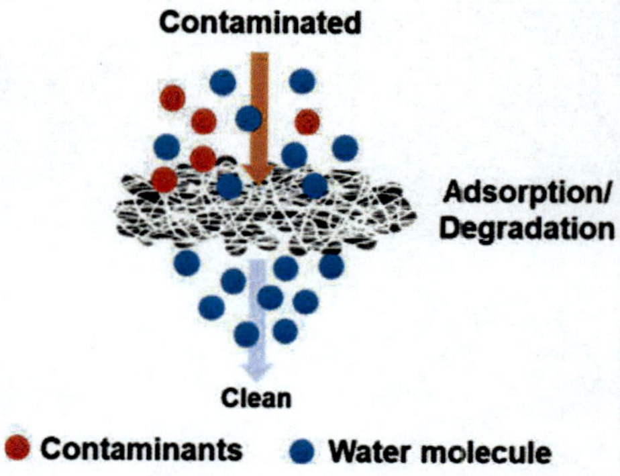

Fig. 3.14 Schematic illustration of the general principle of wastewater treatment using membrane filtration

3.3.2.3 BNC-Based Nanocomposite for Efficient Wastewater Treatment

Water contamination caused by heavy metals and organic dyes released from the ever-growing socioeconomic development pose significant threats to environment [109].

Organic dyes have negative impacts on the health of ecosystem and humans because of their carcinogenic and mutagenic nature [110–112]. Heavy metals can bind to and interfere with vital cellular components, and have been listed as a major public concern by the World Health Organization [113]. Typical membrane filtration methods to tackle these contaminants rely on either adsorption of the contaminants or the degradation of the contaminants (Fig. 3.14) [114].

A recent work has demonstrated a robust, scalable membrane made of PDA and BNC which can effectively remove multiple heavy metal ions and organic dyes (Fig. 3.15) [55]. PDA has rich catechol and amine functional groups, which can allow covalent/noncovalent binding with organic and inorganic species. The incorporation of PDA particles during BNC bio-fabrication can lead to a scalable and robust membrane. As shown in Fig. 3.11b, a high density of PDA particles can be easily loaded within the BNC network, demonstrating the unique advantage of the bio-fabrication method compared to other BNC functionalization methods. The high loading of PDA resulted in the effective removal of organic dyes and heavy metal ions at high concentrations (40–60 ppm) from feedwater. The PDA/BNC membrane was successfully deployed to remove even a cocktail of pollutants (Fig. 3.11i). The mechanical and chemical stability of BNC enabled robust handling of the PDA/BNC membrane, which can be easily regenerated and reused for multiple cycles without performance degradation. The PDA/BNC composite membrane represents a powerful material platform for wastewater treatment due to its applications in both adsorption-based water treatment and filtration with tunable porosity. Moreover, the versatile surface modification of PDA by more specific high contaminant binding functional species can be realized in future PDA/BNC-based membranes.

Besides relying on the traditional adsorption-based active materials, a catalytically active BNC-based ultrafiltration membrane has been developed to remove organic pollutants (Fig. 3.16) [56]. Based on the previously reported GO/BNC membrane,

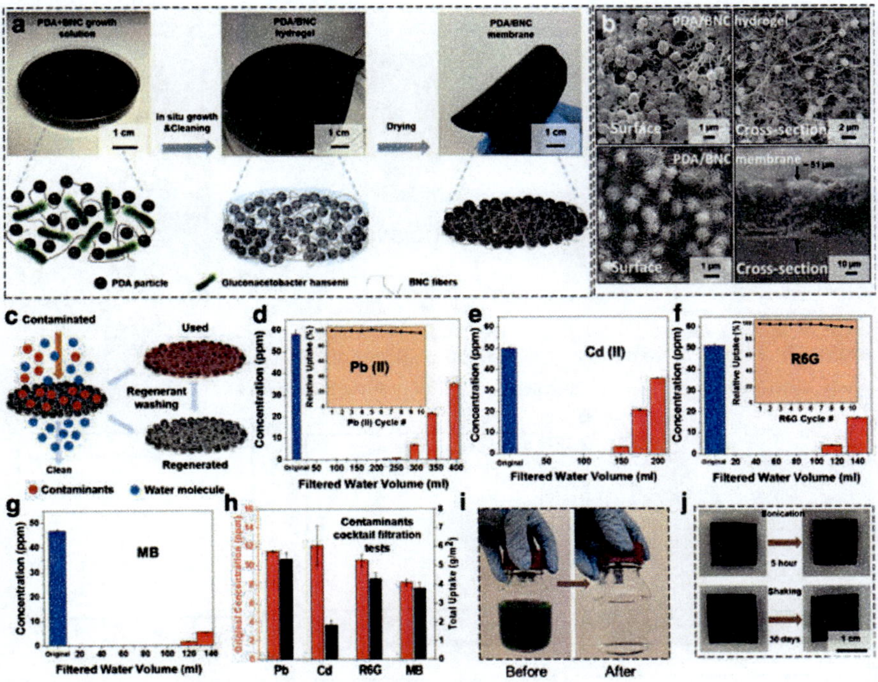

Fig. 3.15 **a** Schematic and photographs showing the bio-fabrication of PDA/BNC membranes. **b** SEM images of the surface and cross section of PDA/BNC hydrogel and membrane. **c** Schematic illustrating the filtration experiment. Heavy metal ions (**d**), (**e**) and organic dyes (**f**), **g** Removal performance via filtration/regeneration. **h** Single filtration tests of contaminants cocktail (left bar shows original concentration; the right bar shows the total uptake by PDA/BNC membrane). **i** Photographs showing the mixture of pollutants cocktail before and after filtration test. **j** Stability tests of PDA/BNC membrane under extensive sonication and mechanical agitation [55]

this work presented a catalytically active membrane realized via the in situ formation of palladium nanoparticles (PdNPs) on a biofabricated GO/BNC membrane. Taking advantage of the mechanical, chemical, stability, and ultrahigh surface area of GO, uniform and high density of PdNPs can be immobilized on the GO sheets within the BNC network to maximize the catalytic activity and prevent aggregation. The layered nature of Pd/GO/BNC forms nanocapillaries across the membrane that can ensure the maximum contact between organic pollutants and PdNPs decorated on the GO sheets. Due to the above advantages, the Pd/GO/BNC membrane exhibited excellent catalytic degradation of organic compounds (up to 99.3% pollutants degradation at a concentration of 60 mg/mL and the ability to treat multiple pollutants simultaneously). The flux of Pd/GO/BNC membrane reached 33.1 L/m^{2}h under 58 psi, distinguishing it from the previously developed PdNPs/GO composites. The novel membrane design reported in this work can be easily adapted for potential applications in water treatment, catalysis and energy harvesting. Moreover, the combination

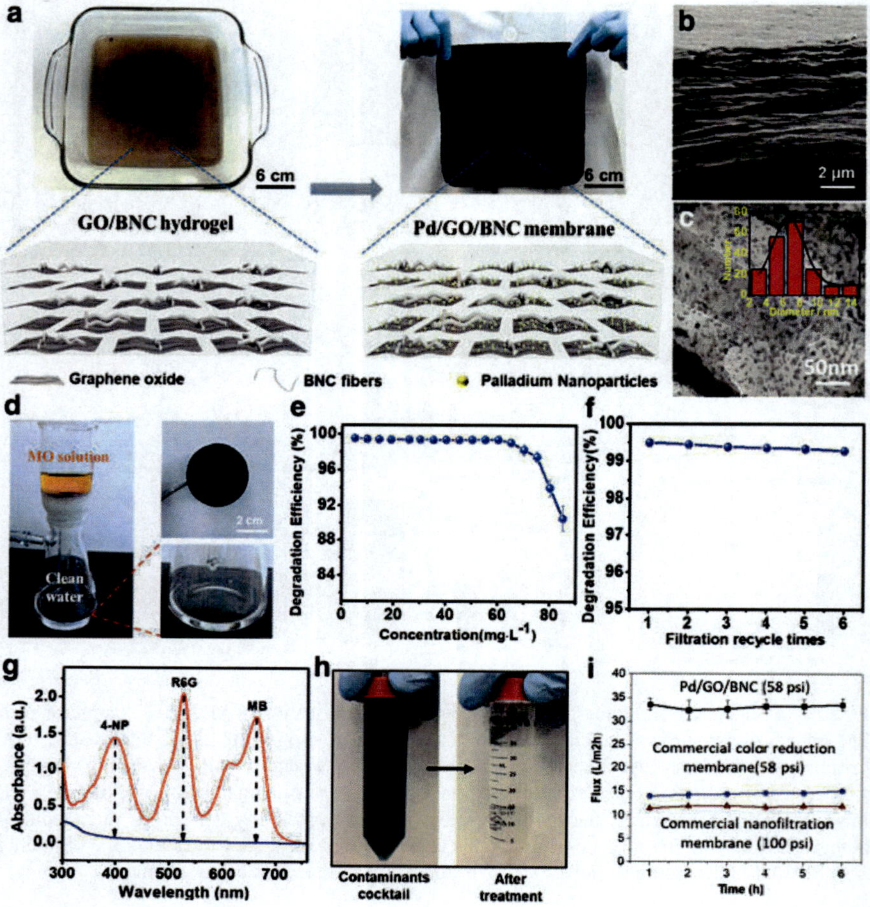

Fig. 3.16 **a** Schematic illustration and photographs showing the fabrication of Pd/GO/BNC membrane. **b** SEM image of the cross section of Pd/GO/BNC membrane. **c** TEM image of Pd/Go/BNC membrane (inset: size distribution of PdNPs). **d** Photograph of a bench-top filtration setup with Pd/GO/BNC filtration membrane. **e** Dye degradation efficiency of Pd/GO/BNC membrane at different methylene orange (MO) concentrations up to 100 mg/m. **f** Dye degradation efficiency of Pd/GO/BNC over multiple cycles of regeneration. **g** UV–vis spectra showing the degradation of a pollutant cocktail: 4-nitrophenol (4-NP), rhodamine 6G (R6G), methylene blue (MB) in the presence of $NaBH_4$. **h** Images showing the contaminants cocktail before and after filtration. **i** Water flux of Pd/GO/BNC membranes, commercial color reduction membranes (under 58 psi), and commercial nanofiltration membranes (under 100 psi) [56]

of bio-fabrication and in situ nanoparticle formation strategies demonstrated in this work can provide inspiration for future BNC-based multifunctional nanocomposites.

3.3.3 BNC-Based Flexible Energy Storage Devices

The rapid development and portable, personalized demand of flexible electronics such as bendable mobile phones, flexible displays, and wearable devices have driven the need for high performance, flexible energy storage devices (FESD) to a new level [115–121]. FESDs based on traditional metallic current collectors (e.g. nickel, copper foam, and stainless steel welded mesh) suffer from high gravimetric and volumetric densities despite their mechanical resilience and high electrical conductivity [122]. Carbon-based materials (e.g. carbon cloth, carbon paper, and graphene) lack electrochemical activity and are usually expensive in spite of having low density and stability [123]. Due to the limitation of conventional materials, applying structural, compositional design of electrochemically active materials with novel polymeric materials, especially cellulosic materials, has become an attractive trend, aiming to balance performance and flexibility [124].

In the past 10 years, BNC and BNC-based materials have attracted increasing interests in the preparation of FESDs (either electrode materials or gel electrolytes) (Fig. 3.17) [7]. From 2009 to 2021, a notably 26% of the publications in various applications of BNC-based materials are energy-related. The unique nanostructures resulted from the bio-assembly of nanoscale fibers can be inherited after specific chemical and physical treatment, which can increase the total active volume of the electrodes for FESDs, such as supercapacitors, secondary batteries (lithium-ion batteries, sodium-ion batteries, potassium-ion batteries, lithium–sulfur batteries), and fuel cells [24]. More importantly, electrochemical active materials can be easily incorporated with BNC and BNC-based materials to yield highly porous, versatile composites attractive for FESDs.

To realize high-performance BNC-based FESDs, the fabrication techniques typically involve vacuum filtration [28, 125], pyrolysis activation (or called carbonization) [40, 126, 127].

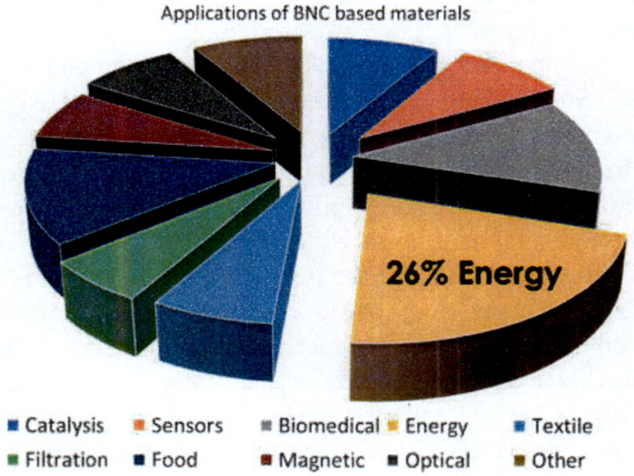

Fig. 3.17 Relative proportion of publications in various applications of BNC-based materials from 2009 to 2021. The data was obtained from the Web of Science

Hu and coworkers demonstrated a lightweight MXene/BNC FESD electrode for supercapacitor (Fig. 3.18) [128]. The MXene/BNC electrode was realized using all solution-based paper-making process via vacuum filtration of MXene solution and BNC fiber solution. Laser-cutting was used to achieve kirigami patterning on the MXene/BNC electrodes and stretchable, bendable micro-supercapacitor arrays (MSCAs) can be easily assembled. It exhibited excellent mechanical flexibility and electrochemical performance. This work demonstrated that, combining novel 2D materials with high specific volumetric capacitance, BNC, and additional laser cutting techniques, versatile FESDs electrode for micro-supercapacitor arrays (MSCAs) can be efficiently obtained.

Liu and coworkers reported a free-standing N-doped BNC-derived aerogel electrode [129]. BNC hydrogel was first immersed with metal–organic framework (MOF) precursors for a ZIF-8/BNC composites which then undergone freeze-drying and pyrolysis activation (Fig. 3.19). The pyrolysis activated N-self-doped carbon nanofiber aerogels can be used as electrodes for all-solid-state supercapacitor. The micro- and mesoporous structure can greatly increase the electrochemical surface area and decrease the ion diffusion pathways, thus achieved high specific capacity (224 F/g at 0.5 A/g and 612 mF/cm^2 at 1.37 mA/cm^2), outstanding rate capability and excellent cycling stability. This work is very representative to show the preparation of BNC-derived aerogel-based materials for energy storage, first by doping the pristine BNC with various precursors followed by freeze-drying and pyrolysis. The obtained BNC-derived carbon aerogels typically have high electrochemical performance but it needs additional polymer-based substrate to achieve FESDs.

Gao and coworkers developed an interesting layer-by-layer method to achieve high-performance BNC-based electrodes for lithium-Ion batteries [130]. BNC was firstly activated by 2,2,6,6-tetramethylpiperidine-1-oxyl (TEMPO) and then went through multiple cycles of incubated with functional materials and freeze-drying (Fig. 3.20). Finally, the BNC aerogel was compressed to obtain a laminar nanomat which can be used as a high-performance LIB anode. This method opens up more possibilities to realize multifunctional BNC-based materials despite their complexity and lengthy processes.

A simplified method was adopted by Yang and coworkers to develop an ultra-robust gel electrolyte of lithium metal batteries [131]. BNC hydrogel was mixed, immersed with $Li_{0.33}La_{0.557}TiO_3$ (LLTO) nanowires and then undergone freeze-drying/compression. The obtained gel electrolyte showed excellent mechanical properties and ability to immobilize anions, which significantly enhanced the capacity of assembled lithium metal battery.

Besides using vacuum filtration, pyrolysis activation and freeze-drying/compression, which offer either poor prospects in terms of scalability and/or compromise the mechanical properties of BNC, a bio-fabrication/post-modification hybrid method was recently developed to prepare BNC-based FESDs (Fig. 3.21) [57, 58]. GO flakes, poly(3,4-ehylene dioxythiophene)-poly(styrene sulfonate) (PEDOT:PSS), and tin oxide particles were added to the BNC growth solution and eventually hybrid composites electrodes were obtained. Notably, the non-conductive BNC network can be coated with conducting polymers, and the fiber

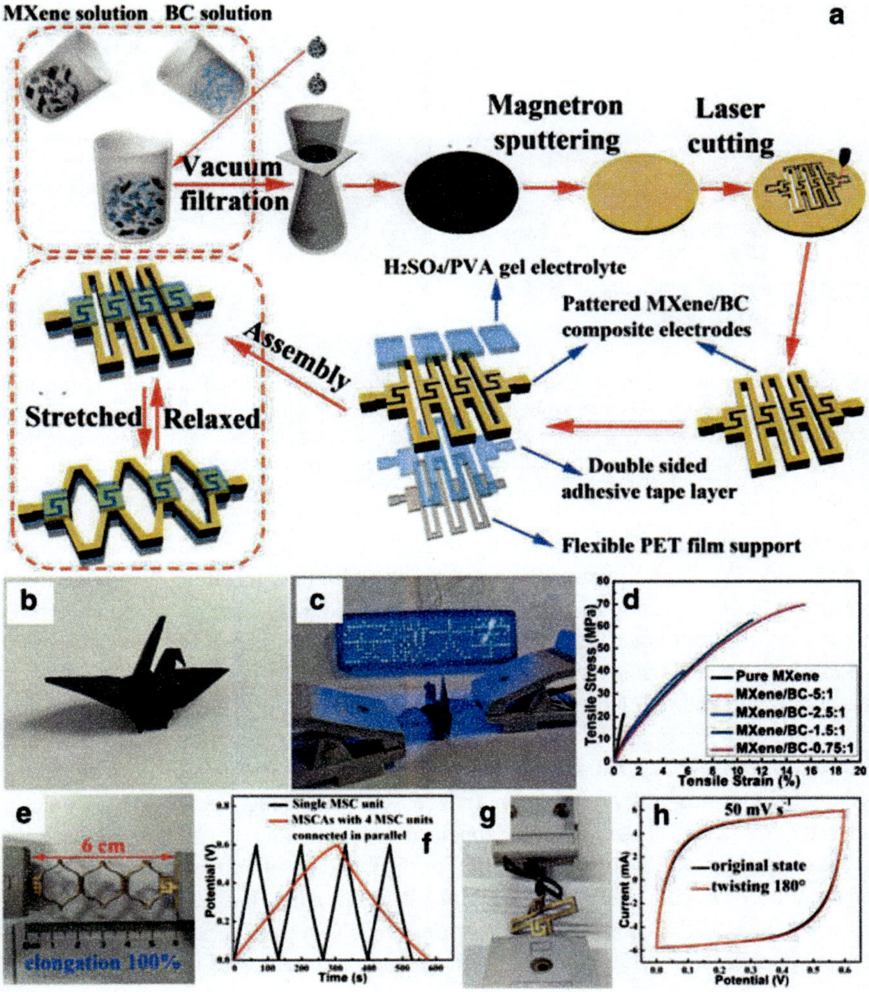

Fig. 3.18 **a** Flowchart of the protocol for preparing MXene/BNC composite electrodes and their kirigami patterning with laser cutting for all-solid-state stretchable supercapacitors. **b** A photograph of a paper crane folded from the MXene/BNC composite electrode. **c** A photograph of paper crane serving as a conductor to light an LED. **d** Typical stress–strain curves of the pure MXene paper and MXene/BNC composite prepared with mass ratios ranging from 5:1 to 0.75:1. **e** A photograph of an MSCA consisting of 4 MSC units connected in parallel under an applied strain of 100% elongation. **f** GCD curves for the MSCA and a single MSC unit at a high current density of 2.0 mA cm^{-2}. **g** A photograph of an MSCA containing four MSC units connected parallel when twisted to 180°. **h** CV curves of the MSCA with and without the twist [128]

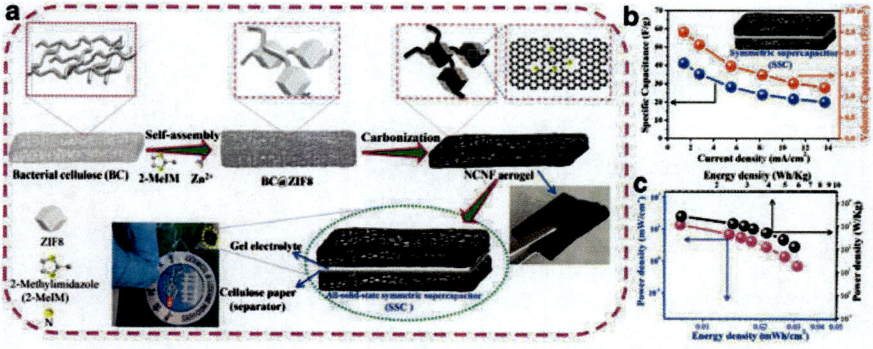

Fig. 3.19 **a** Fabrication of an all-solid-state symmetric supercapacitor (SSC) based on free-standing N-self-doped carbon nanofiber aerogels. **b** Rate performance based on the GCD curves at different current densities (the inset is a picture of the assembled all-solid-state symmetric supercapacitor). **c** Energy and power densities of the SSC based on different scan rates [129]

themselves can prevent the restacking of RGO which leads to loss of active area during reduction. Consequently, these electrodes exhibited excellent electrochemical performance and cycling stability. While assembled with electrolyte-coated BNC separators, all solid-state flexible supercapacitor devices can be formed. Through bio-fabrication and post modification, the RGO/PEDOT:PSS/BNC electrodes possessed outstanding electrochemical performance with cycling stability, holding great potential for long-term stable performance. After assembling the obtained electrodes with a BNC membrane as a separator, an all solid-state supercapacitor with excellent flexibility can be realized. Without using additional polymeric substrate, these BNC-based supercapacitors possess low density and excellent mechanical properties, allowing stable performance under extreme mechanical deformation. The bio-fabrication and post-modification hybrid techniques may provide a new insight into the preparation of BNC-based FESDs materials with optimal performance while preserving all the unique features of BNC.

3.4 Comments on the Technical and Economic Problems and Future Trends

In this chapter, we have discussed the fabrication strategies and design principles to fabricate various BNC-based nanocomposites, especially the innovative combination of bio-fabrication and post-modification methods. The versatile post-modification options allow the incorporation of various functional materials (such as GO, PDA, Mxenes, MOF, etc.), and the as-prepared BNC-based multifunctional nanocomposites are highly promising for water treatment, energy harvesting, chemical sensing, catalysis, and life sciences. However, there are several challenges associated with BNC for its real-world application.

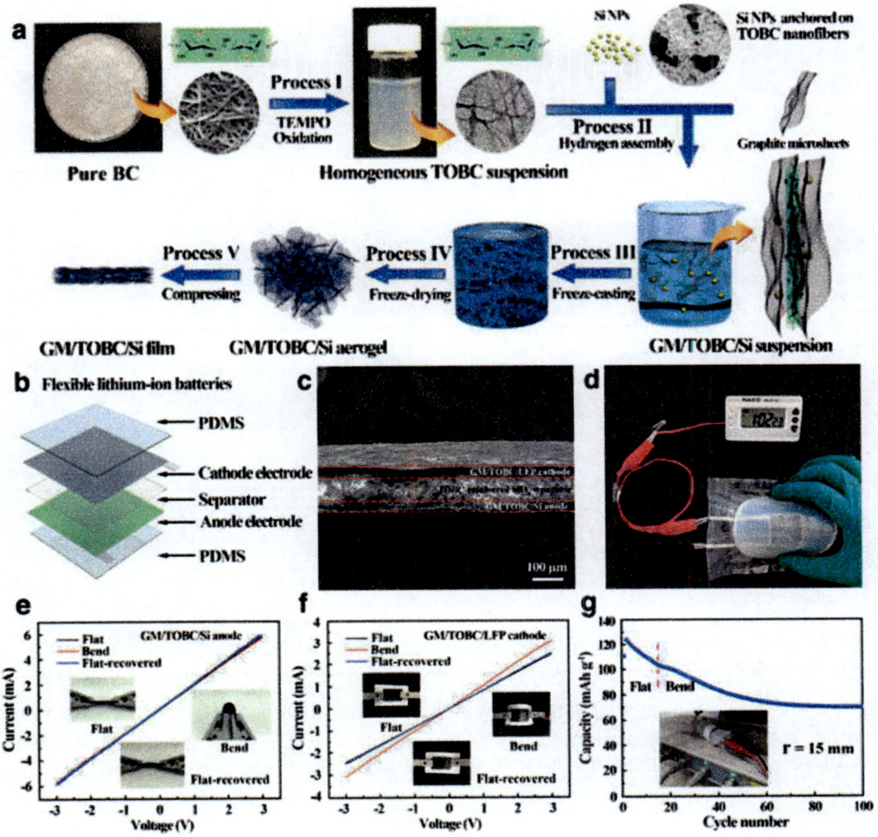

Fig. 3.20 **a** Schematic illustration of the preparation of freestanding GM/TOBC/Si nanomat. **b** Component units of flexional Li-ion full battery. **c** Cross-sectional SEM image of flexible full cell. **d** Photograph of digital watch lightened by the assembled full LIB under bending state. Conductivity of GM/TOBC/Si (**e**) and (**f**) GM/TOBC/$LiFePO_4$ at flat and bending states. **g** Cycle performance of the flexible full-cell at the bending state [130]

So far, the applications of BNC-related products fall into two polarized regimes: (i) industrial-scale produced food products at a low price; (ii) niche products designed for biotechnology with a high price [132, 133]. The situation in the second regime is due to the high capital investment and high operating costs, which impose severe economic constraints on the commercialization of BNC in industries other than food industry. Thus, increasing efforts have been devoted to designing economical processes for the production of BNC such as the evaluation of culture media compositions, fermentation systems, genetic engineering, and post-production modification processes.

Secondly, to utilize the highly interconnected porous network of BNC aerogel, freeze-drying is generally required to maintain the high porosity. However, freeze-drying involves high energy consumption, which increases the manufacturing cost

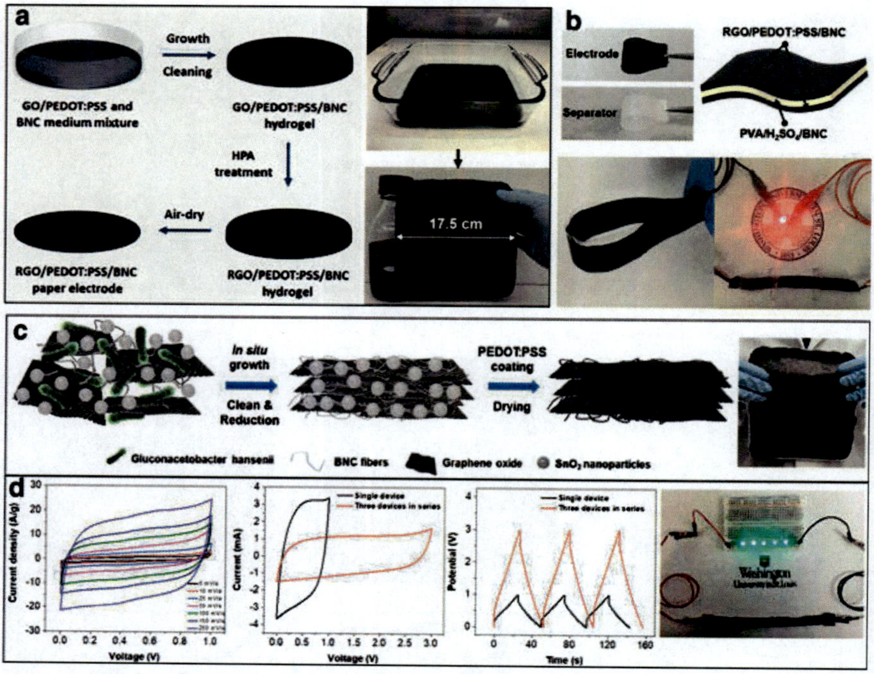

Fig. 3.21 **a** Fabrication of RGO/PEDOT:PSS/BNC electrodes. **b** The RGO/PEDOT:PSS/BNC electrode can be easily assembled with a BNC film to obtain flexible supercapacitor devices [57]. **c** Fabrication of PEDOT:PSS/SnO_2/RGO/BNC electrode. **d** Electrochemical performance of the electrodes and the assembled devices [58]

and is not friendly for large-scale production. Seeking a cost-effective drying method to preserve the highly porous structure of BNC aerogel is significantly important for scalable and cost-effective production.

Thirdly, BNC exhibits excellent biocompatibility which causes concerns about its durability during outdoor implementation. Utilized by various organisms as a carbon source, cellulose is known to be decomposed by cellulase, which commonly exists in various microorganisms, such as fungi, bacteria, and protozoans [134]. Thus, for real-world application, such as BNC-based solar evaporator in river and seawater, the degradation of BNC might occur, which eventually leads to lower evaporation after long-time operation. Thus, understanding the durability of BNC during real world is highly important. In addition, future studies about improving the resistance against decomposition should also be included.

Despite the challenges in advancing to practical applications, it is foreseeable that increasing efforts will be dedicated to boost BNC-based nanocomposites to various applications, due to the unique properties of BNC and the progress in the design and realization of functional BNC-based nanocomposites.

References

1. H. Jin, Y. Nishiyama, M. Wada, S. Kuga, Nanofibrillar cellulose aerogels. Colloids Surf. A **240**(1–3), 63–67 (2004)
2. D. Klemm, F. Kramer, S. Moritz, T. Lindström, M. Ankerfors, D. Gray, A. Dorris, Nanocelluloses: a new family of nature-based materials. Angew. Chem. Int. Ed. **50**(24), 5438–5466 (2011)
3. H. Valo, S. Arola, P. Laaksonen, M. Torkkeli, L. Peltonen, M.B. Linder, R. Serimaa, S. Kuga, J. Hirvonen, T. Laaksonen, Drug release from nanoparticles embedded in four different nanofibrillar cellulose aerogels. Eur. J. Pharm. Sci. **50**(1), 69–77 (2013)
4. L. Falk, H. Emmerich, P. Antje, L. Dieter, T. Stefanie, N. Marie-Alexandra, W. Martin, R. Thomas, Cellulosic aerogels as ultra-lightweight materials. Part 2: Synthesis and properties 2ndICC 2007, Tokyo, Japan, October 25–29, 2007. Holzforschung: Int. J. Biol. Chem. Phys. Technol. Wood **63**(1), 3–11 (2009)
5. P. Gatenholm, D. Klemm, Bacterial nanocellulose as a renewable material for biomedical applications. MRS Bull. **35**(3), 208–213 (2010)
6. F.M. Portela da Gama, F. Dourado, Bacterial nanocellulose: what future? BioImpacts: BI **8**(1), 1–3 (2018)
7. Z.-Y. Wu, H.-W. Liang, L.-F. Chen, B.-C. Hu, S.-H. Yu, Bacterial cellulose: a robust platform for design of three dimensional carbon-based functional nanomaterials. Acc. Chem. Res. **49**(1), 96–105 (2016)
8. M. Shoda, Y. Sugano, Recent advances in bacterial cellulose production. Biotechnol. Bioproc. E **10**(1), 1–8 (2005)
9. S. Yamanaka, K. Watanabe, N. Kitamura, M. Iguchi, S. Mitsuhashi, Y. Nishi, M. Uryu, The structure and mechanical properties of sheets prepared from bacterial cellulose. J Mater. Sci. **24**(9), 3141–3145 (1989)
10. S. Salmon, S.M. Hudson, Crystal morphology, biosynthesis, and physical assembly of cellulose, chitin, and chitosan. J. Macromol. Sci. Part C: Polym. Rev. **37**(2), 199–276 (1997)
11. D. Klemm, B. Heublein, H.-P. Fink, A. Bohn, Cellulose: fascinating biopolymer and sustainable raw material. Angew. Chem. Int. Ed. **44**(22), 3358–3393 (2005)
12. Y. Huang, C. Zhu, J. Yang, Y. Nie, C. Chen, D. Sun, Recent advances in bacterial cellulose. Cellulose **21**(1), 1–30 (2014)
13. W. Hu, S. Chen, J. Yang, Z. Li, H. Wang, Functionalized bacterial cellulose derivatives and nanocomposites. Carbohyd. Polym. **101**, 1043–1060 (2014)
14. L. Tian, Q. Jiang, K.K. Liu, J. Luan, R.R. Naik, S. Singamaneni, Bacterial nanocellulose-based flexible surface enhanced Raman scattering substrate. Adv. Mater. Interf. **3**(15), n/a-n/a (2016)
15. L. Tian, J. Luan, K.-K. Liu, Q. Jiang, S. Tadepalli, M.K. Gupta, R.R. Naik, S. Singamaneni, Plasmonic biofoam: a versatile optically active material. Nano Lett. **16**(1), 609–616 (2016)
16. L.-F. Chen, Z.-H. Huang, H.-W. Liang, H.-L. Gao, S.-H. Yu, Three-dimensional heteroatom-doped carbon nanofiber networks derived from bacterial cellulose for supercapacitors. Adv. Func. Mater. **24**(32), 5104–5111 (2014)
17. L.-F. Chen, X.-D. Zhang, H.-W. Liang, M. Kong, Q.-F. Guan, P. Chen, Z.-Y. Wu, S.-H. Yu, Synthesis of nitrogen-doped porous carbon nanofibers as an efficient electrode material for supercapacitors. ACS Nano **6**(8), 7092–7102 (2012)
18. S. Tong, M. Zheng, Y. Lu, Z. Lin, X. Zhang, P. He, H. Zhou, Binder-free carbonized bacterial cellulose-supported ruthenium nanoparticles for Li-O2 batteries. Chem. Commun. **51**(34), 7302–7304 (2015)
19. B. Wang, X. Li, B. Luo, J. Yang, X. Wang, Q. Song, S. Chen, L. Zhi, Pyrolyzed bacterial cellulose: a versatile support for lithium ion battery anode materials. Small **9**(14), 2399–2404 (2013)
20. H.-W. Liang, Q.-F. Guan, Z. Zhu, L.-T. Song, H.-B. Yao, X. Lei, S.-H. Yu, Highly conductive and stretchable conductors fabricated from bacterial cellulose. NPG Asia Mater. **4**, e19 (2012)

21. R.T. Olsson, M.A.S. Azizi Samir, G. Salazar Alvarez, L. Belova, V. Strom, L.A. Berglund, O. Ikkala, J. Nogues, U.W. Gedde, Making flexible magnetic aerogels and stiff magnetic nanopaper using cellulose nanofibrils as templates. Nat. Nano **5**(8), 584–588 (2010)
22. Q. Jiang, L. Tian, K.-K. Liu, S. Tadepalli, R. Raliya, P. Biswas, R.R. Naik, S. Singamaneni, Bilayered biofoam for highly efficient solar steam generation. Adv. Mater. n/a-n/a (2016)
23. D. Ghim, Q. Jiang, S. Cao, S. Singamaneni, Y.-S. Jun, Mechanically interlocked 1T/2H phases of MoS2 nanosheets for solar thermal water purification. Nano Energy **53**, 949–957 (2018)
24. W. Lei, D. Jin, H. Liu, Z. Tong, H. Zhang, An overview of bacterial cellulose in flexible electrochemical energy storage. Chemsuschem **13**(15), 3731–3753 (2020)
25. P. Gatenholm, D. Klemm, Bacterial nanocellulose as a renewable material for biomedical applications. MRS Bull. **35**(3), 208–213 (2011)
26. L. Tian, Q. Jiang, K.-K. Liu, J. Luan, R.R. Naik, S. Singamaneni, Bacterial nanocellulose-based flexible surface enhanced raman scattering substrate. Adv. Mater. Interf. **3**(15), 1600214 (2016)
27. Y.J. Kang, S.-J. Chun, S.-S. Lee, B.-Y. Kim, J.H. Kim, H. Chung, S.-Y. Lee, W. Kim, All-solid-state flexible supercapacitors fabricated with bacterial nanocellulose papers, carbon nanotubes, and triblock-copolymer ion gels. ACS Nano **6**(7), 6400–6406 (2012)
28. S. Li, D. Huang, B. Zhang, X. Xu, M. Wang, G. Yang, Y. Shen, Flexible supercapacitors based on bacterial cellulose paper electrodes. Adv. Energy Mater. **4**(10), 1301655 (2014)
29. M. Yuan, Q. Jiang, K. Liu, S. Singamaneni, S. Chakrabartty, Towards an integrated qr code biosensor: light-driven sample acquisition and bacterial cellulose paper substrate. IEEE Trans. Biomed. Circ. Syst. **12**(3), 452–460 (2018)
30. R.T. Olsson, M.A.S. Azizi Samir, G. Salazar-Alvarez, L. Belova, V. Ström, L.A. Berglund, O. Ikkala, J. Nogués, U.W. Gedde, Making flexible magnetic aerogels and stiff magnetic nanopaper using cellulose nanofibrils as templates. Nat. Nanotechnol. **5**, 584 (2010)
31. E. Morales-Narváez, H. Golmohammadi, T. Naghdi, H. Yousefi, U. Kostiv, D. Horák, N. Pourreza, A. Merkoçi, Nanopaper as an optical sensing platform. ACS Nano **9**(7), 7296–7305 (2015)
32. M. Chen, H. Kang, Y. Gong, J. Guo, H. Zhang, R. Liu, Bacterial cellulose supported gold nanoparticles with excellent catalytic properties. ACS Appl. Mater. Interf. **7**(39), 21717–21726 (2015)
33. F. Lai, Y.-E. Miao, Y. Huang, Y. Zhang, T. Liu, Nitrogen-doped carbon nanofiber/molybdenum disulfide nanocomposites derived from bacterial cellulose for high-efficiency electrocatalytic hydrogen evolution reaction. ACS Appl. Mater. Interf. **8**(6), 3558–3566 (2016)
34. H. Wei, W. Leng, J. Song, C. Liu, M.R. Willner, Q. Huang, W. Zhou, P.J. Vikesland, Real-time monitoring of ligand exchange kinetics on gold nanoparticle surfaces enabled by hot spot-normalized surface-enhanced raman scattering. Environ. Sci. Technol. **53**(2), 575–585 (2019)
35. G. Yang, J. Xie, Y. Deng, Y. Bian, F. Hong, Hydrothermal synthesis of bacterial cellulose/AgNPs composite: a "green" route for antibacterial application. Carbohyd. Polym. **87**(4), 2482–2487 (2012)
36. T. Zhang, W. Wang, D. Zhang, X. Zhang, Y. Ma, Y. Zhou, L. Qi, Biotemplated synthesis of gold nanoparticle-bacteria cellulose nanofiber nanocomposites and their application in biosensing. Adv. Func. Mater. **20**(7), 1152–1160 (2010)
37. X. Li, S. Chen, W. Hu, S. Shi, W. Shen, X. Zhang, H. Wang, In situ synthesis of CdS nanoparticles on bacterial cellulose nanofibers. Carbohyd. Polym. **76**(4), 509–512 (2009)
38. H. Li, Z. Cheng, Q. Zhang, A. Natan, Y. Yang, D. Cao, H. Zhu, Bacterial-derived, compressible, and hierarchical porous carbon for high-performance potassium-ion batteries. Nano Lett. **18**(11), 7407–7413 (2018)
39. Z.-Y. Wu, C. Li, H.-W. Liang, J.-F. Chen, S.-H. Yu, Ultralight, flexible, and fire-resistant carbon nanofiber aerogels from bacterial cellulose. Angew. Chem. Int. Ed. **52**(10), 2925–2929 (2013)
40. L.-F. Chen, Z.-H. Huang, H.-W. Liang, Q.-F. Guan, S.-H. Yu, Bacterial-cellulose-derived carbon nanofiber@MnO2 and nitrogen-doped carbon nanofiber electrode materials: an asymmetric supercapacitor with high energy and power density. Adv. Mater. **25**(34), 4746–4752 (2013)

41. L.-F. Chen, Z.-H. Huang, H.-W. Liang, W.-T. Yao, Z.-Y. Yu, S.-H. Yu, Flexible all-solid-state high-power supercapacitor fabricated with nitrogen-doped carbon nanofiber electrode material derived from bacterial cellulose. Energy Environ. Sci. **6**(11), 3331–3338 (2013)
42. H.-W. Liang, Z.-Y. Wu, L.-F. Chen, C. Li, S.-H. Yu, Bacterial cellulose derived nitrogen-doped carbon nanofiber aerogel: an efficient metal-free oxygen reduction electrocatalyst for zinc-air battery. Nano Energy **11**, 366–376 (2015)
43. C. Long, D. Qi, T. Wei, J. Yan, L. Jiang, Z. Fan, Nitrogen-doped carbon networks for high energy density supercapacitors derived from polyaniline coated bacterial cellulose. Adv. Func. Mater. **24**(25), 3953–3961 (2014)
44. Z.-Y. Wu, H.-W. Liang, C. Li, B.-C. Hu, X.-X. Xu, Q. Wang, J.-F. Chen, S.-H. Yu, Dyeing bacterial cellulose pellicles for energetic heteroatom doped carbon nanofiber aerogels. Nano Res. **7**(12), 1861–1872 (2014)
45. S. Yamanaka, J. Sugiyama, Structural modification of bacterial cellulose. Cellulose **7**(3), 213–225 (2000)
46. D.R. Ruka, G.P. Simon, K.M. Dean, In situ modifications to bacterial cellulose with the water insoluble polymer poly-3-hydroxybutyrate. Carbohyd. Polym. **92**(2), 1717–1723 (2013)
47. L.L. Zhou, D.P. Sun, L.Y. Hu, Y.W. Li, J.Z. Yang, Effect of addition of sodium alginate on bacterial cellulose production by Acetobacter xylinum. J. Ind. Microbiol. Biotechnol. **34**(7), 483 (2007)
48. H.-C. Huang, L.-C. Chen, S.-B. Lin, C.-P. Hsu, H.-H. Chen, In situ modification of bacterial cellulose network structure by adding interfering substances during fermentation. Biores. Technol. **101**(15), 6084–6091 (2010)
49. H. Luo, G. Xiong, Z. Yang, S.R. Raman, H. Si, Y. Wan, A novel three-dimensional graphene/bacterial cellulose nanocomposite prepared by in situ biosynthesis. RSC Adv. **4**(28), 14369–14372 (2014)
50. W.-I. Park, H.-S. Kim, S.-M. Kwon, Y.-H. Hong, H.-J. Jin, Synthesis of bacterial celluloses in multiwalled carbon nanotube-dispersed medium. Carbohyd. Polym. **77**(3), 457–463 (2009)
51. Z. Yan, S. Chen, H. Wang, B. Wang, J. Jiang, Biosynthesis of bacterial cellulose/multi-walled carbon nanotubes in agitated culture. Carbohyd. Polym. **74**(3), 659–665 (2008)
52. Q. Jiang, L. Tian, K.-K. Liu, S. Tadepalli, R. Raliya, P. Biswas, R.R. Naik, S. Singamaneni, Bilayered biofoam for highly efficient solar steam generation. Adv. Mater. **28**(42), 9400–9407 (2016)
53. Q. Jiang, H. Gholami Derami, D. Ghim, S. Cao, Y.-S. Jun, S. Singamaneni, Polydopamine-filled bacterial nanocellulose as a biodegradable interfacial photothermal evaporator for highly efficient solar steam generation. J. Mater. Chem. A **5**(35), 18397–18402 (2017)
54. Q. Jiang, D. Ghim, S. Cao, S. Tadepalli, K.-K. Liu, H. Kwon, J. Luan, Y. Min, Y.-S. Jun, S. Singamaneni, Photothermally active reduced graphene oxide/bacterial nanocellulose composites as biofouling-resistant ultrafiltration membranes. Environ. Sci. Technol. **53**(1), 412–421 (2019)
55. H. Gholami Derami, Q. Jiang, D. Ghim, S. Cao, Y.J. Chandar, J.J. Morrissey, Y.-S. Jun, S. Singamaneni, A robust and scalable polydopamine/bacterial nanocellulose hybrid membrane for efficient wastewater treatment. ACS Appl. Nano Mater. **2**(2), 1092–1101 (2019)
56. T. Xu, Q. Jiang, D. Ghim, K.-K. Liu, H. Sun, H.G. Derami, Z. Wang, S. Tadepalli, Y.-S. Jun, Q. Zhang, S. Singamaneni, Catalytically active bacterial nanocellulose-based ultrafiltration membrane. Small **14**(15), 1704006 (2018)
57. Q. Jiang, C. Kacica, T. Soundappan, K.-K. Liu, S. Tadepalli, P. Biswas, S. Singamaneni, An in situ grown bacterial nanocellulose/graphene oxide composite for flexible supercapacitors. J. Mater. Chem. A **5**(27), 13976–13982 (2017)
58. K.-K. Liu, Q. Jiang, C. Kacica, H.G. Derami, P. Biswas, S. Singamaneni, Flexible solid-state supercapacitor based on tin oxide/reduced graphene oxide/bacterial nanocellulose. RSC Adv. **8**(55), 31296–31302 (2018)
59. S. Cao, P. Rathi, X. Wu, D. Ghim, Y.-S. Jun, S. Singamaneni, Cellulose nanomaterials in interfacial evaporators for desalination: a "natural" choice. Adv. Mater. n/a (n/a), 2000922

60. S. Cao, Q. Jiang, X. Wu, D. Ghim, H. Gholami Derami, P.-I. Chou, Y.-S. Jun, S. Singamaneni, Advances in solar evaporator materials for freshwater generation. J. Mater. Chem. A **7**(42), 24092–24123 (2019)
61. J.P. Camden, J.A. Dieringer, J. Zhao, R.P. Van Duyne, Controlled plasmonic nanostructures for surface-enhanced spectroscopy and sensing. Acc. Chem. Res. **41**(12), 1653–1661 (2008)
62. H. Ko, S. Singamaneni, V.V. Tsukruk, Nanostructured surfaces and assemblies as SERS media. Small **4**(10), 1576–1599 (2008)
63. A. Campion, P. Kambhampati, Surface-enhanced Raman scattering. Chem. Soc. Rev. **27**(4), 241–250 (1998)
64. J.F. Li, Y.F. Huang, Y. Ding, Z.L. Yang, S.B. Li, X.S. Zhou, F.R. Fan, W. Zhang, Z.Y. Zhou, D.Y. Wu, B. Ren, Z.L. Wang, Z.Q. Tian, Shell-isolated nanoparticle-enhanced Raman spectroscopy. Nature **464**(7287), 392–395 (2010)
65. T.Y. Liu, K.T. Tsai, H.H. Wang, Y. Chen, Y.H. Chen, Y.C. Chao, H.H. Chang, C.H. Lin, J.K. Wang, Y. Wang, Functionalized arrays of Raman-enhancing nanoparticles for capture and culture-free analysis of bacteria in human blood. Nat. Commun. **2** (2011)
66. J.N. Anker, W.P. Hall, O. Lyandres, N.C. Shah, J. Zhao, R.P. Van Duyne, Biosensing with plasmonic nanosensors. Nat. Mater. **7**(6), 442–453 (2008)
67. N. Gandra, A. Abbas, L.M. Tian, S. Singamaneni, Plasmonic planet-satellite analogues: hierarchical self-assembly of gold nanostructures. Nano Lett. **12**(5), 2645–2651 (2012)
68. L. Tian, K.-K. Liu, M. Fei, S. Tadepalli, S. Cao, J.A. Geldmeier, V.V. Tsukruk, S. Singamaneni, Plasmonic nanogels for unclonable optical tagging. ACS Appl. Mater. Interf. **8**(6), 4031–4041 (2016)
69. M. Moskovits, Surface-enhanced Raman spectroscopy: a brief retrospective. J. Raman Spectrosc. **36**(6–7), 485–496 (2005)
70. R.S. Golightly, W.E. Doering, M.J. Natan, Surface-enhanced raman spectroscopy and homeland security: a perfect match? ACS Nano **3**(10), 2859–2869 (2009)
71. J. Homola, Surface Plasmon resonance sensors for detection of chemical and biological species. Chem. Rev. **108**(2), 462–493 (2008)
72. X. Fang, S.R. Ahmad, Detection of explosive vapour using surface-enhanced Raman spectroscopy. Appl. Phys. B **97**(3), 723–726 (2009)
73. K. Hering, D. Cialla, K. Ackermann, T. Dörfer, R. Möller, H. Schneidewind, R. Mattheis, W. Fritzsche, P. Rösch, J. Popp, SERS: a versatile tool in chemical and biochemical diagnostics. Anal. Bioanal. Chem. **390**(1), 113–124 (2008)
74. M.E. Stewart, C.R. Anderton, L.B. Thompson, J. Maria, S.K. Gray, J.A. Rogers, R.G. Nuzzo, Nanostructured plasmonic sensors. Chem. Rev. **108**(2), 494–521 (2008)
75. C.H. Lee, L.M. Tian, S. Singamaneni, Paper-based SERS swab for rapid trace detection on real-world surfaces. ACS Appl. Mater. Inter. **2**(12), 3429–3435 (2010)
76. L. Tian, J.J. Morrissey, R. Kattumenu, N. Gandra, E.D. Kharasch, S. Singamaneni, Bioplasmonic paper as a platform for detection of kidney cancer biomarkers. Anal. Chem. **84**(22), 9928–9934 (2012)
77. D. Martinak, A. Rudolph, In Explosives detection using an ion mobility spectrometer for airport security, security technology, in *Proceedings. The Institute of Electrical and Electronics Engineers 31st Annual 1997 International Carnahan Conference on*, 15–17 Oct 1997, pp. 188–189
78. W.W. Yu, I.M. White, Inkjet printed surface enhanced raman spectroscopy array on cellulose paper. Anal. Chem. **82**(23), 9626–9630 (2010)
79. J. Langer, D. Jimenez de Aberasturi, J. Aizpurua, R.A. Alvarez-Puebla, B. Auguié, J.J. Baumberg, G.C. Bazan, S.E.J. Bell, A. Boisen, A.G. Brolo, J. Choo, D. Cialla-May, V. Deckert,L. Fabris, K. Faulds, F.J. García de Abajo, R. Goodacre, D. Graham, A.J. Haes, C.L. Haynes, C. Huck, T. Itoh, M. Käll, J. Kneipp, N.A. Kotov, H. Kuang, E.C. Le Ru, H.K. Lee, J.-F. Li, X.Y. Ling, S.A. Maier, T. Mayerhöfer, M. Moskovits, K. Murakoshi, J.-M. Nam, S. Nie, Y. Ozaki, I. Pastoriza-Santos, J. Perez-Juste, J. Popp, A. Pucci, S. Reich, B. Ren, G.C. Schatz, T. Shegai, S. Schlücker, L.-L. Tay, K.G. Thomas, Z.-Q. Tian, R.P. Van Duyne, T. Vo-Dinh, Y. Wang, K.A. Willets, C. Xu, H. Xu, Y. Xu, Y.S. Yamamoto, B. Zhao, L.M. Liz-Marzán, Present and future of surface-enhanced raman scattering. ACS Nano **14**(1), 28–117 (2020)

80. A. Abbas, L. Tian, J.J. Morrissey, E.D. Kharasch, S. Singamaneni, Hot spot-localized artificial antibodies for label-free plasmonic biosensing. Adv. Func. Mater. **23**(14), 1789–1797 (2013)
81. E.C. Le Ru, P.G. Etchegoin, *Principles of Surface-Enhanced Raman Spectroscopy: and Related Plasmonic Effects* (Elsevier, Amsterdam; Boston, 2009)
82. M. Elimelech, The global challenge for adequate and safe water. J. Water Supply: Res. Technol. - Aqua **55**(1), 3–10 (2006)
83. P.-H. Gleick, ed., *The World's Water: The Biennial Report on Freshwater Resources,* vol. 8 (Island Press, Washington, DC, 2014)
84. J. Eliasson, The rising pressure of global water shortages. Nature **517** (7532) (2014)
85. S. Cao, X. Wu, Y. Zhu, R. Gupta, A. Tan, Z. Wang, Y.-S. Jun, S. Singamaneni, Polydopamine/hydroxyapatite nanowire-based bilayered membrane for photothermal-driven membrane distillation. J. Mater. Chem. A **8**(10), 5147–5156 (2020)
86. X. Wu, S. Cao, D. Ghim, Q. Jiang, S. Singamaneni, Y.-S. Jun, A thermally engineered polydopamine and bacterial nanocellulose bilayer membrane for photothermal membrane distillation with bactericidal capability. Nano Energy **79**, 105353 (2021)
87. W. Shang, T. Deng, Solar steam generation: steam by thermal concentration **1**, 16133 (2016)
88. H. Ghasemi, G. Ni, A.M. Marconnet, J. Loomis, S. Yerci, N. Miljkovic, G. Chen, Solar steam generation by heat localization **5**, 4449 (2014)
89. Y.-S. Jun, X. Wu, D. Ghim, Q. Jiang, S. Cao, S. Singamaneni, Photothermal membrane water treatment for two worlds. Acc. Chem. Res. **52**(5), 1215–1225 (2019)
90. W. Li, Z. Li, K. Bertelsmann, D.E. Fan, Portable low-pressure solar steaming-collection unisystem with polypyrrole origamis. Adv. Mater. **31**(29), 1900720 (2019)
91. A. Politano, P. Argurio, G. Di Profio, V. Sanna, A. Cupolillo, S. Chakraborty, H.A. Arafat, E. Curcio, Photothermal membrane distillation for seawater desalination. Adv. Mater. **29**(2), 1603504 (2017)
92. X. Wu, Q. Jiang, D. Ghim, S. Singamaneni, Y.-S. Jun, Localized heating with a photothermal polydopamine coating facilitates a novel membrane distillation process. J. Mater. Chem. A **6**(39), 18799–18807 (2018)
93. M.A. Shannon, P.W. Bohn, M. Elimelech, J.G. Georgiadis, B.J. Marinas, A.M. Mayes, Science and technology for water purification in the coming decades. Nature **452**(7185), 301–310 (2008)
94. R. Singh, N. Hankins, *Emerging Membrane Technology for Sustainable Water Treatment* (Elsevier, 2016)
95. M.M. Pendergast, E.M.V. Hoek, A review of water treatment membrane nanotechnologies. Energy Environ. Sci. **4**(6), 1946–1971 (2011)
96. P. Le-Clech, V. Chen, T.A.G. Fane, Fouling in membrane bioreactors used in wastewater treatment. J. Membr. Sci. **284**(1–2), 17–53 (2006)
97. R. Zhang, Y. Liu, M. He, Y. Su, X. Zhao, M. Elimelech, Z. Jiang, Antifouling membranes for sustainable water purification: strategies and mechanisms. Chem. Soc. Rev. (2016)
98. H.C. Flemming, G. Schaule, T. Griebe, J. Schmitt, A. Tamachkiarowa, Workshop on membranes in drinking water production technical innovations and health aspects biofouling—the achilles heel of membrane processes. Desalination **113**(2), 215–225 (1997)
99. R. Komlenic, Rethinking the causes of membrane biofouling. Filtr. Sep. **47**(5), 26–28 (2010)
100. J.S. Baker, L.Y. Dudley, Biofouling in membrane systems—A review. Desalination **118**(1–3), 81–89 (1998)
101. R.J. Barnes, J.H. Low, R.R. Bandi, M. Tay, F. Chua, T. Aung, A.G. Fane, S. Kjelleberg, S.A. Rice, Nitric oxide treatment for the control of reverse osmosis membrane biofouling. Appl. Environ. Microbiol. **81**(7), 2515–2524 (2015)
102. H. Lade, D. Paul, J.H. Kweon, Quorum quenching mediated approaches for control of membrane biofouling. Int. J. Biol. Sci. **10**(5), 550–565 (2014)
103. H.-L. Yang, J.C.-T. Lin, C. Huang, Application of nanosilver surface modification to RO membrane and spacer for mitigating biofouling in seawater desalination. Water Res. **43**(15), 3777–3786 (2009)

104. A. Kumar, P.K. Vemula, P.M. Ajayan, G. John, Silver-nanoparticle-embedded antimicrobial paints based on vegetable oil. Nat. Mater. **7**(3), 236–241 (2008)
105. F. Perreault, M.E. Tousley, M. Elimelech, Thin-film composite polyamide membranes functionalized with biocidal graphene oxide nanosheets. Environ. Sci. Technol. Lett. **1**(1), 71–76 (2014)
106. F. Perreault, A. Fonseca de Faria, M. Elimelech, Environmental applications of graphene-based nanomaterials. Chem. Soc. Rev. **44** (16), 5861–5896 (2015)
107. V. Kochkodan, N. Hilal, A comprehensive review on surface modified polymer membranes for biofouling mitigation. Desalination **356**, 187–207 (2015)
108. J.R. Ray, S. Tadepalli, S.Z. Nergiz, K.-K. Liu, L. You, Y. Tang, S. Singamaneni, Y.-S. Jun, Hydrophilic, bactericidal nanoheater-enabled reverse osmosis membranes to improve fouling resistance. ACS Appl. Mater. Interf. **7**(21), 11117–11126 (2015)
109. Z.J. Ren, A.K. Umble, Water treatment: recover wastewater resources locally. Nature **529**(7584), 25–25 (2016)
110. A.R. Khataee, M.B. Kasiri, Photocatalytic degradation of organic dyes in the presence of nanostructured titanium dioxide: Influence of the chemical structure of dyes. J. Mol. Catal. A: Chem. **328**(1), 8–26 (2010)
111. K. Yu, S. Yang, C. Liu, H. Chen, H. Li, C. Sun, S.A. Boyd, Degradation of organic dyes via bismuth silver oxide initiated direct oxidation coupled with sodium bismuthate based visible light photocatalysis. Environ. Sci. Technol. **46**(13), 7318–7326 (2012)
112. V.A. Sakkas, M.A. Islam, C. Stalikas, T.A. Albanis, Photocatalytic degradation using design of experiments: a review and example of the Congo red degradation. J. Hazard. Mater. **175**(1), 33–44 (2010)
113. R.P. Schwarzenbach, B.I. Escher, K. Fenner, T.B. Hofstetter, C.A. Johnson, U. von Gunten, B. Wehrli, The challenge of micropollutants in aquatic systems. Science **313**(5790), 1072 (2006)
114. F. Chen, A.S. Gong, M. Zhu, G. Chen, S.D. Lacey, F. Jiang, Y. Li, Y. Wang, J. Dai, Y. Yao, J. Song, B. Liu, K. Fu, S. Das, L. Hu, Mesoporous, three-dimensional wood membrane decorated with nanoparticles for highly efficient water treatment. ACS Nano **11**(4), 4275–4282 (2017)
115. J.A. Rogers, T. Someya, Y. Huang, Materials and mechanics for stretchable electronics. Science **327**(5973), 1603–1607 (2010)
116. D.-H. Kim, J. Song, W.M. Choi, H.-S. Kim, R.-H. Kim, Z. Liu, Y.Y. Huang, K.-C. Hwang, Y.-W. Zhang, J.A. Rogers, Materials and noncoplanar mesh designs for integrated circuits with linear elastic responses to extreme mechanical deformations. Proc. Natl. Acad. Sci. **105**(48), 18675–18680 (2008)
117. S. Bae, H. Kim, Y. Lee, X. Xu, J.-S. Park, Y. Zheng, J. Balakrishnan, T. Lei, H. Ri Kim, Y.I. Song, Y.-J. Kim, K.S. Kim, B. Ozyilmaz, J.-H. Ahn, B.H. Hong, S. Iijima, Roll-to-roll production of 30-inch graphene films for transparent electrodes. Nat. Nano **5**(8), 574–578 (2010)
118. G.H. Gelinck, H.E.A. Huitema, E. van Veenendaal, E. Cantatore, L. Schrijnemakers, J.B.P.H. van der Putten, T.C.T. Geuns, M. Beenhakkers, J.B. Giesbers, B.-H. Huisman, E.J. Meijer, E.M. Benito, F.J. Touwslager, A.W. Marsman, B.J.E. van Rens, D.M. de Leeuw, Flexible active-matrix displays and shift registers based on solution-processed organic transistors. Nat. Mater. **3**(2), 106–110 (2004)
119. J.A. Rogers, Z. Bao, K. Baldwin, A. Dodabalapur, B. Crone, V.R. Raju, V. Kuck, H. Katz, K. Amundson, J. Ewing, P. Drzaic, Paper-like electronic displays: large-area rubber-stamped plastic sheets of electronics and microencapsulated electrophoretic inks. Proc. Natl. Acad. Sci. **98**(9), 4835–4840 (2001)
120. M.L. Hammock, A. Chortos, B.C.K. Tee, J.B.H. Tok, Z. Bao, 25th anniversary article: the evolution of electronic skin (e-skin): a brief history, design considerations, and recent progress. Adv. Mater. **25**(42), 5997–6038 (2013)
121. Y. Shao, M.F. El-Kady, L.J. Wang, Q. Zhang, Y. Li, H. Wang, M.F. Mousavi, R.B. Kaner, Graphene-based materials for flexible supercapacitors. Chem. Soc. Rev. **44**(11), 3639–3665 (2015)

122. H. Li, Z. Tang, Z. Liu, C. Zhi, Evaluating flexibility and wearability of flexible energy storage devices. Joule **3**(3), 613–619 (2019)
123. Y. Wang, Q. Yang, Y. Zhao, S. Du, C. Zhi, Recent advances in electrode fabrication for flexible energy-storage devices. Adv. Mater. Technol. **4**(7), 1900083 (2019)
124. Z. Liu, F. Mo, H. Li, M. Zhu, Z. Wang, G. Liang, C. Zhi, Advances in flexible and wearable energy-storage textiles. Small Methods **2**(11), 1800124 (2018)
125. F. Wang, H.-J. Kim, S. Park, C.-D. Kee, S.-J. Kim, I.-K. Oh, Bendable and flexible supercapacitor based on polypyrrole-coated bacterial cellulose core-shell composite network. Compos. Sci. Technol. **128**, 33–40 (2016)
126. X. Wang, D. Kong, Y. Zhang, B. Wang, X. Li, T. Qiu, Q. Song, J. Ning, Y. Song, L. Zhi, All-biomaterial supercapacitor derived from bacterial cellulose. Nanoscale **8**(17), 9146–9150 (2016)
127. X. Wang, D. Kong, B. Wang, Y. Song, L. Zhi, Activated pyrolysed bacterial cellulose as electrodes for supercapacitors. Sci. China Chem. **59**(6), 713–718 (2016)
128. S. Jiao, A. Zhou, M. Wu, H. Hu, Kirigami patterning of mxene/bacterial cellulose composite paper for all-solid-state stretchable micro-supercapacitor arrays. Adv. Sci. **6**(12), 1900529 (2019)
129. H. Chen, T. Liu, J. Mou, W. Zhang, Z. Jiang, J. Liu, J. Huang, M. Liu, Free-standing N-self-doped carbon nanofiber aerogels for high-performance all-solid-state supercapacitors. Nano Energy **63**, 103836 (2019)
130. X. Zhou, Y. Liu, C. Du, Y. Ren, R. Xiao, P. Zuo, G. Yin, Y. Ma, X. Cheng, Y. Gao, Layer-by-layer engineered silicon-based sandwich nanomat as flexible anode for lithium-ion batteries. ACS Appl. Mater. Interf. **11**(43), 39970–39978 (2019)
131. C. Ding, X. Fu, H. Li, J. Yang, J.-L. Lan, Y. Yu, W.-H. Zhong, X. Yang, An ultrarobust composite gel electrolyte stabilizing ion deposition for long-life lithium metal batteries. Adv. Func. Mater. **29**(43), 1904547 (2019)
132. H.M.C. Azeredo, H. Barud, C.S. Farinas, V.M. Vasconcellos, A.M. Claro, Bacterial cellulose as a raw material for food and food packaging applications. Front. Sustain. Food Syst. **3** (7) (2019)
133. M. Gama, F. Dourado, S. Bielecki, *Bacterial Nanocellulose : From Biotechnology to Bio-Economy* (2016)
134. A.A. Klyosov, Trends in biochemistry and enzymology of cellulose degradation. Biochemistry **29**(47), 10577–10585 (1990)

Chapter 4
Nanocellulose Aerogels

Wenshuai Chen, Haipeng Yu, Ming Dai, Chaoji Chen, Shengjie Ling, Li-Feng Chen, and Qing Li

Abstract Nanocellulose aerogels have attracted attention in various fields because they have distinctive structures and properties, and are produced from abundant and sustainable precursors. In this chapter, we will summarize the development of aerogels using nanocellulose as a building block. First, we will describe the fabrication of nanocellulose aerogels by freeze-drying, supercritical drying, and atmospheric pressure drying methods. Then, we will discuss how to control the structures, properties, and functions of the aerogels via various strategies. Next, we will discuss the application of aerogels in energy storage, triboelectric nanogenerators, sensors, water purification, thermal insulation and scaffolds, according to the characteristics of the aerogels. Finally, we will describe the challenges and future trends associated with nanocellulose aerogels from our personal perspective.

W. Chen (✉) · H. Yu (✉) · M. Dai · Q. Li
Key Laboratory of Bio-Based Material Science and Technology, Ministry of Education, Northeast Forestry University, Harbin 150040, P. R. China
e-mail: chenwenshuai@nefu.edu.cn

H. Yu
e-mail: yuhaipeng20000@nefu.edu.cn

C. Chen
School of Resource and Environmental Science, Wuhan University, Wuhan 430079, P. R. China

S. Ling
School of Physical Science and Technology, ShanghaiTech University, Shanghai 201210, P. R. China

L.-F. Chen
CAS Key Laboratory of Mechanical Behavior and Design of Materials (LMBD), Department of Thermal Science and Energy Engineering, Hefei National Laboratory for Physical Sciences at the Microscale, University of Science and Technology of China, Hefei 230026, Anhui, P. R. China

L. Hu et al. (eds.), *Emerging Nanotechnologies in Nanocellulose*, NanoScience and Technology, https://doi.org/10.1007/978-3-031-14043-3_4

4.1 Introduction

Aerogels are ultralight and highly porous materials [1, 2]. They are produced from wet gels, suspensions, dispersions or solutions, in which the liquid is replaced by gas without serious structural shrinkage. Aerogels are mainly divided as organic-, inorganic-, and composite aerogels. Some inorganic aerogels, such as those made from silica, suffer from brittleness. Some organic aerogels and composite aerogels are flexible, but their production process is complex. The fabrication cost of most aerogels is high, and sometimes toxic and harmful reagents are used. Thus, novel building blocks for the sustainable fabrication of environmentally friendly aerogels are still required.

Because nanocellulose has a one-dimensional nanofiber structure with abundant hydroxyl groups on its surface [3–5], it can be easily assembled into aerogels with entangled web-like structures. The structure and performance of an aerogel can be optimized by carefully manipulating the structure and assembly of nanocellulose [6]. Nanofibrillated cellulose (NFC), cellulose nanocrystal (CNC), and bacterial cellulose (BC) can all be used as building blocks to produce aerogels. NFC and CNC are generally dispersed in water to form suspensions or hydrogels, whereas BC exists as pellicles. Nanocellulose aerogels can be produced by freeze-drying, supercritical drying, or atmospheric pressure drying methods.

Nanocellulose aerogels can be fabricated in a variety of shapes and sizes, depending on the demand. Their structures and properties can be optimized via many methods, resulting in multiple types of aerogels. Nanocellulose aerogels have broad application potential in energy, environmental, and biomedical areas. In the past 20 years, great progress has been made in the development of nanocellulose aerogels, ranging from fundamental research to practical applications. In this chapter, we summarize the development of nanocellulose aerogels, with specific focus on their fabrication, structures, and properties, the control of their functions, and their applicability in some key areas. Finally, we will elaborate our views and prospects on the technical and economic issues and future trends relating to nanocellulose aerogels.

4.2 Constructing Nanocellulose Aerogels

Generally, nanocellulose is dispersed in water or an organic solvent to form a suspension, dispersion, wet gel, or pellicle. To fabricate an aerogel, the liquid should be removed while retaining the nanofiber skeleton without collapse. Lyophilization, supercritical drying, and atmospheric pressure drying are the main methods of nanocellulose aerogel fabrication. Although the mechanisms by which aerogels are constructed differ, the as-obtained nanocellulose aerogels are all lightweight and have porous structures.

4.2.1 Freeze-Drying

Freeze-drying, which is also called lyophilization, is a versatile technique that is widely used for the production of porous nanocellulose aerogels using a freeze-drier. It works by first freezing the nanocellulose colloid/dispersion/suspension/gel/pellicle, then reducing the pressure and heating, and direct sublimating the frozen water in the sample to a vapor (Fig. 4.1a, b) [7]. This produces a lightweight aerogel with a highly porous structure (Fig. 4.1c) [7].

Before sublimating, the precursor is frozen to transform the liquid into ice crystals. Some freeze-driers have the functions for freezing the samples, whereas others require the sample to be frozen in advance, then placed in the freeze-drier for lyophilization. The nanocellulose samples are generally frozen at approximately −18, −56, −80, and −196 °C. As the ice crystals form, they squeeze the nanocellulose to transfer to their boundaries. The nanocellulose then self-aggregates as the ice crystals sublimate. Finally, porous aerogels with web-like entangled nanofiber structures are generated (Fig. 4.1d, e) [7]. Aerogels fabricated by freeze-drying aqueous suspensions of NFC are lightweight and flexible, have bulk densities in the range of 1.3×10^{-3} to 17.0×10^{-3} g cm^{-3}, and a porosity of 98%.

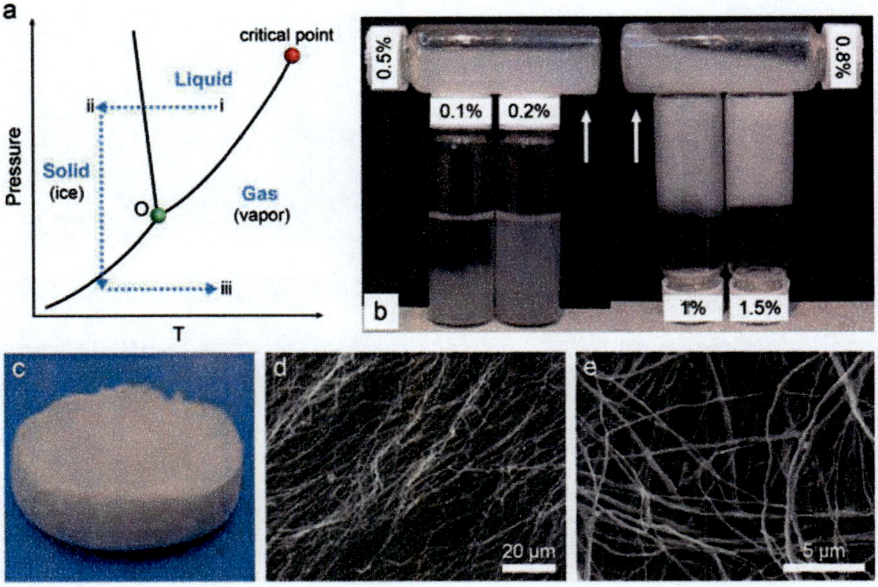

Fig. 4.1 Fabrication of nanocellulose aerogels by freeze-drying method. **a** Phase diagram of the water. **b** NFC aqueous suspensions with solid contents of 0.1–1.5 wt%. **c** Digital photo of a 23.6 mg NFC aerogel. **d, e** SEM images of the NFC aerogels [7]. **b–e** are reproduced with permission from [7]. Copyright 2011 Royal Society of Chemistry

4.2.2 Supercritical Drying

Supercritical drying, which is also called critical point drying, is another efficient technique for nanocellulose aerogel production.

The technique works by allowing the liquid to reach its own critical point during the drying process, enabling the complete transformation from the liquid phase to a supercritical fluid by controlling the pressure and temperature (Fig. 4.2a). During the supercritical drying process, the solvent has no apparent surface tension, so the skeleton structure is maintained without collapsing as the wet sample is transformed into an aerogel. It is essential to remove the solvent from the nanocellulose/solvent mixture without generating a two-phase system, eliminating capillary forces (Fig. 4.2b) [8]. This can be achieved by first compressing and heating the sample to above the critical pressure and temperature of the solvent. The sample is then decompressed and cooled to room conditions. The generated low-pressure solvent vapor is replaced by air via diffusion, and the nanocellulose self-aggregates and forms a porous aerogel (Fig. 4.2c, d) [8]. Aerogels produced by applying the

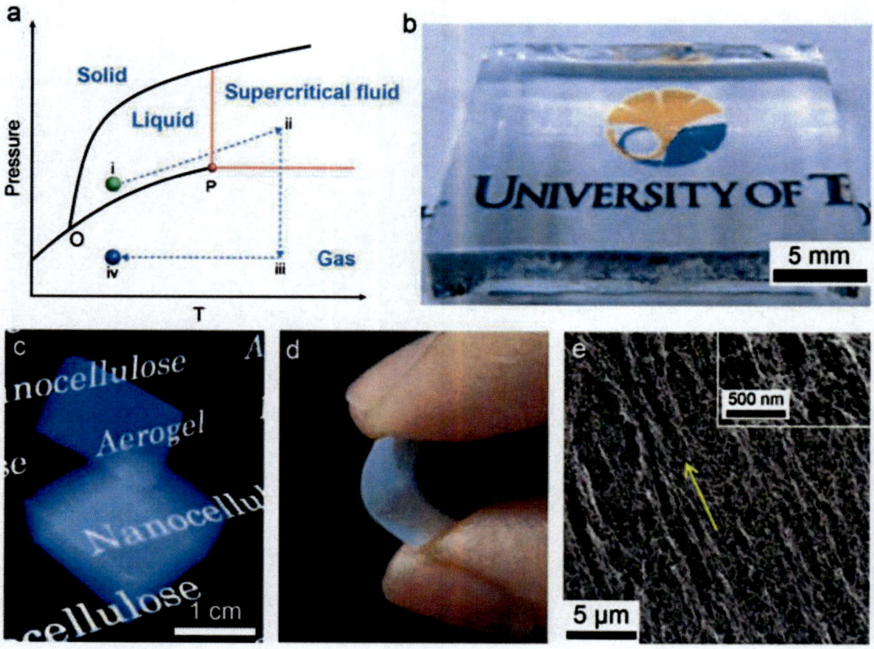

Fig. 4.2 Fabrication of nanocellulose aerogels by supercritical drying method. **a** Phase diagram of the carbon dioxide. **b** Digital photo of a TEMPO-oxidized NFC hydrogel. **c** Digital photo of TEMPO-oxidized NFC aerogels. **d** Digital photo of a densified TEMPO-oxidized NFC aerogel after compression. **e** SEM image of the TEMPO-oxidized NFC aerogel [8]. **b–e** are reproduced with permission from [8]. Copyright 2014 Wiley-VCH

supercritical drying technique have porous entangled nanofiber structures, and show no clear signs of aggregation to sheets (Fig. 4.2e) [8].

4.2.3 Atmospheric Pressure Drying

Compared with freeze-drying and supercritical drying, atmospheric pressure drying does not need extreme pressure, low temperature or high temperature. It is a simple operation, energy-saving and environment-friendly method. Generally, the solvent in the gel structure is mainly water, and the surface tension is large [6]. Thus, during the drying process, the pore wall will collapse due to capillary pressure. Assuming that the pores are evenly distributed in the material species, F_1, F_2, F_3 and F_4 represent the capillary pressure acting on the pore wall, and the resultant force generated by the capillary pressure in the horizontal direction is represented by F_1', F_2', F_3' and F_4' (Fig. 4.3a) [9]. The capillary pressure (P) is calculated refer to reference [10]:

$$P = \frac{-2\gamma \cos\theta}{r} \tag{4.1}$$

where γ means the surface tension, θ represents the contact angle between the wall of pore and the solvent, and r is the pore radius. According to Young's Laplace equation, several methods to reduce the capillary pressure can be classified as [11]: (1) use a solvent with low surface tension (to reduce the γ value); (2) increase the pore diameter to reduce the contact angle between the solvent and the pore wall; (3) make uniform pores to balance the capillary pressure on the pore wall; (4) improve the mechanical properties of the hole wall.

Enhancing the mechanical properties of porous materials and using low surface tension solvents to replace water or alcohol in nanocellulose gels are feasible strategies to achieve atmospheric drying. Glycidoxypropyltrimethoxysilane and branched polyethyleneimine are used to improve the mechanical properties of the pore walls, and then the nanocellulose-based aerogels with the high specific surface area are prepared by solvent replacement and atmospheric drying (Fig. 4.3b, c) [9]. A nanocellulose aerogel film is prepared by vacuum filtration, solvent exchange, and atmospheric drying (Fig. 4.3d–f) [12]. The nanocellulose aerogel film with a permeability of >90%, tensile strength of 97 MPa and density of ~0.6 g cm^{-3} is obtained. Nanocellulose-based aerogels with a density of 2–20 kg m^{-3}, specific surface area of 90 m^2 g^{-1}, specific compression modulus of 26.8 ± 6.1 kPa m^3 kg^{-1}, good wet stability and shape recovery of 80–90%, are prepared by solvent exchange and atmospheric drying of a mixture of nanocellulose, alginate and $CaCO_3$ (Fig. 4.3g–j) [13].

4.2.4 Foaming

Nanocellulose can be compounded with polymers and some additives to prepare aerogels by the foaming process. It has high mechanical strength and a large specific surface area, which enable it to form a large number of nucleation points in the process of polymer foaming to induce heterogeneous nucleation, enhance the nucleation of cells and improve the mechanical, rheological and thermal properties of polymers [14]. However, the abundant hydroxyl groups on the surface of nanocellulose lead to poor compatibility between nanocellulose and hydrophobic polymers [15]. Surface modification of nanocellulose is usually carried out to improve the interface compatibility. The molding process of aerogels mainly includes: (1) raw material dissolution, (2) nucleation of bubbles, (3) bubble growing, and (4) stable solidification of bubbles [16]. The polymer foaming technologies include physical foaming and chemical foaming, among which supercritical fluid foaming is the most common

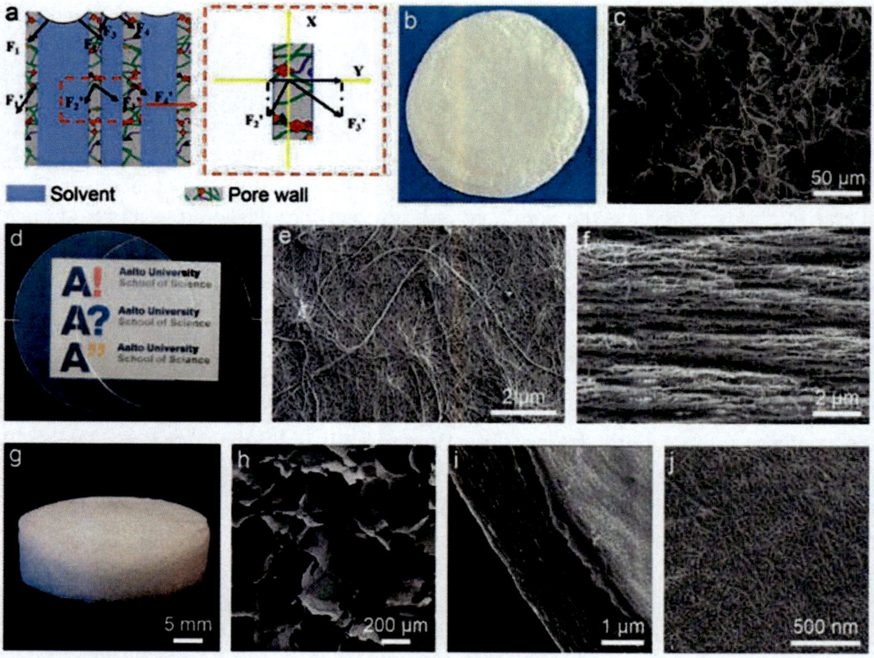

Fig. 4.3 Fabrication of nanocellulose aerogels by atmospheric pressure drying method. **a** Force analysis of the inner wall of gel during drying process. **b** Digital photo of an NFC composite aerogel. **c** SEM image of the NFC composite aerogel [9]. **a–c** are reproduced with permission from [9]. Copyright 2019 Elsevier. **d** Digital photo of an NFC aerogel membrane (left) and a compacted film (right). **e, f** SEM image of the nanocellulose aerogel membrane [12]. **d–f** are reproduced with permission from [12]. Copyright 2015 Wiley-VCH. **g** Digital photo of an NFC composite aerogel. **h–j** SEM images of the NFC composite aerogel [13]. **g–j** are reproduced with permission from [13]. Copyright 2021 Elsevier

physical foaming method. The supercritical fluid physical foaming is environmentally friendly because its foaming agent is carbon dioxide, nitrogen, etc., and the obtained aerogel material has a small cell size and high cell density [17]. The common molding methods include extrusion, injection and intermittent foam molding. The hydrophobic-modified nanocellulose is combined with polypropylene to prepare a composite aerogel with a porosity of 80% by the injection foaming method [15]. The addition of nanocellulose improves the foaming quality of polypropylene, resulting in a small cell size, high cell density and narrow bubble size distribution. When the porosity is less than 80%, the specific flexural modulus and strength of the aerogel increase.

4.3 Controlling the Structures, Properties and Functions of the Aerogels

Nanocellulose aerogels can be produced via various methods from different building blocks. The components and structures of the aerogels determine their properties and functions. The fabrication of a nanocellulose aerogel is affected by many factors. Various nanocellulose aerogels can be produced by carefully designing and manipulating the structure and surface activity of the nanocellulose, the active components addition, the hybrid methods used, the precursor concentration, the drying method, the processing parameters, and the pretreatment and post-treatment stages.

4.3.1 Building Blocks

Various types of NFC, CNC and BC can be used as building blocks for aerogel fabrication.

Different nanocellulose building blocks produce aerogels with different characteristics. Long and high aspect ratio NFC and chemically modified NFC are frequently used to directly produce porous aerogels from an NFC suspension or gel. The NFC self-aggregates and forms hydrogen bonds between the adjacent hydroxyl groups, and generates a strong entangled web-like skeleton to support the highly porous aerogels. The aerogels are flexible and can be repeatedly bent and twisted without damage (Fig. 4.4a–c) [18]. Hydrochloric acid and sulphuric acid hydrolyzed CNC have also been used to fabricate aerogels. Because the CNC is relatively short, it generally self-aggregates into two-dimensional (2D) sheet-like structures during freeze-drying process, which are mutually connected (Fig. 4.4d, e) [19]. CNC aerogels can also be prepared from their partner hydrogels via solvent exchange with ethanol followed by supercritical drying. The aerogels have a density of 78 mg cm^{-3} and specific surface areas of up to 605 m^2 g^{-1} [19].

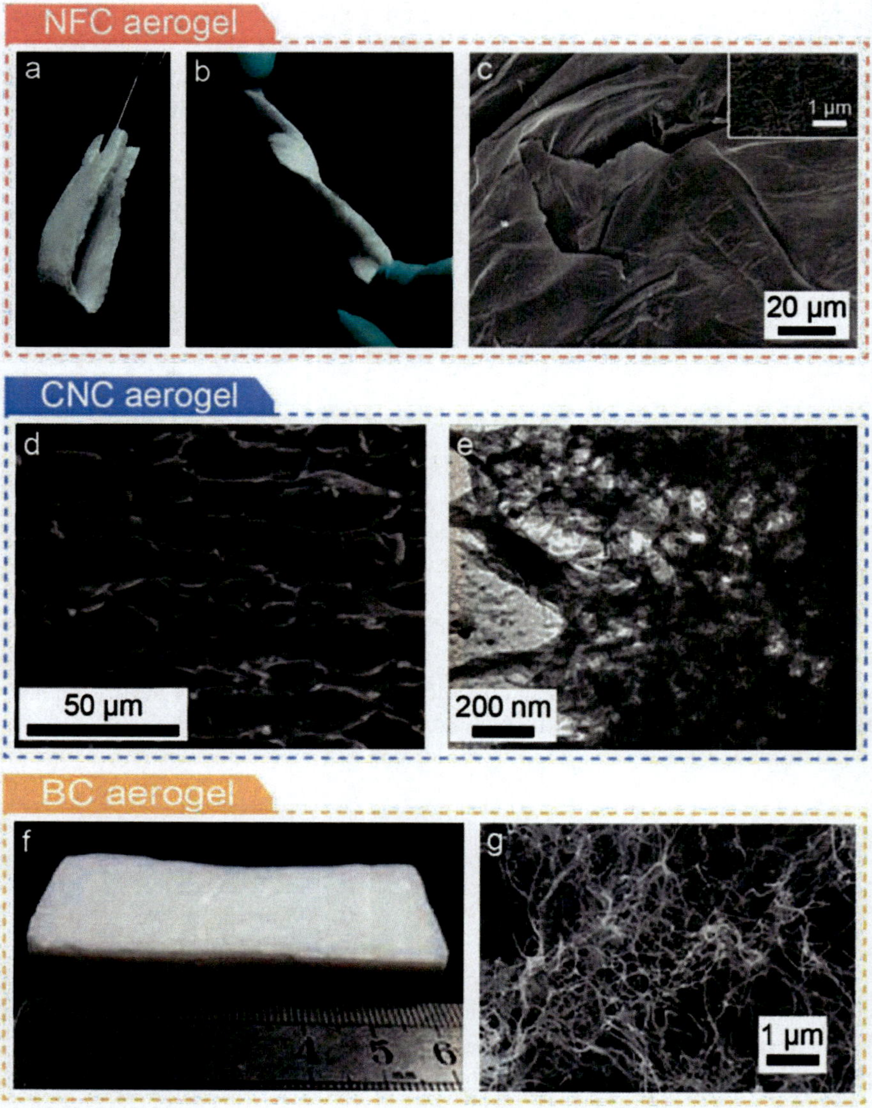

Fig. 4.4 Fabrication of nanocellulose aerogels using various building blocks. **a, b** Digital photos of an NFC aerogel. **c** SEM image of an NFC aerogel [18]. **a–c** are reproduced with permission from [18]. Copyright 2014 Wiley-VCH. **d, e** SEM images of CNC aerogels [19]. **d, e** are reproduced with permission from [19]. Copyright 2010 Royal Society of Chemistry. **f** Digital photo of a BC aerogel. **g** SEM image of a BC aerogel [20]. **f, g** are reproduced with permission from [20]. Copyright 2012 Springer Nature

Aldehyde-modified CNCs and hydrazide-modified CNCs can be mixed with polypyrrole nanofibers, polypyrrole-coated carbon nanotubes, or spherical manganese dioxide nanoparticles via an aqueous sol–gel process to produce hybrid aerogels with capacitive nanoparticles distributed within the CNC network. BC pellicles can also be used directly to fabricate aerogels (Fig. 4.4f) [20]. The BC aerogels have nanofibrous structures owing to the long ribbon-like structure of BC and its self-aggregation during drying (Fig. 4.4g) [20]. The aerogels are flexible and have an apparent density of ~7 mg cm^{-3} and a porosity of ~99.6%. BC can also be disintegrated into NFC via mechanical nanofibrillation, and into CNC via strong acid hydrolysis for aerogel fabrication.

4.3.2 Precursor Concentration

The structure, density, porosity, and mechanical properties of the nanocellulose aerogels can be controlled by simply manipulating the precursor concentration before drying (Fig. 4.5a) [7]. Freeze-dried NFC and TEMPO-oxidized NFC aqueous suspensions produce aerogels with porosities of >99% [7, 8]. The NFC self-aggregates into long nanofibers, and forms 3D web-like structures when the concentration of the suspension is <0.2 wt% (Fig. 4.5b, c) [7]. If the concentration of the suspension is increased to >0.5 wt%, the NFC forms interconnected 2D sheet-like structures (Fig. 4.5d, e) [7]. The densities of the aerogels increased from $\sim 1.1 \times 10^{-3}$ to 8.8×10^{-3} g cm^{-3} when the concentration of suspensions is increased from 0.1 to 0.8 wt%. Transparent, mechanically tough aerogels are produced by the supercritical drying of TEMPO-oxidized NFC hydrogels. As the hydrogel concentration increases, the density of the aerogel linear increased from 4 to 40 mg cm^{-3}, and the porosities decreases from 98.1 to 99.7% [8]. Because the density and the porosity of aerogels are critically dependent on the solid contents of the suspensions and hydrogels from which they are formed, ultralight aerogels can be produced from ultralow-concentration suspension if the nanocellulose is well dispersed in water. When a transparent NFC suspension with a concentration of ~0.018% is freeze-dried, aerogel with an ultralow density of 0.2×10^{-3} g cm^{-3} and a high porosity of >99.9% is obtained (Fig. 4.5f) [7]. Owing to the self-aggregation of single nanofibers and their bundles along the longitudinal direction, the ultralight aerogels are organized into entangled nanofibers with lengths above 1 mm and widths in the range of 40–180 nm (Fig. 4.5g) [7].

4.3.3 Ice Template

The nanocellulose dispersion is frozen in advance to fabricate aerogels by freeze-drying. Therefore, the pore and cell wall structure of the aerogels are critically determined by the ice structure.

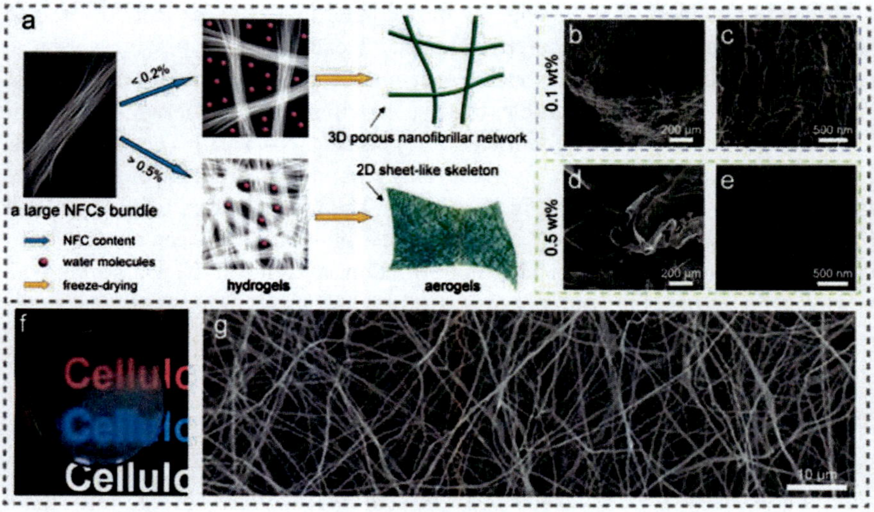

Fig. 4.5 Fabrication of nanocellulose aerogels with various solid contents. **a** Scheme illustrating the effect of the solid content of NFC aqueous suspension on the microstructure of the aerogels. **b**, **c** SEM images of an aerogel produced from a 0.1 wt% NFC aqueous suspension. **d**, **e** SEM images of an aerogel produced from a 0.5 wt% NFC aqueous suspension. **f** Digital photo and **g** SEM image of ultralight NFC aerogel produced from a ~0.018 wt% NFC aqueous suspension [7]. **a–g** are reproduced with permission from [7]. Copyright 2011 Royal Society of Chemistry

Aerogels with various structures are generated using ice as a template. The cooling rate and freeze-casting direction determine the formation of ice crystals. Therefore, the structure of the ice crystals can be modified by manipulating the cooling parameters. When a >0.5 wt% aqueous suspension of NFC is frozen in a refrigerator at −18 °C, aerogels with interpenetrated 2D sheet-like structures are produced after lyophilization [7]. Freezing NFC hydrogels in liquid nitrogen also generates large sheet-like aggregates (Fig. 4.6a) [21]. Such sheets and pores result from the slow growth of ice crystals that gradually push the slender nanofibers into sheets. Liquid propane is also used as a medium for NFC suspension freezing. Fast cooling is realized because liquid propane cools well below the boiling point for heat transferring; this yields smaller ice crystals, and the original nanofibrillar network is retained during lyophilization (Fig. 4.6b) [21].

Freeze-casting is an efficient method for fabricating nanocellulose aerogels with aligned pores [25]. It involves the controlled solidification of a nanocellulose dispersion/gel followed by the sublimation of the ice crystal under reduced pressure. Unidirectional freezing, bidirectional freezing, and radial freezing have been frequently used to control the pore structures of nanocellulose aerogels. During unidirectional freezing, nanocellulose suspension starts to freeze under a single temperature gradient. The ice grows along the direction of freezing, resulting in aerogels with honeycomb structures of anisotropic tubular pores (Fig. 4.6c–g) [22, 23]. During bidirectional freezing of a nanocellulose dispersion/gel, the nucleation and growth

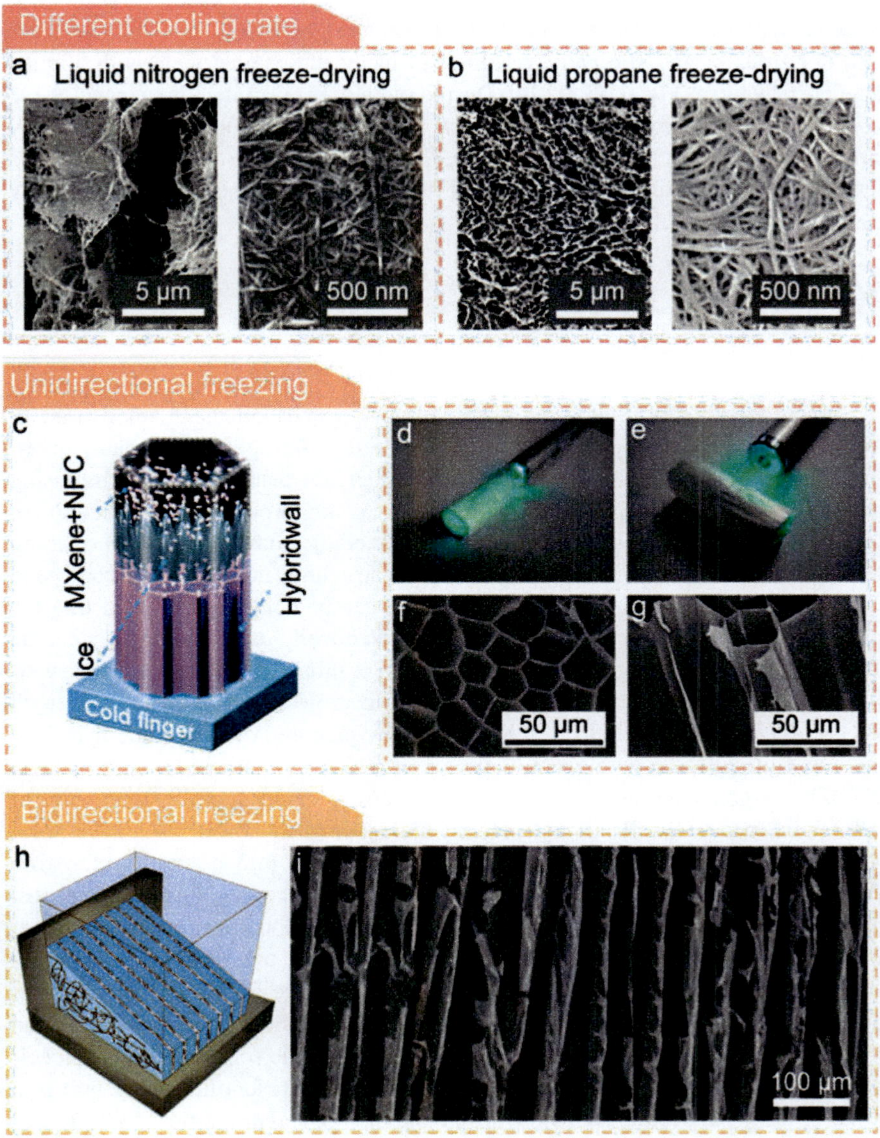

Fig. 4.6 Fabrication of nanocellulose aerogels using ice template method. Production of NFC aerogels by: **a** liquid nitrogen freeze-drying and **b** liquid propane freeze-drying [21]. **a**, **b** are reproduced with permission from [21]. Copyright 2011 American Chemical Society. **c** Production of NFC aerogel by unidirectional freezing [22]. **c** is reproduced with permission from [22]. Copyright 2020 Wiley-VCH. **d**, **e** Digital photos showing the laser penetration tests of TEMPO-oxidized NFC aerogel. SEM images of **f** cross- and **g** longitudinal-sectional of a TEMPO-oxidized NFC aerogel prepared by unidirectional freezing [23]. **d**–**g** are reproduced with permission from [23]. Copyright 2016 American Chemical Society. **h** Production of an NFC aerogel by bidirectional freezing. **i** SEM image of an NFC/silver nanowire composite aerogel prepared by bidirectional freezing [24]. **h**, **i** are reproduced with permission from [24]. Copyright 2020 American Chemical Society

of ice crystals are controlled under dual temperature gradients. Thus, the ice grows both horizontally and vertically, resulting in nanocellulose aerogels with aligned lamellar structures (Fig. 4.6h, i) [24]. During radial freeze-drying, the nanocellulose dispersion is subjected to two temperature gradients: along both the longitudinally and radial directions. The as-generated aerogels have radially oriented pore structures. Thus, nanocellulose aerogels with different pore structures can be obtained by carefully controlling the freezing process via lyophilization.

4.3.4 Processing Before Drying

The macrostructure of a nanocellulose aerogel depends on the method used to process the nanocellulose suspension before drying. Aerogels with bead-, fiber-, film-, and bulk-like macrostructures can be generated using different processing methods. A bead-like nanocellulose aerogel can be produced by injecting an NFC/polyethylenimine/3-glycidyloxypropyl trimethoxysilane mixture into liquid nitrogen, then carrying out lyophilization. The structure, mechanical strength and Cu (II) adsorption capacity of the aerogel bead can be controlled by adjusting the mass ratio of the precursors (Fig. 4.7a–c) [26]. Nanocellulose aerogel fibers can be produced by freeze-drying the wet injecting fiber, or infusing nanocellulose into wet-spinning hollow fibers, then carrying out lyophilization. Porous coaxial fibers with NFC aerogel cores and cellulose-rich sheaths were prepared by wet-spinning hollow fibers and filling them with an NFC precursor before lyophilization (Fig. 4.7d–f) [27]. The as-generated coaxial fibers have a density of 0.2 g cm^{-3}, a porosity of 85%, and a specific tensile strength of 23.5 MPa g cm^{-3}.

A nanocellulose dispersion can be vacuum-filtered into a wet sheet before lyophilization to fabricate a nanocellulose aerogel film (Fig. 4.7g–i) [28]. An aerogel film can also be produced by simply compressing the bulk aerogel into a film under some pressure. With regard to bulk aerogel fabrication, the mold into which the nanocellulose dispersion is poured has a critical influence on the bulk structure of the aerogel (Fig. 4.7j, k) [29]. Aerogels of various sizes and shapes can be obtained from different molds. 3D printing is an effective technique for fabricating nanocellulose aerogels with complex geometries. A nanocellulose monolith can be prepared by 3D printing nanocellulose inks with suitable rheological properties. After freeze-drying, 3D printing aerogels are obtained (Fig. 4.7l–o) [30]. The topography of such an aerogel can be manipulated by designing its structure and controlling the interactions of its building blocks.

4.3.5 Integration with Active Components

Owing to their 1D nanofiber structures, and the physical/chemical activity of their surfaces, nanocellulose and chemically modified nanocellulose can be integrated with

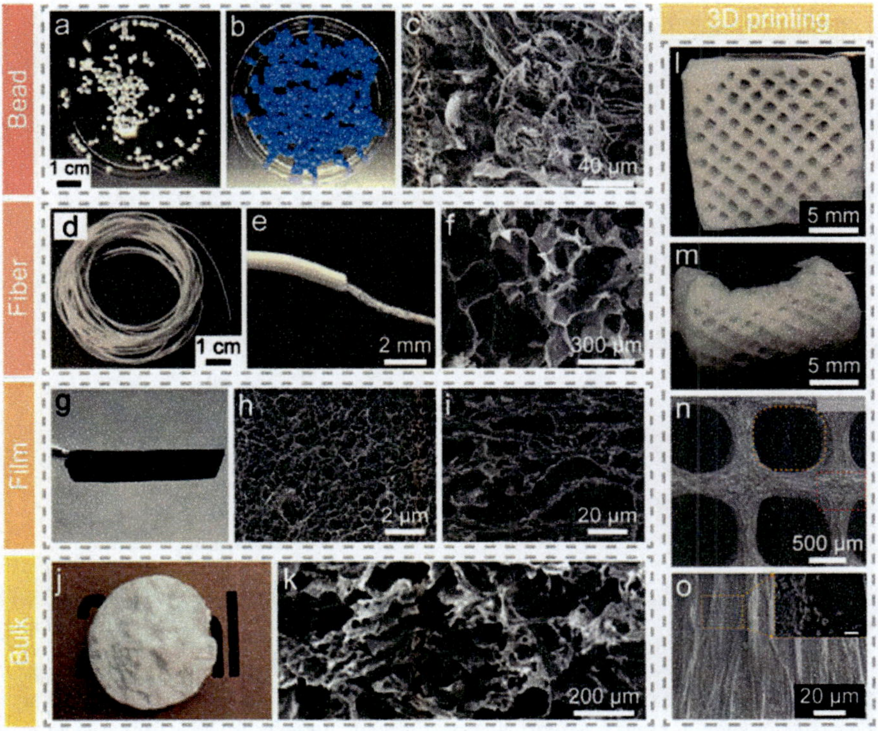

Fig. 4.7 Fabrication of nanocellulose aerogels with various macrostructures. Digital photos of NFC composite aerogel beads **a** before and **b** after adsorption of copper ions. **c** SEM image of an NFC composite aerogel bead [26]. **a–c** are reproduced with permission from [26]. Copyright 2020 Elsevier. **d, e** Digital photos of NFC aerogel-based coaxial fibers. **f** SEM image of the NFC aerogel core of the coaxial fiber [27]. **d–f** are reproduced with permission from [27]. Copyright 2019 Elsevier. **g** Digital photo of an NFC/carbon nanotube (CNT) composite aerogel film. **h, i** SEM images of the NFC/CNT composite aerogel film [28]. **g–i** are reproduced with permission from [28]. Copyright 2021 American Chemical Society. **j** Digital photo and **k** SEM image of NFC/NaCl composite aerogel [29]. **j, k** are reproduced with permission from [29]. Copyright 2018 American Chemical Society. **l, m** Digital photos and **n, o** SEM images of 3D printed NFC aerogels [30]. **l–o** are reproduced with permission from [30]. Copyright 2021 American Chemical Society

various substances to fabricate composite aerogels. The active components used and their interactions with the nanocellulose determine the structures and properties of the aerogels. Polymers, carbon, metals, metallic oxides, salt, and dyes can all be hybridized with nanocellulose in the form of solutions, particles, wires, and sheets on different scales to produce multiple types of aerogels with different properties. Various polymers have been mixed with nanocellulose to construct aerogels. The integration of amylopectin and NFC produces aerogels with microcellular structures that have cell dimensions in the range of 20–70 mm [31]. The cooperation of NFC with a chitosan matrix produces soft and flexible aerogels (Fig. 4.8a, b) [32].

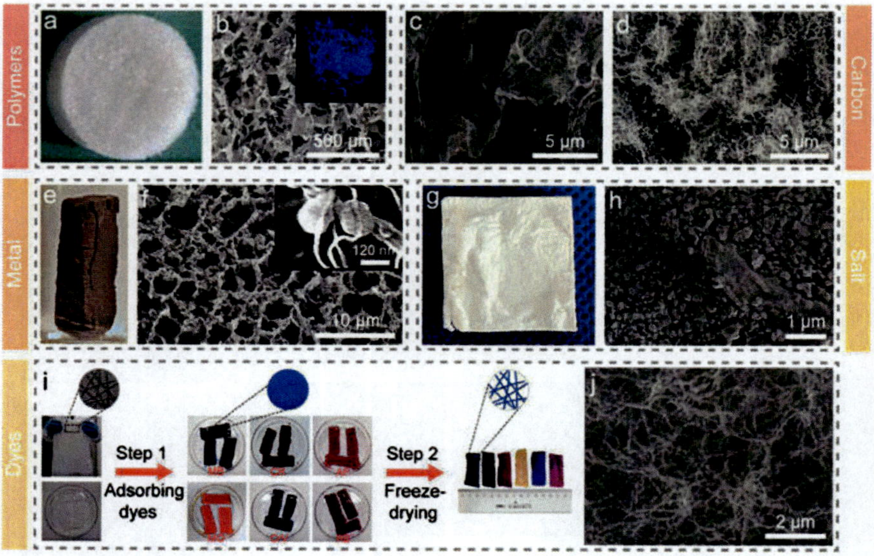

Fig. 4.8 Fabrication of various nanocellulose composite aerogels. **a** Digital photo and **b** SEM image of an NFC/chitosan composite aerogel [32]. **a**, **b** are reproduced with permission from [32]. Copyright 2017 Wiley-VCH. **c**, **d** SEM images of an NFC/few-walled carbon nanotube (FWCNT) composite aerogel [33]. **c**, **d** are reproduced with permission from [33]. Copyright 2013 Wiley-VCH. **e** Digital photo and **f** SEM image of a BC/cobalt ferrite nanoparticle composite aerogel [34]. **e**, **f** are reproduced with permission from [34]. Copyright 2010 Springer Nature. **g** Digital photo and **h** SEM image of an NFC/LiCl composite aerogel [35]. **g**, **h** are reproduced with permission from [35]. Copyright 2021 Elsevier. **i** Process used to produce BC/dye composite aerogels. **j** SEM image of an N, S doped, BC-derived carbon composite aerogel [36]. **i**, **j** are reproduced with permission from [36]. Copyright 2014 Springer

Nanocellulose can also be hybridized with various forms of carbon such as carbon nanotubes and graphene to produce aerogels. The integration of NFC with few-walled carbon nanotubes (FWCNTs) produces porous and conductivity aerogels (Fig. 4.8c, d) [33]. Metals, such as gold nanorods [37], and metal oxidize, such as cobalt ferrite nanoparticles [34], have been added to or synthesized within BC networks to produce composite aerogels with plasmonic activity and ferromagnetic properties (Fig. 4.8e, f) [34]. Salts such as lithium chloride have also been combined with NFC via solution exchanging followed by lyophilization (Fig. 4.8g, h) [35]. The as-generated composite aerogels have a strong moisture-absorbing ability. Nanocellulose can also be adsorbed or dyed with different dyes for aerogel preparation. BC hydrogels are immersed in various organic dye solutions. The dyes are absorbed by the BC, and the resulting hydrogels are freeze-dried to produce various colored BC/dye hybrid aerogels (Fig. 4.8i) [36]. Heteroatom doped carbon composite aerogels can be obtained by pyrolyzing such BC/dye hybrid aerogels (Fig. 4.8j) [36]. Therefore, the integration of active components can apparently provide nanocellulose aerogels with novel structures, properties and functions.

4.3.6 Carbonization

Cellulose is a good carbon source and nanocellulose aerogels can be pyrolyzed to produce carbon aerogels. Generally, the aerogels are pyrolyzed in a tubular furnace and protected by an argon or nitrogen atmosphere. The hydrophilic functional groups of cellulose, such as hydroxyl groups, gradually disappear as the temperature increases. The crystalline structure of nanocellulose is destroyed and amorphous carbon is formed. After carbonization, the shape of the aerogels is retained although they shrink in size owing to the evaporation of the volatile components. Carbon aerogels are low in density, became hydrophobic, and exhibit compression resilience.

NFC-derived carbon aerogels have 3D networks that are interconnected by 2D sheet-like structures (Fig. 4.9a, b) [38]. BC-derived carbon aerogels have a porous, interconnected 3D structure (Fig. 4.9c, d) [39], the carbon nanofibers of which are 10–20 nm wide. NFC-derived carbon aerogels pyrolyzed at 850 °C have a density of 7.8 mg cm^{-3} and a specific surface area of 554.8 m^2 g^{-1}. BC-derived carbon aerogels that have been treated at 1300 °C have a density of 4–6 mg cm^{-3} and a conductivity of 20.6 S m^{-1}. Various precursors such as dyes, polymers, carbon, acids, alkalis, and salts can also be hybridized with nanocellulose before carbonization to produce carbon-based composite aerogels with various properties and functions. The pyrolysis treatment essentially changes the characteristics of nanocellulose aerogels.

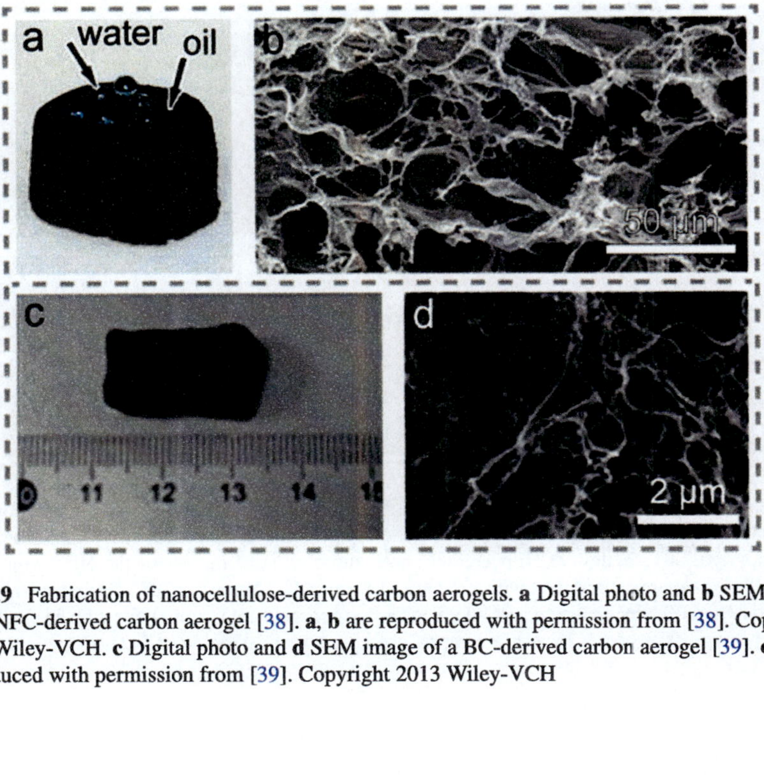

Fig. 4.9 Fabrication of nanocellulose-derived carbon aerogels. **a** Digital photo and **b** SEM image of an NFC-derived carbon aerogel [38]. **a, b** are reproduced with permission from [38]. Copyright 2016 Wiley-VCH. **c** Digital photo and **d** SEM image of a BC-derived carbon aerogel [39]. **c, d** are reproduced with permission from [39]. Copyright 2013 Wiley-VCH

The structures and properties of the carbon aerogels can be manipulated by integrating nanocellulose with various active components before carbonization.

4.3.7 Post-treatment

After fabrication, nanocellulose aerogels and nanocellulose-derived carbon aerogels can be further decorated with active components via post-treatment to improve their performance.

Conducting polymers, biomolecules, and carbon nanotubes can be coated onto a cross-linked NFC aerogel via layer-by-layer (LbL) assembly (Fig. 4.10a) [40]. Such an aerogel has a porosity of 99%, and exhibits enhanced compressive strength,

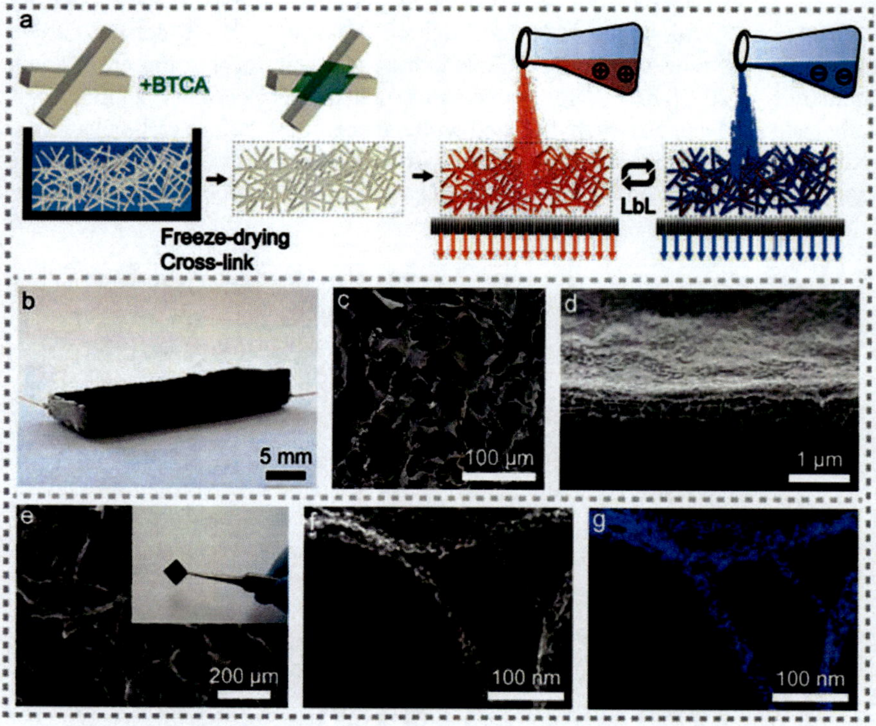

Fig. 4.10 Fabrication of nanocellulose composite aerogels via post-treatment. **a** Production of cross-linked NFC aerogel via layer-by-layer assembly [40]. **a** is reproduced with permission from [40]. Copyright 2013 Wiley-VCH. **b** Digital photo and **c**, **d** SEM images of the NFC composite aerogel [41]. **b–d** are reproduced with permission from [41]. Copyright 2015 Springer Nature. **e** SEM image and **f** STEM image of a BC-derived carbon aerogel/MnO_2 composite with an elemental mapping image of **g** Mn [42]. **e–g** are reproduced with permission from [42]. Copyright 2013 Wiley-VCH

elasticity, fluorescence, and energy storage capacity. In one experiment, an electrode, a separator, and another electrode were coated step-by-step onto an NFC aerogel skeleton. The resulting compressible aerogel supercapacitor comprised carbon nanotube electrodes, while the as-produced aerogel battery is organized with an intercalated copper hexacyanoferrate ion cathode and a carbon nanotube anode (Fig. 4.10b–d) [41].

Nanocellulose-derived carbon aerogels can also be treated with various materials to improve their performances. BC-derived carbon aerogels can be coated with MnO_2 (Fig. 4.10e–g) [42], doped with heteroatoms such as nitrogen, or activate with KOH to optimize their structures and electrochemical properties for energy storage. Post-treatment can introduce novel functionality and further improve the properties of nanocellulose aerogels and carbon aerogels.

4.4 Applications

Nanocellulose aerogels with various structures, properties, and functions have been developed. Nanocellulose aerogels, which are 3D porous networks constructed from sustainable polymer nanofibers, have great potential for use in many fields. In the following section, we will highlight some representative applications of nanocellulose aerogels, including in energy storage, energy harvesting, sensor, water purification, thermal insulation, and bio-scaffold.

4.4.1 Energy Storage

Nanocellulose aerogels have been widely exploited to contracture various energy storage systems such as supercapacitors, lithium ion batteries (LIBs), sodium ion batteries (SIBs), and lithium sulfur batteries [43, 44]. Aerogels that have been decorated with an active component have favorable aerogels characteristics and electrochemical activity. Porous structures of the aerogels are beneficial for constructing aerogels with high active components loading, and interconnected porous networks provide efficient paths for electrons and ions to move and diffuse.

LbL-assembled, single wall carbon nanotubes (SWCNT)-decorated NFC aerogels have been used as the electrodes of an aqueous supercapacitor, which have a specific electrode capacitance of 419 F g^{-1} [40]. A 3D LbL-assembled supercapacitor has a capacitance of 25 F g^{-1} over 400 cycles, and a hybrid battery on an NFC aerogel substrate has a specific charge capacity of 5.9 mA h g^{-1} (Fig. 4.11a) [41]. A chemically cross-linked CNC aerogel loaded with high-capacitive nanoparticles has been used as an electrode in a symmetric coin-cell supercapacitor (Fig. 4.11b–e) [45]. The as-assemble device has low internal resistance and good capacitance retention at high charge–discharge rates. Flexible CNC/multiwalled carbon nanotubes/polypyrrole

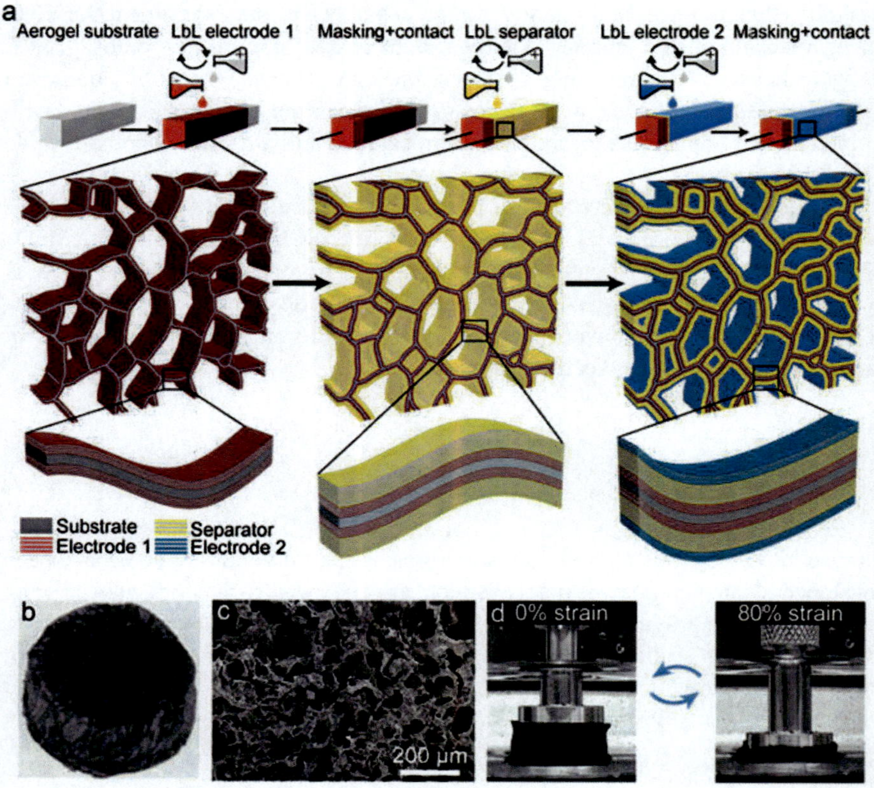

Fig. 4.11 Nanocellulose aerogels for energy storage systems. **a** Fabrication of NFC composite aerogels for an energy storage device [41]. **a** is reproduced with permission from [41]. Copyright 2015 Springer Nature. **b** Digital photo and **c** SEM image of a CNC composite aerogel for a supercapacitor. **d** Digital photos of cyclic compression of a CNC composite aerogel in a saturated Na_2SO_4 solution [45]. **b–d** are reproduced with permission from [45]. Copyright 2015 Wiley-VCH

composite aerogels have been used as electrodes in a supercapacitor. The electrodes and supercapacitor exhibit compression stability at 80% compression [45].

Nanocellulose-derived carbon aerogels have been used as substrates for active components, or have been doped with heteroatoms and used as electrodes in various energy storage devices. An asymmetric supercapacitor with a positive electrode comprising a BC carbon aerogel coated with MnO_2 and a negative electrode comprising a nitrogen-doped BC carbon aerogel (Fig. 4.12a) can be reversibly charged/discharged at a voltage of 2.0 V in 1.0 M Na_2SO_4 solution [42]. The supercapacitor achieves an energy density of 32.9 Wh kg^{-1} and a maximum power density of 284.6 kW kg^{-1}. A flexible all-solid-state supercapacitor can be created using nitrogen-doped BC carbon aerogels as electrodes (Fig. 4.12b) [46]. Such a supercapacitor has a maximum power density of 390.5 kW kg^{-1} and is stable after 5000 cycles. A supercapacitor with N, P-co-doped BC carbon aerogels as electrodes

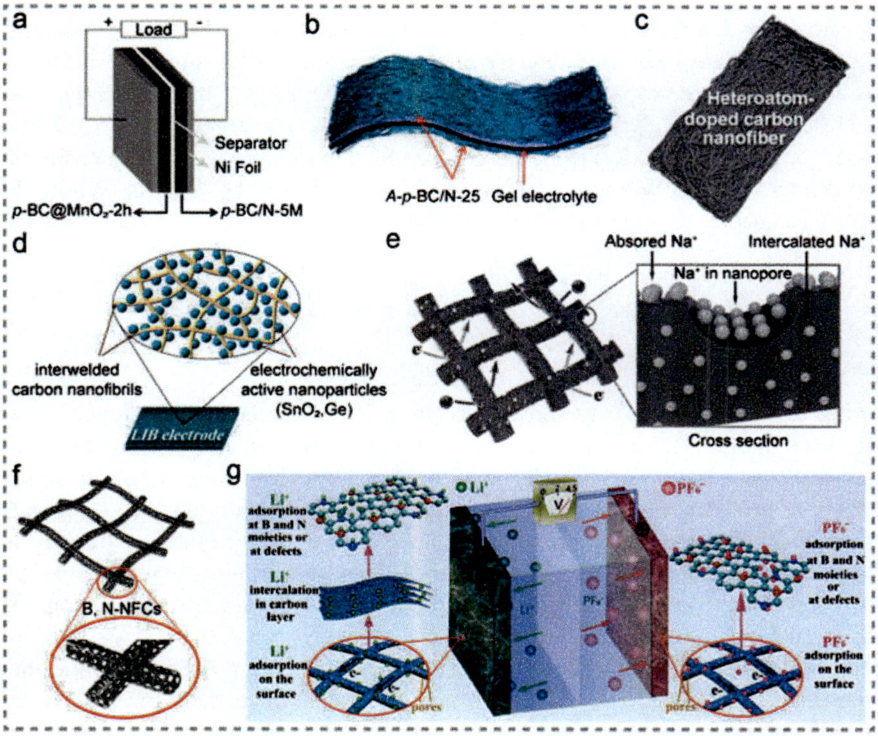

Fig. 4.12 Nanocellulose-derived carbon aerogels for the energy storage system. **a** Assembling asymmetric supercapacitor using BC-derived carbon composite aerogels as electrodes [42]. **a** is reproduced with permission from [42]. Copyright 2013 Wiley-VCH. **b** Assembling an all-solid-state supercapacitor using BC-derived carbon composite aerogels as electrodes [46]. **b** is reproduced with permission from [46]. Copyright 2013 Royal Society of Chemistry. **c** BC-derived, heteroatom doped carbon aerogels for supercapacitors [47]. **c** is reproduced with permission from [47]. Copyright 2014 Wiley-VCH. **d** BC-derived carbon composite aerogels for LIBs [48]. **d** is reproduced with permission from [48]. Copyright 2013 Wiley-VCH. **e** BC-derived N, O dual-doped carbon aerogels for SIBs [49]. **e** is reproduced with permission from [49]. Copyright 2016 Wiley-VCH. **f** BC-derived B, N dual-doped carbon aerogels for SIBs [50]. **f** is reproduced with permission from [50]. Copyright 2017 Wiley-VCH. **g** BC-derived B, N dual-doped carbon aerogels for LICs [51]. **g** is reproduced with permission from [51]. Copyright 2017 Wiley-VCH

has a maximum power density of 186.0 kW kg^{-1} (Fig. 4.12c) [47]. Nanocellulose-derived carbon aerogels can also be used as substrates to construct anodes for LIBs (Fig. 4.12d) [48]. A SnO_2 nanoparticle-decorated BC carbon aerogel maintains a reversible capacity of 600 mAh g^{-1} after 100 cycles at 100 mA g^{-1}. A Ge nanoparticle-decorated BC carbon aerogel exhibits a specific capacity of 500 mAh g^{-1} after 100 cycles at 100 mA g^{-1}. Heteroatom doped nanocellulose carbon aerogels have also been used as electrodes for SIBs and lithium-ion capacitors (LICs). An N, O-dual-doped BC carbon aerogel is mixed with polyvinylidene fluoride and acetylene black, and used as an anode for SIBs (Fig. 4.12e) [49]. It has a specific capacity of 545 mAh g^{-1} at 100 mA g^{-1} after 100 cycles. A B, N co-doped BC carbon aerogel has been used directly as an electrode for SIBs (Fig. 4.12f) [50]. It has a specific charge capacity of 581 mAh g^{-1} at 100 mA g^{-1} after 120 cycles. A B, N dual-doped BC carbon aerogel has been used to construct binder-free electrodes for LICs (Fig. 4.12g) [51], the as-assembled device has an energy density of 220 Wh kg^{-1} at 225 W kg^{-1}, a power density of 22 500 W kg^{-1} at 104 Wh kg^{-1}, and 81% capacitance retention after 5000 cycles.

4.4.2 Triboelectric Nanogenerator

Owing to their high porosity and surface roughness, nanocellulose aerogels have been used in triboelectric nanogenerators (TENGs) to harvest energy from the environment. The energy is harvested by coupling contact electrification and electrostatic induction. When it contacts other materials, cellulose generally donates electrons. Thus, nanocellulose aerogels are usually considered as positive materials of TENG. Compared with densely packed films, highly porous nanocellulose aerogels provide increased contact area and electrostatic induction, resulting in greater charge on the porous surface.

Active components such as rabbit fur (Fig. 4.13a–d) [52], polyethylenimine and aminosilane have been integrated into aerogels to improve their tribopositivity. An NFC/rabbit fur composite aerogel-based TENG exhibits a power density of 3.4 W m^{-2} realized on a 4.7 MΩ load at a pressure of 30 kPa. A TENG assembled by pairing an NFC/polyethylenimine aerogel with four layers of PVDF mats has a power density of 13.3 W m^{-2} at a load resistance of 106 Ω (Fig. 4.13e) [53], and a TENG assembled from an aminosilane-functionalized NFC aerogel film paired with polyimide aerogel (Fig. 4.13f, g) has an output voltage of 57.2 V and an output current of 8.6 μA [54]. Owing to its favorable rheological properties, an NFC suspension can be printed on substrates with various patterns before freeze-drying to produce a printed hierarchically structured 3D aerogel for a TENG (Fig. 4.13h) [55]. The printed NFC aerogel/Ag/PET layer is paired with a printed PDMS/Ag/PET layer. The as-assembled TENGs can harvest mechanical energy to drive 88 light-emitting diodes.

4.4.3 Sensor

Sensors are detection devices that can sense the measured information, and can transform the sensed information into electrical signals or other required forms of information output according to certain rules. Based on the sensitive elements, sensors can be used to detect thermal, light, gas, force, magnetic, humidity, sound, radiation, color, and taste [6]. Usually, a sensor is composed of sensing materials and conversion elements [56]. The performance of the sensor mainly depends on the

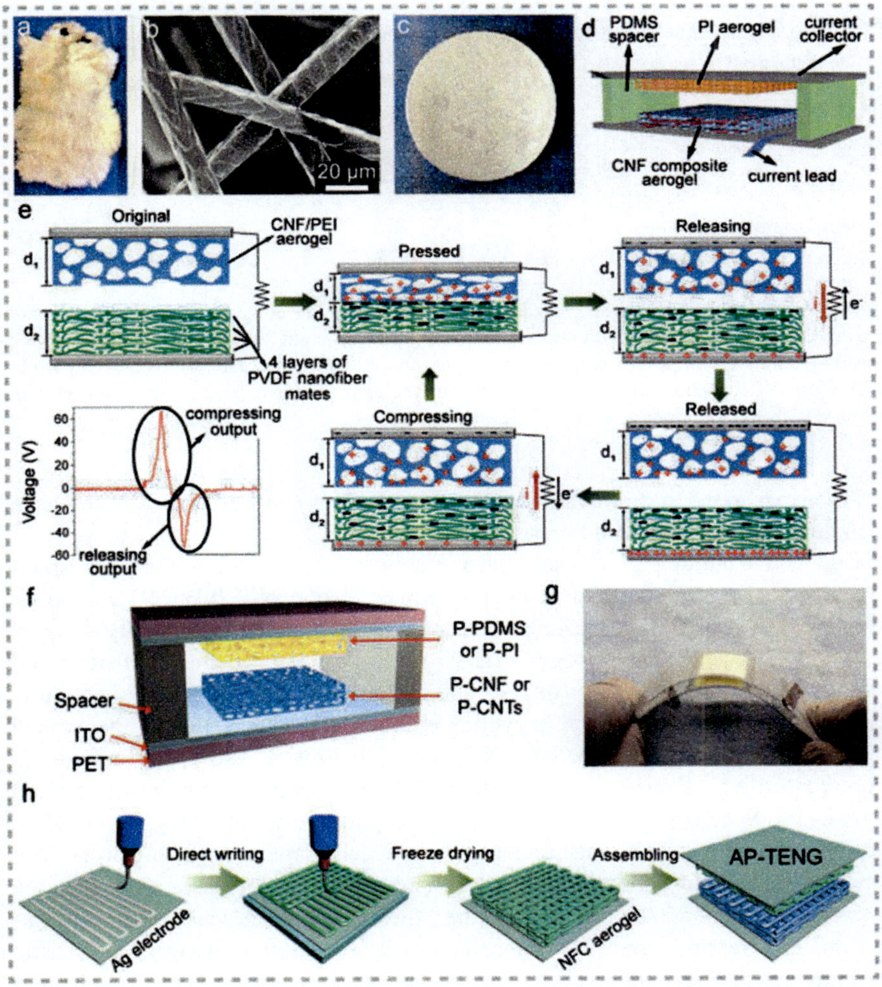

Fig. 4.13 Nanocellulose aerogels for TENG. **a** Digital photo and **b** SEM image of rabbit fur. **c** Digital photo of an NFC/rabbit fur composite aerogel film. **d** Assembling TENG from an NFC composite aerogel [52]. **a–d** are reproduced with permission from [52]. Copyright 2018 Royal Society of Chemistry. **e** Schematic illustrating the working principle of an NFC composite aerogel-based TENG [53]. **e** is reproduced with permission from [53]. Copyright 2018 Elsevier. **f** Schematic and **g** digital photo of an NFC aerogel-based TENG [54]. **f, g** are reproduced with permission from [54]. Copyright 2018 Wiley-VCH. **h** Schematic illustrating the preparation of a 3D printed NFC aerogel-based TENG [55]. **h** is reproduced with permission from [55]. Copyright 2019 Elsevier

chemical composition of the sensing material, surface modification, the microstructure and morphology of the sensing layer and some other factors [57]. Nanocellulose aerogels are highly porous, have favorable mechanical properties, and are easily decorated with active components. Therefore, they have been designed as sensors

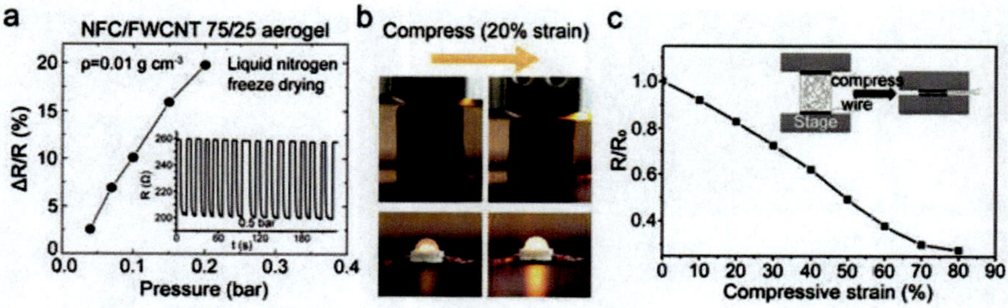

Fig. 4.14 Nanocellulose aerogels for sensors. **a** An NFC/FWCNT composite aerogel for pressure sensing [33]. **a** is reproduced with permission from [33]. Copyright 2013 Wiley-VCH. **b** Pressure sensitivity of a BC composite aerogel [59]. **b** is reproduced with permission from [59]. Copyright 2021 Wiley-VCH. **c** Pressure response of a BC-derived carbon aerogel [39]. **c** is reproduced with permission from [39]. Copyright 2013 Wiley-VCH

and used to respond to various stimuli. Recently, various nanocellulose aerogels have been developed for strain and pressure sensing [6, 58].

Conductive building blocks have been integrated into aerogels to provide responsive resistance during mechanical compression or tension. NFC/FWCNT composite aerogels have been used as sensors to detect the compression exerted by dry nitrogen gas (Fig. 4.14a) [33]. An aerogel with 25 wt% FWCNT content is sensitive and exhibits a 10% relative change of resistance when subjected to 0.1 bar of pressure. It is also stable when subjected to cycling at 0.5 bar. The sensibility of the aerogel decreases when decreasing the FWCNT content to 20 wt%, and ~5 bar pressure is required to affect a 10% change in resistance. BC/polymethylsilsesquioxane/reduced graphite oxide aerogels have also been used as pressure sensors (Fig. 4.14b) [59]. During compressing, the contact among the reduced graphite oxide increases. The conductivity therefore increases as the compressive strain increases. Owing to the intrinsic alignment of their porous structures, these aerogels are sensitive and responsive to low-pressure. Because nanocellulose-derived carbon aerogels are elastic and conductive, they can also be exploited as pressure sensors. The combination of compressible and conductive properties endows aerogels with responsiveness to pressure. BC-derived carbon aerogels exhibit good pressure response behavior (Fig. 4.14c) [39]. Upon compression, the resistance decreases almost linearly as the strain increases from 0 to 70%. Because the aerogel has a robust interconnected carbon nanofiber network, it is stable during cyclic compression, and exhibits a continuous and reversible change of current with strain.

4.4.4 Water Purification

Clean water is very important for people's health and environmental safety. However, due to industrial production and human daily activities, the problem of water pollution is becoming more and more serious. Therefore, it is critical to develop water purification materials and technologies for generating clean water. Owing to the activity of nanocellulose and high porosity, nanocellulose aerogels, nanocellulose composite aerogels, and nanocellulose-derived carbon aerogels have been developed for water purification [60]. The purification is mainly achieved by adsorption and filtration methods. For adsorbing purification, the contaminants are adsorbed by the aerogels, resulting in clean water after taking out the adsorbed aerogels. For filtrating purification, the impurity substance and water are separated by filtration through the aerogels.

Nanocellulose aerogels and nanocellulose hybrid aerogels are used as adsorbers to remove toxic ions and organic pollutants from dirty water. The hydroxyl groups of nanocellulose and the active components introduced to the aerogels effectively adsorb pollutants, and the interconnected pores of the aerogels can store large amounts of pollutants. Purification can be terminated by simply removing the aerogel from the water.

NFC aerogels can be used directly as adsorbents to remove methylene blue (MB) and toluidine blue (TB) from water (Fig. 4.15a–c) [18]. NFC aerogels and TEMPO oxidized aerogels have adsorption capacities for MB of 2.9 and 3.7 mg g^{-1}, and for TB of 3.0 and 4.2 mg g^{-1}, respectively. TEMPO-oxidized NFC/chitin nanofiber hybrid aerogels have adsorption capacities of 217 and 531 mg g^{-1} for As (III) and MB, respectively [62]. TEMPO-oxidized NFC/zeolitic imidazolate framework-8 hybrid aerogels has an equilibrium adsorption capacity of 83.3 mg g^{-1} for Rhodamine B solution [63]. CNC/metal–organic framework particles (MOFs) aerogels can be used to remove potassium dichromate. An aerogel comprising 50 wt% MOFs can adsorb 85% of the Cr (VI) within 24 h [64].

Hydrophobic nanocellulose aerogel and carbon aerogels can be used to adsorb oil from water. Because the surface of nanocellulose contains hydrophilic hydroxyl groups, it should be modified or converted to carbon before the adsorption of oil. Oil/water separation can be achieved by fabricating a silylated NFC aerogel by integrating NFC with a methyltrimethoxysilane solution. The resulting aerogels are hydrophobic and oleophilic, and can adsorb organic solvents and oils with capacities up to 100 times their own weight [65]. Nanocellulose carbon aerogels are lightweight, porous and oleophilic; they rapidly adsorb oil and organic solvents from water and store them within themselves. When a piece of nanocellulose derived carbon aerogel is introduced into a contaminated water, the aerogels rapidly adsorbs the contaminants, and yield clean water (Fig. 4.15d) [39]. With regard to the adsorption of various oils and organic solvents, BC-derived carbon aerogels can adsorb the liquids up to 106–312 times their own weight [39], whereas NFC-derived carbon aerogels display a weight gain of 70–225 times of their own weight [38]. Aerogels that have adsorbed contaminants can be recycled by various methods including squeezing, combustion, and distillation.

Oil can also be separated from water using nanocellulose aerogels filters. Because the aerogels are highly porous and contain numerous hydroxyl groups, they are superhydrophilic and oleophobic under water. They are easily wetted; water penetrates rapidly into their whole network, and is therefore readily stored. Because water is oleophobic, oil is rejected by wet aerogels. Thus, nanocellulose aerogel filters can

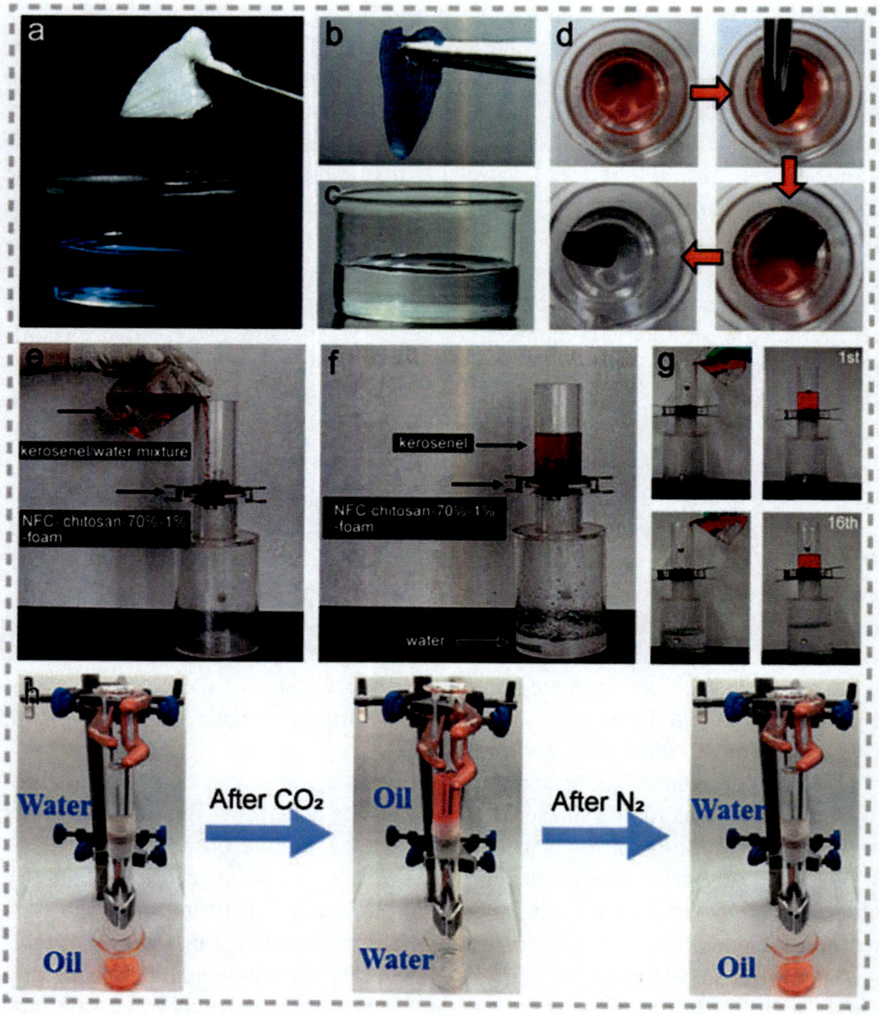

Fig. 4.15 Nanocellulose aerogels for water purification. **a**–**c** Digital photos showing the removal of MB from water using an NFC aerogel [18]. **a**–**c** are reproduced with permission from [18]. Copyright 2014 Wiley-VCH. **d** Digital photos showing the removal of gasoline from water using a BC-derived carbon aerogel [39]. **d** is reproduced with permission from [39]. Copyright 2013 Wiley-VCH. **e, f** Digital photos showing the separation of kerosene/water mixture using an NFC/chitosan composite aerogel filter. **g** Sixteen cycles of kerosene/water separation using the NFC/chitosan composite aerogel filter [32]. **e**–**g** are reproduced with permission from [32]. Copyright 2017 Wiley-VCH. **h** Digital photos illustrating the separation of hexane/water mixture using an NFC composite aerogel [61]. **h** is reproduced with permission from [61]. Copyright 2019 Wiley-VCH

be used to separate oil/water layered mixtures by gravity: oil is rejected and water passes through. High-porosity aerogels with numerous micron-scale pores promote the rapid flow of water. A wetted NFC/chitosan composite aerogel filter can be used to separate a kerosene/water mixture (40% v/v). After pouring the mixture into the top of a container, the water is gradually penetrating and passes through the filter, whereas the kerosene is retained above the aerogel (Fig. 4.15e, f) [32]. Such aerogels can sustain stable water flux over 16 repeated cycles (Fig. 4.15g) [32], which indicates their robust oil/water separation capability. The surface wettability of nanocellulose aerogels can be modified by integration with stimuli-response components. NFC aerogel grafted with poly(*N*,*N*-dimethylamino-2-ethyl methacrylate) polymer brushes can remove the oil from an oil/water mixture. When exposed to CO_2, the aerogel becomes hydrophilic and the separation behavior is reversed (Fig. 4.15h) [61].

4.4.5 Thermal Insulation

Thermal management plays an important role in the fields of architecture, environment and energy. An important point to realize thermal management is to develop thermal insulation materials. Thermal insulation limits or impairs heat exchange between substances by increasing the overall thermal resistance to heat transfer. The heat transfer process mainly contains three basic modes: heat conduction, convection and radiation [66]. Because nanocellulose aerogels are highly porous, they are an ideal platform for thermal insulation. The thermal conductivity of nanocellulose aerogels is mainly determined by solid-state thermal conductivity and gas-phase thermal conductivity, which is closely related to its density, pore size distribution and structure. When the pore size inside of the aerogels is smaller than the average free path of gas molecules, the pores will restrict gas movement and inhibit convection, and the thermal conductivity of the material will decrease [67]. The pore size of the nanocellulose aerogel can be adjusted according to the actual requirements.

The nanocellulose aerogels have a thermal conductivity of <0.017 W m^{-1} K^{-1}. When such aerogels are compressed, the decrease in porosity results in an increase in thermal conductivity, although the value remains below 0.025 W m^{-1} K^{-1} (Fig. 4.16a) [18]. Because they contain some large pores, the thermal conductivity of NFC/chitosan composite aerogels is a little higher than that of air (approximately 0.025 W m^{-1} K^{-1}), but they still have good thermal insulation performance. Regardless of whether such an aerogel is placed on a cold plate at 0 °C or on a hot plate at 70 °C for 5 min, the aerogel remains stable and no apparent temperature gradient occurs in the height direction of the aerogel (Fig. 4.16b) [32]. This demonstrates the favorable insulation properties of the aerogel over a wide temperature range.

An NFC/graphene oxide/boric acid/sepiolite nanorod composite aerogel with aligned tubular pores (Fig. 4.16c–e) has a thermal conductivity of 0.015 W m^{-1} K^{-1} in the radial orientation [68]. However, the thermal conductivity of such an aerogel in the axial direction (along the pores) is apparently higher than that in the radial

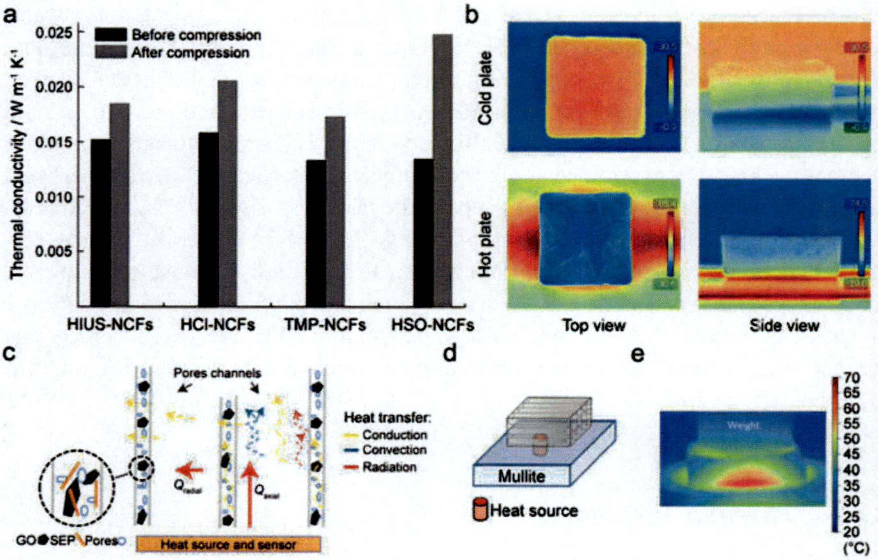

Fig. 4.16 Nanocellulose aerogels used for thermal insulation. **a** Thermal conductivity of various nanocellulose aerogels and compressed nanocellulose aerogels [18]. **a** is reproduced with permission from [18]. Copyright 2014 Wiley-VCH. **b** Thermographic images of an NFC/chitosan composite aerogel in contact with plates of various temperatures [32]. **b** is reproduced with permission from [32]. Copyright 2017 Wiley-VCH. **c** Schematic illustrating thermal conductivity in the radial and axial directions of an aerogel with aligned pores. **d** Schematic and **e** thermographic image of NFC/graphene oxide/boric acid/sepiolite nanorod composite aerogels in contact with a heat source [68]. **c–e** are reproduced with permission from [68]. Copyright 2014 Springer Nature

direction. Aerogel fibers are also used for thermal management. The coaxial fibers organized into an NFC aerogel core and a cellulose-rich sheath can be used as thermal insulators [27]. They are effective at −20 to 150 °C. Both the NFC aerogel core and the sheath are critical for the thermal insulating property of the fibers.

4.4.6 Scaffold

In the process of tissue cell reconstruction, the three-dimensional network structure, called scaffold, is used for the survival and attachment of cells before regenerating the tissue. Through scaffolds, cells can be fixed in a specific location. Scaffold materials provide a place for cell growth, reproduction, metabolism and other activities [69]. With the continuous proliferation of cells, they gradually disappear, and the tissues or organs that guide cell regeneration form a geometry similar to that of scaffolds [70]. Because the scaffold materials are in direct contact with animal tissue, they should have the characteristics of biodegradability, non-toxicity, good biocompatibility and mechanical supporting ability.

Nanocellulose aerogels are biocompatible and have advantageous mechanical properties. Therefore, highly porous nanocellulose aerogels have been developed as scaffolds for tissue engineering. After the absorption of aqueous liquids, such aerogels provide an environment that is analogous to the extracellular matrix and promote the adhesion and proliferation of cells. L929 cells adhere to and proliferate on the NFC/chitosan composite aerogel, and a subconfluent layer of metabolically active cells is observed after 96 h (Fig. 4.17a–c) [32]. BC aerogels can be used as a scaffold for auricular cartilage tissue engineering (Fig. 4.17d–f) [71]. A bilayer BC scaffold with a porosity of 75% and a pore size of approximately 50 μm, organized as a dense BC layer and a macroporous BC/alginate layer, provides a platform for culturing human nasoseptal chondrocytes, and freshly isolated human nasoseptal chondrocytes and mononuclear cells can form cartilage on it. TEMPO-oxidized NFC and carboxymethylated NFC aerogels can be used to evaluate inflammatory responses and tissue reactions (Fig. 4.17g) [72]. NFC aerogel scaffolds are non-toxic, both in vitro or in vivo, and can induce repair and wound healing. However, the degradation rate of nanocellulose is very low, which results in a foreign body reaction.

4.5 Conclusion

In this chapter, we focused on the development of nanocellulose aerogels and their applications. NFC, CNC and BC aerogels are mainly produced by freeze-drying, supercritical drying, and atmospheric pressure drying methods. Such aerogels are lightweight and highly porous. The characteristics of aerogels can be controlled by carefully manipulating certain parameters. These include, but are not limited to, the type of nanocellulose building blocks, the concentration of the nanocellulose suspension, the growth direction of the ice template, the processing methods used prior to the drying process, the integration with active components, carbonization and the conversion to carbon aerogels, and the post-treatment of the aerogels. Therefore, multiple types of nanocellulose aerogels with different structures, properties and functions have been developed and can be used for various applications such as energy storage and harvest, sensor, water purification, thermal management and scaffold.

4.5.1 Technical and Economic Problems

Although numerous nanocellulose aerogels have been developed, and they have great potential for use in many fields, some technical and economic challenges remain. These challenges include: (1) The fabrication of aerogels by freeze-drying or supercritical drying method is usually time-consuming. (2) The mass production of large aerogels remains a challenge. (3) It remains difficult to prepare ultralight and robust

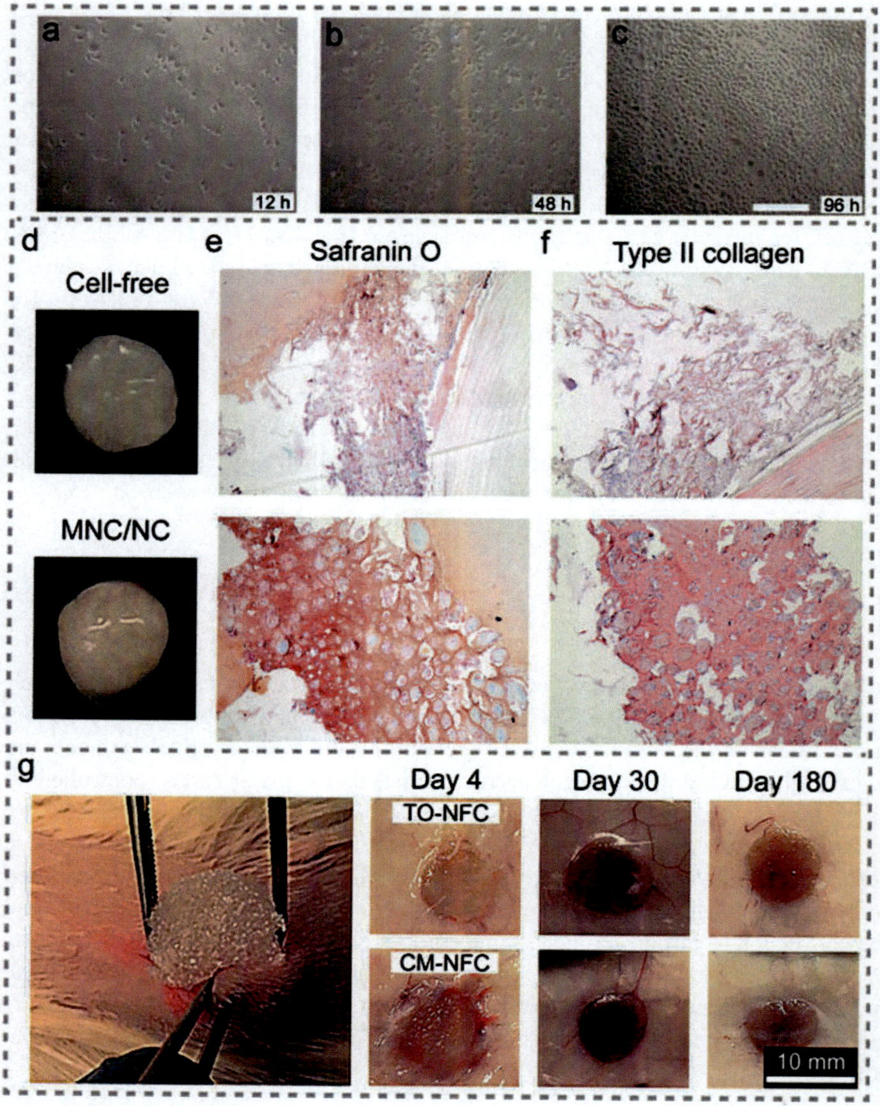

Fig. 4.17 Nanocellulose aerogels for scaffold. **a–c** Images showing the morphology of L929 cells grown on an NFC/chitosan composite aerogel for various periods of time [32]. **a–c** are reproduced with permission from [32]. Copyright 2017 Wiley-VCH. **d–f** A bilayer BC aerogel scaffold supports the formation of neocartilage [71]. **d–f** are reproduced with permission from [71]. Copyright 2014 Elsevier. **g** Fate of NFC aerogel scaffolds over various periods of time [72]. **g** is reproduced with permission from [72]. Copyright 2018 Elsevier

aerogels with densities of less than 1.0×10^{-3} g cm^{-3} because when the concentration of the nanocellulose suspension is too low, the aerogels shrink during the fabrication process. (4) Some methods used to improve the performance of aerogels are very complicated; they sometimes introduce expensive or unwholesome components, and adversely affect other properties of the aerogels. (5) In some applications, nanocellulose aerogels still have obvious defects. For example, when used as tissue engineering scaffolds, the degradation rate of nanocellulose is too low. (6) The industrialization of nanocellulose aerogels remains immature. (7) Compared with widely used commercial foam or gel, nanocellulose aerogels are still expensive.

4.5.2 Future Trends

In the future, more novel nanocellulose aerogels will be developed with an increasing focus on sustainability. The fabrication of nanocellulose and aerogels will become more mature on the laboratory scale, and the structures, properties and functions of the aerogels will be further optimized. Some aerogels will have new properties, and there will be more new applications. Therefore, nanocellulose aerogels will have an ideal market prospect. In the future, more attention should be paid to the following: (1) The cost of nanocellulose and aerogels should be reduced. (2) Industrial-grade aerogel mass production technology should be optimized. (3) The green production and environment friendly aerogels should be maintained. (4) Rapid and low-energy consumption technologies for the production of nanocellulose aerogels and carbon aerogels need to be developed. (5) The structures and properties of various NFC, CNC and BC aerogels need to be compared in depth. (6) More attention should be paid to the sustainability of nanocellulose aerogels. (7) More attention should be paid to the synergistic mechanisms associated with various building blocks used to fabricate aerogels. (8) The application of nanocellulose aerogels should be expanded. (9) Aerogels and the devices and systems they are used to create should be designed more intelligently.

References

1. A. Soleimani Dorcheh, M.H. Abbasi, Silica aerogel; synthesis, properties and characterization. J. Mater. Process. Technol. **199**(1), 10–26 (2008)
2. J. Zou, J. Liu, A.S. Karakoti, A. Kumar, D. Joung, Q. Li, S.I. Khondaker, S. Seal, L. Zhai, Ultralight multiwalled carbon nanotube aerogel. ACS Nano **4**(12), 7293–7302 (2010)
3. Y. Habibi, L.A. Lucia, O.J. Rojas, Cellulose nanocrystals: chemistry, self-assembly, and applications. Chem. Rev. **110**(6), 3479–3500 (2010)
4. R.J. Moon, A. Martini, J. Nairn, J. Simonsen, J. Youngblood, Cellulose nanomaterials review: structure, properties and nanocomposites. Chem. Soc. Rev. **40**(7), 3941–3994 (2011)
5. I. Siró, D. Plackett, Microfibrillated cellulose and new nanocomposite materials: a review. Cellulose **17**(3), 459–494 (2010)

6. Y. Chen, L. Zhang, Y. Yang, B. Pang, W. Xu, G. Duan, S. Jiang, K. Zhang, Recent progress on nanocellulose aerogels: preparation, modification, composite fabrication, applications. Adv. Mater. **33**(11), 2005569 (2021)
7. W. Chen, H. Yu, Q. Li, Y. Liu, J. Li, Ultralight and highly flexible aerogels with long cellulose I nanofibers. Soft Matter **7**(21), 10360–10368 (2011)
8. Y. Kobayashi, T. Saito, A. Isogai, Aerogels with 3D ordered nanofiber skeletons of liquid-crystalline nanocellulose derivatives as tough and transparent insulators. Angew. Chem. Int. Ed. **126**(39), 10562–10565 (2014)
9. Y. Li, N. Grishkewich, L. Liu, C. Wang, K.C. Tam, S. Liu, Z. Mao, X. Sui, Construction of functional cellulose aerogels via atmospheric drying chemically cross-linked and solvent exchanged cellulose nanofibrils. Chem. Eng. J. **366**(15), 531–538 (2019)
10. B. Wang, W. Zhang, W. Zhang, A.S. Mujumdar, L. Huang, Progress in drying technology for nanomaterials. Dry. Technol. **23**(1–2), 7–32 (2005)
11. P.R. Aravind, P. Shajesh, G.D. Soraru, K.G.K. Warrier, Ambient pressure drying: a successful approach for the preparation of silica and silica based mixed oxide aerogels. J. Sol-Gel Sci. Technol. **54**(1), 105–117 (2010)
12. M.S. Toivonen, A. Kaskela, O.J. Rojas, E.I. Kauppinen, O. Ikkala, Ambient-dried cellulose nanofibril aerogel membranes with high tensile strength and their use for aerosol collection and templates for transparent, flexible devices. Adv. Funct. Mater. **25**(42), 6618–6626 (2015)
13. J. Rostami, K. Gordeyeva, T. Benselfelt, E. Lahchaichi, S.A. Hall, A.V. Riazanova, P.A. Larsson, G. Cinar Ciftci, L. Wågberg, Hierarchical build-up of bio-based nanofibrous materials with tunable metal–organic framework biofunctionality. Mater. Today **48**, 47–58 (2021)
14. M.V.G. Zimmermann, M.P. da Silva, A.J. Zattera, R.M. Campomanes Santana, Effect of nanocellulose fibers and acetylated nanocellulose fibers on properties of poly(ethylene-co-vinyl acetate) foams. J. Appl. Polym. Sci. **134**(17), 44760 (2017)
15. L. Wang, M. Ando, M. Kubota, S. Ishihara, Y. Hikima, M. Ohshima, T. Sekiguchi, A. Sato, H. Yano, Effects of hydrophobic-modified cellulose nanofibers (CNFs) on cell morphology and mechanical properties of high void fraction polypropylene nanocomposite foams. Compos. Part A Appl. Sci. Manuf. **98**, 166–173 (2017)
16. Y. Luo, S. Xiao, S. Li, Effect of initial water content on foaming quality and mechanical properties of plant fiber porous cushioning materials. Bioresources **12**(2), 4259–4269 (2017)
17. H.-C. Zhang, C.-N. Yu, Y. Liang, G.-X. Lin, C. Meng, Foaming behavior and microcellular morphologies of incompatible SAN/CPE blends with supercritical carbon dioxide as a physical blowing agent. Polymers **11**(1), 89 (2019)
18. W. Chen, Q. Li, Y. Wang, X. Yi, J. Zeng, H. Yu, Y. Liu, J. Li, Comparative study of aerogels obtained from differently prepared nanocellulose fibers. ChemSusChem **7**(1), 154–161 (2014)
19. L. Heath, W. Thielemans, Cellulose nanowhisker aerogels. Green Chem. **12**(8), 1448–1453 (2010)
20. H.-W. Liang, Q.-F. Guan, Z. Zhu, L.-T. Song, H.-B. Yao, X. Lei, S.-H. Yu, Highly conductive and stretchable conductors fabricated from bacterial cellulose. NPG Asia Mater. **4**(6), e19 (2012)
21. J.T. Korhonen, P. Hiekkataipale, J. Malm, M. Karppinen, O. Ikkala, R.H.A. Ras, Inorganic hollow nanotube aerogels by atomic layer deposition onto native nanocellulose templates. ACS Nano **5**(3), 1967–1974 (2011)
22. Z. Zeng, C. Wang, G. Siqueira, D. Han, A. Huch, S. Abdolhosseinzadeh, J. Heier, F. Nüesch, C. Zhang, G. Nyström, Nanocellulose-MXene biomimetic aerogels with orientation-tunable electromagnetic interference shielding performance. Adv. Sci. **7**(15), 2000979 (2020)
23. Z.-Z. Pan, H. Nishihara, S. Iwamura, T. Sekiguchi, A. Sato, A. Isogai, F. Kang, T. Kyotani, Q.-H. Yang, Cellulose nanofiber as a distinct structure-directing agent for xylem-like micro-honeycomb monoliths by unidirectional freeze-drying. ACS Nano **10**(12), 10689–10697 (2016)
24. Z. Zeng, T. Wu, D. Han, Q. Ren, G. Siqueira, G. Nyström, Ultralight, flexible, and biomimetic nanocellulose/silver nanowire aerogels for electromagnetic interference shielding. ACS Nano **14**(3), 2927–2938 (2020)

25. G. Shao, D.A.H. Hanaor, X. Shen, A. Gurlo, Freeze casting: from low-dimensional building blocks to aligned porous structures—a review of novel materials, methods, and applications. Adv. Mater. **32**(17), 1907176 (2020)
26. C. Tang, P. Brodie, Y. Li, N.J. Grishkewich, M. Brunsting, K.C. Tam, Shape recoverable and mechanically robust cellulose aerogel beads for efficient removal of copper ions. Chem. Eng. J. **392**(15), 124821 (2020)
27. J. Zhou, Y.-L. Hsieh, Nanocellulose aerogel-based porous coaxial fibers for thermal insulation. Nano Energy **68**, 104305 (2020)
28. L. Bai, J. Ma, H. Song, Y. Yang, C. Zhi, S.-Y. Lee, H. Yu, S. Liu, J. Li, M. Yu, W. Chen, Flexible, electrically conductive, nanostructured, asymmetric aerogel films for lithium-sulfur batteries. ACS Appl. Mater. Interfaces **13**(49), 59174–59184 (2021)
29. J. Fan, S. Ifuku, M. Wang, K. Uetani, H. Liang, H. Yu, Y. Song, X. Li, J. Qi, Y. Zheng, H. Wang, J. Shen, X. Zhang, Q. Li, S. Liu, Y. Liu, Q. Wang, J. Li, P. Lu, Z. Fan, W. Chen, Robust nanofibrillated cellulose hydro/aerogels from benign solution/solvent exchange treatment. ACS Sustain. Chem. Eng. **6**(5), 6624–6634 (2018)
30. Y. Chen, Z. Yu, Y. Ye, Y. Zhang, G. Li, F. Jiang, Superelastic, hygroscopic, and ionic conducting cellulose nanofibril monoliths by 3D printing. ACS Nano **15**(1), 1869–1879 (2021)
31. A.J. Svagan, M.A.S.A. Samir, L.A. Berglund, Biomimetic foams of high mechanical performance based on nanostructured cell walls reinforced by native cellulose nanofibrils. Adv. Mater. **20**(7), 1263–1269 (2008)
32. Y. Wang, K. Uetani, S. Liu, X. Zhang, Y. Wang, P. Lu, T. Wei, Z. Fan, J. Shen, H. Yu, S. Li, Q. Zhang, Q. Li, J. Fan, N. Yang, Q. Wang, Y. Liu, J. Cao, J. Li, W. Chen, Multifunctional bionanocomposite foams with a chitosan matrix reinforced by nanofibrillated cellulose. ChemNanoMat **3**(2), 98–108 (2017)
33. M. Wang, I.V. Anoshkin, A.G. Nasibulin, J.T. Korhonen, J. Seitsonen, J. Pere, E.I. Kauppinen, R.H.A. Ras, O. Ikkala, Modifying native nanocellulose aerogels with carbon nanotubes for mechanoresponsive conductivity and pressure sensing. Adv. Mater. **25**(17), 2428–2432 (2013)
34. R.T. Olsson, M.A.S. Azizi Samir, G. Salazar-Alvarez, L. Belova, V. Ström, L.A. Berglund, O. Ikkala, J. Nogués, U.W. Gedde, Making flexible magnetic aerogels and stiff magnetic nanopaper using cellulose nanofibrils as templates. Nat. Nanotechnol. **5**(8), 584–588 (2010)
35. Y. Wang, M. Dai, H. Wu, L. Xu, T. Zhang, W. Chen, Z.L. Wang, Y. Yang, Moisture induced electricity for self-powered microrobots. Nano Energy **90**, 106499 (2021)
36. Z.-Y. Wu, H.-W. Liang, C. Li, B.-C. Hu, X.-X. Xu, Q. Wang, J.-F. Chen, S.-H. Yu, Dyeing bacterial cellulose pellicles for energetic heteroatom doped carbon nanofiber aerogels. Nano Res. **7**(12), 1861–1872 (2014)
37. L. Tian, J. Luan, K.-K. Liu, Q. Jiang, S. Tadepalli, M.K. Gupta, R.R. Naik, S. Singamaneni, Plasmonic biofoam: a versatile optically active material. Nano Lett. **16**(1), 609–616 (2016)
38. W. Chen, Q. Zhang, K. Uetani, Q. Li, P. Lu, J. Cao, Q. Wang, Y. Liu, J. Li, Z. Quan, Y. Zhang, S. Wang, Z. Meng, H. Yu, Sustainable carbon aerogels derived from nanofibrillated cellulose as high-performance absorption materials. Adv. Mater. Interfaces **3**(10), 1600004 (2016)
39. Z.-Y. Wu, C. Li, H.-W. Liang, J.-F. Chen, S.-H. Yu, Ultralight, flexible, and fire-resistant carbon nanofiber aerogels from bacterial cellulose. Angew. Chem. Int. Ed. **52**(10), 2925–2929 (2013)
40. M. Hamedi, E. Karabulut, A. Marais, A. Herland, G. Nyström, L. Wågberg, Nanocellulose aerogels functionalized by rapid layer-by-layer assembly for high charge storage and beyond. Angew. Chem. Int. Ed. **52**(46), 12038–12042 (2013)
41. G. Nyström, A. Marais, E. Karabulut, L. Wågberg, Y. Cui, M.M. Hamedi, Self-assembled three-dimensional and compressible interdigitated thin-film supercapacitors and batteries. Nat. Commun. **6**(1), 7259 (2015)
42. L.-F. Chen, Z.-H. Huang, H.-W. Liang, Q.-F. Guan, S.-H. Yu, Bacterial-cellulose-derived carbon nanofiber@MnO_2 and nitrogen-doped carbon nanofiber electrode materials: an asymmetric supercapacitor with high energy and power density. Adv. Mater. **25**(34), 4746–4752 (2013)
43. J.-H. Kim, D. Lee, Y.-H. Lee, W. Chen, S.-Y. Lee, Nanocellulose for energy storage systems: beyond the limits of synthetic materials. Adv. Mater. **31**(20), 1804826 (2019)

44. W. Chen, H. Yu, S.-Y. Lee, T. Wei, J. Li, Z. Fan, Nanocellulose: a promising nanomaterial for advanced electrochemical energy storage. Chem. Soc. Rev. **47**(8), 2837–2872 (2018)
45. X. Yang, K. Shi, I. Zhitomirsky, E.D. Cranston, Cellulose nanocrystal aerogels as universal 3D lightweight substrates for supercapacitor materials. Adv. Mater. **27**(40), 6104–6109 (2015)
46. L.-F. Chen, Z.-H. Huang, H.-W. Liang, W.-T. Yao, Z.-Y. Yu, S.-H. Yu, Flexible all-solid-state high-power supercapacitor fabricated with nitrogen-doped carbon nanofiber electrode material derived from bacterial cellulose. Energy Environ. Sci. **6**(11), 3331–3338 (2013)
47. L.-F. Chen, Z.-H. Huang, H.-W. Liang, H.-L. Gao, S.-H. Yu, Three-dimensional heteroatom-doped carbon nanofiber networks derived from bacterial cellulose for supercapacitors. Adv. Funct. Mater. **24**(32), 5104–5111 (2014)
48. B. Wang, X. Li, B. Luo, J. Yang, X. Wang, Q. Song, S. Chen, L. Zhi, Pyrolyzed bacterial cellulose: a versatile support for lithium ion battery anode materials. Small **9**(14), 2399–2404 (2013)
49. M. Wang, Z. Yang, W. Li, L. Gu, Y. Yu, Superior sodium storage in 3D interconnected nitrogen and oxygen dual-doped carbon network. Small **12**(19), 2559–2566 (2016)
50. M. Wang, Y. Yang, Z. Yang, L. Gu, Q. Chen, Y. Yu, Sodium-ion batteries: improving the rate capability of 3D interconnected carbon nanofibers thin film by boron, nitrogen dual-doping. Adv. Sci. **4**(4), 1600468 (2017)
51. Q. Xia, H. Yang, M. Wang, M. Yang, Q. Guo, L. Wan, H. Xia, Y. Yu, High energy and high power lithium-ion capacitors based on boron and nitrogen dual-doped 3D carbon nanofibers as both cathode and anode. Adv. Energy Mater. **7**(22), 1701336 (2017)
52. H.-Y. Mi, X. Jing, Z. Cai, Y. Liu, L.-S. Turng, S. Gong, Highly porous composite aerogel based triboelectric nanogenerators for high performance energy generation and versatile self-powered sensing. Nanoscale **10**(48), 23131–23140 (2018)
53. H.-Y. Mi, X. Jing, Q. Zheng, L. Fang, H.-X. Huang, L.-S. Turng, S. Gong, High-performance flexible triboelectric nanogenerator based on porous aerogels and electrospun nanofibers for energy harvesting and sensitive self-powered sensing. Nano Energy **48**, 327–336 (2018)
54. Q. Zheng, L. Fang, H. Guo, K. Yang, Z. Cai, M.A.B. Meador, S. Gong, Highly porous polymer aerogel film-based triboelectric nanogenerators. Adv. Funct. Mater. **28**(13), 1706365 (2018)
55. C. Qian, L. Li, M. Gao, H. Yang, Z. Cai, B. Chen, Z. Xiang, Z. Zhang, Y. Song, All-printed 3D hierarchically structured cellulose aerogel based triboelectric nanogenerator for multi-functional sensors. Nano Energy **63**, 103885 (2019)
56. Y. Ren, X. Sun, J. Liu, Advances in liquid metal-enabled flexible and wearable sensors. Micromachines **11**(2), 200 (2020)
57. C. Wang, L. Yin, L. Zhang, D. Xiang, R. Gao, Metal oxide gas sensors: sensitivity and influencing factors. Sensors **10**(3), 2088–2106 (2010)
58. L. Dai, Y. Wang, X. Zou, Z. Chen, H. Liu, Y. Ni, Ultrasensitive physical, bio, and chemical sensors derived from 1-, 2-, and 3-D nanocellulosic materials. Small **16**(13), 1906567 (2020)
59. J. Zhang, Y. Cheng, C. Xu, M. Gao, M. Zhu, L. Jiang, Hierarchical interface engineering for advanced nanocellulosic hybrid aerogels with high compressibility and multifunctionality. Adv. Funct. Mater. **31**(19), 2009349 (2021)
60. Y. Fang, C. Jing, G. Li, S. Ling, Z. Wang, P. Lu, Q. Li, C. Dai, S. Gao, B. Chen, L. Bai, H. Zhang, T. Liu, Q. Wang, J. Li, H. Yu, S. Liu, W. Chen, Wood-derived systems for sustainable oil/water separation. Adv. Sustain. Syst. **5**(7), 2100039 (2021)
61. Y. Li, L. Zhu, N. Grishkewich, K.C. Tam, J. Yuan, Z. Mao, X. Sui, CO_2-responsive cellulose nanofibers aerogels for switchable oil-water separation. ACS Appl. Mater. Interfaces **11**(9), 9367–9373 (2019)
62. X. Zhang, I. Elsayed, C. Navarathna, G.T. Schueneman, E.I.B. Hassan, Biohybrid hydrogel and aerogel from self-assembled nanocellulose and nanochitin as a high-efficiency adsorbent for water purification. ACS Appl. Mater. Interfaces **11**(50), 46714–46725 (2019)
63. L. Zhu, L. Zong, X. Wu, M. Li, H. Wang, J. You, C. Li, Shapeable fibrous aerogels of metal–organic-frameworks templated with nanocellulose for rapid and large-capacity adsorption. ACS Nano **12**(5), 4462–4468 (2018)

64. H. Zhu, X. Yang, E.D. Cranston, S. Zhu, Flexible and porous nanocellulose aerogels with high loadings of metal–organic-framework particles for separations applications. Adv. Mater. **28**(35), 7652–7657 (2016)
65. Z. Zhang, G. Sèbe, D. Rentsch, T. Zimmermann, P. Tingaut, Ultralightweight and flexible silylated nanocellulose sponges for the selective removal of oil from water. Chem. Mater. **26**(8), 2659–2668 (2014)
66. V. Apostolopoulou-Kalkavoura, P. Munier, L. Bergström, Thermally insulating nanocellulose-based materials. Adv. Mater. **33**(28), 2001839 (2021)
67. L.-Y. Long, Y.-X. Weng, Y.-Z. Wang, Cellulose aerogels: synthesis, applications, and prospects. Polymers **10**(6), 623 (2018)
68. B. Wicklein, A. Kocjan, G. Salazar-Alvarez, F. Carosio, G. Camino, M. Antonietti, L. Bergström, Thermally insulating and fire-retardant lightweight anisotropic foams based on nanocellulose and graphene oxide. Nat. Nanotechnol. **10**(3), 277–283 (2015)
69. W. Zhang, C. Gao, Morphology transformation of self-assembled organic nanomaterials in aqueous solution induced by stimuli-triggered chemical structure changes. J. Mater. Chem. A **5**(31), 16059–16104 (2017)
70. Y. Chen, S. Zhou, Q. Li, Microstructure design of biodegradable scaffold and its effect on tissue regeneration. Biomaterials **32**(22), 5003–5014 (2011)
71. H. Martínez Ávila, E.-M. Feldmann, M.M. Pleumeekers, L. Nimeskern, W. Kuo, W.C. de Jong, S. Schwarz, R. Müller, J. Hendriks, N. Rotter, G.J.V.M. van Osch, K.S. Stok, P. Gatenholm, Novel bilayer bacterial nanocellulose scaffold supports neocartilage formation in vitro and in vivo. Biomaterials **44**, 122–133 (2015)
72. A. Rashad, S. Suliman, M. Mustafa, T.Ø. Pedersen, E. Campodoni, M. Sandri, K. Syverud, K. Mustafa, Inflammatory responses and tissue reactions to wood-based nanocellulose scaffolds. Mater. Sci. Eng. C **97**, 208–221 (2019)

Chapter 5
Nanocellulose-Based Materials with Superior Mechanical Performance

Upamanyu Ray, Shuze Zhu, Zhenqian Pang, and Teng Li

Abstract The intrinsic mechanical properties of nanocellulose, with a specific stiffness and specific strength higher than most metals and alloys, are superior engineering materials. This makes the renewable yet abundantly available nanocellulose a promising material in multi-fold stimulating applications involving high strength and toughness. This chapter sheds light on the principle mechanistic design features leading to the superb mechanical properties in nanocellulose-based materials. Some of them, such as nanocellulose-based films, macrofibers, bulk wood and lightweight nanocomposites, are discussed. Different modeling approaches focusing on the underlying mechanism behind these fascinating mechanical properties are also reviewed. Finally, some broader future perspectives that can enrich this discipline are analyzed, which hopefully will stimulate the readers to explore the fertile opportunities of nanocellulose-based sustainable materials engineering in the foreseeable future.

5.1 Introduction

Natural cellulose-based materials, such as wood and cotton have been used by mankind ever since the advent of civilization. For example, papermaking in China and use of cotton in India started more than 2000 years ago. In recent decades, with the advent of nanotechnology and the invention of material characterization instruments such as scanning electron microscopy (SEM), transmission electron microscopy (TEM) and atomic force microscopy (AFM), and traditional processing techniques of cellulose started evolving. New techniques to modify cellulose were gradually devised, among which extracting cellulose at the nanoscale was the most

U. Ray · Z. Pang · T. Li (✉)
Department of Mechanical Engineering, University of Maryland, College Park, MD 20742, USA
e-mail: lit@umd.edu

S. Zhu
Department of Engineering Mechanics, Zhejiang University, Hangzhou 310027, China

L. Hu et al. (eds.), *Emerging Nanotechnologies in Nanocellulose*, NanoScience and Technology, https://doi.org/10.1007/978-3-031-14043-3_5

fundamental one. Nanocellulose became a new building block with exceptional properties and enabled a wide range of potential applications. Also, to fulfill the ever-increasing needs of the growing population of planet Earth in a sustainable manner, factors such as low weight, biodegradability, recyclability, low carbon footprint and abundant availability became extremely crucial. Cellulose happened to offer each one of them. Since the first cellulose nanofibril was identified in 1977 [1], the impact of nanocellulose on our daily life has been steadily growing. For example, market estimates influenced by nanocellulose in the field of medical and life sciences alone were around $97 billion globally in 2017 [2]. The market potential for nanocellulose in USA is around 6.4 million metric tons and globally it is 35 million metric tons, a number bound to increase with more novel applications [3, 4]. In the academic front, the number of scientific literature involving nanocellulose by the end of 2018 has increased by over 19,000% from 2000 [5]. Research efforts are in progress to utilize nanocellulose in many promising applications such as structural materials, lightweight composites, barrier films, aircraft and wind turbine blades, antimicrobial films, flexible displays, next-generation batteries, reinforcing fillers for polymers, biomedical implants, electronic components, biodegradable straws and separation membranes.

Since mechanical performance is crucial for the practical use of materials in many applications, understanding and designing high-performance cellulose-based materials have received tremendous attention. Two of the most predominant mechanical properties considered in engineering material design are strength and toughness. While the material strength is its ability to resist the plastic deformation, the toughness of a material is its ability to absorb energy and undergo plastic deformation before fracture. Unfortunately, these two properties are often mutually exclusive. Materials with higher strength are usually less tough. An optimal combination of these two properties is highly desirable but challenging to achieve. A simultaneous increase of strength and toughness had been achieved in some materials such as nanotwin boundaries [6] or bulk metallic glass-dendrite composites [7] but such efforts were rather material specific, complex and expensive. A general strategy in materials science which can defeat the perpetual conflict of strength and toughness is in great need. Intriguingly, cellulose already possesses some superb intrinsic mechanical properties. For example, the specific modulus and specific strength of cellulose are indeed higher than most metals, composites and ceramics as can be seen in any standard Ashby plot [8]. If enhanced toughness can be incorporated with the existing appealing mechanical properties of cellulose, it has the potential to make cellulose an appropriate material in the quest for advanced high-performance renewable materials.

Cellulose is available from multiple sources in nature [8, 9]. Among them, the main natural source is the plant cell wall [10]. Figure 5.1a shows the intrinsically hierarchical structure of wood fibers. A 20–50 μm thick wood fiber comprises of nanofibrillated cellulose (5–50 nm in diameter), which can further be disintegrated into finer elementary fibrils (1.5–3.5 nm in diameter). These fibrils are made of cellulose molecular chains, the fundamental building blocks. Cellulose molecule is a linear chain of ringed glucose molecules, with a repeat unit (Fig. 5.1b) comprising

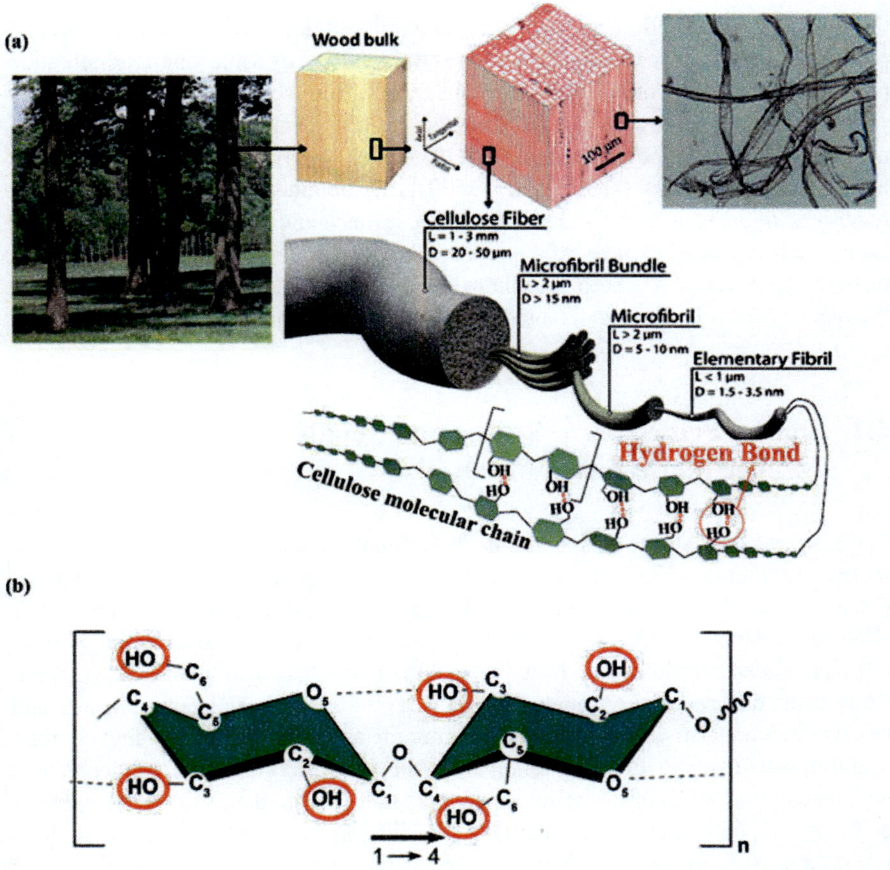

Fig. 5.1 **a** Wood hierarchical structure shown from a top-down perspective, starting from a tree trunk and ending at the fundamental building blocks of cellulose molecular chains. Image in the top panel reproduced with permission [17]. Copyright 2013, American Chemical Society. Image titled "Cellulose molecular chain" at the bottom panel reproduced with permission [18]. Copyright 2015, National Academy of Sciences. **b** A schematic of one repeat unit in a molecular cellulose chain. Six hydroxyl groups, responsible for hydrogen bonding within the same chain and/or neighboring chains, of each unit are shown in red circles

of two anhydroglucose rings ($C_6H_{10}O_5$) linked through a C–O–C covalent bond. Rich hydroxyl groups in the cellulose molecule (six in each repeat unit) enable formation of hydrogen bonds, both inside (intra-chain) and outside (inter-chain) the chain. The intra-chain hydrogen bonding stabilizes the linkage and results in the linear configuration of the cellulose chain. On the other hand, inter-chain hydrogen bonding among neighboring cellulose molecules plays a pivotal role in the deformation and failure behaviors of cellulose-based materials [11]. The facile hydrogen bonding between cellulose molecules plays a pivotal role in the superb mechanical properties of nanocellulose-based materials, as to be detailed in later sections. Apart from the plants, another cellulosic source is bacterial cellulose (BC) [12–14], which is produced by certain types of bacteria such as *Acetobacter* by a microbial fermentation process. The length of nanocellulose chains from various sources and its morphology vary slightly depending on the variable sources. For example, the length of a BC nanofiber (degree of polymerization (DOP) ~ 8000 [15]) is often longer than that of cellulose nanofibers (DOP ~ 500–2500 [16]) obtained from plants. However, regardless of the source, the underlying cellulose mechanics is expected to remain the same and it is crucial to have a complete mechanistic understanding on the interplay of the nanocellulose fibers to be able to fully extract the maximum potential from the mechanical properties in a nanocellulose-based material.

5.2 Design Principles of Strong and Tough Nanocellulose-Based Materials

The history of designing nanocellulose-based materials as a novel concept gained increasing attention from the final stages of the twentieth century. For example, Revol et al. [19] first fabricated a nanopaper comprised of cellulose nanocrystals (CNCs) as early as 1992 by concentrating an aqueous CNC suspension. Similar efforts to obtain cellulose-based nanopapers using cellulose nanofiber (NFC) suspensions from different source materials [20, 21] such as sugar beet, tunicates and wood pulp were also demonstrated just before the advent of the twenty-first century. Simultaneous investigations spanning from the year 2000 onwards, addressed the mechanical properties of the cellulose nanocrystals, the fundamental building block of these nanocellulose-based materials. Various experimental techniques such as Raman spectroscopy, inelastic X ray scattering and Atomic Force microscopy, as well as different molecular modeling techniques reported values of mechanical properties of CNCs such as, but not limited to, the longitudinal elastic modulus to be ranging from 124 to 220 GPa [22–26]. Such impressive mechanical properties indicated that the fundamental nanocellulose building blocks were likely stronger than steel and comparable to strong synthetic fibers such as Kevlar, making them ideal candidates for reinforcements to engineer strong and tough materials. With gradual improvement in sophisticated nanoscale experimental design techniques, focus is gradually increased to utilize the superior mechanics of the constituent building blocks and fabricate strong and tough novel cellulosic materials (e.g. nanopaper, macrofibers, composites etc.), which will cause minimal harm to the environment. The primary mechanics design principles adopted to achieve this encompass modifications involving (a) structural, (b) interfacial and (c) topological features, which are discussed in this section.

(a) *Modification of structural features*

The modification of structural features by utilizing the unique morphology of cellulose may result in distinct deformation and failure processes in its molecular chains,

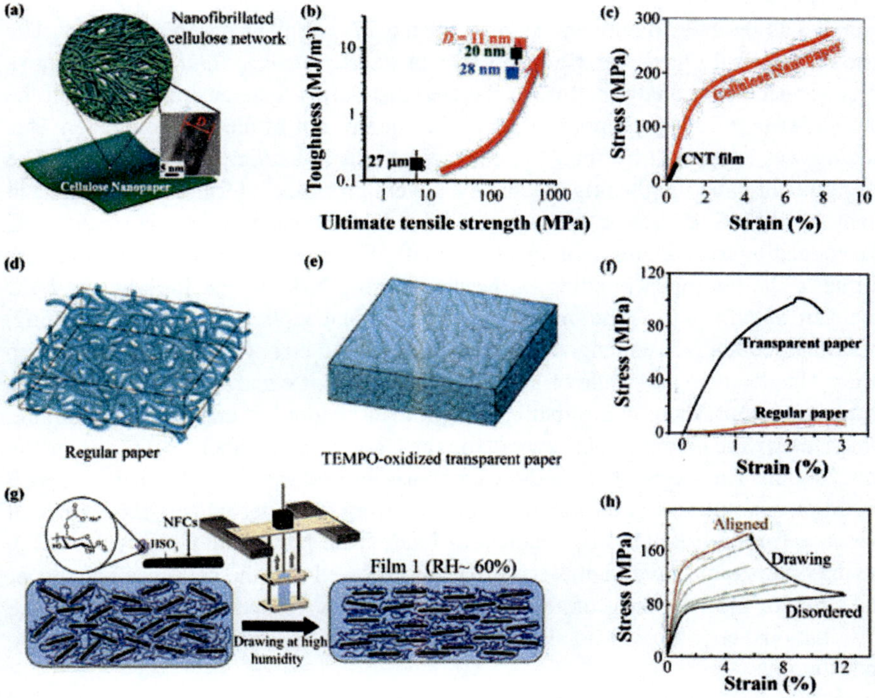

Fig. 5.2 Design principles involving modification of structural features. **a** A schematic and a TEM image (inset) of cellulose nanopaper with NFC fibers ~10 nm in diameter. **b** Simultaneous increase in toughness and tensile strength with decrease in mean NFC fiber diameter from 27 μm to 11 nm gives rise to the anomalous scaling law—the smaller, stronger and tougher. **c** Experimental stress–strain curves showing that high strength materials such as CNTs lack hydrogen bonds resulting in strength much lower than cellulose nanopaper. **a–c** Reproduced with permission [18]. Copyright 2015, National Academy of Sciences. **d, e** Schematic of **d** regular and **e** transparent paper, where the TEMPO treatment unzips the original tubular fibers resulting in higher density of packing. **f** Stress–strain curves showing higher strength in transparent paper. **d–f** Reproduced with permission [27]. Copyright 2013, American Chemical Society. **g** Design technique involving wet-drawing to introduce alignment among NFCs in CNC-carboxymethyl cellulose (CMC) film. **h** Engineering stress–strain curves showing superior mechanical properties due to alignment in the composite film. **g–h** Reproduced with permission [28]. Copyright 2015, American Chemical Society

leading to the extraordinary mechanical performance of nanocellulose-based materials. Different methods such as reducing the fiber size (diameter) of the NFCs, collapsing the wood cell wall or introducing anisotropy can be applied under this design principle. One such study by Zhu et al. [18] checked how the cellulose fiber size (diameter) affects the overall mechanical performance. A cellulose nanopaper (30–50 μm thickness) (Fig. 5.2a) was fabricated from bleached wood pulp through a filtration process. The mean diameters of the cellulose nanofibers (NFC) in the nanopaper were scaled down from 27 μm to 11 nm by chemical and mechanical treatment. It was seen that the desirable but hard-to-achieve property of simultaneous increase of strength (40 times) and toughness (130 times) (Fig. 5.2b) was obtained as the NFC diameters decreased from 27 μm to 11 nm, respectively. The stimulating results gave rise to a scaling law of *smaller, stronger and tougher*. From a fracture mechanics analysis, it was predicted that the ultimate tensile strength of the nanopaper is inversely proportionally to the square root of the NFC diameter. This finding was supported by robust experimental tensile-loading measurements. The strengthening and toughening mechanisms were proposed as follows. As the mean diameter of NFC decreases, there is a reduction in the intrinsic defect size in the NFC fibers, leading to an increase of tensile strength. However, the increase in toughness is due to the hydrogen bonding in the neighboring NFC fibers. Initially the NFC fibers are entangled in a random fashion. With gradual application of tensile loading, the entangled fibers disentangle, become aligned and then begin to slide against each other. This leads to a cascade of hydrogen bond breaking and reforming among the hydroxyl groups, thereby dissipating a significant amount of energy before failure, which leads to an increase in the overall toughness of the nanopaper. To understand the contributions from hydrogen bonding, comparisons were made with a material, such as carbon nanotubes (CNT), which lacked such a bonding mechanism. Experimental evidence from tensile tests of nanopapers made from NFCs and CNTs (Fig. 5.2c), having comparable fiber diameters of ~11 nm showed that the cellulose nanopaper had strength and toughness of around 9 and 281 times, respectively, higher than the CNT nanopaper, which also validated the hypothesis about the hydrogen bonding contribution.

Another method to modify the structural features was demonstrated by collapsing the wood cell walls and thereby amplifying the mechanical performance in a transparent paper, as reported by Fang et al. [27]. Regular paper (Fig. 5.2d) is synthesized without the microfluidizer treatment and thus is comprised of microsized fibers (~27 μm) with many porous cavities. However, for the wood fibers in the transparent paper (Fig. 5.2e), TEMPO treatment and filtration swells up the fiber diameter from ~5–30 nm to an average fiber diameter of 26 μm, whereas their degree of polymerization decreases. Unzipping of the wood fibers takes place in the axial direction, resulting in a high density packing of ribbon-like unzipped wood fibers. This paper possesses a combination of the usually conflicting properties such as an ultrahigh optical transparency of 96% and an ultrahigh haze of 60% at room temperature. This is in stark contrast to common transparent papers which have a high transparency of 90% but a low haze of around 20%. Mechanical properties of this transparent paper are also superior. Experimental tensile testing (Fig. 5.2f) suggests that it is much stronger (tensile strength ~105 MPa) and tougher (toughness ~1.8 J/m^3) than the regular paper (strength ~8 MPa; toughness ~0.15 J/m^3). This is because the TEMPO treatment unzips the hollow cellulose fibers of the regular paper and significantly enhances the surface area of contact between the neighboring fibers. The numerous hydroxyl groups on the contact surface area of the unzipped fibers are in closer proximity, leading to the much-increased spatial density of interfacial hydrogen bonds. The improvement of mechanical properties in the TEMPO treated fiber is another example of the effectiveness of increasing the spatial density of interfacial hydrogen bonds in enhancing mechanical properties.

Other papers have been reported to possess much higher strength and toughness such as the one demonstrated by Sehaqui et al. [29], where a composite paper comprised of wood fibers and NFCs was designed. On increasing the weight content of NFCs from 0 to 10%, the strength increased from 98 to 160 MPa, and the toughness increased from 1.7 to 4.42 MJ/m^3. Incorporation of NFCs into the paper suppressed the internal voids and increased the interfibrillar hydrogen bond density, which resulted in this enhanced mechanical performance. Different processing techniques may result in cellulose nanopapers of even higher strength (~350 MPa [30]) and toughness (15.1 MJ/m^3 [31]). These studies on cellulose-based papers suggest many unexplored possibilities that can be included in multifunctional material design strategies to create desirable mechanistic features such as enhanced strength and toughness [32].

Manipulation of alignment among the cellulose nanofibers is another frequently used structural feature design strategy which has been used to develop nanocellulose-based materials with superior mechanical performance. A distinctive study to understand the effect on alignment on the mechanical properties was reported by Wang et al. [28]. Here, a wet-drawing technique was applied (Fig. 5.2g) on a CNC-carboxymethyl (CMC) (1:1 by weight) composite film to align the CNCs embedded in the matrix. As a result, the strength of the drawn composite film having aligned CNCs increased to ~187 MPa from ~93 MPa (Fig. 5.2h) in comparison to the non-aligned (disordered) film. The greater the alignment, the higher was the interfacial overlap between the neighboring fibers. This demanded a greater energy to overcome the considerably increased facile reformation of hydrogen bonds before eventual failure, which was the reason behind such superlative mechanical performance. The Young's modulus of the aligned film also increased from ~6.5 GPa (disordered) to ~15 GPa. The drawing procedure did not have any prominent effect on the toughness as the work of fracture for both the aligned (~10 MJ/m^3) and disordered (~11.8 MJ/m^3) was not largely compromised. This suggests a similar deformation mechanism in both the aligned and the disordered films, which makes the design principle of inducing alignment very appealing in novel nanocellulose materials. The principles revealed in this sub-section lays out the foundation to realize many intriguing strong and tough nanocellulose-based materials (e.g.: bulk wood, films), exhibiting high-mechanical-performance, as will be discussed in the next section.

(b) *Modification of interfacial features*

One of the main mechanics design principles in nanocellulose-based composites is to functionalize nanocellulose with other materials so that its mechanical properties is much superior than its constituents. This design principle leverages the unique structure of cellulose since the –OH groups of cellulose molecules can form hydrogen bonds with its neighboring molecular chains as well as with other functional groups of the composites. The interfacial space of the cellulose molecules is thus tuned to maximize the effective stress transfer between the cellulose and the matrix. The synergistic effects of cellulose and the other constituent materials in the composite elevate the overall mechanical performance of the material. Some

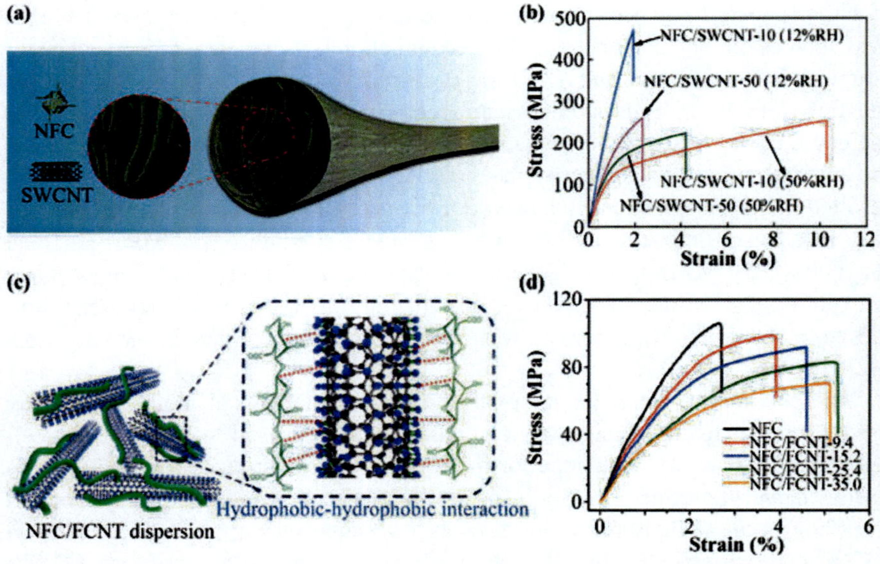

Fig. 5.3 Design principles involving modification of interfacial features. **a** Schematic illustrating the interfacial engineering of SWCNT and NFC by improved dispersion. **b** Stress–strain plots of NFC/SWCNT filaments at different humidity. **a, b** Reproduced with permission [33]. Copyright 2019, American Chemical Society. **c** Schematic showing the fabrication of fluorinated CNT-based NFC film. **d** Stress–strain plots of the composite NFC/FCNT film with varying mass fraction of FCNT. **c, d** Reproduced with permission [34]. Copyright 2018, American Chemical Society

approaches reporting modifications of interfacial features of nanocellulose-based composites are discussed below.

One approach to engineer the interface is to evenly spread out the reinforcing agent by refining the dispersion of the cellulose in the composite suspension and obtain nanocellulose-based filaments with better mechanical properties as demonstrated by Wan et al. [33]. Here, the TEMPO-treated NFC suspension was mixed with single-walled CNT (SWCNT) powder followed by ultrasonication and further dispersion by a three-roll mill. Sonication broke the agglomerates in the suspension and milling firmly wrapped the NFCs with the SWCNT through application of intensified stress. Then, the suspension was wet-spinned through a spinneret which aligned the building blocks (NFC and SWCNT) (Fig. 5.3a) in addition to densifying the resulting filament. The NFC/SWCNT-10 and NFC/SWCNT-50 filaments obtained in this procedure had 10% and 50% weight ratio of SWCNT, respectively, in the composite filament. At 12% RH, the strength (~472.17 MPa) and modulus (28.28 GPa) of NFC/SWCNT-10 was much higher than the corresponding strength (289.12 MPa) and modulus (19.55 GPa) of pure NFC filaments (Fig. 5.3b) clearly highlighting the importance of designing the interfacial features of the composite in increasing the mechanical performance. The lower strength (~263.12 MPa) and modulus (17.19 GPa) for NFC/SWCNT-50 probably occurred due to the formation of agglomerates in the network and increasing defects. At a higher humidity (RH

~ 50%), the strength of the filaments was lower than that for RH ~ 10% due more slippage caused by liquid water at the NFC/SWCNT interface. Interestingly, the stiffness increased (Fig. 5.3b, green line) with the increasing SWCNT content at higher humidity mostly because of the ability of water to form hydrogen bonds stabilizing the network. The filament obtained here was found to possess a higher strength and modulus than most NFC-based films, membranes, microfibers and nanopapers.

Application of the design principle of interfacial feature modification can also improve the toughness of the bulk material, as demonstrated by Wang and Wu [34]. Here, a fluorinated CNT (FCNT) and NFC-based composite film was reported (Fig. 5.3c) with strong hydrophobic-hydrophobic interactions between NFC and FCNT, as confirmed by Fourier transform infrared (FTIR) spectroscopy. The strong interfacial interactions ensured a uniform dispersion of the composite solution. On increasing the FCNT mass fraction from 9.4 to 15.2%, the toughness and elongation at break for the composite film increased relative to the pure NFC film (Fig. 5.3d). Increasing the FCNT content beyond 15.2% decreased the toughness, possibly due to agglomeration defects. The decrease of tensile strength with increased loading and the resulting increasing toughness (till 15.2 wt% of CNT) might be due to hydrophobic interactions because FTIR confirmed negligible hydrogen bonding between FCNT and NFCs. Given that NFCs are traditionally hydrophilic, this is captivating and demands future study on the underlying mechanism. The higher toughness makes these kinds of films suitable for a wide range of applications where usage will involve considerable bending and folding, as in sensors. All the studies mentioned above depict that the effectiveness of tuning the interfacial features in obtaining diverse but beneficial mechanical properties are paramount.

(c) *Modification of topological features*

The modification of topological features is another predominant design principle to fabricate strong and tough nanocellulose-based materials. This involves physical and chemical cross-linking. Physical cross-linking involves altering the structure of nanocellulose in a manner so that the hydrogen bonding between adjacent cellulose nanofibers is enhanced. More facile reformation of hydrogen bonds of the resulting nanofibers undergoing deformation leads to superior mechanical performance as already shown in Figs. 5.2 and 5.3. Chemical cross-linking involves introducing a receptor molecule, commonly called the cross-linker, which connects the functional groups (–OH) of cellulose molecular chains by covalent bonding, making the network much more stronger than physical cross-linking. Ye et al. [35] demonstrated a dual cross-linked film formed by both physical and chemical cross-linking (Fig. 5.4a). Chemical cross-linkers (epichlorohydrin) were added in a regenerated cellulose suspension to form a chemical cross-linked gel, which was subsequently stretched to induce physical cross-linking by aligning the constituent nanofibers. The dual cross-linked film thus obtained had a reasonable strength (~98.5 MPa) and high toughness (~36.1 MJ/m^3), as shown in Fig. 5.4b (black line) and Fig. 5.4c (labelled 'IF-0%'), respectively. If a pre-stretching strain as high as 120% was introduced before formation of the chemical cross-linked gel, then the strength of the anisotropic dual cross-linked film (ACF) was much higher (~253.2 MPa) (Fig. 5.4b, cyan line) with a modest reduction in toughness (~29.7 MJ/m^3) (Fig. 5.4c, labelled 'ACF-120%'). The mechanistic understanding of the high strength and toughness in the dual cross-linked film can be explained as follows. Pre-stretching aligned partially the nanofibers in the film and helped minimize the defects in the porous structure. Chemical cross-linking then improved the connectivity between the nanofibers and enhanced the interfacial stress transfer process under tensile loading. Further physical cross-linking on the chemical cross-linked film aligned the nanofibers to a greater extent and minimized the defect size even more. This maximized the facile reformation of hydrogen bonds in the dual cross-linked film upon tensile loading. More pre-stretching (from 0 to 120%) thus minimized the defects even more, leading to the trend of higher strength and toughness as shown in Fig. 5.4c. The cumulative synergy provided by modification of the topological features by both physical and chemical cross-linking resulted in a novel dual cross-linked film which had a toughness, higher than most nanocellulose and wood-based films as well as densified wood (Fig. 5.4d).

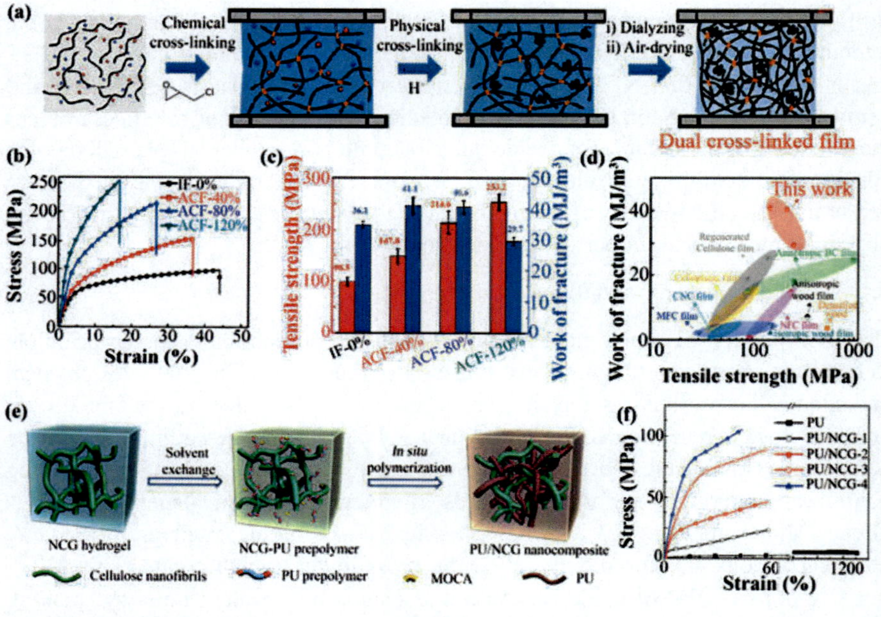

Fig. 5.4 Design principles involving modification of topology. **a** Schematic illustrating the modification of topological features by introducing both physical and chemical cross-linking. **b** Stress–strain plots of isotropic (black) film (IF), and anisotropic dual cross-linked cellulose films (ACFs). **c** Comparison of work of fracture, a parameter indicating toughness, and tensile strength in IF (black) and ACFs of varying pre-stretching strains. **d** Comparison of work of fracture and tensile strength of the obtained dual cross-linked film with different cellulose and wood-based films, densified wood etc. **a–d** Reproduced with permission [35]. Copyright 2019, American Chemical Society. **e** A schematic showing a double network formation in a porous nanocellulose gel (NCG) and polyurethane (PU) nanocomposite. **f** Stress–strain curves of the pristine PU and NCG-PU nanocomposite with increasing NCG content from NFC-1 (13 wt%) to NCG-4 (56 wt%). **e–f** Reproduced with permission [36]. Copyright 2019, American Chemical Society

The chemical cross-linking as described above can further be tuned to introduce multiple functional groups and form a double-network, as demonstrated by Li et al. [36]. In this study, a nanoporous cellulose gel (NCG), and acetone and tetrahydrofuran were first solvent-exchanged (Fig. 5.4e). Then the resulting material was infused with polyurethane (PU) and 3,3-dichloro-4,4-diamino-diphenylmethanelithium (MOCA) to initiate an in situ polymerization reaction. The –NCO groups of the PU could form polyurethane linkage with the –OH groups of the NCG. The double network PU/NCG nanocomposite thus formed, displayed an improved mechanical performance as shown in Fig. 5.4f. As the NCG content increased from ~13% (NCG-1) to ~56% (~NCG-4), the strength increased from 5.5 MPa to 105 MPa, respectively. The important role of NCG acting as a nanofiller is evident here as the strength of the nanocomposite is higher than most PU/cellulose nanocrystals, PU/microfibrillated cellulose and other PU/regenerated cellulose nanocomposites (maximum strength ~27.7 MPa [37–40]). The topological feature modification by forming a double network may have some future potential in other nanocellulose-based strong and tough materials which need to be implemented in applications involving considerable twisting.

5.3 Superb Mechanical Performance of Nanocellulose-Based Materials

In this section, some nanocellulose-based materials, which are fabricated using the design principles described in the above section, are discussed. The materials reviewed here are (a) micro/macro fibers, (b) films, (c) bulk wood and (d) nanocomposites.

(a) *Micro/macro fibers*

Strong and lightweight cellulose-based micro or macro-sized fibers are designed to enrich the field of renewable eco-composites as a suitable replacement for steels, alloys and plastics, thus paving the way for a sustainable future. Li et al. [41] demonstrated a low-cost, hydrid microfiber made of NFC and graphene oxide (GO) nanosheets (Fig. 5.5a–c), 1:1 by weight. Here, NFCs (~100–400 nm in length) were prepared by TEMPO treatment [27] and the GO (~1.2 μm average diameter) was prepared by Hummer's method [42]. The hybrid microfiber (~20 mm long) was prepared by manual wet-spinning of the hybrid solution in water and then dried by pulling it out under tension. The evidence of alignment between the fibers was demonstrated by images from polarized optical microscopy. These microfibers are extremely flexible (Fig. 5.5b) and possess remarkable twistability (Fig. 5.5c).

This material design exploited the synergistic interactions between NFCs and GOs described as follows. Due to the high strength of the GO nanosheets (3–5 layers), the tensile failure of this hybrid is expected to occur during inter-fiber sliding and not by fracturing of the GO sheets. NFCs offer extra bonding options between the

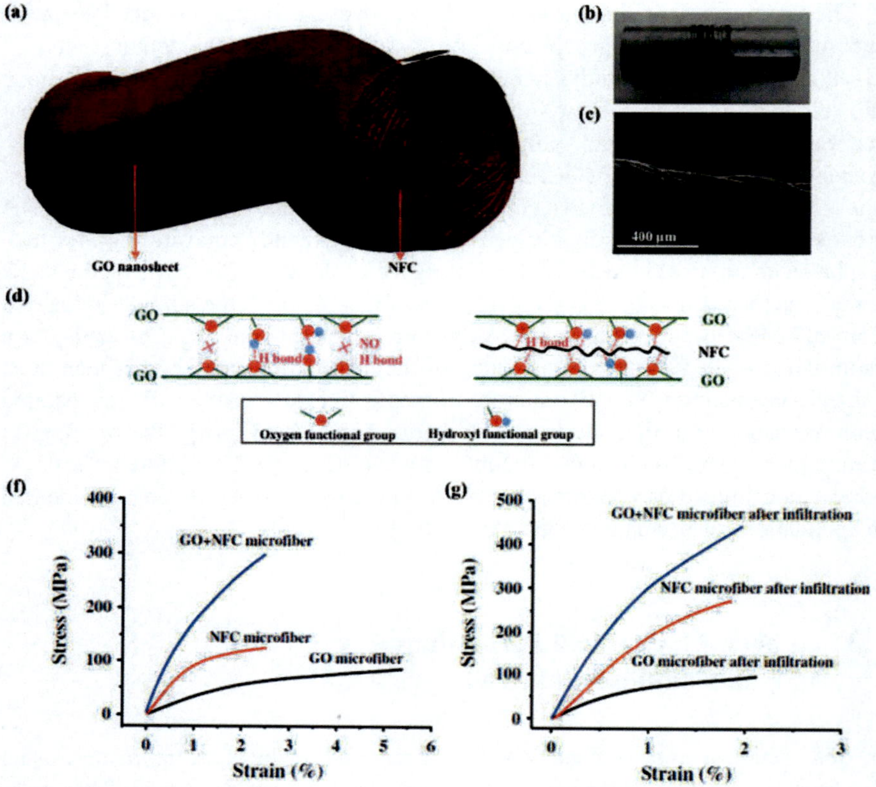

Fig. 5.5 **a** Schematic illustration of a GO-NFC hybrid microfiber comprised of aligned GO nanosheets and NFC fibers. **b** Photograph of the coiled hybrid fiber and **c** SEM image of the GO and NFC fiber twisted against each other. **d** Schematic showing how the incorporation of NFC renders additional hydrogen bonding in the GO-NFC hybrid. Experimental stress–strain curves of GO microfiber, NFC microfiber and the GO-NFC hybrid microfiber before (**e**) and after (**f**) metal ion (Ca^{2+}) infiltration. Reproduced with permission [41]. Copyright 2015, Nature Publishing Group

GO nanosheets (Fig. 5.5d). Hydrogen bonds are formed when –OH or –O– groups in the NFCs interact with the –OH of GOs. The abundant –OH groups in NFCs increased the hydrogen bonding options, resulting in a cascade of breaking and reformation of hydrogen bonding under tensile loading. This helped increase the strength and toughness of these microfibers. On the other hand, the GO sheets are much longer than NFCs. They linked the NFC chains and provided extra bonding sites over increasing distances. This linking of NFCs was not possible just by interactions of NFCs with each other, owing to their small lengths. Because of this bridging effect of GO nanosheets and the weaving and wrapping of NFCS around the GO sheets, inter-sheet hydrogen bonding of the system further intensified leading to greater fracture energy under tensile deformation and resulted in a modified microfiber with higher strength and toughness. Pure GO sheets only have –OH groups of one sheet forming hydrogen bond in closer proximity with another –OH group of a neighboring sheet, along with weak vdW forces. There was no synergistic effect because of the much limited hydrogen bonding sites.

The proposed analysis was validated with experimental observations. These well-aligned hybrid microfibers have ultimate tensile strength ~274.6 ± 22.4 MPa, which was higher than both the NFC microfibers (139.1 ± 28.7 MPa) and the GO microfibers (84.0 ± 2.8 MPa) separately, as evidenced in Fig. 5.5e. The average elastic modulus of GO-NFC microfiber (20.6 ± 0.9 GPa) was also substantially higher than both the NFC microfibers (15.5 ± 4.5 GPa) and the GO microfibers (2.3 ± 2 GPa). Incorporation of metal ions like Ca^{2+} further enhanced the tensile strength of the hybrid (Fig. 5.5f) by bringing the NFCs and GOs in closer proximity. The infiltration was done by putting the dried fiber in a 10% wt. $CaCl_2$ solution and pulling the fibers out by drying under tension. This study opened multiple probable possibilities where synergistic effects can be used. For example, including larger GO nanosheets and experimenting with other metal ions, can be applied in creating other high-performance hybrid microfibers with much superior mechanical properties.

Cellulose-based macrofibers can also be extremely useful to the field of high-performance biocomposites, as well as novel applications in biomedicine and nanofluidics. Considering the higher degree of polymerization and industrial-scale production potential of the microbial fermentation processes, BC-derived materials can be produced at a large scale with lower cost, opening greater benefits in using them to fabricate macrofibers. One such study was conducted by Wang et al. [43] where BC from Acetobacter xylinum was used to make BC pellicles, which were then wet-twisted to form macrofibers. Then the macrofibers were subjected to drawing till 30% strain. This formed a BC-30 macrofiber, which possessed a high strength of 826 MPa and a high Young's modulus (50 GPa). In comparison to low-density metals such as lightweight steel, natural cellulose fibers, jute, cotton and sisal fibers, the specific Young's modulus (36.4 GPa cm^3 gm^{-1}) and specific tensile strength (598 MPa cm^3 gm^{-1}) of this macrofiber was much higher. Moreover, these ultra-strong and ultra-stiff macrofibers showed excellent dyeability, which other high-performance, stronger and stiffer Kevlar aramid macrofibers lack. This may expand their usability in multitudes of other domains such as textiles and highlights the efficacy of BC to design macrofibers with appealing mechanical properties.

Efforts intended to pre-orient the cellulose fibers gradually while undergoing modification by chemical cross-linking, can also represent an effective course of action to fabricate cellulose macrofibers using bottoms-up strategies. This was demonstrated by Mittal et al. [44] where the challenge of reproducing nanoscale cellulose fiber mechanics to macroscale was addressed and partially overcome by designing a double flow-assisted self-assembly of the nanocellulose fibers, enabling an almost linear chain arrangement of NFCs in the final macrofibers. Using a mixture of spruce and pine bleached wood fibers as the starting material, cellulose nanofiber suspension was prepared by a sequence of steps involving TEMPO-mediated oxidation, homogenization, sonication and centrifugation. The flow-focusing device (Fig. 5.6a), where the prepared cellulose suspension was then injected, contained a tapered central channel with two sheaths, orthogonally connected to it along its length. When the cellulose suspension was first injected in the core channel, fibrillar

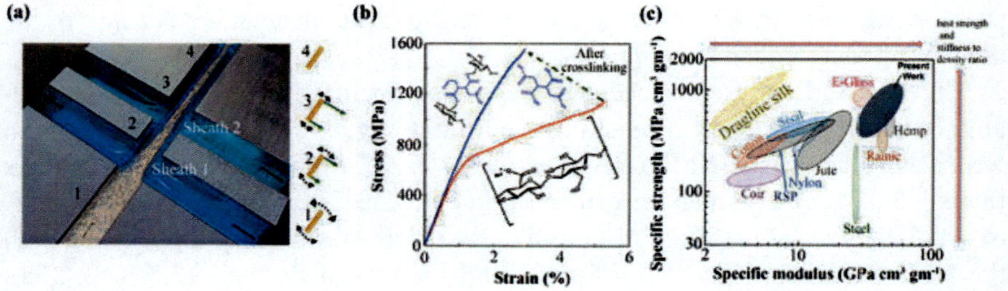

Fig. 5.6 **a** Schematic of the flow-focusing device to design cellulose macrofibers having near unidirectional alignment. The right panel shows the type of interactions at the positions labelled 1–4. **b** Engineering stress–strain curves of the macrofiber with (blue) and without (red) cross-linking at high humidity conditions (RH = 50%). **c** Comparison of specific strength (MPa cm^3 gm^{-1}) and specific modulus (GPa cm^3 gm^{-1}) of the obtained macroscale fiber with major existing bio and synthetic fibers. **a–c** Reproduced with permission [44]. Copyright 2018, American Chemical Society

alignment was poor due to extensive Brownian diffusion and electrostatic repulsions (Position 1, Fig. 5.6a) from the –COOH groups of NFCs. As the core flow further accelerated along the tapered center channel, deionized (DI) water was sprayed from the first sheath (Sheath 1, Fig. 5.6a). This formed a temporary alignment in the fibrils and simultaneously prevented the suspension to stick to the channel walls. With the flow accelerating along the channel, HCl at low pH was injected from the second sheath (Sheath 2, Fig. 5.6a) which arrested the electrostatic repulsion (Position 2, Fig. 5.6a) due to protonation of carboxyl groups leading to the formation of $–COO^-$ ions. This resulted in denser stacking and superior alignment in the fibrils. Then, with continual flow downstream (Position 3–4, Fig. 5.6a) along the gradually narrowed section of the central channel, the fiber suspension slowly immobilized, thus arresting the Brownian diffusion and formed continuous cellulosic threads which were subsequently manually removed and air-dried.

Nearly perfect unidirectional nanofiber alignment (orientation index ~0.92) in these lightweight macrofibers was observed and the relative size (diameter ~6.8 μm; length ~9–12 mm) increased 300 times along the process, leading to the formation of macroscale fibers. In presence of humidity (50% RH), one sample fiber displayed a strength of ~1200 MPa (Fig. 5.6b, red line). Furthermore, when chemical cross-linking was induced in the fibers by injecting 1,2,3,4-butane tetracaboxylic acid (BTCA), instead of HCl, through Sheath 2 (Fig. 5.6a), the average strength of the fibers at similar 50% RH increased to a maximum of 1570 MPa (Fig. 5.6b, blue line). The macroscale fibers also demonstrated an ultra-high Young's modulus (~80 GPa) at similar high humidity (RH = 50%) conditions. Similar macrofibers conditioned at a much lower humidity (RH = 14%) displayed a lower strength (maximum ~1406 MPa) and comparable Young's modulus (maximum ~86 GPa). The flow-focused assembly thus enhanced the interfacial stress transfer by aligning the constituent nanofibers of the macroscale fiber. Chemical cross-linking refined the stress transfer mechanism pre-aligned nanofibers by stimulating the connections between cellulose nanofibers through covalent bond formation, thus restricting the movement of water and boosting the mechanical performance of the macrofiber even at high humidity. The macroscale fiber demonstrated the highest specific strength to a specific modulus ratio when compared with multiple bio-based and synthetic fibers such as cotton, jute, nylon, steel, hemp and sisal, as shown in Fig. 5.6c.

(b) *Films*

An important category of strong and tough nanocellulose-based materials is represented by films. A progress in mechanics design strategies of films may cause significant breakthrough in the many fields such as, but not limited to, sensors and optical devices. There are stimulating demonstrations involving bacterial cellulose (BC) [45] to fabricate ultra-strong, ultra-tough and highly flexible BC films with high structural alignment. Here, BC films were produced by cutting BC pellicles of size 20 × 70 mm obtained from *Acetobacter xylinum* and wet drawing them upto 40% (Fig. 5.7a) using a tensile testing machine. This was then followed by hot pressing. Wet drawing loosened the entangled cellulose nanofibrils, made them longer and narrower, and enhanced the degree of alignment along the drawing direction. Hot pressing brought the fibrils in closer proximity to each other, indicating extended facile hydrogen bond formation. As a result, the mechanical strength of the BC films increased with a higher degree of wet drawing (Fig. 5.7b). Given the light-weightiness (low-density) of cellulose, the specific tensile strength of the BC film was substantially higher than that of titanium and aluminum alloys, which are some of the lightest and strongest known metallic materials (Fig. 5.7c). As illustrated in Fig. 5.7d, both the strength and the work of fracture increased simultaneously as the alignment in BC nanofibers increased, clearly showing the same anomalous scaling law of increased strength (~1.0 GPa) and toughness (~25 MJ·m^{-3}) for these films. Remarkably, these BC films were also foldable (Fig. 5.7e) into desirable shapes, thus increasing their usability in various applications.

Nanocellulose-based films fabricated from natural wood can also be envisioned by imposing greater degree of alignment of cellulose chains along the fiber axis direction, thus aiding in enhancement of mechanical properties. This was demonstrated by Zhu et al. [46] through an extremely straightforward process starting with natural wood (Basswood, Pine). Alignment, indicating increasing structural anisotropy, of the cellulose fibers in the wood could be tuned by changing the cutting directions of raw wood right at the beginning (Fig. 5.8a). The film was then fabricated through a series of processes involving removal of hemicellulose and lignin from the natural wood, followed by mechanical pressing which resulted in transparent films (Fig. 5.8a). All these processes preserved the alignment of NFCs in the wood. The anisotropy of such films could be tailored by changing the cutting direction in the wood raw materials (Fig. 5.8b) by varying the angle of cutting (θ) by 22.5 and 45° and changing the fiber alignment (ϒ) accordingly. This anisotropic film exhibits high transparency (≈90% transmittance) and superb intensity ratio (~350%) of transmitted light. The mechanical strength of the aligned fibers in the anisotropic film was also high (~350 MPa), around three times more than that of conventional cellulose nanopaper consisting of randomly (isotropic) oriented nanofibers, as illustrated

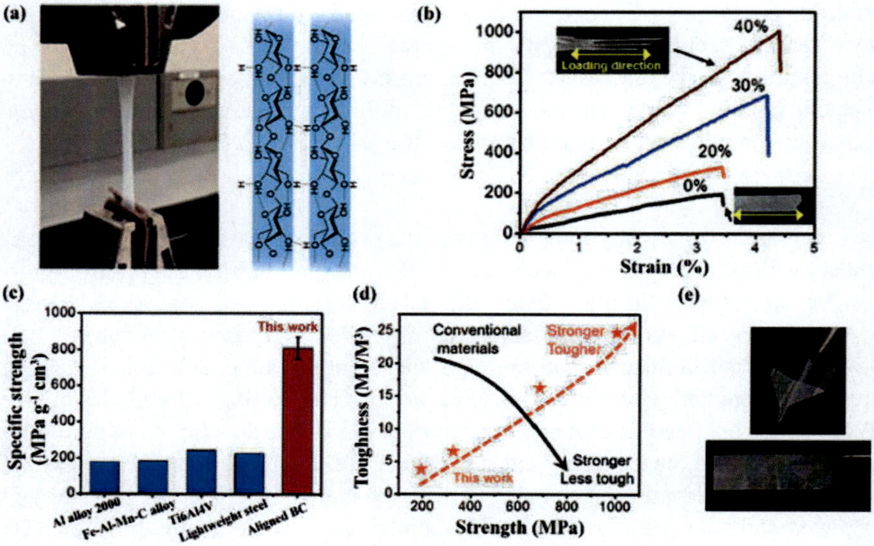

Fig. 5.7 **a** Transparent and anisotropic films made of aligned bacterial cellulose (BC) nanofibers fabricated by wet drawing of BC gel. **b** Stress–strain curves of the original (0%) and the wet-drawn (20, 30 and 40%) BC films after drying. **c** Comparison of the specific strength of the BC film with other existing high-performance materials. **d** Stronger and tougher simultaneously, a Holy Grail in materials engineering. **e** Optical image showing the high degree of flexibility (folding and unfolding) in the BC films. **a–e** Reproduced with permission [45]. Copyright 2018, Wiley-VCH

in Fig. 5.8c. Additionally, the strength of anisotropic films in a direction perpendicular to the fiber axis was much smaller (~23.2 MPa). This was because the aligned fibers in the anisotropic film offered a large number of parallel interfaces which, in unison, had a much better mechanical load transfer capacity causing the high strength. Inter-fiber sliding in the direction perpendicular to fiber axis minimized the number of amenable interfaces, thus reducing their load transfer capacity, indicating a lower strength. The breaking and reformation of hydrogen bonds while sliding in the aligned orientation due to larger available interfaces resulted in more energy dissipation and thus an increased film toughness of 7.38 MJ m^{-3}, around 3.3 times higher than the isotropic film. Experimental tensile testing is performed for different ϒ such as 0°, 4.2° and 10.3° corresponding to cutting angles of $\theta = 0°$, 22.5° and 45°, respectively, as shown in Fig. 5.8d. It is seen that with decrease in ϒ (indicating aligned fibers), the mechanical strength of the film increased. This was because of the fact that with decrease in the alignment angle, there is an additive effect of the hydrogen bonds at the interface along the loading direction, which was balanced by force F_2 as in Fig. 5.8e, and also of the covalent bonds within the same cellulose fibers which was balanced by force F_1 (Fig. 5.8e). This cumulative effect was much larger in a low alignment angle. With increase in ϒ, contribution from covalent bonds (balancing F_2) decreased and only the hydrogen bonding between the adjacent fibers (balancing F_1) gained prominence, thus resulting in low strength. Such transparent

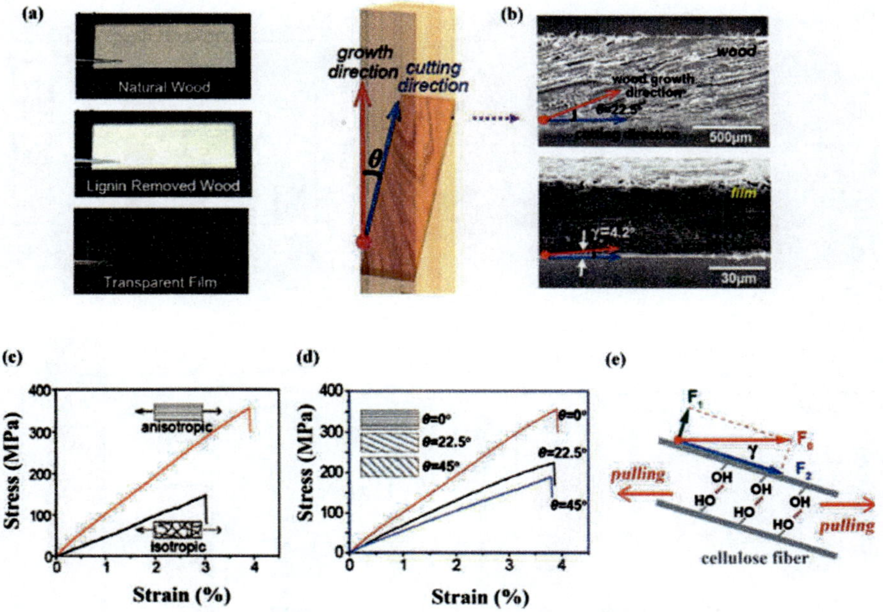

Fig. 5.8 **a** Photographs of natural wood undergoing delignification and pressing to directly transform to an anisotropic transparent film. **b** Films with tunable anisotropy can be obtained by cutting wood in different directions. SEM images of the wood being cut into slices at 22.5° and its corresponding transparent film are shown. **c** Experimental stress–strain curves of the anisotropic film ($\theta = 0°$) in this work and an isotropic nanopaper. **d** Stress–strain curves of the tunable anisotropic films with change in cutting angles (0, 22.5, and 45°). **e** Schematic of the forces acting along and perpendicular to the fibers. Reproduced with permission [46]. Copyright 2017, Wiley-VCH

films with tunable anisotropy can be used in many targeted applications in the field of green electronics.

(c) *Bulk Wood*

Plant wood contains hemicellulose (~19.5%), lignin (~20.8%) and cellulose (~44.01%) [47], the latter being the primary component. An important goal to engineer wood is to remove the hemicellulose and lignin by an adequate proportion that can transform the bulk natural wood into high performance materials. Song et al. [47] presented such a universal top-down strategy to process natural woods such as basswood, oak, poplar, cedar and pine into a densified wood through a simple yet effective two-step process (Fig. 5.9a). Here, natural wood (~120 mm by 44 mm by 44 mm) was chemically treated followed by boiling in deionized water. This was followed by mechanical pressing at 100 °C for 1 day, which leads to collapse of cell walls and total densification, resulting in a piece of densified wood (~115.6 mm by 46.4 mm by 9.5 mm). Adjusting the boiling time is the key here because it produces densified wood with partial degrees of lignin removal. Total absence of lignin crushes the wood during pressing and too much lignin does not allow the wood to be fully

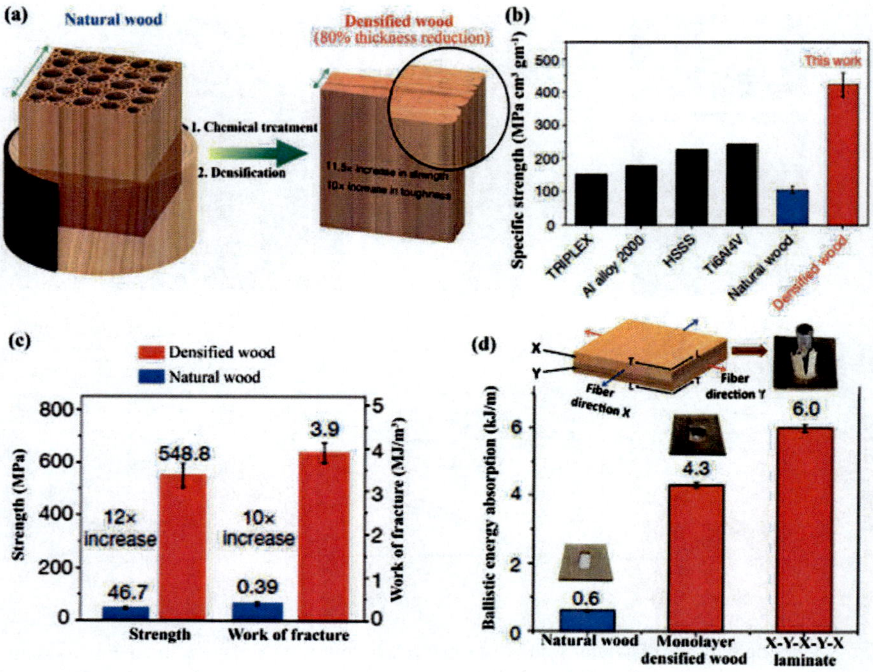

Fig. 5.9 **a** Schematic illustration of the processing approach of densified wood from bulk natural wood. **b** Comparison of the specific tensile strength of the densified wood with other high-performance structural materials. **c** Comparison of work of fracture of natural and densified wood. **d** Ballistic energy absorption comparison of natural wood, monolayer densified wood and the X–Y–X–Y–X laminated densified wood. Inset showing L shows the tree growth (fiber alignment) direction and T shows the direction perpendicular to it. Reproduced with permission [47]. Copyright 2018, Macmillan Publishers Limited

densified. So, an optimum balance of the cellulose, lignin and hemicellulose was required to obtain the densified wood of maximum strength (586.8 MPa), which was found to be 38.7%, 11.3% and 5.2%, respectively. The specific strength of the densified wood was much higher than typical metals (e.g., high strength steel HSSS, Al alloys) and titanium alloys (Fig. 5.9b). The enhanced hydrogen bond formation between the well aligned neighboring cellulose nanofibers in the densified wood was the primary reason for this favorable increase, along with the low density (1.3 gm/cm^3) of the densified wood. In comparison to natural wood, the densified wood had 12 times increase in strength and over 10 times increase in work of fracture as shown in Fig. 5.9c. The large number of repetitive hydrogen bond breaking and reforming events among the rich hydroxyl groups of the densely packed cell walls in the densified wood required a much higher energy to fracture the surface, making this wood much tougher than natural wood.

The above approach showed promising future potential in making low-cost armor as shown by the ballistic testing (Fig. 5.9d). For its demonstration, a five-layer laminated wood (X–Y–X–Y–X laminate as in inset of Fig. 5.9d) was constructed, where

5 pieces of monolayer densified wood was placed on top of each other in alternate 90° orientation and then pressed. X is the fiber alignment direction and Y is perpendicular to the fiber alignment. When comparing the ballistic absorption per unit sample thickness of natural wood, monolayer densified wood and X–Y–X–Y–X laminated wood, it was observed that the ballistic energy absorption of the monolayer densified wood and X–Y–X–Y–X laminate was around 4.3 times and 6 times higher than natural wood (Fig. 5.9d), respectively. The projectile, fired at around 120 miles/hour, was actually trapped by the X–Y–X–Y–X laminated wood (inset Fig. 5.9d) and the ballistic resistance of this wood was found approximately the same, both along the fiber alignment (5.6 ± 0.2 kJ m^{-1}) and perpendicular to the fiber alignment (6.0 ± 0.1 kJ m^{-1}) direction. Substantial energy dissipation because of the greater resistance, provided by the highly packed wood cell walls, was the main reason for obstruction of projectile perforation, as evidenced by the severe chapping of the densified wood surfaces (inset: Fig. 5.9d). Even under moisture attack (~95% RH), there was only a minor drop in densified wood strength (~493.1 MPa). On applying a surface painting (Polyurethane, Minwax) on the surface, the strength (~535.9 MPa) showed no prominent drop. This material is expected to exhibit a remarkable potential in low-cost, lightweight yet high-performance applications.

(d) *Nanocomposites*

Nanocellulose-based composites or nanocomposites [48] are manufactured by primarily applying the design principle of interfacial feature modification which predominantly utilizes the secondary hydrogen bonding in the cellulose molecules to reinforce it with other polymers. One intriguing study to integrate cellulose into high performance structural composites was demonstrated by Zhou et al. (Fig. 5.10) [49]. Their investigation reported a high-performance graphite-NFC composite (1:1 by mass) fabricated using bottoms-up approach by leveraging the hydrogen bonds and displayed promising potential to be scalably manufactured with minimal carbon footprint. Sonication enabled the exfoliation of commercial graphite powder into fewer-layered graphite. A uniform dispersion of the NFC on the graphite (Fig. 5.10a) followed. No harsh chemical treatment was involved. The resulting slurry was cast-dried (Fig. 5.10a) and hot-pressed (60 °C) to obtain the final composite having a density of around 1.2 g cm^{-3} and 20–50 μm thickness. Figure 5.10b shows the fabulous integration of low-cost, recyclability, processability, specific toughness and specific strength that this composite offers with respect to existing structural materials. Due to the low graphite and cellulose density, the specific strength (794 MPa g^{-1} cm^{-3}) is also significantly greater than prominent engineering materials such as Al alloys, steel, TRIPLEX and titanium alloys as summarized in Fig. 5.10c. As graphite flakes are hybridized into the NFC matrix, the tensile strength of the composite (901 ± 129 MPa) increased by ~3.7-times than that of NFC. The work of fracture (up to 30.0 MJ/m^3) (Fig. 5.10d) also increased nearly 5.9-times than NFCs and the ballistic energy absorption increased dramatically by ~4.6 and ~5.4 times (Fig. 5.10e) when compared with NFC nanopaper and commercial graphite, respectively. Approaches such as this to make high-performance cellulose-based

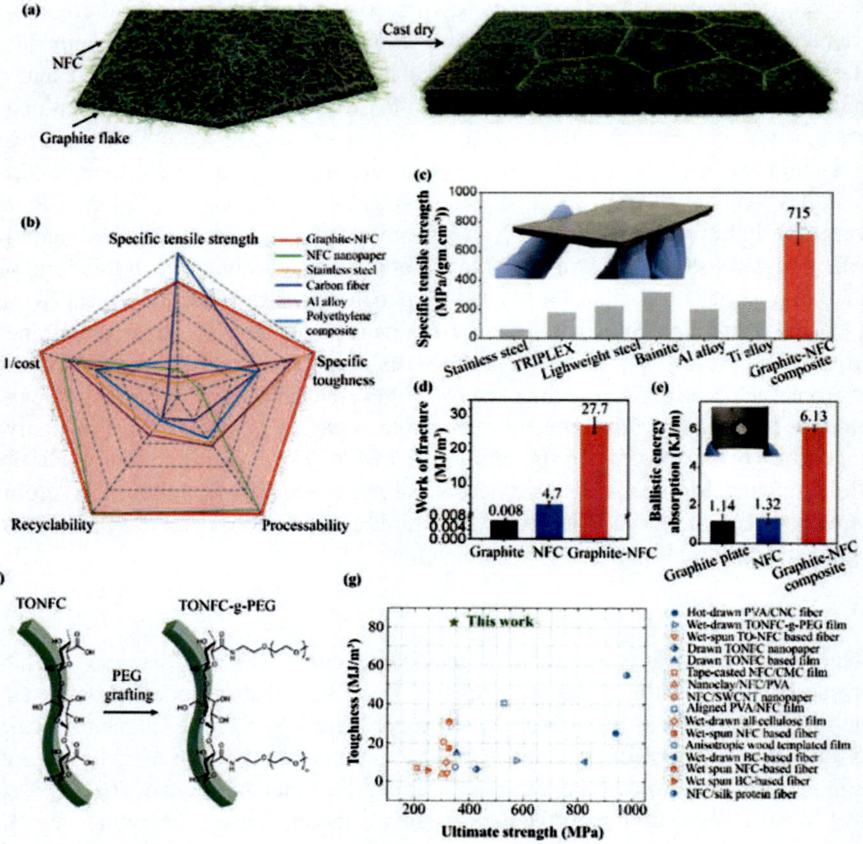

Fig. 5.10 Schematic of the processing approach for the graphite-NFC nanocomposite (1:1 by mass) using NFC-enable dispersion, cast-drying and pressing. **a** Radar plot comparing the mechanical traits, normalized by the maximum value, of graphite-NFC composites and other materials. **c** Comparison of specific tensile strength of the composite with high-performance metallic alloys. **d** Bar diagrams showing the work of fracture of the graphite-NFC composite, NFC nanopaper, and graphite paper. **e** Ballistic energy absorption comparison in NFC, commercial graphite and the graphite-NFC composite. **a–e** Reproduced with permission [49]. Copyright 2019, ELSEVIER. **f** Schematic showing the grafting of poly-ethylene glycol (PEG) on TEMPO-oxidized cellulose nanofibers (TO-NFC). **g** Comparison of the ultimate strength (MPa) and toughness (MJ/m^3) of the anisotropic TONFC-g-PEG nanocomposite with respect to other cellulose-based nanocomposites reported from previous literature. **f–g** Reproduced with permission [50]. Copyright 2018, American Chemical Society

composites, involving only mild mechanical sonication and minimal chemical treatments, have the remarkable potential to be adopted in industry-scale processes for commercialization.

Nanocomposites having improved mechanical performance can also be designed by grafting nanocellulose with other polymers, as reported by Geng et al. [50] where poly-ethylene glycol (PEG) was grafted on NFCs (Fig. 5.10f) in a matrix of poly-lactic acid (PLA). A TEMPO-mediated oxidation was performed on a softwood pulp to introduce carboxyl groups in the cellulose molecules and make them suitable for grafting. The TEMPO oxidized cellulose nanofiber (TO-NFC) suspension was then functionalized through a carbodiimide moderated amidation reaction to introduce the grafting, the presence of which was verified by Fourier-transform infrared spectroscopy. This resulted in the grafted TONFC-PEG (TONFC-g-PEG) suspension, which was then dissolved in N,N-dimethylformamide (DMF) with further addition of PLA. After stirring to ensure uniform disperse of the suspension followed by oven-drying and compression, an isotropic TONFC-g-PEG nanocomposite was obtained. The isotropic nanocomposite film was then stretched at 100 °C and cooled down to 20 °C rapidly to obtain an anisotropic TONFC-g-PEG nanocomposite film. The strength and toughness of this anisotropic TONFC-g-PEG nanocomposite film was 342 MPa and 82.5 MJ/m^3, respectively. Both the mechanical properties were considerably enhanced in comparison to the isotropic TONFC-g-PEG nanocomposite (strength—66.9 MPa; toughness—2.9 MJ/m^3) and the non-grafted isotropic TONFC nanocomposite (strength—63.5 MPa; toughness—1.6 MJ/m^3). The superior mechanical performance of the nanocomposite revealed how the combination of the design principles of structural (through controlled alignment) and interfacial (through grafting) modifications can be used to realize strong and tough cellulosic materials. Furthermore, this nanocomposite demonstrated a reasonable strength and considerably high toughness when compared to other existing cellulose-based nanocomposites (Fig. 5.10g). Both PLA and NFCs, being biopolymers make this nanocomposite very conducive for scalable green engineering.

Investigations related to interfacial mechanics involving the interaction of cellulose with water are limited but demonstrated [30, 51–53]. The relation of cellulose strength and water is counterintuitive. Presence of water is found to deteriorate [30, 51] the strength of cellulose but the factors responsible for this decrease are the same ones which promote fiber alignment and thereby enhance the dry strength. Till date, the detailed interplay of water with the cellulose fibril interface throughout the entire fabrication process remains an open question and there is a lack of experimental data. In places where there are difficulties in obtaining experimental results, modeling approaches may offer insights on the mechanism study. Many of the findings reported in the preceding two sections can further be enriched by investigating them through materials modeling, as described in the following section.

5.4 Modeling Approaches Explaining the Multi-scale Mechanics

Modeling approaches augment the experimental understanding by offering a fundamental qualitative insight on the failure and deformation behavior in nanocellulose-based materials. Mechanical behaviors across different length scales, ranging from nanometers to millimeters, can be investigated through distinct modeling techniques connecting the structure–mechanics relationships of different forms of nanocellulose-based high performance materials. This section discusses the different modeling approaches adopted to guide the novel design of cellulose nanomaterials. The different modeling strategies reviewed here are (a) atomistic modeling at the nanoscale, (b) modeling at micron (coarse-grained modeling) and continuum (finite element modeling) scales, and (c) theoretical modeling.

(a) *Atomistic modeling*

Atomistic modeling, performed by implementing molecular dynamics (MD) simulations using open source software such as LAMMPS [54], is an effective computational tool to investigate the mechanical behavior involving deformation and failure at the nanoscale. These simulations reveal the mechanical performance of the material model at the nanoscale and predict mechanical properties at the macroscale. It has long been used to predict the tensile strength [55] and elastic modulus [56] of crystalline cellulose. In one of the fundamental studies performed by Zhu et al. [18], already described in Fig. 5.2a–c, MD simulations were used to validate the envisioned mechanism of how inter-fiber sliding, facilitating the breaking and reforming of hydrogen bonds in the NFCs, plays an instrumental role to simultaneously increase both strength and toughness in the cellulose nanopaper. Such studies are important as interfacial sliding is expected to be a primary failure mode in cellulose nanopaper, because the strength of individual nanofibers is very high (~7.5 GPa [57]) and they are difficult to fail. Here, a nanocellulose bundle model was constructed with seven nanocellulose chains in a pattern as found in the crystalline cellulose nanostructure. The interactions between different molecules in the NFC chains were mapped by the ReaxFF force field [58]. The center chain, surrounded by six neighboring chains, was pulled out (only three chains shown in Fig. 5.11a for visual clarity). Figure 5.11b plots the variation of the total potential energy due to chain sliding at the fiber interface as a function of the pulling-out displacement. The zigzag profile of the curve clearly captures the hydrogen bond breaking and reforming events. When the hydrogen bond is stretched the most, the potential energy rises to a local maximum (inset of Fig. 5.11b corresponding to the first peak) and then the energy drops because the hydrogen bond is broken. The curve picks up again as the hydrogen bond reformed with the neighboring hydroxyl group (inset of Fig. 5.11b corresponding to second peak) during sliding. This further continues with the stretching of the hydrogen bond again, giving rise to the ramping up of the energy curve (third peak of Fig. 5.11b), thereby leading to a cascade of hydrogen bond breaking and reforming events until two surfaces are no longer in contact, signifying failure. The combined strength of hydrogen bonds and the cascade of energy dissipation required to completely separate the cellulose fibers gave rise to the enhanced toughness for the nanopaper. To understand the contributions from hydrogen bonding, comparisons were made with a material, such as carbon nanotubes (CNT), which lacked such a bonding mechanism. Atomistic simulation results of NFC (red lines, Fig. 5.11c) and CNT bundle (blue lines, Fig. 5.11c) models undergoing analogous inter-fiber sliding, showed that the energy dissipated to cause the failure was much higher in cellulose bundles than in

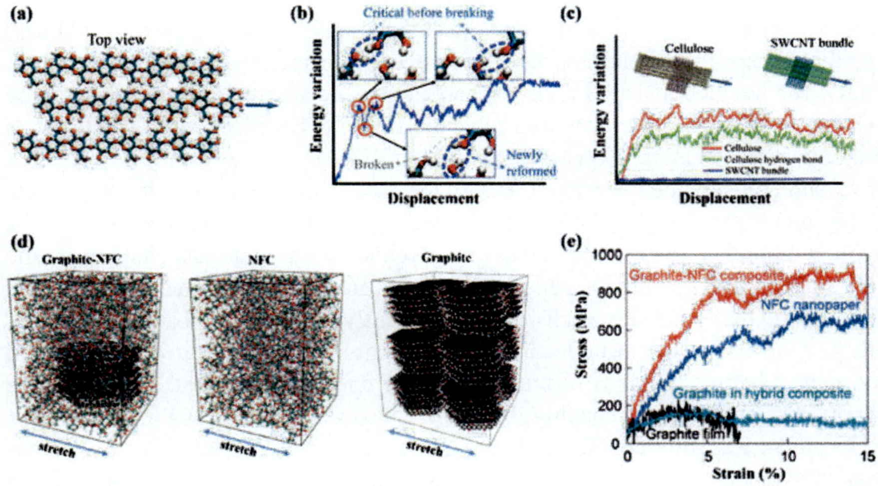

Fig. 5.11 Atomistic modeling of nanocellulose-based materials. **a** An atomistic model of cellulose chain pulling out to study the effect of sliding. **b** Total potential energy variation versus the pulling displacement obtained from molecular dynamics simulations as shown in (**a**). Prominent ups and downs in the curve represent the hydrogen bond breaking and forming events, respectively, for cellulose. **c** Total potential energy variation versus pulling-out displacement is plotted for cellulose fibers (in red) and CNT (in blue) bundles. **a–c** Reproduced with permission [18]. Copyright 2015, National Academy of Sciences. **d** Initial molecular models of the graphite-NFC hybrid, NFC and pure graphite. **e** Stress–strain curves for models in (**d**), obtained from molecular dynamics simulations, to qualitatively demonstrate the contributions of the graphite stack and the NFC in rendering high strength and toughness to the graphite-NFC composite. **d–e** Reproduced with permission [49]. Copyright 2019, ELSEVIER

CNT, with the primary energy resisting the sliding in cellulose being contributed by hydrogen bonds (green lines, Fig. 5.11c). Modeling results validate the experimental evidence from tensile tests of cellulose and CNTs.

Atomistic modeling can also provide other comprehensive mechanistic interpretations related to structural interfaces. For example, MD simulations [59, 60] predicted that the hydrogen bonds are more dominant in the (110)–(110) interface of the most commonly occurring cellulose 1β structure, in comparison to other (200)–(200) interfaces, which lacked adequate spatial hydrogen bonding density. The role of amorphous NFC interface, reinforced with a polymer matrix, in enhancing the mechanical performance of the composite by structural feature modification was also reported [61] through atomistic modeling. Such analysis is critical in designing nanocomposites because nanocellulose frequently acts as a filler by penetrating between the stacks of the reinforcing material. In one such study, NFCs filled up the spaces between the graphite stacks to yield a nanocellulose-graphite hybrid (Fig. 5.10a–e) having much superior mechanical performance in comparison to the pristine films based on its constituent materials, graphite and NFC. A mechanistic explanation of simultaneous strengthening and toughening in this nanocomposite was provided by molecular dynamics simulations of graphite-NFC (1:1 by mass), NFC and graphite models

(Fig. 5.11d). As in Fig. 5.11e, the strength of the graphite-NFC composite (red lines) was the highest, followed by the NFC nanopaper (blue lines) and graphite film (black lines). The interactions of NFCs in the nanocomposite are two-fold such as—(a) the –OH groups instigating hydrophilic behavior on the NFC surface and (b) –CH bonds promoting hydrophobic behavior with graphite. NFC, in between the flake gaps, stuck the graphite layers together by increasing the load transferring capacity of graphite flakes, making the graphite an active load bearing agent (cyan lines, Fig. 5.11e). The hydrogen bonding in between cellulose molecular chains simultaneously made the NFCs better load bearing agents. These coupled interactions between graphite and NFCs due to the presence of each other consequently enhanced the load bearing capability of the entire composite leading to a high strength. The repeated breaking and forming of hydrogen bonds of the NFCs while under deformation, caused greater energy dissipation in the loading direction, causing increased toughness. Without NFCs, there was only bonding at the graphite flake edges while weak vdW forces existed in the inter-flake regions, making pure graphite weak and brittle. In addition to being in reasonable agreement with the tensile testing measurements (0.91 GPa), the mechanistic insights provided by molecular modeling received further experimental validation from in situ atomic force microscopy (AFM) pull-out tests, which reported the average fracture strength of the graphite flake in the composite at around 0.83 GPa.

(b) *Modeling at the micron and continuum scales*

Mechanics of hierarchical nanocellulose-based materials transpire at length scales ranging from nanometers in molecules to meters in real-world samples. These multi-scale mechanics cannot be explained by just one technique and requires different strategies and fundamental principles to represent the mechanics at each of the length scales. This section covers the modeling approaches at the micron scale (e.g. coarse-grained modeling) and the continuum scale (e.g. finite element modeling).

Atomistic modeling can be employed in the nanometer and nanoseconds length and time scale, respectively. However, modeling systems in the micron scale are computationally demanding using MD simulations. For this reason, a coarse-grained (CG) modeling scheme is used to bridge the gap between the nano and the continuum level mechanical properties. In CG modeling, a group of atoms is mapped as beads by parametrizing the forces from MD simulations. This approximation reduces the computational expenses by manifolds, making this scheme very suitable to capture the mechanical properties dependent on length-scale effects unable to be addressed by MD-based investigations. A bottoms-up, scalable CG model reported by Ray et al. [62] used an all atomistic model (Fig. 5.12a) where 1 bead was represented by a single anhydro-glucose monomer ($C_6H_{10}O_5$), which comprised of 21 atoms of carbon, hydrogen and oxygen. This 1 bead formed the Level-1 in the CG model, which was then scaled up to Level-2. The Level-2 CG model was comprised of 36 chains of Level-1 beads (Fig. 5.12b), which was further scaled up to form a Level-3 CG bead, that consisted of 7 chains consisting of Level-2 beads (Fig. 5.12c). The distance between two Level-3 CG beads was ~20.304 nm. Thus, a random network of cellulose fibers formed using the Level-3 CG beads (Fig. 5.12d) was able to

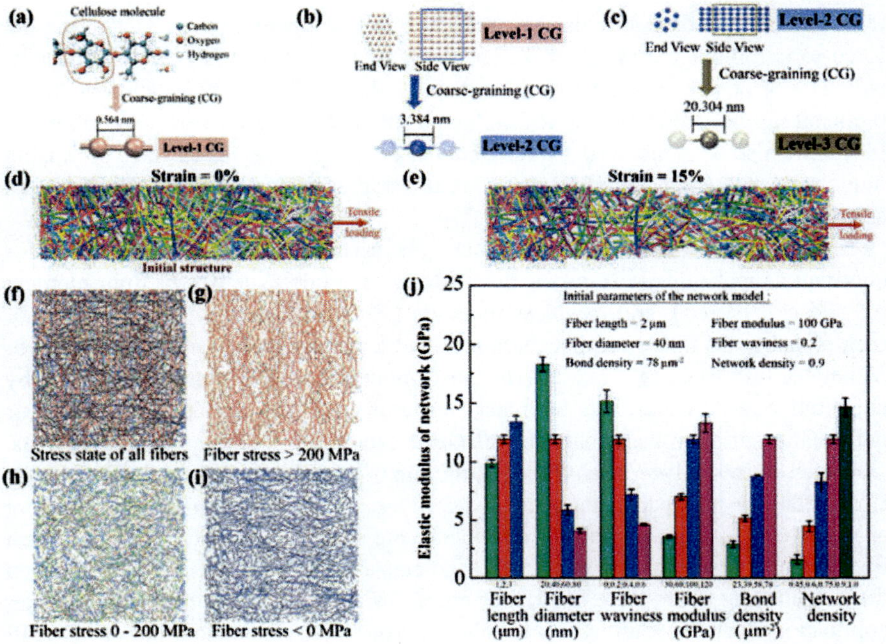

Fig. 5.12 Coarse-grained (CG) and finite element modelling (FEM) at the micron and continuum length scale, respectively. **a** Level-1 of a CG model comprising 21 atoms of carbon, hydrogen and oxygen. **b** Level-2 in the CG scheme obtained from mapping 36 chains in the Level-1 model, each consisting of 6 Level-1 CG beads. **c** Level-3 CG in the scheme obtained from mapping 7 chains in the Level-2 model, each consisting of 6 Level-2 CG beads. **d–e** Random network of chains comprised of Level-3 CG beads to model a representative structure of cellulose nanopaper undergoing tensile deformation. The unstretched model (strain = 0% as in (**d**)) and the final stretched model (strain = 15% as in (**e**)) shown here. Reproduced with permission [62]. Copyright 2021, Springer. **f** Stress distribution of all the fibers in the network. **g** Stress distribution of the fibers having tensile stresses greater than 200 MPa. **h** Stress distribution of the fibers having tensile stresses between 0 and 200 MPa. **i** Stress distribution of the fibers having tensile stresses less than 0 MPa, indicating compressive stress. **j** Variation of parameters such as fiber length, fiber diameter, bond density, fiber modulus, fiber waviness and network density on the elastic modulus of the random network. **f–j** Reproduced with permission [67]. Copyright 2017, ELSEVIER

model a micron-scale cellulose nanopaper, shedding light on the in situ deformation and failure mechanisms while the nanopaper underwent tensile deformation (Fig. 5.12e). Such insights are challenging to observe in experiments and almost impossible using atomistic simulations. Design strategies of proposing longer cellulose fibers to increase the overall nanopaper strength transfer, as suggested in this study, can also be very effective to shed deeper insights in the mechanics design.

Other CG modeling schemes explaining the reason why natural wood with porous cell walls display a higher strength and toughness (Fig. 5.9c) than densified wood with collapsed channels [47], have also aided in the mechanistic understanding. The enhanced isotropic ballistic resistance (Fig. 5.9d) of the laminated densified wood from the same study could also be explained by CG modeling after comparing the force variation due to pulling of the cell walls, as a function of separation distance. Novel CG schemes to investigate the bending failure of NFC bundles and fracture of bioinspired staggered NFC-based Bouligand structures [63], as well as hybrid fibers [64], have also been reported. Overall, coarse-graining strategies can be a promising modeling scheme in nanocellulose, and future studies can establish the CG approach as a powerful modeling methodology.

Modeling approaches at the continuum scale rest on the fundamental framework of continuum mechanics and apply various numerical models such as the finite element method (FEM) [65], and meshless methods [66] to investigate the connection of factors such as NFC size and distribution, network density on the mechanical behavior of nanocellulose-based materials. In these models, the NFCs are represented by constitutive laws, which may be linear elastic or nonlinear. Breakable and fixed or deformable spring linkages between the NFCs capture the hydrogen bonding effects. One such study analyzed the stress distribution in a 2D fibrous network (Fig. 5.12f–i) of cellulose nanopaper and modeled its elastic properties using finite element analysis [67]. Here, the nanocellulose fibers were modelled as rigid bodies with each fiber segment connected to other fixed segments at its two ends. The bonds between the neighboring fibers were fixed, thus not allowing fiber sliding or rotation during application of tensile loading. At a tensile strain of 1%, the stress distribution of (a) all fibres (Fig. 5.12f), fibers having tensile stress greater than 200 MPa (Fig. 5.12g), fibers having tensile stresses between 0 and 200 MPa (Fig. 5.12h) and fibers having tensile stresses lesser than 0 MPa (compressive stress) (Fig. 5.12i), are shown. This was done to verify that around two-thirds of the fibers in the network were under tensile loading while the remaining one-third was contracting laterally due to Poisson's effects. The effect of varying parameters such as fiber length, fiber diameter, bond density, fiber modulus, fiber waviness and network density were investigated with respect to the elastic modulus of the random network (Fig. 5.12j). With increase in fiber length, the elastic modulus increased because the average number of bonds per fiber also increased, thus elevating their stress transfer capacity. With increase in fiber diameter, the elastic modulus decreased because of the reduction in the number of load-bearing fibers, since the network volume is constant. On increasing the fiber waviness, the elastic modulus decreased because a substantial number of coiled (wavy) fibers had to be unstretched and straightened before being able to transfer the tensile load. The elastic modulus was also found to increase proportionally with increase in elastic modulus of the individual fibers. In addition, the modulus increased linearly with a higher bond and network density due to the presence of more inter-fiber bonds. The model predicted that the elastic modulus will converge to zero at a network density less than 0.39.

Separate investigations such as those by Shishehbor et al. [68] analyze different loading conditions such as bending, shearing and torsion and report their dependency on bonded and non-bonded interactions, using a simplified continuum-based structural model. This helped in recording the accurate, linear elastic response of NFCs on the stiffness of the crystalline nanocellulose and may provide a framework to study the inelastic behavior of NFCs in future. In summary, the numerical models are very useful to explore the underlying nanocellulose mechanics but before adopting them to analyze the mechanics, a rigorous analysis needs to be done by checking the available computational resources and/or developing a clear understanding of the governing experimental phenomenon, to take full advantage of their potential.

(c) *Theoretical modeling*

Theoretical modeling, based on the mathematical formulations defined by the theories of continuum mechanics, may also offer salient clarifications in mechanics study of nanocellulose which agree well with applicable experimental results. This approach, unlike the modeling techniques described before, cannot visualize the deformation and distribution of stress in nanocellulose networks. But here, the material parameters (e.g. nanofibril length, nanofibril radius) and the inter-fiber hydrogen bond interactions are incorporated in a mathematical model in order to estimate the mechanical properties of the nanocellulose-based material under study. Hence, theoretical modeling strategies can bridge the molecular scale mechanics of cellulose nanofibrils with that at the continuum scale containing the cellulose fiber network. For example, Meng et al. [69] conducted a theoretical study to explain the effect of aligned and random nanofiber orientations. The established three-dimensional crack-bridging model in shown in Fig. 5.13a.

The crack-bridging model of the cellulose nanopaper (Fig. 5.13a) was assumed to be an elastic, homogenous and isotropic solid structure. Tensile deformation was applied (y direction) to the model (arrows in Fig. 5.13a) by a uniform stress (σ_∞). During the deformation, cellulose nanofibers start sliding relative to each other resulting in the formation of a crack after a certain threshold. The crack gradually propagated as the nanofibers were continually pulled out from the crack surfaces. This formed the crack bridging zone (Fig. 5.13a) behind the tip of the crack. The nanofibers in the crack bridging zone were modeled in three ways: (a) aligned, (b) at a specific angle and (c) in random orientation to the tensile loading direction. For randomly oriented fibers in the crack bridging zone, the fracture toughness decreased with increase in nanofiber radius (R), as shown in Fig. 5.13b. This is accordance with the experimental results already described in Fig. 5.2a–c [18] because with increase in R, the interspatial hydrogen bonding decreased, leading to reduced breaking and reformation of hydrogen bonds during tensile deformation, thus accelerating the failure of the nanopaper. A distribution parameter (A) was introduced to study the effect of aligned ($A = 100$), perpendicularly misaligned ($A = -100$) and random ($A = 0$) orientations of the nanofibers on fracture toughness (Fig. 5.13c). As the nanofibers became more aligned (e.g. A increasing from 0 to 100), the nanopaper was able to delay the crack propagation by resisting a larger deformation force, which resulted in the maximum fracture toughness of the nanopaper. This study provided a qualitative understanding of the experimental data already described in Fig. 5.8a–c, thus ensuring the validity of the proposed model.

Unlike the variable nanofiber orientations in the crack-bridging zone as described above, a separate multiscale theoretical study [70] considered nanofibers parallel to each other and perpendicular to the crack surface. A correlation was developed (Fig. 5.13d) between fracture toughness and physical parameters such as nanofiber

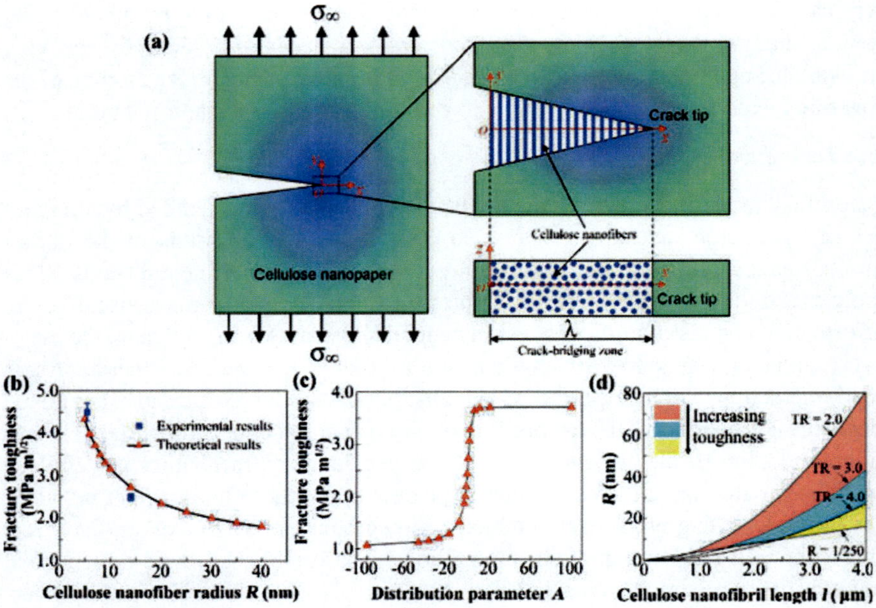

Fig. 5.13 Theoretical modeling to study the cellulose nanopaper mechanics. **a** Schematic of the developed three-dimensional crack-bridging model. **b** Plot of fracture toughness versus nanofiber radius (*R*). **c** Plot of fracture toughness versus distribution parameter (*A*), indicating the nanofiber orientation. **a–c** Reproduced with permission [69]. Copyright 2018, ELSEVIER. **d** Design map of cellulose nanopaper with respect to the toughening ratio (*TR*), *R* and *l* using a crack-bridging model. Reproduced with permission [70]. Copyright 2017, ELSEVIER

length (*l*) and radius (*R*), by proposing a unified cohesive law which considered the self-healing property of the hydrogen bonds on the fiber interfaces. A toughening ratio (TR) was formulated, the higher value of which signified reduced crack-opening displacement and thus superior bridging effects. When *R* was fixed, the *TR* increased with increase in *l*. This indicated that for longer nanofibers, the crack-bridging zone will also be much longer which will subdue the rate of crack propagation and increase the fracture toughness of the nanopaper. In addition, when *l* was fixed, TR increased with decrease in *R*. This was because, for a constant nanofiber length, the thinner nanofibers in the crack-bridging zone offered an increased interfacial area to be involved in relative nanofiber sliding which lead to an increase in fracture toughness. The design map (Fig. 5.13d), thus created, could help the fabrication of the cellulose nanopaper with tailored mechanical performance. Thus, robust theoretical models, complimented by experiments, have an exceptional potential in enriching the nanocellulose mechanics framework by assessing the underlying structure–mechanics relationships.

5.5 Future Perspectives and Limitations

Invention of materials such as steel and plastics have long revolutionized the human society. But currently, the environmental pollution ensuing from manufacture of steel and plastic wastes are increasing alarmingly. For example, current predictions suggest that the plastic waste in the ocean will triple by 2040 [71] if globally controlled interventions are not imposed. Hence there is an urgent need to shift from traditional petrochemical-based products to carbon neutral green materials. Independent networks, exploring prospects of an improved Earth, such as the Millennium Project [72] also identifies "Sustainable Development and Climate Change" as the first of the 15 global challenges that needs to be immediately addressed through integrated research efforts and inter-governmental policies. Thus, replacement of steel and plastics by advanced nanocellulose-based materials may offer promising solutions to guide the mankind toward a sustainable future. This section discusses an array of ongoing efforts which may help realize the future vision of strong and tough nanocellulose-based green materials unfolding multifaceted opportunities toward sustainable materials engineering [73].

Strong and tough nanocellulose-based fibers, designed using innovative mechanistic principles, have already displayed a wide range of applications in multiple domains, making them potential candidates for future green materials engineering. Macroscopic wood fibers [74] obtained by delignification of natural wood and subsequent twisting, have a high mechanical strength (~106.5 MPa, in comparison to 54.9 MPa of natural wood) and are ~20 fold tougher (~7.7 vs. 0.36 MJ m^{-3}) than natural wood because of their retained aligned cellulose microfibers. In addition, they are highly flexible, bendable and exhibit excellent knitting properties and dyeability. This makes them suitable for future smart textile applications. Hybrid composite microfibers based on carbon nanotube and nanocellulose [75], developed via scalable 3D printing methods possess a combination of high mechanical strength (~247 MPa) and electrical conductivity (~216 S cm^{-1}), making them promising candidates for wearable electronic devices. Helically coiling smaller fibers may lead to large BC-based macrofibers [76] having a high strength (~535.4 MPa) and toughness (44.8 MJ/m^3) which can also demonstrate possible applications in the textile industry. BC macrofibers [43] may record a high tensile strength (~826 MPa) due to the large length and the alignment of nanofibers and are dyeable. This enhances the possibility of utilizing them for future production of high-performance biocomposites and textiles, as well as impacting the field of biomedicine and nanofluidics. Cellulose macrofibers [44], fabricated by multiscale flow control of nanocellulose, can effectively transfer the fiber mechanics of nanoscale to the macroscale. These macroscale fibers possess high Young modulus (~86 GPa) and tensile strength (1.57 GPa), making them extremely effective as load-bearing, lightweight structural materials.

Strong and lightweight nanocellulose-based films may also have the potential to replace the traditional plastics and metals through innovative mechanics design using the natural resources. Transparent films [27, 77] with their exquisite mechanical properties, coupled with ultrahigh transparency and ultrahigh haze, may enhance their utility as packaging materials and also in photovoltaic applications, such as solar panel, solar roof, or solar windows. Ultra-strong and tough, super clear cellulose films fabricated by Ye et al. [35] using a dual cross-linking approach (Fig. 5.4a–d) might be adapted in next-generation opto-electronic applications. Anisotropic transparent films (Fig. 5.8) with tunable anisotropy [46] may have the potential to unlock a range of novel green electronics and fluidic devices. Ultra-thin (thickness ~8.3 μm) films [78] with aligned NFCs have also been reported to exhibit a superior strength (~342 MPa) and modulus (~43.6 GPa) in addition to being foldable into various shapes, which manifests their remarkable potential as diaphragm in wood speakers, which may have future relevance in the arena of acoustics. Environmentally friendly enzyme assisted processing techniques may result in cellulose nanopaper [31] possessing a maximum degree of polymerization (DOP) of ~1100, having a high strength (214 MPa), modulus (13.2 GPa) and toughness (15.1 MJ/m^3). This demonstrates the correlation between DOP and mechanical performance and emphasizes on the need to obtain cellulose nanofibers of higher DOP to obtain films of superior mechanical properties. Fang et al. [79] elucidated a DOP protection strategy by cooking the initial wood blocks in an aqueous alkaline sulfite-anthraquinone-methanol (ASAM) liquor. The novel ASAM pulping approach resulted in anisotropic, low-density (~1.37 gm/cm^3) nanocellulose films having a high DOP of ~1944, in addition to a high strength of ~1014.7 MPa and modulus of ~60.2 GPa along with a reasonably high toughness of ~15.2 MJ/m^3. The specific strength (820 MPa cm^3 gm^{-1}) of the film was reported to be higher than known natural polymers and many existing metals (e.g.: Aluminum, Monels, Cast iron, Carbon steel, Mg_2Zn etc.). Plethora of diverse implementations establish the future prospects of such nanocellulose-based films in high-performance structural applications as well as in electronics and intensify their future likelihood in gradually replacing petroleum and synthetic polymer-based products. Computational studies, adopting machine-learning assisted approaches [80, 81] and using experimental studies as training data, can also compliment the experimental design techniques to design tailored nanocellulose-based films with superior mechanical properties.

Novel nanocellulose-based materials, fabricated by optimizing the structure–mechanics [82] relationships in bulk wood through a top-down approach, can also be employed to develop next-generation renewables to be used in the field of high-performance structural materials, energy storage, nanoionics, light management and many more. Partial delignification followed by densification of bulk wood (e.g.: oak, poplar, pine, cedar) may yield materials such as super wood (Fig. 5.9) [47], which possesses a low density and has the potential to serve as a lightweight high-performance material, urgently needed in the automobile industry. Strong, tough and lightweight engineered wood from the fast-growing bamboo [83], possessing a high tensile strength (~1 GPa) and modulus (59.6 GPa), can also be appropriate for similar applications. This is because carbon fiber reinforced polymers currently used in advanced vehicles have high costs. Current lightweight materials such as advanced high-strength steels (AHSS), magnesium (Mg) alloys and aluminum (Al) alloys are replacing cast iron and conventional mild steel components; but AHSS is too heavy, Al alloys are still expensive and Mg alloys are susceptible to corrosion.

Densified woods [47, 83], having a strength higher than most known structural materials along with high modulus, ballistic resistance and superb dimensional stability under moisture, offer a promising alternative. Incorporation of these lightweight materials will increase vehicle efficiency by manifolds. For example, a 10% reduction in vehicle weight lead to a 6–8% fuel economy improvement. This will reduce the planet's carbon footprint by significant amounts, considering that 79 million vehicles were supposedly sold in 2018 globally. Similar approaches to modify natural wood into cross-laminated timber are already being used in skyscraper construction, thus reducing a building's carbon footprint by 75%. Complete delignification followed by densification of bulk wood may result in an engineered wood [84] having reasonably high strength (~270 MPa) and elastic modulus (~35 GPa), in addition to being moldable into different shapes, thus increasing their adaptability in multiple sustainable structural applications. Finally, a close collaboration between plant and materials scientists can also effectively utilize gene editing technologies [85] to obtain engineered wood having targeted hierarchical structures, which may open a new paradigm in wood sciences with very attractive functionalities.

Nanocellulose-based composites attract considerable attention, because the different polymers and nanomaterials used to reinforce these nanocomposites magnify the mechanics design space and help in fabricating tunable devices for selective applications. For example, nanocomposite filaments based on SWCNT/NFC (Fig. 5.3a, b) can be used to design sensors, which can lift a weight more than 520 times their own and detect strain variations occurring from contractions of muscles in a human body. Such strain sensors may display novel cross-disciplinary applications in clinical managements by effectively monitoring patients. Methods such as grating NFCs with PEG can be successfully used to envision nanocellulose-based composite macro-ribbons [86] having a high mechanical strength (~576 MPa), modulus (17.2 GPa) and toughness (~17 MJ/m^3). Such high-strength macroscopic ribbons may demonstrate extensive applications in the field of biomedical devices or in optoelectronics. Two-dimensional materials such as MXene [87] invented as recently as 2011, have also been reported to reinforce cellulose nanofibrils to fabricate nanocomposites having a reasonably high mechanical strength (~136.5 MPa) [87] and modulus (9.3 GPa). The presence of cellulose nanofibrils aided in the superior mechanical performance of the nanocomposite whereas the exceptional photothermal, electrothermal and strain sensing properties due to the presence of MXenes unveiled its resourcefulness for prospective applications in soft robotics and flexible electronics. Other nanocomposites based on graphite and nanocellulose [49], fabricated by a bottom-up approach, already showed a high mechanical tensile strength (up to 1.0 GPa) and toughness (up to 30.0 MJ/m^3). The lightweight composite surpasses most conventional structural materials in terms of specific strength and toughness, including various steels, alloys, and even some carbon fibers. The scalable manufacture of this composite required little energy and produced minimal harmful waste, makes its applications more lucrative [88], thus opening doors for developing high-performance nanocomposites in a more sustainable manner.

Although various bottoms-up and top-down mechanics design strategies have already generated many applications for nanocellulose based high-performance materials, this field of material science is still at a nascent stage. Since most of the innovations described above have been performed in the academic laboratories, certain bottlenecks need to be tackled before advancing its commercialization. Economic feasibility is a predominant obstacle for possible large-scale industrial implementation. For example, TEMPO-mediated oxidation performed to extract TO-NFCs are expensive and may waste large amounts of water, if scaled up. Research efforts restricted to laboratory, rarely focus on performing a life-cycle analysis to assess the environmental impacts at every stage of the manufacturing process, but this is extremely important across the industrial value chain. Hence, future dialogues and collaborative efforts between industry and academia are crucial to accelerate the impact and widespread commercialization of nanocellulose-based materials. From a research perspective, a fundamental understanding of the interactions between water and cellulose nanomaterials is key to design a novel material with mechanical performance consistent across variable ambient conditions. For this, high throughput screening techniques, shedding light on how the bound water influences the conformation of cellulose nanocrystals, could also be very insightful. Supporting experimental characterizations provided by Nuclear Magnetic Resonance spectroscopy and/or dielectric measurements, may validate the simulation-predicted results and provide a holistic understanding. Secondary hydrogen bonding has already been established to influence the mechanical performance of strong and tough nanocellulose-based materials, a detailed study describing the stage of the materials processing from when the contributions from hydrogen bonding start affecting the material behavior, is still absent. Visualizations of hydrogen bonding through experimental 3D tomography techniques, which can already characterize the nanoscale reinforcement morphology in carbon nanomaterials [89], may reveal significant results enhancing our understanding of the mechanism. However, in spite of the grand challenges, current research progress kindles future hopes that the emerging research technologies [90] may well be reproduced in the industrial scale. This, when possible, is anticipated to bring to fruition the idea of green and renewable nanocellulose in prospective high-tech applications involving lightweight structures, biomedicine, electronics, energy storage and energy-efficient buildings.

5.6 Conclusion

In summary, an array of nanocellulose-based materials with superior mechanical performance have been discussed. In this chapter, significant mechanistic studies of materials related to cellulose derived both from plants and bacteria, the two most common sources of this abundantly available and bio-renewable polymer, are included. Regardless of the source, the fundamental design principles driving the counterintuitive properties of higher strength and higher toughness in nanocellulose are noted and discussed in detail. Then, some examples of cellulose-based materials (e.g., fibers, films, bulk wood, nanocomposites), synthesized either by bottoms-up or top-down approach, have been reported along with the key findings encompassing their superior mechanical properties. Complementary modeling approaches which support experimental evidences and elucidate the mechanism behind multiple appealing mechanical properties of cellulose, are also described. Future perspectives of the emerging areas requiring mechanically strong and tough high-performance cellulosic materials are outlined, along with the grand challenges that need to be addressed in the immediate future to advance this research realm. This chapter emphasizes on the potential impact of cellulose-based materials and hopes to inspire current or future researchers to nurture the appealing mechanical performance of this next-generation material through relevant avenues, that can address and thereby mitigate the significant challenges in the field.

References

1. K. Esau, *Anatomy of Seed Plants*, 2nd edn. (Wiley, 1977)
2. N. Lin, A. Dufresne, Nanocellulose in biomedicine: current status and future prospect. Eur. Polym. J. **59**, 302–325 (2014). https://doi.org/10.1016/j.eurpolymj.2014.07.025
3. J.O.A. Shatkin, T.H. Wegner, E.M. (TED) Bilek, Market projections of cellulose nanomaterial-enabled products—part 1: applications. TAPPI J. **13**, 9–16 (2014)
4. J.O.H.N. Cowie, E.M. (TED) Bilek, T.H. Wegner, Market projections of cellulose nanomaterial-enabled products—part 2: volume estimates. TAPPI J. **13**, 57–69 (2014)
5. Z. Dai, V. Ottesen, J. Deng, R.M.L. Helberg, L. Deng, A brief review of nanocellulose based hybrid membranes for CO_2 separation. Fibers **7**, 1–18 (2019)
6. D. Jang, X. Li, H. Gao, J.R. Greer, Deformation mechanisms in nanotwinned metal nanopillars. Nat. Nanotechnol. **7**, 594–601 (2012). https://doi.org/10.1038/nnano.2012.116
7. M.D. Demetriou, M.E. Launey, G. Garrett, J.P. Schramm, D.C. Hofmann, W.L. Johnson, R.O. Ritchie, A damage-tolerant glass. Nat. Mater. **10**, 123 (2011). https://doi.org/10.1038/NMAT2930
8. R.J. Moon, A. Martini, J. Nairn, J. Youngblood, Cellulose nanomaterials review: structure, properties and nanocomposites. Chem. Soc. Rev. **40**, 3941–3994 (2011). https://doi.org/10.1039/c0cs00108b
9. D. Klemm, F. Kramer, S. Moritz, T. Lindström, M. Ankerfors, D. Gray, A. Dorris, Nanocelluloses: a new family of nature-based materials. Angew. Chem. Int. Ed. **50**, 5438–5466 (2011). https://doi.org/10.1002/anie.201001273
10. G. Chinga-carrasco, Cellulose fibres, nanofibrils and microfibrils: the morphological sequence of MFC components from a plant physiology and fibre technology point of view. Nanoscale Res. Lett. **6**, 1–7 (2011)
11. U. Ray, S. Zhu, Z. Pang, T. Li, Mechanics design in cellulose-enabled high-performance functional materials. Adv. Mater. **33**, 2002504 (2020). https://doi.org/10.1002/adma.202002504
12. M. Iguchi, S. Yamanaka, A. Budhiono, Bacterial cellulose—a masterpiece of nature's arts. J. Mater. Sci. **35**, 261–270 (2000)
13. B.V. Mohite, S.V. Patil, A novel biomaterial: bacterial cellulose and its new era applications. Biotechnol. Appl. Biochem. 101–110 (2013). https://doi.org/10.1002/bab.1148
14. Z. Wu, H. Liang, L. Chen, B. Hu, S. Yu, Bacterial cellulose: a robust platform for design of three dimensional carbon-based functional nanomaterials. Acc. Chem. Res. **49**, 96–105 (2016). https://doi.org/10.1021/acs.accounts.5b00380

15. E.J. Foster, R.J. Moon, U.P. Agarwal, M.J. Bortner, J. Bras, S. Camarero-Espinosa, K.J. Chan, M.J.D. Clift, E.D. Cranston, S.J. Eichhorn, D.M. Fox, W.Y. Hamad, L. Heux, B. Jean, M. Korey, W. Nieh, K.J. Ong, M.S. Reid, S. Renneckar, R. Roberts, J.A. Shatkin, J. Simonsen, K. Stinson-Bagby, N. Wanasekara, J. Youngblood, Current characterization methods for cellulose nanomaterials. Chem. Soc. Rev. **47**, 2609–2679 (2018). https://doi.org/10.1039/c6cs00895j
16. K. Nelson, T. Retsina, M. Lakovlev, A. van Heiningen, American process: production of low cost nanocellulose for renewable, advanced materials applications, in *Catalytic Materials Nanofibers—From Research to Manufacture* (2016), pp. 267–302
17. H. Zhu, Z. Jia, Y. Chen, N. Weadock, J. Wan, O. Vaaland, X. Han, T. Li, L. Hu, Tin anode for sodium-ion batteries using natural wood fiber as a mechanical buffer and electrolyte reservoir. Nano Lett. **13**, 3093–3100 (2013). https://doi.org/10.1021/nl400998t
18. H. Zhu, S. Zhu, Z. Jia, S. Parvinian, Y. Li, O. Vaaland, L. Hu, T. Li, Anomalous scaling law of strength and toughness of cellulose nanopaper. PNAS **112**, 8971–8976 (2015). https://doi.org/10.1073/pnas.1502870112
19. J.F. Revol, H. Bradford, J. Giasson, R.H. Marchessault, D.G. Gray, Helicoidal self-ordering of cellulose microfibrils in aqueous suspension. Int. J. Biol. Macromol. **14**, 170–172 (1992). https://doi.org/10.1016/S0141-8130(05)80008-X
20. A. Dufresne, J. Cavaillé, M.R. Vignon, Mechanical behavior of sheets prepared from sugar beet cellulose microfibrils. J. Appl. Polym. Sci. **64**, 1185–1194 (1997). https://doi.org/10.1002/(sici)1097-4628(19970509)64:6%3c1185::aid-app19%3e3.3.co;2-2
21. T. Taniguchi, K. Okamura, New films produced from microfibrillated natural fibres. Polym. Int. **47**, 291–294 (1998). https://doi.org/10.1002/(SICI)1097-0126(199811)47:3%3c291::AID-PI11%3e3.0.CO;2-1
22. F. Tanaka, T. Iwata, Estimation of the elastic modulus of cellulose crystal by molecular mechanics simulation. Cellulose **13**, 509–517 (2006). https://doi.org/10.1007/s10570-006-9068-x
23. A. Šturcová, G.R. Davies, S.J. Eichhorn, Elastic modulus and stress-transfer properties of tunicate cellulose whiskers. Biomacromol **6**, 1055–1061 (2005). https://doi.org/10.1021/bm049291k
24. S.J. Eichhorn, G.R. Davies, Modelling the crystalline deformation of native and regenerated cellulose. Cellulose **13**, 291–307 (2006). https://doi.org/10.1007/s10570-006-9046-3
25. I. Diddens, B. Murphy, M. Krisch, M. Muller, Anisotropic elastic properties of cellulose measured using inelastic X-ray scattering. Macromolecules **41**, 9755–9759 (2008)
26. S. Iwamoto, W. Kai, A. Isogai, T. Iwata, Elastic modulus of single cellulose microfibrils from tunicate measured by atomic force microscopy. Biomacromol **10**, 2571–2576 (2009). https://doi.org/10.1021/bm900520n
27. Z. Fang, H. Zhu, Y. Yuan, D. Ha, S. Zhu, C. Preston, J. Huang, L. Hu, Novel nanostructured paper with ultrahigh transparency and ultrahigh haze for solar cells. Nano Lett. **14**, 765–773 (2014). https://doi.org/10.1021/nl404101p
28. B. Wang, J.G. Torres-Rendon, J. Yu, Y. Zhang, A. Walther, Aligned bioinspired cellulose nanocrystal-based nanocomposites with synergetic mechanical properties and improved hygromechanical performance. ACS Appl. Mater. Interfaces **7**, 4595–4607 (2015). https://doi.org/10.1021/am507726t
29. H. Sehaqui, M. Allais, Q. Zhou, L.A. Berglund, Wood cellulose biocomposites with fibrous structures at micro- and nanoscale. Compos. Sci. Technol. **71**, 382–387 (2011). https://doi.org/10.1016/j.compscitech.2010.12.007
30. A.J. Benítez, J. Torres-Rendon, M. Poutanen, A. Walther, Humidity and multiscale structure govern mechanical properties and deformation modes in films of native cellulose nanofibrils. Biomacromol **14**, 4497–4506 (2013). https://doi.org/10.1021/bm401451m
31. M. Henriksson, L.A. Berglund, P. Isaksson, T. Lindström, T. Nishino, Cellulose nanopaper structures of high toughness. Biomacromol **9**, 1579–1585 (2008). https://doi.org/10.1021/bm800038n
32. Q. Meng, T.J. Wang, Mechanics of strong and tough cellulose nanopaper. Appl. Mech. Rev. **71**, 040801 (2019). https://doi.org/10.1115/1.4044018

33. Z. Wan, C. Chen, T. Meng, M. Mojtaba, Y. Teng, Q. Feng, D. Li, Multifunctional wet-spun filaments through robust nanocellulose networks wrapping to single-walled carbon nanotubes. ACS Appl. Mater. Interfaces **11**, 42808–42817 (2019). https://doi.org/10.1021/acsami.9b15153
34. X. Wang, P. Wu, Fluorinated carbon nanotube/nanofibrillated cellulose composite film with enhanced toughness, superior thermal conductivity, and electrical insulation. ACS Appl. Mater. Interfaces **10**, 34311–34321 (2018). https://doi.org/10.1021/acsami.8b12565
35. D. Ye, X. Lei, T. Li, Q. Cheng, C. Chang, L. Hu, L. Zhang, Ultrahigh tough, super clear, and highly anisotropic nanofiber-structured regenerated cellulose films. ACS Nano **13**, 4843–4853 (2019). https://doi.org/10.1021/acsnano.9b02081
36. K. Li, P. Wei, J. Huang, D. Xu, Y. Zhong, L. Hu, L. Zhang, J. Cai, Mechanically strong shape-memory and solvent-resistant double-network polyurethane/nanoporous cellulose gel nanocomposites. ACS Sustain. Chem. Eng. **7**, 15974–15982 (2019). https://doi.org/10.1021/acssuschemeng.9b02341
37. C. Cai, Z. Wei, X. Wang, C. Mei, Y. Fu, W.H. Zhong, Novel double-networked polyurethane composites with multi-stimuli responsive functionalities. J. Mater. Chem. A **6**, 17457–17472 (2018). https://doi.org/10.1039/c8ta05969a
38. X. Cao, H. Dong, C.M. Li, New nanocomposite materials reinforced with flax cellulose nanocrystals in waterborne polyurethane. Biomacromol **8**, 899–904 (2007). https://doi.org/10.1021/bm0610368
39. X. Yao, X. Qi, Y. He, D. Tan, F. Chen, Q. Fu, Simultaneous reinforcing and toughening of polyurethane via grafting on the surface of microfibrillated cellulose. ACS Appl. Mater. Interfaces **6**, 2497–2507 (2014). https://doi.org/10.1021/am4056694
40. P. Yu, H. He, Y. Luo, D. Jia, A. Dufresne, Reinforcement of natural rubber: the use of in situ regenerated cellulose from alkaline-urea-aqueous system. Macromolecules **50**, 7211–7221 (2017). https://doi.org/10.1021/acs.macromol.7b01663
41. Y. Li, H. Zhu, S. Zhu, J. Wan, Z. Liu, O. Vaaland, S. Lacey, Z. Fang, H. Dai, T. Li, L. Hu, Hybridizing wood cellulose and graphene oxide toward high-performance fibers. NPG Asia Mater. **7**, e150 (2015). https://doi.org/10.1038/am.2014.111
42. D.C. Marcano, D.V. Kosynkin, J.M. Berlin, A. Sinitskii, Z. Sun, A. Slesarev, L.B. Alemany, W. Lu, J.M. Tour, Improved synthesis of graphene oxide. ACS Nano **4**, 4806 (2010)
43. S. Wang, F. Jiang, X. Xu, Y. Kuang, K. Fu, E. Hitz, L. Hu, Super-strong, super-stiff macrofibers with aligned, long bacterial cellulose nanofibers. Adv. Mater. **29**, 1702498 (2017). https://doi.org/10.1002/adma.201702498
44. N. Mittal, F. Ansari, V. Krishne Gowda, C. Brouzet, P. Chen, P.T. Larsson, S.V. Roth, F. Lundell, L. Wa, N.A. Kotov, L.D. Soderberg, Multiscale control of nanocellulose assembly : transferring remarkable nanoscale fibril mechanics to macroscale fibers. ACS Nano **12**, 6378–6388 (2018). https://doi.org/10.1021/acsnano.8b01084
45. S. Wang, T. Li, C. Chen, W. Kong, S. Zhu, J. Dai, A.J. Diaz, E. Hitz, S.D. Solares, T. Li, L. Hu, Transparent, anisotropic biofilm with aligned bacterial cellulose nanofibers. Adv. Funct. Mater. **28**, 1707491 (2018). https://doi.org/10.1002/adfm.201707491
46. M. Zhu, Y. Wang, S. Zhu, L. Xu, C. Jia, J. Dai, J. Song, Y. Yao, Y. Wang, Y. Li, D. Henderson, W. Luo, H. Li, M.L. Minus, T. Li, L. Hu, Anisotropic, transparent films with aligned cellulose nanofibers. Adv. Mater. **29**, 1606284 (2017). https://doi.org/10.1002/adma.201606284
47. J. Song, C. Chen, S. Zhu, M. Zhu, J. Dai, U. Ray, Y. Li, Y. Kuang, Y. Li, N. Quispe, Y. Yao, A. Gong, U.H. Leiste, H.A. Bruck, J.Y. Zhu, A. Vellore, H. Li, M.L. Minus, Z. Jia, A. Martini, T. Li, L. Hu, Processing bulk natural wood into a high-performance structural material. Nature **554**, 224–228 (2018). https://doi.org/10.1038/nature25476
48. C.M. Clarkson, S. El Awad Azrak, E. Forti, G.T. Schueneman, R.J. Moon, J.P. Youngblood, Progress report on recent developments in cellulose nanomaterial composites. Adv. Mater. 2000718 (2020). https://doi.org/10.1002/adma.202000718R1
49. Y. Zhou, C. Chen, S. Zhu, C. Sui, C. Wang, Y. Kuang, U. Ray, D. Liu, A. Brozena, U.H. Leiste, N. Quispe, H. Gui, A. Vellore, H.A. Bruck, A. Martini, B. Foster, J. Lou, T. Li, L. Hu, A. Printed, Recyclable, ultra-strong and ultra-tough graphite structural material. Mater. Today **30**, 17–25 (2019)

50. S. Geng, K. Yao, Q. Zhou, K. Oksman, High-strength, high-toughness aligned polymer-based nanocomposite reinforced with ultralow weight fraction of functionalized nanocellulose. Biomacromol **19**, 4075–4083 (2018). https://doi.org/10.1021/acs.biomac.8b01086
51. M.J. Lundahl, A.G. Cunha, E. Rojo, A.C. Papageorgiou, L. Rautkari, J.C. Arboleda, O.J. Rojas, Strength and water interactions of cellulose I filaments wet-spun from cellulose nanofibril hydrogels. Sci. Rep. **30695**, 1–13 (2016). https://doi.org/10.1038/srep30695
52. R. Sinko, S. Keten, Effect of moisture on the traction-separation behavior of cellulose nanocrystal interfaces. Appl. Phys. Lett. **105** (2014). https://doi.org/10.1063/1.4904708
53. A.G. Cunha, M. Lundahl, M.F. Ansari, L. Johansson, J.M. Campbell, O.J. Rojas, Surface structuring and water interactions of nanocellulose filaments modified with organosilanes toward wearable materials. ACS Appl. Nano Mater. **1**, 5279–5288 (2018). https://doi.org/10.1021/acsanm.8b01268
54. S. Plimpton, Fast parallel algorithms for short-range molecular dynamics. J. Comput. Phys. **117**, 1–19 (1995)
55. X. Wu, R.J. Moon, A. Martini, Tensile strength of Iβ crystalline cellulose predicted by molecular dynamics simulation. Cellulose **21**, 2233–2245 (2014). https://doi.org/10.1007/s10570-014-0325-0
56. X. Wu, R.J. Moon, A. Martini, Crystalline cellulose elastic modulus predicted by atomistic models of uniform deformation and nanoscale indentation. Cellulose **20**, 43–55 (2013). https://doi.org/10.1007/s10570-012-9823-0
57. R.M.A. Domingues, M.E. Gomes, R.L. Reis, The potential of cellulose nanocrystals in tissue engineering strategies. Biomacromol **15**, 2327–2346 (2014). https://doi.org/10.1021/bm500524s
58. T.R. Mattsson, J.M.D. Lane, K.R. Cochrane, M.P. Desjarlais, A.P. Thompson, F. Pierce, G.S. Grest, First-principles and classical molecular dynamics simulation of shocked polymers. Phys. Rev. B **81**, 054103 (2010). https://doi.org/10.1103/PhysRevB.81.054103
59. R. Sinko, S. Mishra, L. Ruiz, N. Brandis, S. Keten, Dimensions of biological cellulose nanocrystals maximize fracture strength. ACS Macro Lett. **3**, 64–69 (2014). https://doi.org/10.1021/mz400471y
60. R. Sinko, S. Keten, Traction–separation laws and stick–slip shear phenomenon of interfaces between cellulose nanocrystals. J. Mech. Phys. Solids **78**, 526–539 (2015). https://doi.org/10.1016/j.jmps.2015.02.012
61. W. Xia, X. Qin, Y. Zhang, R. Sinko, S. Keten, Achieving enhanced interfacial adhesion and dispersion in cellulose nanocomposites via amorphous interfaces. Macromolecules **51**, 10304–10311 (2018). https://doi.org/10.1021/acs.macromol.8b02243
62. U. Ray, Z. Pang, T. Li, Mechanics of cellulose nanopaper using a scalable coarse-grained modeling scheme. Cellulose **28**, 3359–3372 (2021). https://doi.org/10.1007/s10570-021-03740-x
63. M. Shishehbor, P.D. Zavattieri, Effects of interface properties on the mechanical properties of bio-inspired cellulose nanocrystal (CNC)-based materials. J. Mech. Phys. Solids **124**, 871–896 (2019). https://doi.org/10.1016/j.jmps.2018.12.002
64. X. Wang, Z. Pang, C. Chen, Q. Xia, Y. Zhou, S. Jing, R. Wang, U. Ray, W. Gan, C. Li, G. Chen, B. Foster, T. Li, L. Hu, All-natural, degradable, rolled-up straws based on cellulose micro- and nano-hybrid fibers. Adv. Funct. Mater. **30**, 1–9 (2020). https://doi.org/10.1002/adfm.201910417
65. Q. Liu, Z. Lu, Z. Hu, J. Li, Finite element analysis on tensile behaviour of 3D random fibrous materials: model description and meso-level approach. Mater. Sci. Eng. A **587**, 36–45 (2013). https://doi.org/10.1016/j.msea.2013.07.087
66. J. Persson, P. Isaksson, A particle-based method for mechanical analyses of planar fiber-based materials. Int. J. Numer. Methods Eng. **93**, 1216–1234 (2013). https://doi.org/10.1002/nme.4428
67. R. Mao, S. Goutianos, W. Tu, N. Meng, S. Chen, T. Peijs, Modelling the elastic properties of cellulose nanopaper. Mater. Des. **126**, 183–189 (2017). https://doi.org/10.1016/j.matdes.2017.04.050

68. M. Shishehbor, F.L. Dri, R.J. Moon, P.D. Zavattieri, A continuum-based structural modeling approach for cellulose nanocrystals (CNCs). J. Mech. Phys. Solids **111**, 308–332 (2018). https://doi.org/10.1016/j.jmps.2017.11.006
69. Q. Meng, B. Li, T. Li, X.Q. Feng, Effects of nanofiber orientations on the fracture toughness of cellulose nanopaper. Eng. Fract. Mech. **194**, 350–361 (2018). https://doi.org/10.1016/j.engfracmech.2018.03.034
70. Q. Meng, B. Li, T. Li, X. Feng, A multiscale crack-bridging model of cellulose nanopaper. J. Mech. Phys. Solids **103**, 22–39 (2017). https://doi.org/10.1016/j.jmps.2017.03.004
71. W.W.Y. Lau, Evaluating scenarios toward zero plastic pollution. Science (80-.) (2020). https://doi.org/10.1126/science.aba9475
72. J.C. Glenn, E. Florescu, T.M.P. Team, *State of the Future 19.1*, 1st edn. (The Millennium Project, 2017), http://www.millennium-project.org/state-of-the-future-version-19-1/
73. J. Huang, A. Dufresne, N. Lin, *Nanocellulose: From Fundamentals to Advanced Materials* (2019)
74. C. Jia, C. Chen, Y. Kuang, K. Fu, Y. Wang, Y. Yao, S. Kronthal, J. Song, F. Xu, B. Liu, L. Hu, From wood to textiles: top-down assembly of aligned cellulose nanofibers. Adv. Mater. **30**, 1801347 (2018). https://doi.org/10.1002/adma.201801347
75. Y. Li, H. Zhu, Y. Wang, U. Ray, S. Zhu, J. Dai, C. Chen, K. Fu, S.-H. Jang, D. Henderson, T. Li, L. Hu, Cellulose-nanofiber-enabled 3D printing of a carbon-nanotube microfiber network. Small Methods **1**, 1700222 (2017). https://doi.org/10.1002/smtd.201700222
76. H.-L. Gao, R. Zhao, C. Cui, Y.-B. Zhu, S.-M. Chen, Z. Pan, Y.-F. Meng, S.-M. Wen, C. Liu, H.-A. Wu, S.-H. Yu, Bioinspired hierarchical helical nanocomposite macrofibers based on bacterial cellulose nanofibers. Natl. Sci. Rev. **7**, 73–83 (2019). https://doi.org/10.1093/nsr/nwz077
77. H. Zhu, Z. Fang, C. Preston, Y. Li, L. Hu, Transparent paper: fabrications, properties, and device applications. Energy Environ. Sci. **7**, 269–287 (2014). https://doi.org/10.1039/c3ee43024c
78. W. Gan, C. Chen, H.T. Kim, Z. Lin, J. Dai, Z. Dong, Z. Zhou, W. Ping, S. He, S. Xiao, M. Yu, L. Hu, Single-digit-micrometer thickness wood speaker. Nat. Commun. **10**, 1–8 (2019). https://doi.org/10.1038/s41467-019-13053-0
79. Z. Fang, B. Li, Y. Liu, J. Zhu, G. Li, G. Hou, J. Zhou, X. Qiu, Critical role of degree of polymerization of cellulose in super-strong nanocellulose films. Matter **2**, 1000–1014 (2020). https://doi.org/10.1016/j.matt.2020.01.016
80. M. Özkan, M. Borghei, A. Karakoç, O.J. Rojas, J. Paltakari, Films based on crosslinked TEMPO-oxidized cellulose and predictive analysis via machine learning. Sci. Rep. **8**, 1–9 (2018). https://doi.org/10.1038/s41598-018-23114-x
81. M. Özkan, A. Karakoç, M. Borghei, J. Wiklund, O.J. Rojas, J. Paltakari, Machine learning assisted design of tailor-made nanocellulose films: a combination of experimental and computational studies. Polym. Compos. **40**, 4013–4022 (2019). https://doi.org/10.1002/pc.25262
82. C. Chen, Y. Kuang, S. Zhu, I. Burgert, T. Keplinger, A. Gong, T. Li, L. Berglund, S.J. Eichhorn, L. Hu, Structure–property–function relationships of natural and engineered wood. Nat. Rev. Mater. (2020). https://doi.org/10.1038/s41578-020-0195-z
83. Z. Li, C. Chen, R. Mi, W. Gan, J. Dai, M. Jiao, H. Xie, Y. Yao, S. Xiao, L. Hu, A. Strong, Tough, and scalable structural material from fast-growing bamboo. Adv. Mater. **32**, 1–8 (2020). https://doi.org/10.1002/adma.201906308
84. M. Frey, G. Biffi, M. Adobes-vidal, M. Zirkelbach, Y. Wang, K. Tu, A.M. Hirt, K. Masania, I. Burgert, T. Keplinger, Tunable wood by reversible interlocking and bioinspired mechanical gradients. Adv. Sci. **6**, 1802190 (2019). https://doi.org/10.1002/advs.201802190
85. K. Belhaj, A. Chaparro-Garcia, S. Kamoun, N.J. Patron, V. Nekrasov, Editing plant genomes with CRISPR/Cas9. Curr. Opin. Biotechnol. **32**, 76–84 (2015). https://doi.org/10.1016/j.copbio.2014.11.007
86. H. Tang, N. Butchosa, Q. Zhou, A transparent, hazy, and strong macroscopic ribbon of oriented cellulose nanofibrils bearing poly(ethylene glycol). Adv. Mater. **27**, 2070–2076 (2015). https://doi.org/10.1002/adma.201404565

87. W.T. Cao, C. Ma, D.S. Mao, J. Zhang, M.G. Ma, F. Chen, MXene-reinforced cellulose nanofibril inks for 3D-printed smart fibres and textiles. Adv. Funct. Mater. **29**, 1–12 (2019). https://doi.org/10.1002/adfm.201905898
88. D. Klemm, D. Schumann, F. Kramer, N. Heßler, M. Hornung, S. Marsch, Nanocelluloses as innovative polymers in research and application. Adv. Polym. Sci. **205**, 49–96 (2006)
89. B. Natarajan, N. Lachman, T. Lam, D. Jacobs, C. Long, M. Zhao, B.L. Wardle, R. Sharma, J.A. Liddle, The evolution of carbon nanotube network structure in unidirectional nanocomposites resolved by quantitative electron tomography. ACS Nano **9**, 6050–6058 (2015). https://doi.org/10.1021/acsnano.5b01044
90. A. Isogai, Emerging nanocellulose technologies: recent developments. Adv. Mater. 1–10 (2020). https://doi.org/10.1002/adma.202000630

Chapter 6
Light Management of Nanocellulose Films

Zhiqiang Fang, Guanhui Li, Gaoyuan Hou, and Xueqing Qiu

Abstract Nanocellulose films are emerging as a natural and inexpensive optical material for advanced photonic applications such as optical sensors, photovoltaics, displays, and photocatalysis due to their ideal combination of environmental friendliness and unique optical properties (e.g., light transparency, tunable optical haze, selective reflection of left circularly polarized light, and iridescent appearance). The light scattering and light transparency of nanocellulose films and corresponding light management mechanisms are the focus of our chapter. We first introduce the design principles and theories for the light management of nanocellulose films in terms of the nature of light, structural hierarchy of cellulose, optical properties of multiscale cellulose, and structure of the nanocellulose film. Moreover, up-to-date strategies to manipulate their optical properties (light transmittance, transmission haze (light scattering), vivid structural coloration, circularly polarization, and bright whiteness) are highlighted. After that, we demonstrate their potential uses in various value-added photonic fields (e.g., optoelectronics and smart photonic applications) as photonic materials. Finally, some comments on technical and economic problems regarding nanocellulose films for photonic applications and their future development trends are provided.

6.1 Introduction

Nanocellulose is a fundamental building block in every plant and wood cell wall and has emerged as a promising natural raw material candidate for the development of affordable and sustainable photonic materials [1]. Films made primarily of

Z. Fang (✉) · G. Li · G. Hou
State Key Laboratory of Pulp and Paper Engineering, South China University of Technology, Guangzhou 510640, China
e-mail: mszhqfang@scut.edu.cn

X. Qiu
School of Chemical Engineering and Light Industry, Guangdong Provincial Key Laboratory of Plant Resources Biorefinery, Guangdong University of Technology, Panyu District, Guangzhou 510006, China

L. Hu et al. (eds.), *Emerging Nanotechnologies in Nanocellulose*, NanoScience and Technology, https://doi.org/10.1007/978-3-031-14043-3_6

nanocellulose (nanocellulose films) possess specific optical properties, for example, light transparency, tailored optical haze, iridescent appearance, transmittance of right circularly polarized light, and selective reflection of left circularly polarized light. These superior optical properties coupled with other advantages, for example, strong mechanical properties, broad modification capacities, good thermal stability, low thermal coefficient of expansion, inexpensiveness, biocompatibility, biodegradability, and renewability, make nanocellulose film a sustainable optical material in various advanced photonic applications (e.g., photovoltaic devices, photocatalysis, optical sensors, anticounterfeiting, laser, and flexible electronics).

In this chapter, the optical properties (light scattering and transparency) of nanocellulose films and corresponding light management mechanisms are the main focal point. We first introduce the design principles and theories for the light management of nanocellulose films in terms of the nature of light, structural hierarchy of cellulose, optical properties of multiscale cellulose, and structure of the nanocellulose film. Moreover, up-to-date strategies to manipulate their optical properties (light transmittance, transmission haze (light scattering), vivid structural coloration, circularly polarization, and bright whiteness) are highlighted. Afterward, their potential applications in diverse value-added fields (such as optoelectronics and smart photonic applications) as photonic materials are presented. Finally, we provide some comments on technical and economic problems with respect to nanocellulose films for photonic applications and their future development trends.

6.2 Design Principles and Theories

To better understand the design principles and theories for the light management of nanocellulose films, we will briefly introduce the nature of light, the multiscale hierarchical structure of cellulose, as well as the interaction of light with cellulose at the multiscale level and the nanocellulose film. Note that the nanocellulose film mentioned in this chapter is mainly prepared by a bottom-up method using nanocellulose isolated from the cell wall of plants and woods as a raw material.

6.2.1 The Nature of Light

Light is particularly significant for all life on earth and normally refers to electromagnetic radiation that carries energy but has no mass. It shows an extreme broad wavelength ranging from low energy radio waves with wavelengths that are measured in meters to high energy gamma rays (wavelength < approximately 1×10^{-11} m). Unfortunately, only a tiny fraction of the electromagnetic spectrum (visible light) ranging from 400 to 700 nm can be detected by the human eye. Visible light is composed of seven different colors (violet, indigo, blue, green, yellow, orange, and red).

Light behaves both as particles and waves, which is well verified by modifying Young's double slit experiment. Light behaves as waves when it propagates, while it behaves as particles as it leaves or arrives. Therefore, light can be described in terms of a stream of photons with energy. A photon is the minimum quantum of energy that can be transported. Wavelength and frequency are two critical features of light. Wavelength multiple frequency is equal to the speed of light ($c = f \times \lambda$), where c denotes the velocity with an exact value of 299,792,458 m/s, f represents the frequency, and λ designates the wavelength.

When an incident light interacts with matters, it can be reflected, refracted, scattered, transmitted, diffracted, emitted, polarized and/or absorbed, which dominates a wide range of visual appearances of everything around us. Therefore, tailored optical appearance of matters can be achieved through engineering light–matter interactions.

6.2.2 *Structural Hierarchy of Cellulose*

Natural materials have produced a diversity of properties and functions by forming elaborate and complex hierarchical structures that bridge macroscopic properties with atomistic scale components. Here, we briefly introduce the structural hierarchy of cellulose to better understand the design principles for the nanocellulose film with various optical properties.

Cellulose, a linear natural polymer, consists of repeating glucopyranose units that are covalently linked by β-(1, 4) glycosidic bond (Fig. 6.1a). The number of anhydroglucoses determines the length of the cellulose chain that varies with the origin and extraction methods of the cellulose resource. Figure 6.1b displays the structure of multiscale cellulose in wood. Approximately 36 Linear cellulose chains are packed into one semi-crystalline elementary fibril (3.5–5 nm in diameter) with alternating amorphous regions and ordered crystalline that are stabilized via van der Waals forces and hydrogen bonding. Elementary fibril as the smallest building block of natural cellulose is further bundled into large microfibrils (5–50 nm in lateral dimension) that ultimately constitute the layered cell walls of wood fibers with a hollow structure. Wood fibers align parallelly to the axial of the trunk and bind together by a matrix of hemicellulose and lignin to form wood. The hierarchical structure lays the foundation to manipulate the interactions between cellulose and light to a large extent.

6.2.3 *Optical Properties of Multiscale Cellulose*

Figure 6.1a displays the molecular structure of a linear cellulose chain, where each anhydroglucose has three hydroxyl groups located at position C2, C3, and C6, and there are ether, hydroxyl, carbon–carbon, and carbon-hydrogen bonds existing in the cellulose chain. These bonds are unable to adsorb visible light. Therefore, multiscale

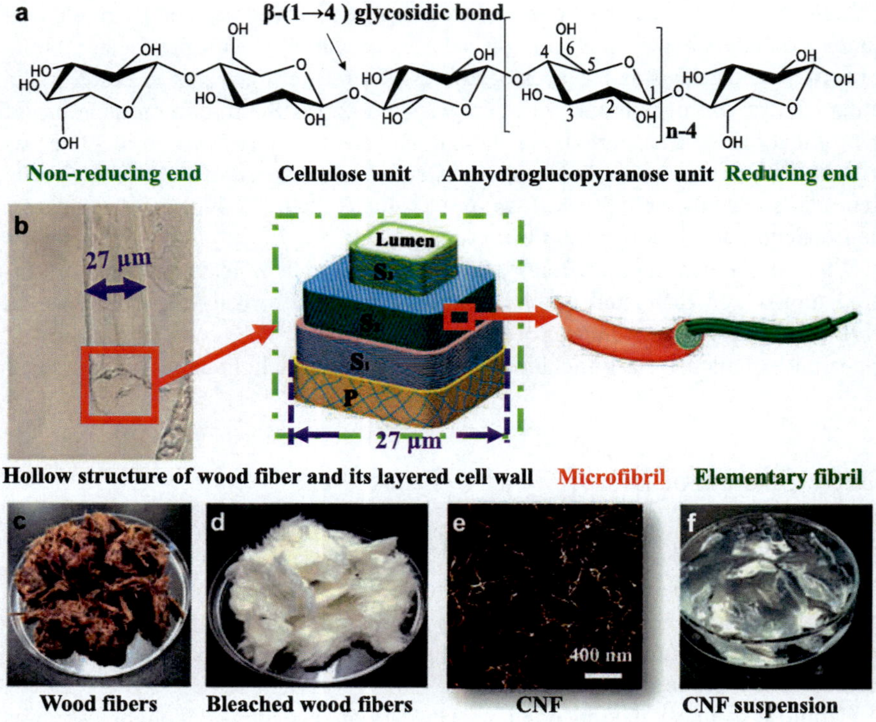

Fig. 6.1 **a** Chemical structure of cellulose. **b** The hierarchical structure of wood cellulose ranging from micrometer scale to nanometer scale. Left part indicating the hollow structure of bleached wood cellulose fiber, central part showing the layered cell wall of wood fiber, right part displaying the schematic of microfibril and elementary fibril and the images of CNFs and their aqueous suspension. Digital images of brown wood cellulose fibers (**c**) and bleached wood fibers (**d**). **e** Atomic force microscope (AFM) image of CNFs. **f** Digital photo of aqueous CNF suspension displaying high optical transparency [2]

cellulose is a non-absorbing media and manifests colorless appearance in theory due to the transmission of incident light. However, micro-sized cellulose fibers (e.g., wood cellulose fibers and cotton) widely used in our daily life manifest an opaqueness. This phenomenon indicates that the physical structure of cellulose also plays a pivotal role in its visual appearance besides the chemical structure.

As shown in Fig. 6.1c, wood fibers disintegrated from chemical pulping usually manifest a brown color due to the existence of light absorbing lignin. After bleaching, wood cellulose fibers become white and opaque (Fig. 6.1d), which is attributed to their hollow structure (left image in Fig. 6.1b) that causes strong light scattering at the interfaces between cellulose and air. Through a combination of carboxymethylation and homogenization, the hollow structure of the wood fiber is destroyed, and its cell wall was disintegrated into cellulose nanofibrils (CNFs) with a width of several nanometers to tens of nanometers (Fig. 6.1e). Given that the size of CNFs is much smaller than the wavelength (400–700 nm) of visible light, the incident light can readily pass through them. Consequently, a transparent CNF suspension in water is obtained (rightmost images in Fig. 6.1f).

Shortly, cellulose is a non-absorbing media because there are no specific groups that can absorb visible light in the molecular structure. Micro-sized cellulose fiber with hollow structure is opaque and white because of strong light scattering between light and cellulose. However, nanocellulose is a colorless nanomaterial due to its nanoscale dimension that allows the visible light to pass through easily.

6.2.4 *Structure of Nanocellulose Film*

In addition to the structural hierarchy of cellulose, the configuration of the nanocellulose film is also a key for designing and manipulating its optical properties. Nanocellulose film is a sheet-like biodegradable material that is composed of CNFs, cellulose nanocrystals (CNCs), bacterial nanocellulose (BNC) or their mixture. Nanocellulose films with completely different optical properties (e.g. selective reflection, high transparency, bright whiteness, and iridescence) can be achieved by rationally modulating structural variations such as the spatial arrangement of nanocellulose, porosity, pore size, thickness, and surface roughness. For instance, by regulating the fibril diameters, nanocellulose can serve as an efficient scattering element for preparing transparent or opaque nanocellulose films. As shown in Fig. 6.2, the nanocellulose film with the finest CNFs exhibits a high optical transparency (Fig. 6.2a) due to the formation of dense structure (Fig. 6.2b, c) that suppresses the light scattering behavior, while the film prepared from coarsest CNFs presents an opaque appearance (Fig. 6.2e) as a result of porous configuration and coarse fibrils (Fig. 6.2f, g) that induces strong multiple light scattering at numerous internal interfaces [3]. Moreover, selective reflection of the nanocellulose film can be achieved by regulating its internal configuration, thus leading to a specific visual appearance [4]. The chiral nematic arrangement of (dimethylmyristylammonio) propanesulfonate (DMAPS):CNC film allows it to be iridescent due to selective reflection (Fig. 6.2h, i).

6.2.5 *Optical Properties of the Nanocellulose Film*

Here, we only describe the optical behaviors of dielectric cellulose materials in the visible spectral region. When light strikes the surface of a nanocellulose film, the interactions of light with the nanocellulose film usually involve absorption, reflection, refraction, and transmission (Fig. 6.3). Strictly speaking, reflection and refraction are a result of scattering that changes the direction of light propagation. These interactions belong to the category of classical geometrical optics and are the fundamental basis of diverse optical functions and devices.

Light adsorption and light scattering are two main physical processes, which determine the visualization of the nanocellulose film. As we mentioned above, the light

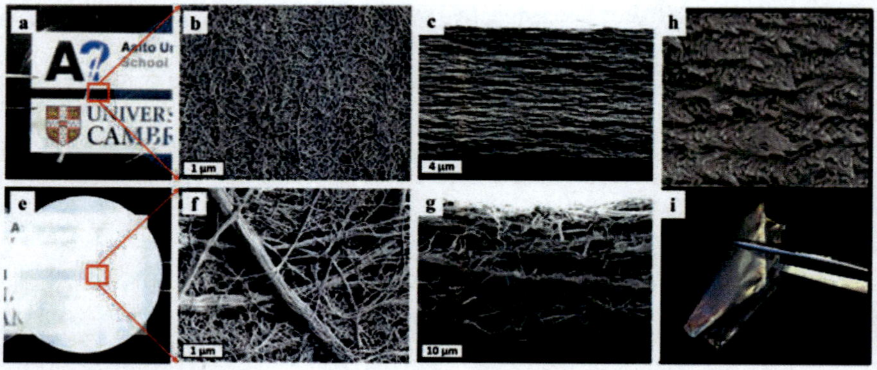

Fig. 6.2 Visual appearance of nanocellulose films with different fibril diameters and corresponding internal structures. Digital photos of **a** transparent and **e** opaque nanocellulose films. Top-view SEM images showing the structures of **b** transparent and **f** opaque nanocellulose films. SEM images indicating cross-sectional fracture surfaces of **c** transparent and **g** opaque nanocellulose films [3]. **h** Cross-sectional SEM image depicting chiral nematic configuration of DMAPS:CNC film. **i** Digital photo of iridescent DMAPS:CNC film [4]

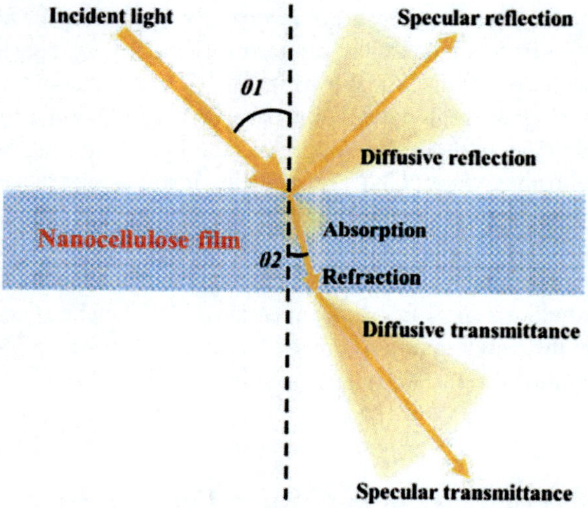

Fig. 6.3 Light-nanocellulose film interactions: light reflection (specular reflection and diffusive reflection), light absorption, light refraction, as well as light transmission (diffusive transmission and specular transmission)

absorption of cellulose is negligible if color components are absent. Therefore, in this chapter, we will discuss the light transmission and light scattering of nanocellulose films in the visible range.

6.2.5.1 Light Transmittance

The transmitted light refers to total light transmittance (T_{total}) that can be generally divided into specular transmittance ($T_{specular}$) and diffusive transmittance ($T_{diffusive}$) according to the propagation direction of the transmitted light (Fig. 6.3). $T_{specular}$ is the light intensity that transmits along the same direction as the incident light wave, while $T_{diffusive}$ refers to the light flux collected at angles deviated from the incident light by an integrating sphere. Their relationship can be described by the following equation: $T_{total} = T_{specular} + T_{diffusive}$.

As a light wave impinges on the surface of a nanocellulose film, the incident light is partially reflected, and the residual light will pass through it. The percentage of transmitted light is highly dependent on the structure of the nanocellulose film. For a dense configuration, the nanocellulose film presents a transparent appearance because most of the incident light can transmit through it. While it possesses a porous structure, light reflection will play a dominant role, thus leading to an opaque visualization.

6.2.5.2 Light Scattering

Light scattering is a physical process because of the interaction of light with the electric field of the scattering particles and can induce the change in the direction of light propagation from an ordered way to a random one, which makes a crucial contribution to the visual appearance of matters. The categories of light scattering and their relationship are shown in Fig. 6.4.

Essentially, three types of scattering take places during light–matter interactions: (1) elastic scattering (the wavelength or frequency of the scattered light is the same as the incident light); (2) inelastic scattering (the emitted radiation has a wavelength or frequency different from that of the incident radiation); (3) quasi-elastic scattering (the wavelength or frequency of the scattered light shifts).

Due to the dielectric properties of the nanocellulose, light scattering of the nanocellulose film refers primarily to elastic scattering where no loss or gain of

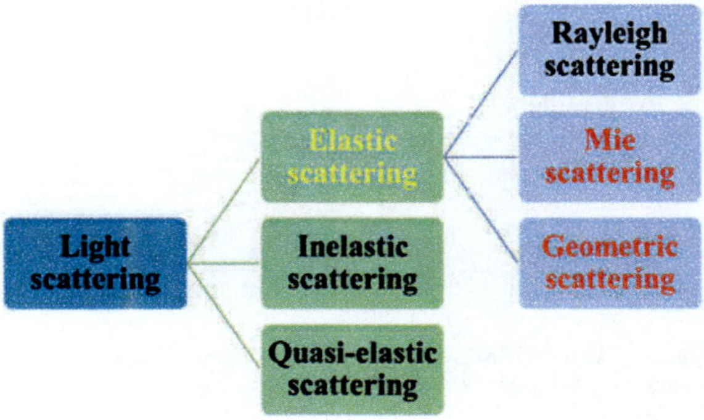

Fig. 6.4 Categories of light scattering and their relationship. The light scattering of the nanocellulose film involves Mie scattering (red) and geometric scattering (red) that belong to elastic scattering (yellow)

light energy (momentum is preserved, no energy loss) occurred. Generally, there are three types of elastic scattering: Rayleigh scattering, Mie scattering, and geometric scattering.

The interaction of light with particles less than the one-tenth wavelength of the incident light leads to Rayleigh scattering and the scattered radiation occurs in all directions uniformly (Fig. 6.5a). The particles may be individual molecules or atoms. Rayleigh scattering is wavelength dependent and increases sharply toward shorter wavelength. It is inversely proportional to the fourth power of the wavelength, i.e., proportional to $1/\lambda^4$.

Mie scattering is caused by particles with diameter larger than one-tenth the wavelength of the incident light but lower than or comparable with the wavelength of the incident light, and the particles have a refractive index different from that of the continuous phase in which they are suspended. Consequently, the resulting scattered radiation is quite random. As shown in Fig. 6.5b, the intensity of forward scattering is much stronger than that of backward scattering. Geometric scattering (also known as nonselective scattering) occurs when the particles are much larger than the wavelength of the radiation (Fig. 6.1c). The larger the scattering cross-section of the particles, the stronger the intensity of forward scattering.

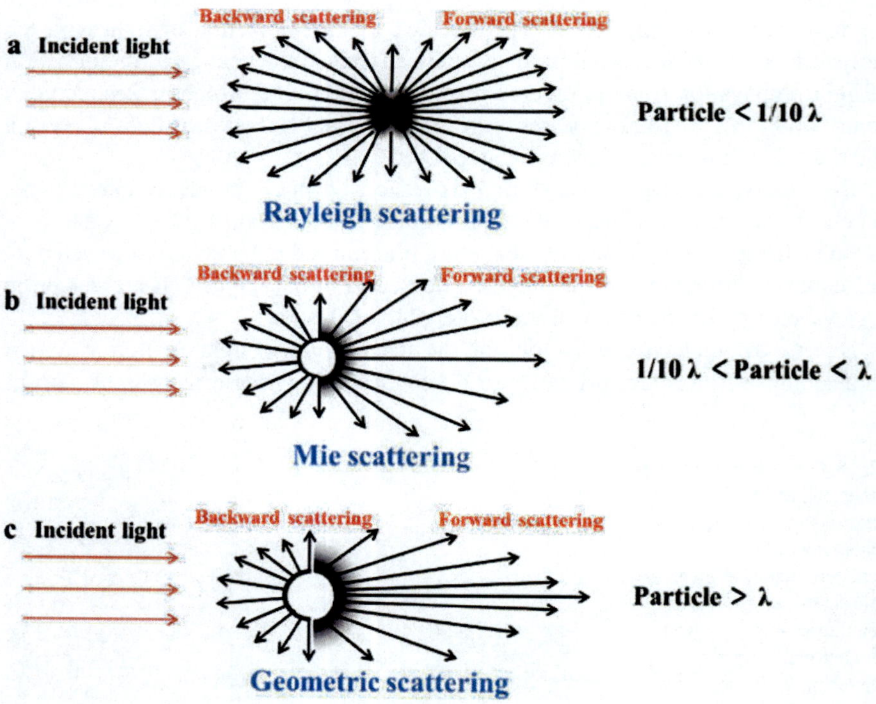

Fig. 6.5 Three types of light scattering (forward scattering and backward scattering): Rayleigh scattering, Mie scattering, and geometric scattering

In short, given that Rayleigh scattering is most prominently observed in gases, we will not discuss the Rayleigh scattering of the nanocellulose film in this chapter. The Mie scattering and Geometric scattering will be our focal point owing to their essential role in the optical properties of the nanocellulose film.

Light transport in a nanocellulose film occurs in several different ways [5], thus allowing a wide range of optical appearance to be observed, for example, from transparency to whiteness, and from haze to clarity. Taking pure CNF film as an example, it can be categorized into transparent and hazy film, transparent and clear film, and opaque and hazy film based on light transmittance and transmission haze (forward light scattering).

When a light encounters a transparent and clear nanocellulose film (Fig. 6.6a), it allows most of the light to be transmitted, with a small part of the light being reflected from the interfaces due to the refractive index contrast between the film and the external medium (e.g., air). Most of the intensity is transmitted in the same direction of the incident light and only a tiny fraction of the transmitted light is scattered due to its homogeneous and dense structure.

Conversely, transparent and hazy nanocellulose film scatters light not only at its interfaces but also within its bulk structure (Fig. 6.6b). Large amount of the transmitted light is scattered in the forward direction, thus leading to a strong forward light scattering behavior (high transmission haze), and only a small part of the intensity propagates through the film in the same direction of the incident light. The amount of transmitted light scattered by the transparent and hazy film is highly dependent on the interplay between the individual scatter properties (e.g. scattering cross-section, aspect ratio, refractive index) and overall properties (e.g. filling fraction, average orientation of the scatters, and structure factor).

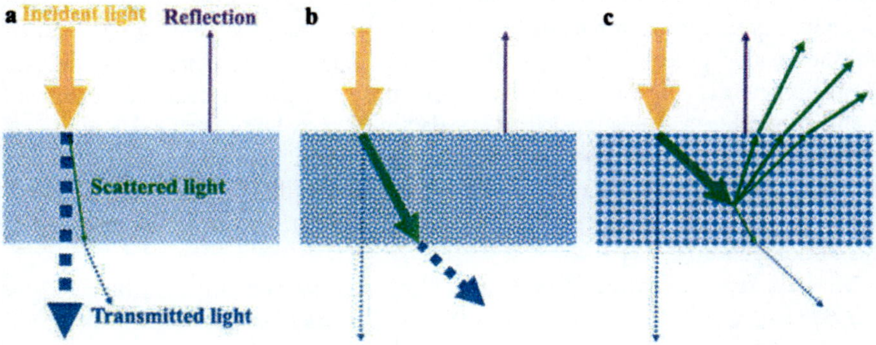

Fig. 6.6 Schematic showing light propagation in various CNF films. **a** Transparent and clear CNF film: most of the incident light directly passes through the CNF film without changing its transport direction. **b** Transparent and hazy CNF film: most of the incident light propagates through the film but vast majority of transmitted light deviates from its original trajectory. **c** Opaque and haze CNF film: light transport in this film undergoes anomalous diffusive behavior. Only a tiny fraction of the incident light can pass through this film, and most light is scattered in a direction that opposes the trajectory of the incident beam

The light propagation in an opaque and hazy nanocellulose film is displayed in Fig. 6.6c. Most of the incident light is elastically scattered multiple times owing to the large refractive index contrast resulting from the microscale nonuniformities (e.g., microscale cellulose crystals, fibril bundles, bubbles, and density fluctuations) within the disordered nanocellulose film. Both surface reflection and backward light scattering contribute to the opaqueness of the nanocellulose film. Furthermore, Mie and/or geometric scattering occurred within heterogeneous nanocellulose film renders its high optical haze (strong forward light scattering), although the light transmittance is quite low. Consequently, the nanocellulose film with refractive index inhomogeneities exhibited a bright whiteness.

6.3 Current Strategies in Light Management of Nanocellulose Film

6.3.1 Light Transmission

Although natural cellulose is a non-absorbing material, the ubiquitous common paper with randomly distributed microscale cellulose fibers manifests an opaque appearance with an optical transparency of less than 50% (Fig. 6.7a). This phenomenon is ascribed not only to severe light scattering at the interfaces between cellulose and air in the bulk of common paper but also to light scattering (surface reflection) at its rough surface (Fig. 6.7b, c). To improve light transmission, tremendous strategies have been employed to minimize refractive index inhomogeneities in the bulk of common paper and/or to reduce its surface roughness: (1) filling its porous structure with transparent materials (e.g., oils, resins [6], molecular cellulose [7], gums, vanishes, nanocellulose [8], and paraffin) with a refractive index similar to cellulose; (2) fibrillating the cell wall of microscale cellulose fibers by mechanical treatments (such as refining and beating [9]); (3) selectively dissolving surface cellulose of microscale cellulose fibers by chemical treatments (e.g., ionic liquid [10, 11], and sulfuric acid), which fills the pores among the paper; and (4) surface modifications of microscale cellulose fibers by chemical methods (e.g., carboxymethylation [12] and carboxylation [2, 13]).

Since the report of optically transparent films made of randomly distributed CNFs in 2009 [14], nanocellulose has been widely used as a raw material to prepare highly transparent films. Thanks to the nanoscale dimension and the rich hydroxyl groups of nanocellulose, the nanocellulose film showed a dense packing nanostructure and a nanoscale surface roughness, thereby significantly suppressing light scattering inside the film that allowed the incident light to be transmitted. Consequently, the nanocellulose film exhibited an optical transparence of ~90%. For example, randomly distributed CNFs are readily inclined to self-assemble into mesoporous CNF films with dense nanostructure during water evaporation (Fig. 6.7e, f). The dense packing structure leads to the high homogeneity of CNF films in terms of refractive index,

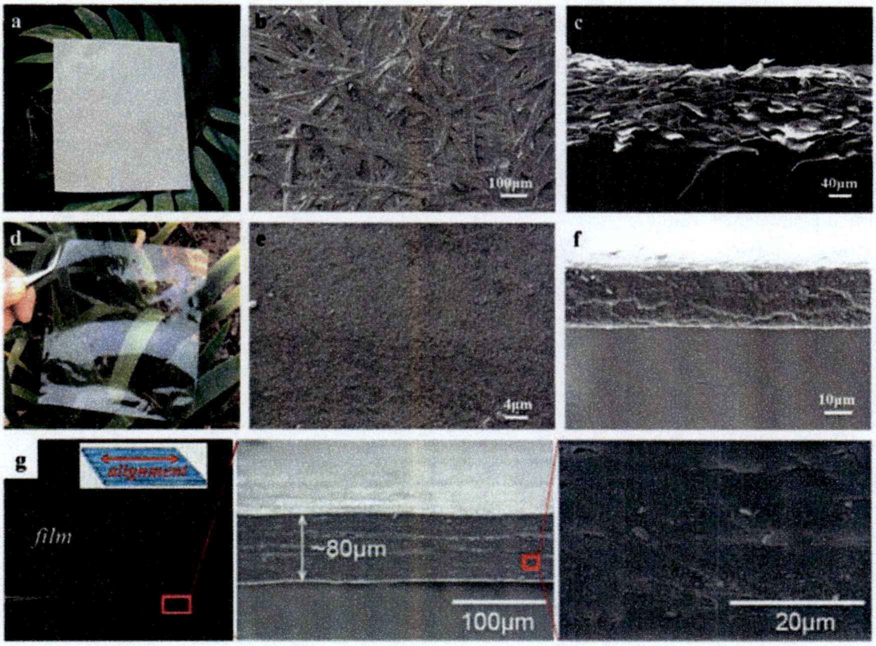

Fig. 6.7 **a** Digital image of common paper and its top-view (**b**) and cross-sectional (**c**) SEM images. **d** Digital photo of isotropic transparent CNF film and its top-view (**e**) and cross-sectional (**f**) SEM images. **h** Digital photo of anisotropic transparent CNF film and its cross-sectional SEM images [18]

thereby significantly suppressing the light scattering in the bulk that renders direct light propagation, with a tiny part of the transmitted light being scattered (as shown in Fig. 6.6a). Consequently, the isotropic CNF film displays a highly transparent and clear appearance (Fig. 6.7d). In addition, surface roughness has a negative influence on the optical transparency of the CNF film [15]. CNF films prepared by casting presented a much smoother surface than those films produced by vacuum filtration, thereby facilitating the transmission of light.

The preparation of isotropic nanocellulose film by a bottom-up method is an energy- and time-consuming procedure owing to the extra-low dewatering of nanocellulose dispersion in water, which tremendously restricts its industrial production and subsequent commercial applications. To address the above challenge, Hu group proposed a top-down approach to prepare anisotropic transparent CNF films directly from wood in a facile and efficient manner [16–18]. The wood block was first delignified by a variety of chemical treatments, followed by hot mechanical pressing of the delignified wood block to densify the porous structure of the delignified wood block. The as-prepared anisotropic transparent CNF film exhibited 90% transparency and a tensile strength of approximately 350 MPa that was much higher than the isotropic transparent CNF film [18]. The superior optical and mechanical properties were ascribed to the dense and well-aligned structure of the CNF film resulting from mechanical pressing (Fig. 6.7g), thereby facilitating light propagation and the formation of hydrogen bonding. However, transparent anisotropic CNF films produced from above top-down method normally demonstrated a relatively high transmission haze of over 60%. Hu's group attempted to reduce the optical haze of anisotropic CNF films to a record-value of 10% while retaining its high transparency by infiltrating transparent polymers into porous delignified wood slices [16, 17].

6.3.2 *Transmission Haze*

Haze represents the visual clarity through a material and can be qualitatively evaluated by the degree of cloudiness observed [19]. Normally, there are two different kinds of haze occurring in materials: reflection haze and transmission haze. Reflection haze occurs when light is reflected from a material, which is associated with the gloss on the material surface. A high gloss surface can lead to a clear reflection. The reflection can appear hazy if there are imperfections existing in the surface of a material. Transmission haze occurs when light passes through a film, which is defined as the ratio of transmitted light scattered by a film that deviates from the incident trajectory at an angle larger than 2.5° to total light transmission. Transparent materials with transmission haze present a milky or fuzzy appearance. The haze can be experimentally measured by a UV–vis spectrometer with an integrated sphere and the corresponding schematic setup of the haze measurement is shown in Fig. 6.8 [20]. The haze value is calculated by the following equation: $\text{Tansmissionhaze}(\%) = \left(\frac{T_4}{T_2} - \frac{T_3}{T_1}\right) \times 100\%$.

Generally, the nanocellulose film exhibits an optically transparent appearance and a nanoscale surface roughness. In this chapter, the transmission haze of the highly transparent nanocellulose film will be the focal point. When the incident light passes through a transparent nanocellulose film, as shown in Fig. 6.9a, the total transmittance is categorized into diffuse transmittance and direct transmittance. For the diffuse transmittance that refers to transmitted light scattered by the nanocellulose film, it also consists of wide-angle scattering (at an angle larger than 2.5° from normal) of light and narrow-angle scattering (at an angle less than 2.5° from normal) of light. Both wide-angle scattering and narrow-angle scattering refer to forward light scattering (Fig. 6.9b). The transmission haze of the nanocellulose film is a result of wide-angle scattering of light. While the narrow angle scattering of the transmitted light is responsible for the clarity of the nanocellulose film (Fig. 6.9c).

Although the hazy visualization of the nanocellulose film was reported in 2009 [14], Cui group initially proposed the concept of optical haze for nanocellulose films and pointed out that this specific optical property may extend their uses to solar cells as an advanced light management layer [21]. Since then, tremendous efforts have been made to manipulate the transmission haze of transparent nanocellulose films by fine-tuning of fibril morphologies and/or film structures (such as packing density and surface roughness).

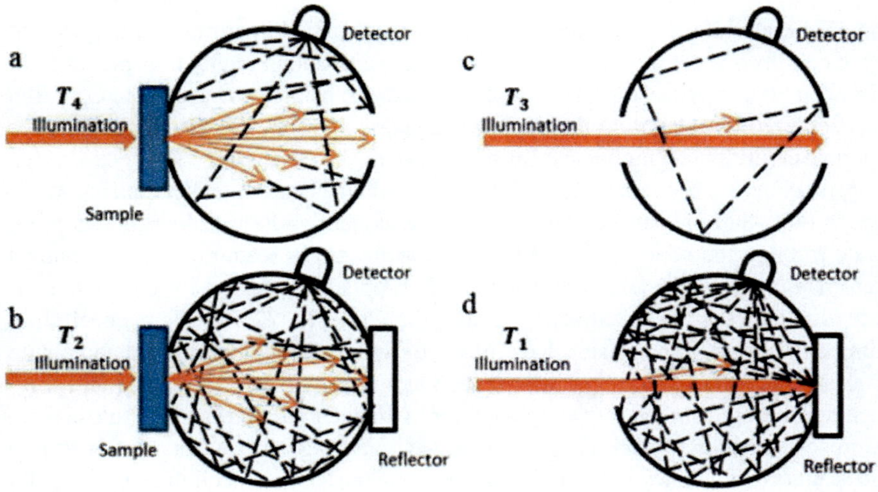

Fig. 6.8 Schematics showing the mechanism and setup for the haze measurement with a UV–vis spectrometer with an integrating sphere. **a** Diffusive transmittance, **b** total transmittance, **c** and **d** are the corrections for the measurement setup [20]

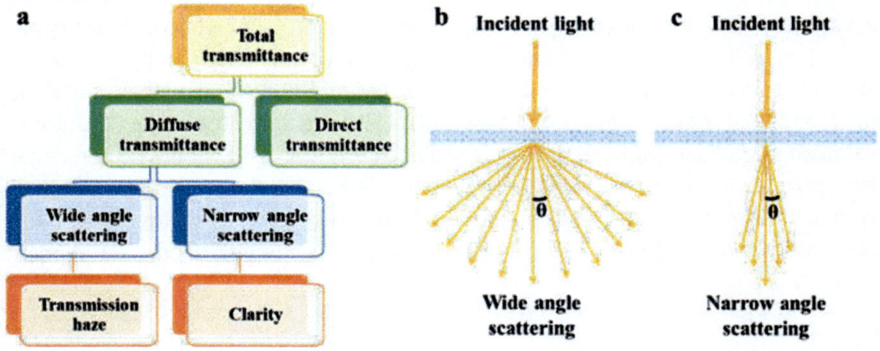

Fig. 6.9 **a** Flowchart showing the category of transparency involving diffuse transmittance and direct transmittance. Diffuse transmittance can be divided into wide angle scattering (transmission haze) and narrow angle scattering (clarity). Schematic indicating the **b** wide angle scattering (at an angle of more than 2.5°) and **c** narrow angle scattering (at an angle of less than 2.5°) of transmitted light

Hu et al. varied the optical properties of the CNF film through rationally tuning the fibril diameter and packing density [22]. When the fibril diameter was less than 50 nm, the optical haze of the CNF film increased with increasing diameter of CNF, while its optical transmittance was well retained. However, at the decrease of packing density, the transmission haze of the CNF film increased, but its optical transmittance showed a declining trend. The incorporation of CNC in the CNF film could also reduce its optical haze while maintaining high transparency [23]. In contrast, adding micro-sized cellulose fibers into CNF dispersion enabled to obtain highly transparent CNF film with high transmission haze [24, 25]. Tunable transmission haze of the CNF film was achieved by regulating the surface roughness as well [15]. The higher surface roughness results in a higher transmission haze of the CNF film.

Nanocellulose has a diameter of 3.5–30 nm, which is much smaller than the wavelength (400–800 nm) of the visible light. Consequently, the visible light can propagate through the nanocellulose film with narrow angle scattering, and an optical transparency but low transmission haze is theoretically observed for the nanocellulose film. However, as reported by previous literature [15, 22, 24–26], nanocellulose films exhibited a transmission haze of up to 86.6%. This abnormal phenomenon is ascribed to the scattering particles such as micro-sized fibril bundles and fragments resulting from incompletely mechanical disintegration of nanocellulose from microscale cellulose fibers. These micro-sized scattering particles increased not only the scattering efficiency but also the surface roughness of nanocellulose film, finally resulting in a high transmission haze of nanocellulose film. Therefore, the ability to control the content of scattering particles is of extremely importance for the tailored transmission haze of nanocellulose film. The higher the content of scattering particles, the higher the transmission haze of nanocellulose film.

In conclusion, nanocellulose film normally exhibited a transparent and clear appearance through removing scattering particles from nanocellulose suspension. In contrast, a transparent and hazy nanocellulose film was obtained by increasing the content of scattering micro-sized particles. The addition of microscale cellulose fibers [8, 24, 27] or cellulose microparticles [25, 28] is an effective route to improve its transmission haze while retaining high light transmittance. It should be noted that micro-sized cellulose fibers were the ideal raw material for the preparation of transparent cellulose film with high transmission haze when taking their low production cost, simple extraction procedure, and strong scattering strength in considerations.

6.3.3 Structural Color

After billions of years' evolution, living organisms have evolved biological materials with a diversity of properties and functions by assembling limited base substances (e.g., proteins, polysaccharides, and minerals) into large varieties of complex structures [29, 30]. For instance, the vibrant structural coloration in a variety of living organisms is generated from coherent or incoherent scattering of light on natural photonic nanostructures made of their constituent components rather than absorption from a pigment.

As an emerging type of natural photonic crystal, CNCs, have recently considered a promising biological building block to produce photonic films with brilliant structural color due to their ability to self-assemble into helicoidal arrangement, nonbleaching, scalability, biocompatibility, and biodegradability, which may offer remarkable potential to substitute traditional, potentially toxic colorants widely used in cosmetics, arts, sensing, security labeling and so on [19, 31, 32]. In comparison to conventional pigmentary color, the structural coloration of photonic CNC films not only shows superior stability under cyclic stimulation, but also exhibits stimuli-responsive feature across the entire visible spectrum.

Thanks to the intrinsic birefringence of crystalline cellulose along the chain axis, CNCs exhibit birefringent optical properties and their intrinsic birefringence can be transmitted to macroscopic CNC materials with ordered arrangements. More interestingly, CNCs in water are inclined to self-assemble into a chiral nematic structure (helical arrangement) that mimics the configuration of chitin widely found in the exoskeletons of arthropods [33, 34]. The inherent birefringence of CNCs along with their ability to self-assemble into a chiral nematic structure enables to produce photonic CNC films with tailored optical properties.

The ability to self-assemble into a liquid crystalline phase of CNC suspension is highly dependent on the balance between orientation-dependent repulsive and attractive interactions. Both short-range hard rod steric repulsions and long-range repulsions of either steric or electrostatic nature are responsible for the repulsive interactions, while attractive interactions are a result of Van der Waals forces. Notably, the surface charge of CNCs has a dominant influence on long-range repulsion.

With the increase in concentration upon the gradual removal of solvent, a colloidal CNC suspension undergoes phase transition from an isotropic phase to a cholesteric phase (as shown in Fig. 6.10a). The charged CNC suspension with low concentration displays an isotropic phase because of the random distribution of the nanorods (Fig. 6.10b). As the CNC suspension reaches a threshold concentration where a phase transition is observed and the randomly distributed CNCs are inclined to locally assemble into the anisotropic phase owing to a compromise between the translational and rotational entropies of individual CNCs. Consequently, both anisotropic and isotropic phases are observed in the intermediate region, and the higher concentrations favor local mutual orientation between individual CNCs, thus leading to a higher ordering of cholesteric liquid crystalline phase with a lower pitch.

Upon further slow solvent evaporation under controlled conditions, a fully cholesteric phase (helicoidal nanostructure) is achieved, and the ordered arrangement inherits the birefringence of the CNCs that presents a cholesteric fingerprint pattern (Fig. 6.10c) between crossed polarizers because of the light propagation at directions normal to the helical axis (**m**). The periodicity of the fingerprint pattern is related to half of the cholesteric pitch, which is typically in the micron range.

Due to the interaction of light with a birefringent cholesteric phase, the cholesteric pitch of the suspension along with the polarization, wavelength, and propagation direction of the incident light significantly determine the optical phenomena of the CNC suspension. When light passes through the **m** axis, the spatial regulation of the refractive index renders the reflection of strongly circularly polarized light for a certain interval of wavelength. The peak wavelength reflected by the cholesteric phase can be given a good approximation by the below equation:

$$\lambda = n_{av} * P * \cos\theta \tag{6.1}$$

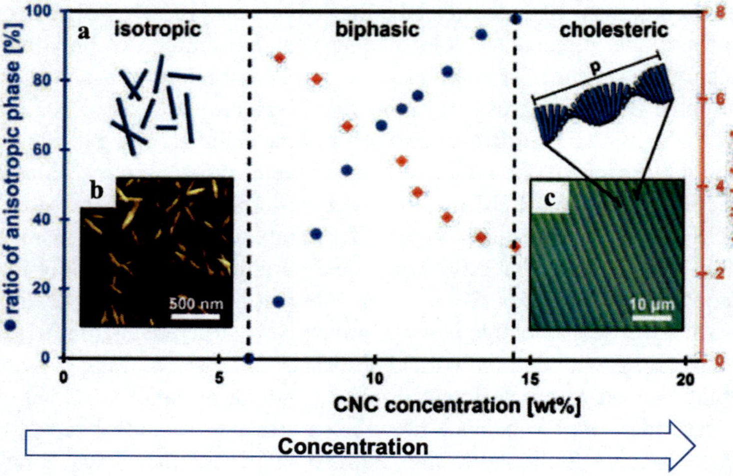

Fig. 6.10 **a** Diagram displaying the first order phase transition of CNC suspension from an isotropic phase to an anisotropic phase (blue dots) with increasing CNC concentration and the corresponding equilibrium pitch (red diamonds). **b** AFM image showing the fibril morphology of CNCs. **c** Optical microscopic photo of a fingerprint pattern of the cholesteric phase under polarized light [31]

where n_{av} is equal to $(n_e + n_o)/2$, n_e and n_o refers to the extraordinary and ordinary optical indices of the phase, respectively. P is the pitch of the helicoid and θ represents the angle of the incident light with respect to **m** inside the CNC suspension. Note that the cholesteric domains in the CNC suspension cannot reflect visible light because their pitch is in the micron range. Fortunately, visible light can still be diffracted in transmission by cholesteric domains oriented almost perpendicular to the incoming light, thus enabling the precise calculation of the average P across samples based on (6.1).

Further increase in the concentration of the cholesteric phase enables the CNC suspension to cross a point of kinetic arrest prior to the formation of the solid film, thereby dramatically restricting the ability to relax and rearrange mutual position and orientation of individual CNCs due to their tightly packed helicoidal nanostructure. From this point onward, the evaporation of the residual water can produce mechanical constraint on the formed gel, where a fully cholesteric phase in CNC suspension is ultimately translated into a chiral nematic ordering nanostructure in the as-prepared solid film.

Due to the decrease in cholesteric pitch changing from micro-sized range in the suspension to submicron range in the solid CNC film, the as-prepared CNC film presents vivid structural coloration in reflection over the visible spectrum owing to cholesteric Bragg reflection (Fig. 6.11a) [35]. As shown in Fig. 6.11b, typical Bouligand arches are observed in the transverse SEM image of the CNC film. Additionally, the optical response of the solid CNC film with chiral nematic ordering arrangement still locally obeys the angular dependence described in (6.1). However, correction

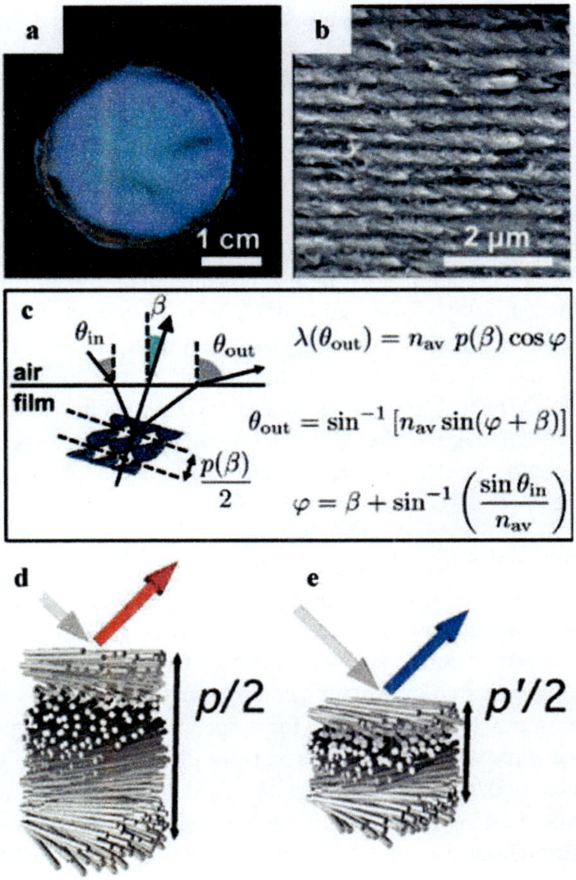

Fig. 6.11 A solid CNC film displaying a brilliant structural color (**a**) and the corresponding cross-sectional SEM image showing its characteristic helicoidal arrangement (**b**). **c** Fergason's law describing the interaction of light with tilted cholesteric domains. **d** and **e** Schematic showing selective reflection of light by regulating the pitch [41]

for Snell's law is required considering the film–air interface, which can be described by Fergason's law (Fig. 6.11c).

The final visual appearance of a CNC film is strongly dependent on many parameters, for example, P [36], the orientation of the helical axis (m), viewing angle, average refractive index, additives (e.g., poly(ethylene glycol) (PEG) [37], poly(vinyl alcohol) (PVA) [38], and surfactants) [4, 39], pH, ionic strength, drying conditions and geometry [40], as well as the polarization, wavelength, and propagation direction of the incident light. For instance, pitch is intrinsically related to the final visible appearance of the CNC-based film and can be modulated by controlling the shape (length and aspect ratio) of CNCs and their interactions, as well as self-assembly conditions (e.g., relative humidity, temperature, and initial concentration). Decreasing the pitch leads to a blue-shift (Fig. 6.11e), while a red-shift is obtained by increasing the pitch (Fig. 6.11d) [41]. Additionally, the alignment of nanorods played an essential role in the optical properties of CNC-based materials. Physical methods such as anchoring, hydrodynamic shear, external electric and magnetic fields, and mechanical stress have been developed to control over the alignment of CNCs across hierarchical length scales: from nanoscale, microscale, to macroscale.

Due to the nonuniform orientation of the helical axis **m** resulting from the broad random distribution of locally cholesteric domains inside the film, pure CNC film typically displays a brilliant, vivid rainbow-like appearance (structure coloration) under polarized light. To achieve tailored optical properties of solid CNC films, many approaches such as external fields, geometrical confinement, and co-assembly were proposed to regulate the pitch and for the orientation of the cholesteric domains [4]. For example, by taking advantage of the intrinsic anisotropic diamagnetic susceptibility of the individual CNCs, Vigolini and coworkers adopted commercial neodymium magnets to control the photonic properties of the CNC film by regulating the orientation of the cholesteric domains in suspension during the self-assembly of CNCs [42]. Three types of designed magnetic environments (no magnetic field, vertical magnetic field, and tilted magnetic field) were separately utilized to control over the final visible appearance of the as-prepared solid films obtained from the identical CNC suspension (Fig. 6.12a–c). As shown in Fig. 6.12d–f, the as-prepared CNC films dried under different magnetic environments demonstrate dramatical differences in optical responses.

The CNC film assembled in a magnetic-free environment has a broad angular wavelength response resulting from a random distribution of helical axis of cholesteric domains (polydomain structure) inside the film (Fig. 6.12g, j). The CNC films prepared in magnetic environments show enhanced intensity of reflected light, however, this only works for a specific range of angles. As shown in Fig. 6.12e, the film, which was prepared from the self-assembly of an aqueous CNC suspension under a vertical magnetic field, presents a strong specular reflection (blue color) owing to the well alignment of the cholesteric domains with uniform pitch throughout the film, with m oriented normal to the film surface (Fig. 6.12h). The cross-sectional SEM image (Fig. 6.12k) further confirms the monodomain structure of the obtained solid film. For the CNC film dried in a tilted magnetic field, a homogeneous cholesteric monodomain structure with a clear tilt with respect to the film–air interface is observed in Fig. 6.12i, l, which enables to explain its unusual optical angular response.

Besides, the structural colors of photonic CNC films can be stabilized and/or enhanced using additives (such as functional polymers, small molecules, and other nanomaterials) and/or optimizing the surface properties and morphology of CNCs [43]. Hydrophilic polymers such as PEG and organic surfactants were added into the CNC film to simultaneously improve its structural color and mechanical flexibility [44]. Yao et al. presented large and flexible photonic CNC/PEG nanocomposite films with uniform structural colorations from blue to red by combining the changes in their weight ratios with the optimization of the surface charge of CNCs [45]. More significantly, the photonic film made of 80 wt% optimized CNC and 20% PEG showed a reversible and smooth structural color shifting from green to transparent upon the increase of relative humidity from 50 to 100%. In a similar method, Zwitterionic surfactants were used by Guidetti et al. to increase the mechanical flexibility of the photonic CNC film while retaining its structural color.

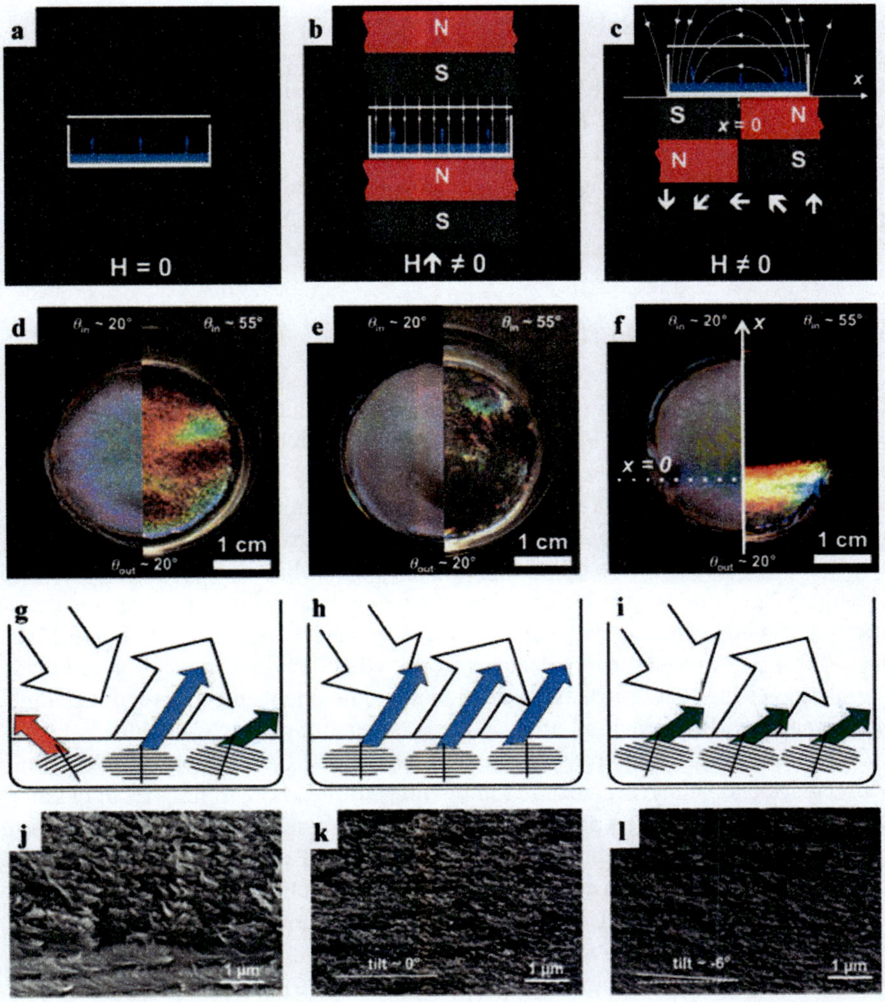

Fig. 6.12 Schematics of three types of magnetic fields for the drying of three identical CNC suspensions: **a** no magnetic field, **b** vertical magnetic field, and **c** tilted magnetic field. **d–f** Digital images showing the visual appearance of the corresponding CNF films obtained from casting under three different magnetic environments. Note that the images were captured from the same observation angle (θ_{out} ~ 20°) using a diffuse white light from either the same angle (θ_{in} ~ 20°, i.e. specular) or different angle (θ_{in} ~ 55°) in order to visualize the cholesteric domains of different orientation. **g–i** Schematic mechanism indicating the orientation of the cholesteric domains responsible for the visual appearance of the corresponding solid CNC films. Note that the black solid lines symbolize the helical axis of the cholesteric domain and the white arrows refer to the incident visible light and specular reflection. **j–l** Cross-sectional SEM images of the corresponding solid CNC films [42]

In short, randomly distributed CNC photonic crystals in the solvent can gradually translate into a fully cholesteric phase upon slow solvent evaporation. Moreover, the helicoidal nanostructure can be well retained in the solid CNC film, which leads to a brilliant, vivid rainbow-like structure coloration. The visual appearance of the as-prepared CNC film could be enhanced by controlling the pitch and orientation of the cholesteric domains using a variety of physical techniques (e.g., external fields, geometrical confinement, and co-assembly) or by using functional materials.

6.3.4 Circularly Polarization

The chiral nematic CNC film behaves as a 1D photonic crystal and presents an inherent capability for circular polarization of light owing to its left-handed (L) helical nanostructure and photonic bandgap, thus having the potential to extend its uses in optical components [46]. When natural light (unpolarized light) propagates by a chiral nematic photonic CNC film, L-circularly polarized light (CPL) is obtained by selective reflection while right-handed (R) CPL is achieved by transmission. As a result, both L- and R- passive CPL are attained on both sides of the film and their intensity can be evaluated by asymmetry factor $|g|$ values. According to the (6.1) in Sect. 6.3.3, the wavelength of selectively reflected L-CPL is as a function of θ and P, thus suggesting that broadband reflection could be obtained because of different pitches and a wide distribution of helical axis orientations in the CNC film.

Xu group exploited the passive circular polarization of the photonic CNC film with left-handed helical organization in the range of near-UV to near-IR to passive L-CPL, thereby rendering a photonic bandgap (PBG)-based CPL with viewing-side-dependent handedness, tunable wavelength by varying PBG, and high $|g|$ values up to 0.87 [47]. The photonic film could transform the incident light by selective reflection to passive right-handed CPL by transmission. Furthermore, the addition of achiral luminophores ranging from molecular to nanometer level into photonic CNC films enabled the conversion of spontaneous emission to R-CPL emission with $|g|$ values up to 0.68.

Generally, the photonic CNC film enables the reflection of left-handed CPL and the transmission of right-handed CPL. Inspired by photonic structures in the cuticle of *Plusiotis resplendens*, photonic CNC films were developed by simply inserting a layer of nematic liquid crystal between two left-handed cholesteric layers with varying pitch values in the thickness direction, thus leading to simultaneous reflection of both L-CPL and R-CPL that could be modulated by temperature variation and an external electric field [48]. Similarly, Wu et al. presented a flexible photonic CNC-based film with sandwich structure. A thin layer of uniaxially oriented polyamine 6, acting as a half-wave retarder, was sandwiched between two CNC/polyethylene glycol diacrylate (PEGDA) layers, thereby enabling simultaneous reflection of L-CPL and R-CPL over the visible spectrum and showing an increased reflectance intensity of up to 77% [49]. Polarization-independent, tunable reflective liquid crystal CNC-based film with sandwich structure, featuring a reflection of approaching 100%

[50], was fabricated by inserting a birefringent half-wave layer of nematic CNC between two reflective chiral nematic CNC-organosilica composite films, which may find applications in security features, solar-gain-regulating technologies, and color filters.

In brief, photonic CNC films demonstrated the intrinsically circular polarization for photonic applications across the visible spectrum, thus enabling their applications in polarizers, optical filters, and coatings.

6.3.5 Bright Whiteness

In comparison to bright and iridescent colorations, achieving a brilliant white appearance is more challenging owing to the same reflection efficiency for broadband visible spectrum [51]. Whiteness is a result of multiple scattering of light in disordered media [52], and the scattering can be measured by transport mean free path (l_t), the smaller l_t renders higher scattering efficiency and less material use to achieve the same whiteness. In spite of the bright whiteness, it is generally realized with thick random systems which multiply the scatter light [51]; nanocellulose has been considered an efficient scattering element for the preparation of various thin and white membranes with minimum optical transmission due to its inherent polydispersity and anisotropy. Progress has been made to prepare thin and white membranes made solely of CNFs [3] or CNCs [53]. Through rationally structuring their configurations (e.g., filling fraction and the physical size of scattering element) at the nanoscale that maximizes refractive index inhomogeneities and scattering efficiency, a bright whiteness was achieved for nanocellulose membranes (Fig. 6.13).

CNCs not only have long been investigated for structural colors, but also can be used to prepare membrane with brilliant whiteness. For example, a templating technique has been developed to prepare thin and white CNC membrane with a l_t of 3–7 μm by using monodisperse polystyrene (PS) spheres with a diameter of 1.27 μm as a template (Fig. 6.13a) [53]. The PS spheres were added into CNC suspension to prepare nanocomposite film by a combination of casting and slow evaporation, and a chemical etching was then applied to remove these PS spherical particles randomly distributed in the dried film. Consequently, a porous structure was obtained for the resulting CNC membrane (Fig. 6.13b), thus improving the light–CNC interactions and finally led to increased scattering efficiency.

In addition to the white CNC membrane, Vignolini and co-workers presented a mechanically stable, thin, and white CNF membrane (Fig. 6.13c) with anomalous diffusion assisted brightness through rationally regulating the fibril morphologies and the distribution of CNFs and/or optimizing the porosity of intertwined CNF network [3]. A sequential centrifugation process was applied to prepare CNFs with different fibril diameters. The as-prepared coarsest CNFs were adopted to prepare a wet gel cake by vacuum filtration, followed by a solvent exchange step. Finally, the gel cake was dried at room temperature to prepare a white membrane with only 9 μm thickness. As we can see from the top-view image in Fig. 6.13d, both coarse CNFs

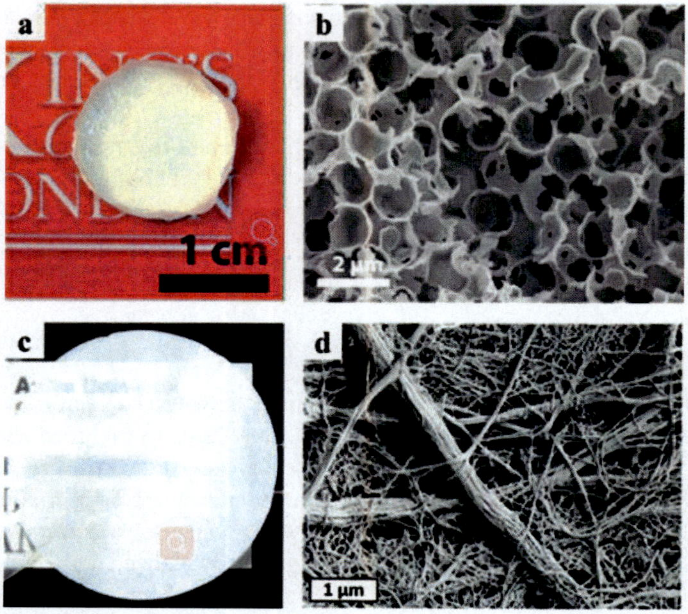

Fig. 6.13 Optical appearance of white membranes made of CNC (**a**) [53] and coarsest CNFs (**c**) [3], respectively, and their corresponding internal structures (**b**, **d**)

and porous structure are observed, which are responsible for the enhanced scattering efficiency. The resulting CNF membrane showed a broadband reflectivity of 60–80% due to its porous configuration, which was much higher than the filter paper (160 µm thickness) that can only reflect approximately 50% of the incident light. Moreover, a white wood with well-aligned CNFs was prepared for future energy-saving and sustainable building applications through a procedure of combining delignification with mechanical densification of natural wood [54]. This white wood exhibited a passive radiative cooling performance due to its ability to backscatter solar radiation and emit strongly in mid-infrared spectrum, thus leading to continuous ambient cooling during both day and night.

In sum, nanocellulose is a promising green nanomaterial for preparing the thin and white membrane with superior scattering efficiency. By designing the structure of the membrane on the scale of the wavelength of light, the transport of light can be controlled to achieve a few micrometers thick membrane with optimal scattering strength. The exploration of nanocellulose for brilliant whiteness may extend its potential uses in solar cells, displays, personal care, paints, and so on.

6.4 Potential Applications of Nanocellulose Film

Many excellent works have reported on the uses of nanocellulose films as a mechanical supporting substrate [19, 32, 55–57] in flexible electronics (e.g., transistors [28, 58–62], solar cells [63–66], and organic light emitting diodes (OLEDs)) [67, 68]. In this part, we focus on the potential high-end applications of the nanocellulose film with tailored optical properties as a functional layer and/or smart component.

6.4.1 *Optoelectronics*

Highly transparent nanocellulose films with high transmission haze can serve as a light management layer for optoelectronic applications, especially for second generation thin film photovoltaics, to enhance device efficiency. Hu et al. first proposed the possibility of applying transparent CNF films with haze in solar cells to enhance their power conversion efficiency by improving light trapping effect [21]. Xu et al. initially reported a CNF film with high transmittance (90.2%) and high optical haze (46.5%) for polysilicon solar cells as a light management layer. As a result, the solar cells attached with the CNF film showed a maximum PCE of 11.48%.

In addition, as we mentioned in Sect. 6.3.2, the haze of the nanocellulose film was ascribed to the cellulose scattering microparticles derived from the incomplete disintegration of the cell wall of the microscale cellulose fibers. To further improve the transmission haze, microscale wood fibers were added into CNFs to improve the scattering strength of the final film. Wu and co-workers presented a high transparent (~90%) CNF/wood fiber film with a haze of up to 78% which can be applied on the surface of the glass substrate for the OLED device as a light diffuser. As shown in Fig. 6.14a, a translucent appearance was observed for the glass substrates attached with the obtained transparent films with different thicknesses. A green OLED device was then successfully fabricated on the glass substrate with a CNF/wood fiber film (Fig. 6.14b), which indicated an enhanced illuminous efficacy of up to 15% due to improved light outcoupling efficiency resulting from the CNF/wood fiber film. Besides, the transparent CNF/wood fiber film with high haze was combined with carbon nanotubes to prepare electrode for the touch screen [8]. The paper touchscreen presents an anti-glare effect when exposed to sunlight (Fig. 6.14c). However, the touchscreen with the PET electrode demonstrated great glaring effects and the underneath letters were fuzzy (Fig. 6.14d).

In fact, transparent cellulose films with microscale cellulose fibers manifested a higher transmission haze than the nanocellulose film, and many efforts have been contributed to preparing transparent and hazy films using micro-sized cellulose fibers as a light scattering source and its applications in optoelectronics [2, 6, 10, 69–72]. Additionally, a transparent wood substrate with a high transmission haze has recently been developed for optoelectronic applications due to its simple and energy-saving fabrication process [71–73]. As the scope of our chapter focuses on using

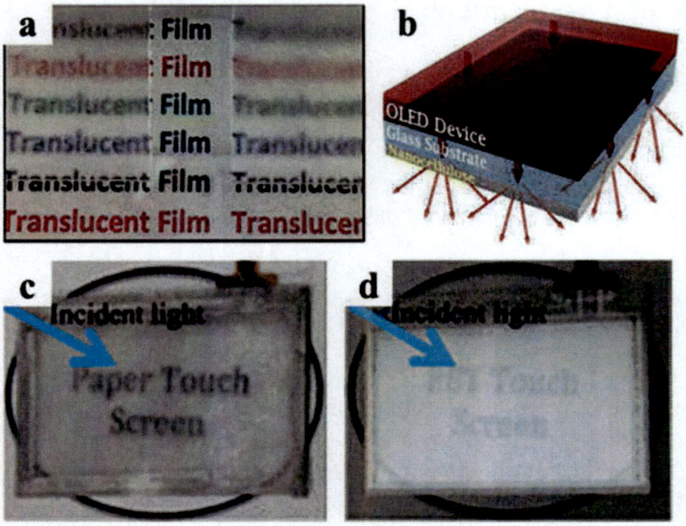

Fig. 6.14 **a** Digital images of translucent glasses attached with CNF/wood fiber films with varying thickness, **b** Schematic structure of green OLED fabricate on glass substrate attached with transparent CNF/wood fiber film with high haze [27]. **c** Touch screen with transparent and hazy CNF/wood fiber film electrode showing anti-glare effect. **d** Touch screen with PET electrode indicating great glare [8]

nanocellulose as a raw material to prepare a transparent and hazy film, we will not discuss it in this section.

6.4.2 *Smart Photonic Applications*

The nanocellulose film also exhibits specific optical properties such as plasmonic or photoluminescent properties and eye-readable and stimuli-responsive structural coloration, which has the potential to apply in colorimetric sensors, anticounterfeiting application, reflective liquid crystal displays, and photonic skins [74].

Zhang and co-workers presented a novel near-infrared (NIR) and visible dual emissive transparent CNF film with high level security codes for security purposes, which had advantages of NIR/visible spectra, nanosecond/microsecond lifetime, and color tuning from blue to yellow [75]. Yb^{3+} doped carbon quantum dots (CQD) showing enhanced fluorescence have been successfully incorporated into the CNF film, where CQD simultaneously acts as visible fluorescence "emitter" and antenna sensitizing Yb^{3+} ions with NIR luminescence. The surface passivation of Yb^{3+}-CQDs with CNFs indicated an eight time improvement in the intensity of NIR luminescence.

Additionally, a BC-based film with plasmonic or photoluminescent properties was fabricated for sensing applications by incorporating various functional nanoparticles

(NPs) such as metal NPs, quantum dots (QDs), and up-conversion NPs into BC [76]. For instance, plasmonic properties can be achieved by the addition of silver and gold NPs into the BC film while the photoluminescent BC film was obtained by conjugating it with CdSe@Zn QDs or amino-functionalized $NaYF_4$:Yb^{3+}@Er^{3+}&SiO_2 up-conversion NPs. The plasmonic and photoluminescent intensity could be regulated by the dosage of graphene oxide. These BC film-based optical sensing platforms have the potential to prove valuable for displaying analytical information in various areas (e.g., diagnostics, environmental monitoring, and food safety).

Photonic CNC films with/without additives (e.g., water-soluble polymers, small molecule, and salts) manifested stimuli-responsive and eye-readable coloration in response to various environmental stimuli (water, solvents, gas, temperature, stress), thereby showing their potential uses in advanced fields such as photonic papers, data security, smart packaging, and sensors) as photonic sensitive functional materials based on structural color [19, 43]. A rewritable photonic CNC/waterborne polyurethane (WPU) composite film with rapid optical response to water and moisture was demonstrated for display application [77]. Through altering the types of stimuli (water, high concentration NaCl solution, and low concentration NaCl solution), temporary, durable, and disguisable patterns were obtained for the photonic CNC/WPU composite film.

Through functionalizing CNCs with amine groups [78] or assembling fluorescent molecules [79] on the surface of CNC, photonic CNC-based films with sensitive responses to small molecules such as toxic formaldehyde can be obtained for sensing uses. Hou and co-workers designed a fluorescent CNC-based film to detect and recognize trace levels of formaldehyde in the moisture conditions by assembling naphthalimide molecules on CNCs. The appearance, fluorescence, and color of the as-prepared CNC-based film were in supersensitive multiresponses to small molecules. More significantly, the dynamic changes in structural color and fluorescence are eye-readable [79].

The chiral nematic CNC-based film with mechanochromic behavior could work as a stress-responsive sensor for immediately visualizing mechanical stresses and monitoring the structural health such as bridges and buildings [80]. As shown in Fig. 6.15a, the CNC elastomer (CNC-E) shows a dynamic change in color from white to blue to yellow to pink to green in response to mechanical stretching when viewed under crossed polarizers. The color change is ascribed to a change in birefringence owing to the reorientation of CNCs in the composite film. As the stretching of CNC-E, the chiral nematic arrangement of CNCs unwinds into a pseudo-nematic structure (Fig. 6.15b). Interestingly, this phenomenon was reversible, and the vivid color disappeared after removing the applied stress.

The photonic CNC film with chiral nematic organization was combined with a hydrogen-bonded network of glucose-polyacrylamide to prepare self-reporting, photonic soft bio-adhesives with specific triple-coupling properties for colorimetric wound healing process monitoring [81]. Photonic bio-adhesives with strong adhesion indicates reddish/transparent appearance as it tightly contacts with wounded biological surfaces. During the healing process, the color of the bio-adhesive will shift from reddish color to bluish color due to the drying of skin when the wound

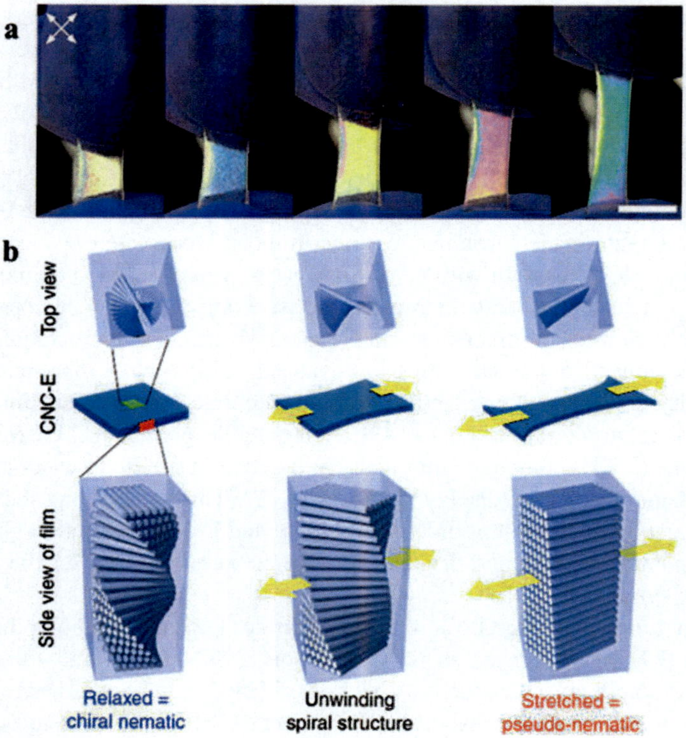

Fig. 6.15 **a** Color change of CNC-E upon mechanical stretching as viewed under crossed polarizers. **b** Schematic illustration of unwinding a chiral nematic structure of CNCs in composite film into a pseudo-nematic arrangement [80]

is healed. Furthermore, bio-adhesives can serve as a colorimetric monitor to detect remote water aerosol transfer on respiratory masks. Bio-adhesive stickers with blue color can be firmly attached to hydrophobic surfaces of N-95 masks, which can be applied to trace the external contamination (e.g., aerogels or droplets containing virus) on a spot. After exposing to spraying water from 1 m distance, the sticker presented a reddish color within 30 s, suggesting that the contaminated mask should be immediately sterilized or disposed.

6.5 Conclusions and Perspective

In conclusion, we have focused on the optical properties (light scattering and transparency) of nanocellulose films and the corresponding mechanisms and latest manipulation strategies of light management. Firstly, the design principles and theories for the light management of nanocellulose films were introduced in terms of the nature of light, structural hierarchy of cellulose, optical properties of multiscale cellulose, and structure of the nanocellulose film. Furthermore, rational manipulation of their optical properties (light transmission, transmission haze, vivid structural coloration, circularly polarization, and bright whiteness) by a variety of advanced strategies was highlighted. Finally, we demonstrated the extensive exploitation of nanocellulose films with tailored optical properties as photonic materials in value-added fields (e.g., optoelectronics and smart photonic applications).

Several challenges in fundamental and applied research must be conquered prior to the commercial applications of the nanocellulose film with tailored optical properties, including:

(1) Fine optimization of the structure (e.g., fibril morphologies and crystallinity) and surface chemistry of the nanocellulose prepared from wood and plants by cost-effective and environmentally friendly top-down methods remain challenging.
(2) Large-scale production of nanocellulose films with tunable optical properties has not been prevalent due to their time- and energy-consuming preparation process, especially for chiral nematic nanocellulose film with brilliant structural colors, CPL reflection or whiteness.
(3) Good weather durability for the nanocellulose film with specific optical properties under harsh indoor/outdoor conditions and thermal stability under high temperature are desirable but remain challenging.
(4) More efforts should be devoted in understanding the underlying mechanism for the light management of the nanocellulose film with specific optical properties.

The light management of the nanocellulose film enables it to be used in novel fields such as optoelectronics, anticounterfeiting applications, colorimetric sensors, and energy-efficient buildings. This booming area is still in its infant stage and more optical properties would be exploited for the nanocellulose film with rapid progress in multi-disciplinary research topics.

References

1. T. Li, C.J. Chen, A.H. Brozena, J.Y. Zhu, L. Xu, C. Driemeier, J.Q. Dai, O.J. Rojas, A. Isogai, L. Wågberg, L.B. Hu, Developing fibrillated cellulose as a sustainable technological material. Nature **590**(7844), 47–56 (2021)
2. Z.Q. Fang, H.L. Zhu, Y.B. Yuan, D.H. Ha, S. Zhu, C. Preston, Q.X. Chen, Y.Y. Li, X.G. Han, S.W. Lee, G. Chen, T. Li, J. Munday, J.S. Huang, L.B. Hu, Novel nanostructured paper with ultrahigh transparency and ultrahigh haze for solar cells. Nano Lett. **14**(2), 765–773 (2014)
3. M.S. Toivonen, O.D. Onelli, G. Jacucci, V. Lovikka, O.J. Rojas, O. Ikkala, S. Vignolini, Anomalous-diffusion-assisted brightness in white cellulose nanofibril membranes. Adv. Mater. **30**(16), E1704050 (2018)
4. G. Guidetti, S. Atifi, S. Vignolini, W.Y. Hamad, Flexible photonic cellulose nanocrystal films. Adv. Mater. **28**(45), 10042–10047 (2016)
5. G. Jacucci, L. Schertel, Y.T. Zhang, H. Yang, S. Vignolini, Light management with natural materials: from whiteness to transparency. Adv. Mater. **33**(28), 2001215 (2020)
6. Y.G. Yao, J.S. Tao, J.H. Zou, B. Zhang, T. Li, J.Q. Dai, M.W. Zhu, S. Wang, K.K. Fu, D. Henderson, E. Hitz, J.B. Peng, L.B. Hu, Light management in plastic-paper hybrid substrate towards high-performance optoelectronics. Energy Environ. Sci. **9**(7), 2278–2285 (2016)

7. W. Hu, G. Chen, Y. Liu, Y.Y. Liu, B. Li, Z.Q. Fang, Transparent and hazy all-cellulose composite films with superior mechanical properties. ACS Sustain. Chem. Eng. **6**(5), 6974–6980 (2018)
8. Z.Q. Fang, H.L. Zhu, C. Preston, X.G. Han, Y.Y. Li, S.W. Lee, X.S. Chai, G. Chen, L.B. Hu, Highly transparent and writable wood all-cellulose hybrid nanostructured paper. J. Mater. Chem. C **1**(39), 6191–6197 (2013)
9. Z.Z. Li, W.X. Liu, F.X. Guan, G.D. Li, Z. Song, D.H. Yu, H.L. Wang, H. Liu, Using cellulose fibers to fabricate transparent paper by microfibrillation. Carbohyd. Polym. **214**, 26–33 (2019)
10. H.L. Zhu, Z.Q. Fang, Z. Wang, J.Q. Dai, Y.G. Yao, F. Shen, C. Preston, W.X. Wu, P. Peng, N. Jang, Q.K. Yu, Z.F. Yu, L.B. Hu, Extreme light management in mesoporous wood cellulose paper for optoelectronics. ACS Nano **10**(1), 1369–1377 (2016)
11. T. Nishino, N. Arimoto, All-cellulose composite prepared by selective dissolving of fiber surface. Biomacromolecules **8**(9), 2712–2716 (2007)
12. F.M. Qin, Z.Q. Fang, J. Zhou, C. Sun, K. Chen, Z.X. Ding, G.H. Li, X. Qiu, Efficient removal of Cu^{2+} in water by carboxymethylated cellulose nanofibrils: performance and mechanism. Biomacromolecules **20**(12), 4466–4475 (2019)
13. P.P. Zhou, P.H. Zhu, G. Chen, Y. Liu, Y.D. Kuang, Y.Y. Liu, Z.Q. Fang, A study on the transmission haze and mechanical properties of highly transparent paper with different fiber species. Cellulose **25**(3), 2051–2061 (2018)
14. M. Nogi, S. Iwamoto, A.N. Nakagaito, H. Yano, Optically transparent nanofiber paper. Adv. Mater. **21**(16), 1595–1598 (2009)
15. W.S. Yang, L. Jiao, D.Y. Min, Z.L. Liu, H.Q. Dai, Effects of preparation approaches on optical properties of self-assembled cellulose nanopapers. RSC Adv. **7**(17), 10463–10468 (2017)
16. C. Jia, C.J. Chen, R.Y. Mi, T. Li, J. Dai, Z. Yang, Y. Pei, S.M. He, H.Y. Bian, S.H. Jang, J.Y. Zhu, B. Yang, L.B. Hu, Clear wood toward high-performance building materials. ACS Nano **13**(9), 9993–10001 (2019)
17. R.Y. Mi, T. Li, D. Dalgo, C.J. Chen, Y.D. Kuang, S.M. He, X.P. Zhao, W.Q. Xie, W.T. Gan, J.Y. Zhu, J. Srebric, R.G. Yang, L.B. Hu, A clear, strong, and thermally insulated transparent wood for energy efficient windows. Adv. Funct. Mater. **30**(1), 1907511 (2020)
18. M.W. Zhu, Y.L. Wang, S.Z. Zhu, L.S. Xu, C. Jia, J.Q. Dai, J.W. Song, Y.G. Yao, Y.B. Wang, Y.F. Li, D. Henderson, W. Luo, H. Li, M.L. Minus, T. Li, L.B. Hu, Anisotropic, transparent films with aligned cellulose nanofibers. Adv. Mater. **29**(21), 1606284 (2017)
19. Z.Q. Fang, H.L. Zhang, S. Qiu, Y.D. Kuang, J. Zhou, Y. Lan, C. Sun, G.H. Li, S.Q. Gong, Z.Q. Ma, Versatile wood cellulose for biodegradable electronics. Adv. Mater. Technol. **6**(2), 2000928 (2021)
20. C. Preston, Z. Fang, J. Murray, H.L. Zhu, J.Q. Dai, J.N. Munday, L.B. Hu, Silver nanowire transparent conducting paper-based electrode with high optical haze. J. Mater. Chem. C **2**(7), 1248–1254 (2014)
21. L.B. Hu, G. Zheng, J. Yao, N. Liu, B. Weil, M. Eskilsson, E. Karabulut, Z.C. Ruan, S.H. Fan, J.T. Bloking, M.D. Mcgehee, L. Wågberg, Y. Cui, Transparent and conductive paper from nanocellulose fibers. Energy Environ. Sci. **6**(2), 513–518 (2013)
22. H.L. Zhu, S. Parvinian, C. Preston, O. Vaaland, Z.C. Ruan, L.B. Hu, Transparent nanopaper with tailored optical properties. Nanoscale **5**(9), 3787 (2013)
23. X.Z. Xu, J. Zhou, L. Jiang, G. Lubineau, T. Ng, B.S. Ooi, H.Y. Liao, C. Shen, L. Chen, J.Y. Zhu, Highly transparent, low-haze, hybrid cellulose nanopaper as electrodes for flexible electronics. Nanoscale **8**(24), 12294–12306 (2016)
24. Z.Q. Fang, H.L. Zhu, W. Bao, C. Preston, Z. Liu, J.Q. Dai, Y.Y. Li, L.B. Hu, Highly transparent paper with tunable haze for green electronics. Energy Environ. Sci. **7**(10), 3313–3319 (2014)
25. M.C. Hsieh, H. Koga, K. Suganuma, M. Nogi, Hazy transparent cellulose nanopaper. Sci. Rep. **7**(1), 41590 (2017)
26. S. Chen, Y.J. Song, F. Xu, Highly transparent and hazy cellulose nanopaper simultaneously with a self-cleaning superhydrophobic surface. ACS Sustain. Chem. Eng. **6**(4), 5173–5181 (2018)
27. W. Wu, N.G. Tassi, H.L. Zhu, Z.Q. Fang, L.B. Hu, Nanocellulose-based translucent diffuser for optoelectronic device applications with dramatic improvement of light coupling. ACS Appl. Mater. Interfaces **7**(48), 26860–26864 (2015)

28. H.L. Ning, Y. Zeng, Y. Kuang, Z.K. Zheng, P.P. Zhou, R.H. Yao, H.K. Zhang, W.Z. Bao, G. Chen, Z.Q. Fang, J.B. Peng, Room-temperature fabrication of high-performance amorphous In-Ga-Zn-O/Al_2O_3 thin-film transistors on ultrasmooth and clear nanopaper. ACS Appl. Mater. Interfaces **9**(33), 27792–27800 (2017)
29. M. Eder, S. Amini, P. Fratzl, Biological composites-complex structures for functional diversity. Science **362**(6414), 543–547 (2018)
30. S. Vignolini, N. Bruns, Bioinspiration across all length scales of materials. Adv. Mater. **30**(19), 1801687 (2018)
31. R.M. Parker, G. Guidetti, C.A. Williams, T.H. Zhao, A. Narkevicius, S. Vignolini, B. Frka-Petesic, The self-assembly of cellulose nanocrystals: hierarchical design of visual appearance. Adv. Mater. **30**(19), 1704477 (2018)
32. Z.Q. Fang, G.Y. Hou, C.J. Chen, L.B. Hu, Nanocellulose-based films and their emerging applications. Curr. Opin. Solid State Mater. Sci. **23**(4), 100764 (2019)
33. R.H. Marchessault, F.F. Morehead, N.M. Walter, Liquid crystal systems from fibrillar polysaccharides. Nature **184**(4686), 632–633 (1959)
34. J.P.F. Lagerwall, C. Schütz, M. Salajkova, J. Noh, P.J. Hyun, G. Scalia, L. Bergström, Cellulose nanocrystal-based materials: from liquid crystal self-assembly and glass formation to multifunctional thin films. NPG Asia Mater. **6**(1), e80 (2014)
35. J.A. Kelly, M. Giese, K.E. Shopsowitz, W.Y. Hamad, M.J. Maclachlan, The development of chiral nematic mesoporous materials. Acc. Chem. Res. **47**(4), 1088–1096 (2014)
36. M.C. Xu, W. Li, C.H. Ma, H.P. Yu, Y.Q. Wu, Y.G. Wang, Z.J. Chen, J. Li, S.X. Liu, Multifunctional chiral nematic cellulose nanocrystals/glycerol structural colored nanocomposites for intelligent responsive films, photonic inks and iridescent coatings. J. Mater. Chem. C **6**(20), 5391–5400 (2018)
37. Y.D. He, Z.L. Zhang, J. Xue, X.H. Wang, F. Song, X.L. Wang, L.L. Zhu, Y.Z. Wang, Biomimetic optical cellulose nanocrystal films with controllable iridescent color and environmental stimuli-responsive chromism. ACS Appl. Mater. Interfaces **10**(6), 5805–5811 (2018)
38. B.C. Wang, A. Walther, Self-assembled, iridescent, crustacean-mimetic nanocomposites with tailored periodicity and layered cuticular structure. ACS Nano **9**(11), 10637–10646 (2015)
39. Y.Y. Cao, W.Y. Hamad, M.J. Maclachlan, Broadband circular polarizing film based on chiral nematic liquid crystals. Adv. Opt. Mater. **6**(17), 1800412 (2018)
40. A.G. Dumanli, G. Kamita, J. Landman, H. Kooij, V. Der, B.J. Glover, J.J. Baumberg, U. Steiner, S. Vignolini, Controlled, bio-inspired self-assembly of cellulose-based chiral reflectors. Adv. Opt. Mater. **2**(7), 646–650 (2014)
41. G. Kamita, B. Frka-Petesic, A. Allard, M. Dargaud, K. King, A.G. Dumanli, S. Vignolini, Biocompatible and sustainable optical strain sensors for large-area applications. Adv. Opt. Mater. **4**(12), 1950–1954 (2016)
42. B. Frka-Petesic, G. Guidetti, G. Kamita, S. Vignolini, Controlling the photonic properties of cholesteric cellulose nanocrystal films with magnets. Adv. Mater. **29**(32), 1701469 (2017)
43. Z.W. Peng, Q.L. Lin, Y.-A.A. Tai, Y.H. Wang, Applications of cellulose nanomaterials in stimuli-responsive optics. J. Agric. Food Chem. **68**(46), 12940–12955 (2020)
44. A. Tran, C.E. Boott, M.J. Maclachlan, Understanding the self-assembly of cellulose nanocrystals-toward chiral photonic materials. Adv. Mater. **32**(41), 1905876 (2020)
45. K. Yao, Q.J. Meng, V. Bulone, Q. Zhou, Flexible and responsive chiral nematic cellulose nanocrystal/poly(ethylene glycol) composite films with uniform and tunable structural color. Adv. Mater. **29**(28), 1701323 (2017)
46. C.L. Duan, Z. Cheng, B. Wang, J.S. Zeng, J. Xu, J.P. Li, W.H. Gao, K.F. Chen, Chiral photonic liquid crystal films derived from cellulose nanocrystals. Small **17**(30), 2007306 (2021)
47. H.Z. Zheng, W.R. Li, W. Li, X.J. Wang, Z.Y. Tang, S.X. Zhang, Y. Xu, Uncovering the circular polarization potential of chiral photonic cellulose films for photonic applications. Adv. Mater. **30**(13), 1705948 (2018)
48. S.N. Fernandes, P.L. Almeida, N. Monge, L.E. Aguirre, D. Reis, C.L.P. Oliveira, N.A.M.F. De, P. Pieranski, M.H. Godinho, Mind the microgap in iridescent cellulose nanocrystal films. Adv. Mater. **29**(2), 1603560 (2017)

49. T.H. Wu, J.D. Li, J.T. Li, S.M. Ye, J. Wei, J.B. Guo, A bio-inspired cellulose nanocrystal-based nanocomposite photonic film with hyper-reflection and humidity-responsive actuator properties. J. Mater. Chem. C **4**(41), 9687–9696 (2016)
50. J.A. De La Cruz, Q.K. Liu, B. Senyuk, A.W. Frazier, K. Peddireddy, I.I. Smalyukh, Cellulose-based reflective liquid crystal films as optical filters and solar gain regulators. ACS Photonics **5**(6), 2468–2477 (2018)
51. L. Cortese, L. Pattelli, F. Utel, S. Vignolini, M. Burresi, D.S. Wiersma, Anisotropic light transport in white beetle scales. Adv. Opt. Mater. **3**(10), 1337–1341 (2015)
52. B.D. Wilts, X.Y. Sheng, M. Holler, A. Diaz, M. Guizar-Sicairos, J. Raabe, R. Hoppe, S.H. Liu, R. Langford, O.D. Onelli, D.Y. Chen, S. Torquato, U. Steiner, C.G. Schroer, S. Vignolini, A. Sepe, Evolutionary-optimized photonic network structure in white beetle wing scales. Adv. Mater. **30**(19), 1702057 (2018)
53. S. Caixeiro, M. Peruzzo, O.D. Onelli, S. Vignolini, R. Sapienza, Disordered cellulose-based nanostructures for enhanced light scattering. ACS Appl. Mater. Interfaces **9**(9), 7885–7890 (2017)
54. T. Li, Y. Zhai, S.M. He, W.T. Gan, Z.Y. Wei, M. Heidarinejad, D. Dalgo, R.Y. Mi, X.P. Zhao, J.W. Song, J.Q. Dai, C.J. Chen, A. Aili, A. Vellore, A. Martini, R.G. Yang, J. Srebric, X.B. Yin, L.B. Hu, A radiative cooling structural material. Science **364**(6442), 760–763 (2019)
55. H.L. Zhu, Z.Q. Fang, C. Preston, Y.Y. Li, L.B. Hu, Transparent paper: fabrications, properties, and device applications. Energy Environ. Sci. **7**(1), 269–287 (2014)
56. Z.Q. Fang, H.L. Zhu, C. Preston, L.B. Hu, Development, application and commercialization of transparent paper. Transl. Mater. Res. **1**(1), 015004 (2014)
57. H.L. Zhu, W. Luo, P.N. Ciesielski, Z. Fang, J.Y. Zhu, G. Henriksson, M.E. Himmel, L.B. Hu, Wood-derived materials for green electronics, biological devices, and energy applications. Chem. Rev. **116**(16), 9305–9374 (2016)
58. J. Huang, H.L. Zhu, Y.C. Chen, C. Preston, K. Rohrbach, J. Cumings, L.B. Hu, Highly transparent and flexible nanopaper transistors. ACS Nano **7**(3), 2106–2113 (2013)
59. Y. Fujisaki, H. Koga, Y. Nakajima, M. Nakata, H. Tsuji, T. Yamamoto, T. Kurita, M. Nogi, N. Shimidzu, Transparent nanopaper-based flexible organic thin-film transistor array. Adv. Funct. Mater. **24**(12), 1657–1663 (2014)
60. D. Gaspar, S.N. Fernandes, A.G. De Oliveira, J.G. Fernandes, P. Grey, R.V. Pontes, L. Pereira, R. Martins, M.H. Godinho, E. Fortunato, Nanocrystalline cellulose applied simultaneously as the gate dielectric and the substrate in flexible field effect transistors. Nanotechnology **25**(9), 094008 (2014)
61. W.Z. Bao, Z.Q. Fang, J.Y. Wan, J.Q. Dai, H.L. Zhu, X.G. Han, X.F. Yang, C. Preston, L.B. Hu, Aqueous gating of van der Waals materials on bilayer nanopaper. ACS Nano **8**(10), 10606–10612 (2014)
62. C.Y. Wang, C. Fuentes-Hernandez, J.C. Liu, A. Dindar, S. Choi, J.P. Youngblood, R.J. Moon, B. Kippelen, Stable low-voltage operation top-gate organic field-effect transistors on cellulose nanocrystal substrates. ACS Appl. Mater. Interfaces **7**(8), 4804–4808 (2015)
63. D.H. Ha, Z.Q. Fang, N.B. Zhitenev, Paper in electronic and optoelectronic devices. Adv. Electron. Mater. **4**(5), 1700593 (2018)
64. Y.H. Zhou, C. Fuentes-Hernandez, T.M. Khan, J.C. Liu, J. Hsu, J.W. Shim, A. Dindar, J.P. Youngblood, R.J. Moon, B. Kippelen, Recyclable organic solar cells on cellulose nanocrystal substrates. Sci. Rep. **3**(1), 1536 (2013)
65. L. Gao, L.F. Chao, M.H. Hou, J. Liang, Y.H. Chen, H.D. Yu, W. Huang, Flexible, transparent nanocellulose paper-based perovskite solar cells. Npj Flex. Electron. **3**(1), 4 (2019)
66. M.H. Jung, N.M. Park, S.Y. Lee, Color tunable nanopaper solar cells using hybrid $CH_3NH_3PbI_{3-X}Br_x$ perovskite. Sol. Energy **139**, 458–466 (2016)
67. H.L. Zhu, Z. Xiao, D.T. Liu, Y.Y. Li, N.J. Weadock, Z.Q. Fang, J.S. Huang, L.B. Hu, Biodegradable transparent substrates for flexible organic-light-emitting diodes. Energy Environ. Sci. **6**(7), 2105–2111 (2013)
68. Y. Okahisa, A. Yoshida, S. Miyaguchi, H. Yano, Optically transparent wood-cellulose nanocomposite as a base substrate for flexible organic light-emitting diode displays. Compos. Sci. Technol. **69**(11), 1958–1961 (2009)

69. D.H. Ha, Z.Q. Fang, L.B. Hu, J.N. Munday, Paper-based anti-reflection coatings for photovoltaics. Adv. Energy Mater. **4**(9), 1301804 (2014)
70. D.H. Ha, J. Murray, Z.Q. Fang, L.B. Hu, J.N. Munday, Advanced broadband antireflection coatings based on cellulose microfiber paper. IEEE J. Photovolt. **5**(2), 577–583 (2015)
71. M.W. Zhu, T. Li, C.S. Davis, Y.G. Yao, J.Q. Dai, Y.B. Wang, F. Alqatari, J.W. Gilman, L.B. Hu, Transparent and haze wood composites for highly efficient broadband light management in solar cells. Nano Energy **26**, 332–339 (2016)
72. C. Jia, T. Li, C.J. Chen, J.Q. Dai, I.M. Kierzewski, J.W. Song, Y.J. Li, C.P. Yang, C.W. Wang, L.B. Hu, Scalable, anisotropic transparent paper directly from wood for light management in solar cells. Nano Energy **36**, 366–373 (2017)
73. Y.Y. Li, M. Cheng, E. Jungstedt, B. Xu, L.C. Sun, L. Berglund, Optically transparent wood substrate for perovskite solar cells. ACS Sustain. Chem. Eng. **7**(6), 6061–6067 (2019)
74. H. Yi, S.-H. Lee, H. Ko, D. Lee, W.-G. Bae, T. Kim, D.S. Hwang, H.E. Jeong, Ultra-adaptable and wearable photonic skin based on a shape-memory, responsive cellulose derivative. Adv. Funct. Mater. **29**(34), 1902720 (2019)
75. Z. Zhang, H. Chang, B.L. Xue, S.F. Zhang, X. Li, W.-K. Wong, K.C. Li, X.J. Zhu, Near-infrared and visible dual emissive transparent nanopaper based on Yb(III)–carbon quantum dots grafted oxidized nanofibrillated cellulose for anti-counterfeiting applications. Cellulose **25**(1), 377–389 (2018)
76. E. Morales-Narváez, H. Golmohammadi, T. Naghdi, H. Yousefi, U. Kostiv, D. Horák, N. Pourreza, A. Merkoçi, Nanopaper as an optical sensing platform. ACS Nano **9**(7), 7296–7305 (2015)
77. H. Wan, X.F. Li, L. Zhang, X.P. Li, P.F. Liu, Z.G. Jiang, Z.Z. Yu, Rapidly responsive and flexible chiral nematic cellulose nanocrystal composites as multifunctional rewritable photonic papers with eco-friendly inks. ACS Appl. Mater. Interfaces **10**(6), 5918–5925 (2018)
78. W.B. Song, J.-K. Lee, M.S. Gong, K. Heo, W.-J. Chung, B.Y. Lee, Cellulose nanocrystal-based colored thin films for colorimetric detection of aldehyde gases. ACS Appl. Mater. Interfaces **10**(12), 10353–10361 (2018)
79. A.Q. Hou, H.H. Chen, C.W. Zheng, K.L. Xie, A.Q. Gao, Assembly of a fluorescent chiral photonic crystal membrane and its sensitive responses to multiple signals induced by small molecules. ACS Nano **14**(6), 7380–7388 (2020)
80. O. Kose, A. Tran, L. Lewis, W.Y. Hamad, M.J. Maclachlan, Unwinding a spiral of cellulose nanocrystals for stimuli-responsive stretchable optics. Nat. Commun. **10**(1), 510 (2019)
81. M. Kim, H. Lee, M.C. Krecker, D. Bukharina, D. Nepal, T.J. Bunning, V.V. Tsukruk, Switchable photonic bio-adhesive materials. Adv. Mater. **2021**, 2103674 (2021)

Chapter 7
Nanocellulose Paper for Flexible Electronic Substrate

Zhaoyang Li, Jun Zhou, and Junwen Zhong

Abstract Paper substrate made with nanocellulose is widely used to fabricate electronics by virtue of the superiorities of flexibility, lightweight, and eco-friendly, etc. Herein, advantages/limitations for nanocellulose paper as electronics substrate, electrodes design and fabrication on nanocellulose paper, and recent advances of various nanocellulose-based electronics, will be systematically reviewed and discussed to provide some insights and inspirations in developing paper electronics.

7.1 Introduction

Modern people are living in a world full of electronics such as computers, cellphone, various sensors, etc. which are part of people's daily life and make work, study, and entertainment easier and more convenient. However, one crisis appears in the life cycle of electronic products, that is the electronic wastes (E-waste) problem [1–3]. One of the reasons is that most of the current electronics are constructed on nondegradable substrates like silicon wafer, glass, and polymer, of which the full degradation time will be as long as hundreds or even thousands of years. It is estimated that 44.7 million tons of E-waste is produced in 2016, which can be placed in line from New York to Bangkok and back [4]. On the other hand, the normal substrate for flexible electronics, polymers, are mainly made from oil, which belongs to non-renewable resource [5]. Thus, developing a recycle, low-cost, and easy-gotten electronic substrate is necessary and has caused lots of interest.

Z. Li · J. Zhong (✉)
Department of Electromechanical Engineering and Centre for Artificial Intelligence and Robotics, University of Macau, Macau SAR 999078, China
e-mail: junwenzhong@um.edu.mo

Z. Li
e-mail: zhaoyangli@um.edu.mo

J. Zhou · J. Zhong
Wuhan National Laboratory for Optoelectronics, Huazhong University of Science and Technology, Wuhan 430074, China
e-mail: jun.zhou@mail.hust.edu.cn

L. Hu et al. (eds.), *Emerging Nanotechnologies in Nanocellulose*, NanoScience and Technology, https://doi.org/10.1007/978-3-031-14043-3_7

Paper composed of cellulose is an appropriate candidate for the next-generation flexible electronic substrate, as cellulose is widely distributed in plant, animal, and bacteria, simultaneously owing advantages of easily-degradable and non-toxic, etc [6–8]. Even though various of paper electronics such as battery, [9] capacitor, [10] energy harvester, [11] sensor, [12] and transistor [13] have been demonstrated, normal paper substrate still exists obvious drawbacks like bad robustness, none-transparent, bad thermal stability. Nanocellulose is one of the strongest natural materials, of which the axial elastic modulus is larger than that of Kevlar fiber [14]. Specialty paper made with nanocellulose is a good choice for constructing next-generation "green electronics". In this chapter, we will discuss the advantages and limitations for nanocellulose paper as electronics substrate, introduce the electrodes design and fabrication on nanocellulose paper, and summarize the most recent advances of various nanocellulose paper-based electronics. Moreover, discussions regarding to newfangled applications, current challenges and future directions will be also presented, which can some useful insights and inspirations for the development of paper electronics.

7.2 Advantages and Limitations for Nanocellulose Paper Electronic Substrate

Nanocellulose is a kind of natural polymer with glucose as monomers and normally can be divided into two types, which are cellulose nanocrystal (CNC) and nanofibrillated cellulose (NFC) [16, 17]. CNC has smaller size, higher crystalline, and better strength, making it a good additive for composite material. NFC is normally fabricated from wood, with diameters in tens of nanometers and length reaching micrometers. Thus, most of the nanocellulose-based electronic substrate is made with NFC. This type of flexible electronic substrate has a lot of advantages, as summarized in Fig. 7.1. These advantages are important for constructing kinds of flexible electronics. In this section, we will discuss the importance of the advantages and current limitations of nanocellulose paper substrate based on NFC.

7.2.1 *Flexible, Light, and Transparent*

Paper substrate composed of nanocellulose keeps the features of normal paper, it is still flexible and light, which belongs to basic requirements for flexible electronic substrate. The density of a typical nanocellulose paper is about 1.53 g/cm^3, and it can be even folded as a paper plane, as shown in Fig. 7.2a [18]. Nanocellulose paper has another outstanding advantage different from normal paper, that is the high transparency. The transparency of paper is affected by the surface morphology, porosity, thickness, and diameter of the cellulose fibers, etc. In nanocellulose paper,

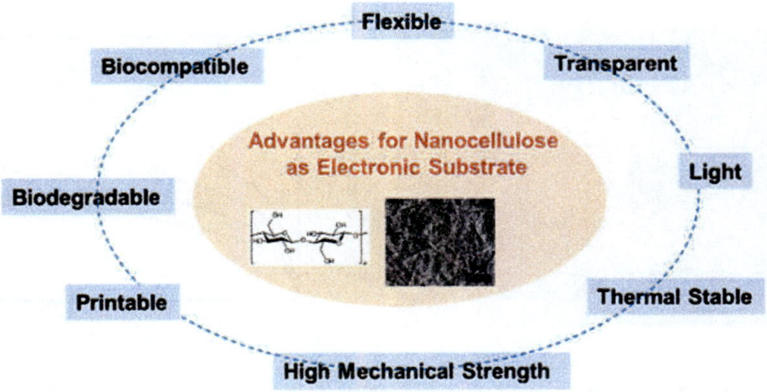

Fig. 7.1 Summary of nanocellulose paper as electronics substrate. Insert shows the molecular structure of nanocellulose and scanning electron microscope (SEM) image of NFC [15]. Copyright 2014, Royal Society of Chemistry

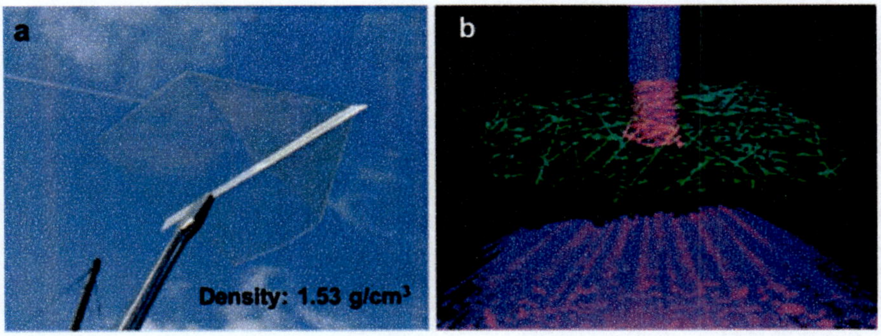

Fig. 7.2 **a** Light transmittance of the nanocellulose paper [18]. Copyright 2009, Wiley–VCH, **b** Schematic of a nanocellulose paper and corresponding light scattering effect [19]. Copyright 2013, Royal Society of Chemistry

the surface morphology is relatively smooth and the porosity is small because of the small diameter of the nanocellulose (nanometer scale). Thus, the light scattering inside the nanocellulose paper is reduced and the light can directly go through it, endowing it the high transparency, as shown in Fig. 7.2b [19]. For example, the transparency for a nanocellulose paper reported by Zhu, et al. can reach as high as 92–93% [20]. High transparent substrate is important for optical electronics, such as solar cell, light-emitting diode, touch screen, transparent sensors, etc.

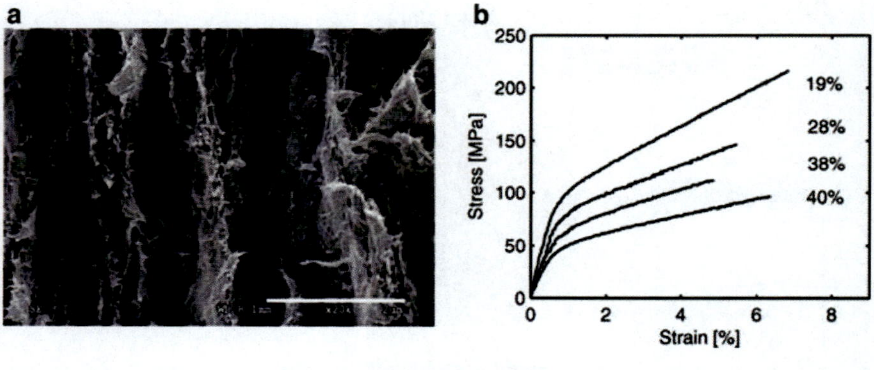

Fig. 7.3 **a** Cross-sectional SEM of a nanocellulose paper, showing the layers structure. **b** Stress–strain curves of nanocellulose papers with different porosities [21]. Copyright 2008, American Chemical Society

7.2.2 *Mechanical Strong*

Mechanical performances of substrate are very key factors for insuring the successful fabrication and long-time usage of flexible electronics. As indicated in the cross-sectional SEM image in Fig. 7.3a, nanocellulose paper is composed of many layers of nanocellulose fibers [21]. Such structure can reduce the strain generating by tension or compression process. Moreover, the intermolecular force like hydrogen bond and the crosslinking between nanocellulose fibers endow the high mechanical strength, which is 10 times better than normal paper [22]. Obviously, the porosity should affect the mechanical strength very much, as shown in Fig. 7.3b, the mechanical strength of nanocellulose paper reduces with the porosity increasing. The mechanical strength for nanocellulose paper nanocellulose paper with porosity of 19% is about 205 MPa, which is strong enough for electronic substrate and can be compared with polymer substrate like PET [21]. Furthermore, electrodes are easier printed on nanocellulose paper than polymer substrate, especially for the aqueous-based electrode materials, which will somehow reduce the fabrication cost and difficulty.

7.2.3 *Thermal Stable*

Many electronics need a heat-treating process with temperature between 150 and 200 °C to improve the performances and electronics will generate heat during working [23]. Natural cellulose will start to degrade when temperature reaching 300 °C [24]. Hsieh et al. report a nanocellulose paper with hardwood dissolving sulfite pulps as raw material, which keeps stable after staying in 200 °C for 30 min [25]. Under the same heat-treating condition, photo paper will be damaged, as shown in Fig. 7.4a. On the other hand, the thermal expansion of substrate will affect the

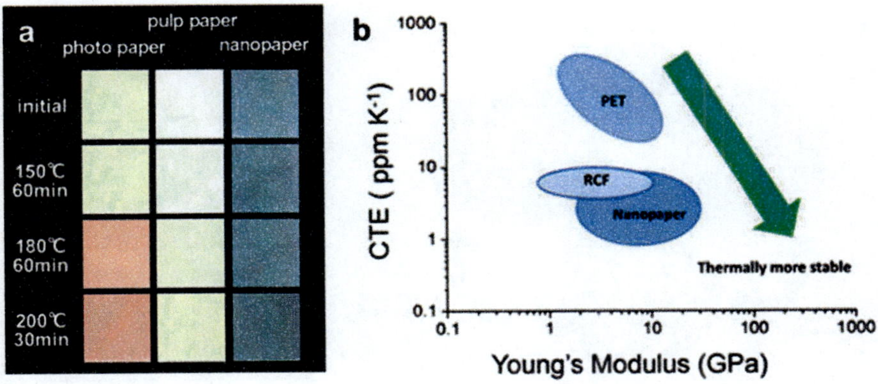

Fig. 7.4 **a** Photographs of nanocellulose paper and photo paper in various conditions after heating [25]. Copyright 2013, Royal Society of Chemistry. **b** Coefficient of thermal expansions versus Young's modulus for nanocellulose papers and PET [22]. Copyright 2013, Royal Society of Chemistry

fabrication process of electronics. The thermal expansion coefficient of nanocellulose paper is normally below 8.5 ppm/K, which is significantly lower than that of polymer substrate like PET, as shown in Fig. 7.4b [22]. Low thermal expansion coefficient will not damage the device structure during the necessary heat-treating process in the fabrication or usage.

7.2.4 *Biodegradable and Biocompatible*

Nanocellulose paper is totally made from wood and can be biodegraded back to the nutrient for nature, making it a perfect recycled bio-material as alternative and ecofriendly substrate for electronics, as indicated in Fig. 7.5a [26]. A lot of researchers have studied on nanocellulose paper-based electronics and demonstrate the biodegradability. For instance, Gao et al. prove that nanocellulose paper substrate can be fully biodegraded in 210 days, as shown in Fig. 7.5b [27]. Moreover, nanocellulose paper is non-toxic, so it's also a good electronic substrate for scenarios that require the biocompatibility, such as wearable electronics and implantable electronics.

7.2.5 *Limitations*

Even though nanocellulose paper has a lot of advantages to be the next-generation substrate for green electronics. There are still some limitations that limit it as a largely used or universal electronic substrate. (1) The energy and money cost for fabricating

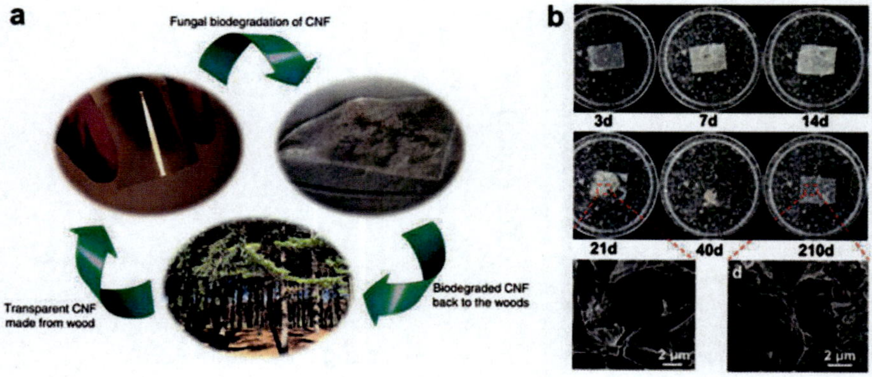

Fig. 7.5 **a** A likely life cycle of the bio-based and biodegradable nanocellulose paper, including extracting cellulose nanofibrils from woods for making nanocellulose paper, and degrading nanocellulose paper through fungal biodegradation as nutrition of woods without adverse environmental effects [26]. Copyright 2015, Nature Publishing Group. **b** Photos and SEM images indicating the bio-degradation process of nanocellulose paper [27]. Copyright 2016, American Chemical Society

nanocellulose are still very high, the energy for producing a ton of nanocellulose is as high as hundreds and even thousands of kWh [28]. (2) Mature technology for fast and large manufacture of nanocellulose paper and electrode is still lacking. (3) The chemical and physical stability of nanocellulose paper are still not as good as polymer substrate, current nanocellulose paper substrate is more appropriate for one-time-used electronics.

7.3 Electrodes for Nanocellulose Paper Substrate

To deposit conductive materials on nanocellulose paper substrate to be the electrodes is the first step for constructing electronics. In general, these electrodes should have reasonable conductive performance according to specific requirement of different electronics, simultaneously will not hamper the advantages of nanocellulose paper substrate. As shown in Fig. 7.6, the electrodes for nanocellulose paper substrate can be divided into four types when classified by physical dimensions: conductive nanoparticles like metal nanoparticles, which is defined as zero-directional electrode; conductive nanowires like carbon nanotubes (CNTs) and metal nanowires, which is defined as one-directional electrode; conductive nanosheets like graphene, which is defined as two-directional electrode; and conductive polymer like PEDOT:PSS, which is defined as thin-film electrode. In this section, we will discuss fabrication methods, performances, and advantages/limitations of each type of electrodes.

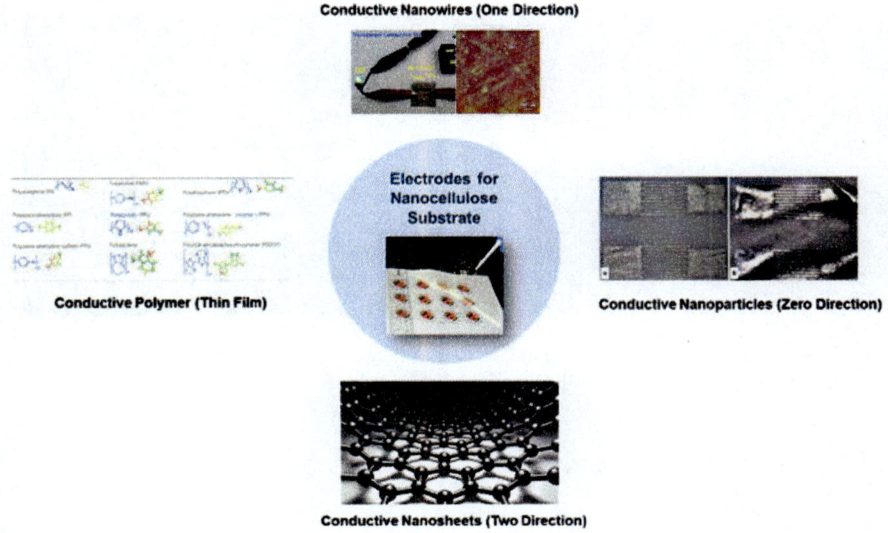

Fig. 7.6 Summary of electrodes for nanocellulose paper substrate, which can be classified by the physical dimensions: zero direction, [29] one direction, [30] two direction, [31] and thin film [32]. Copyright 2012, Springer. Copyright 2013, American Chemical Society. Copyright 2013, Royal Society of Chemistry

7.3.1 Conductive Nanoparticle

Conductive nanoparticles such as silver nanoparticle can be used to make conductive inks that can be directly printed on nanocellulose paper substrate as electrodes, which has the potential to be a quick, large-scale and precise electrode fabrication strategy. Conductive nanoparticle ink has very high level of conductivity and can be easily absorbed by the substrate and finally tightly bond with the substrate when the ink is dry. Chinga-Carrasco, et al., adopt inkjet-printed technology to coat silver nanoparticles on nanocellulose paper substrate [29]. First, the substrate surface is processed by hexamethyldisilazane (HMDS) to reduce its water-wettability and improve the resolution of printing inkjet-printed silver nanoparticles. Second, the silver nanoparticles are printed on the HMDS-modified films, which composes nanofibrils with diameters less than 20 nm, as indicated in Fig. 7.7a. The surface resistivity of the inkjet-printed silver electrode is about 5 Ω/Sq, and the smallest space between the nearby electrodes is about 100–200 μm, as shown in Fig. 7.7b. However, zero-directional electrode needs a dense structure to form the conductive network, as a result, this type of electrode on nanocellulose paper substrate will seriously hamper the transparency.

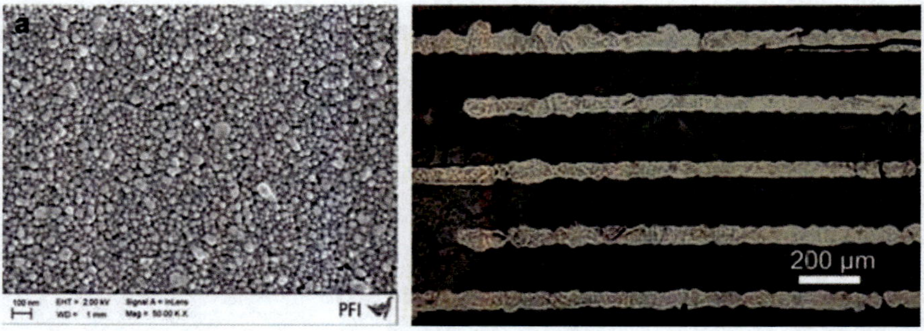

Fig. 7.7 **a** Surface SEM image and **b** optical image of inkjet-printed silver nanoparticles electrode on nanocellulose paper substrate [29]. Copyright 2012, Springer

7.3.2 Conductive Nanowire

One-directional conductive material has good length-diameter ratio and can form a conductive network with small amount of material, which ensures a high transparency of the nanocellulose paper substrate and will not influence the mechanical property of substrate. CNT is a typical one-directional conductive material, simultaneously owing the advantages of lightweight and high mechanical strength, etc [30]. Fang et al. use Meyer rod coating method to deposit CNTs on nanocellulose paper substrate, which has potential as a large-scale industrial method, as shown in Fig. 7.8a [33]. Specifically, 1 mg/ml of CNTs is dispersed in water and 10 mg/ml of sodium dodecyl-benzene-sulfonate (SDBS) serves as the surfactant. Increasing the amount of the CNTs can enhance the conductively, as shown in Fig. 7.8b. Sheet resistance is about 500 Ω/Sq, if the CNTs amount reaches 0.7 $\mu g/cm^2$. By the way, increasing the CNT amount is not an effective way to get better conductively anymore for this work. Zhong et al. test the mechanical and optical performances of CNT electrode on nanocellulose paper substrate [34]. Deformation under bending angle up to 90° and durability tests are realized. During the measurement, sheet resistance of about 2 $k\Omega/cm^2$ almost remains stable (Fig. 7.8c, d). The CNT electrode is also bent continuously one thousand times with a maximum bending angle of ~60°, and the I–V curve remains stable (Fig. 7.8e). These results indicate the good flexibility and stability of CNT electrodes on the nanocellulose paper substrate. On the other hand, the transparency is as high as 90% with CNT electrodes (Fig. 7.8f).

Metal nanowires like silver and coper nanowires are another excellent candidate for one-directional electrode, as it has better conductivity than CNTs. Silver nanowire electrode is usually fabricated with a "transfer method", as indicated in Fig. 7.9a [35]. Specifically, nanocellulose is filtered through tetramethylpiperidine oxide (TEMPO)-treated nanocellulose paper to smooth the surface, and then a filtered silver nanowires network is embedded on the smooth surface with a lamination process, as shown in Fig. 7.9b. Kang et al. prove that such silver nanowires electrode can conformally cover the surface with strong adhesion strength, making it adapt structure deformation even after folding it from +180° to −180°, as shown in Fig. 7.9c [36]. They also

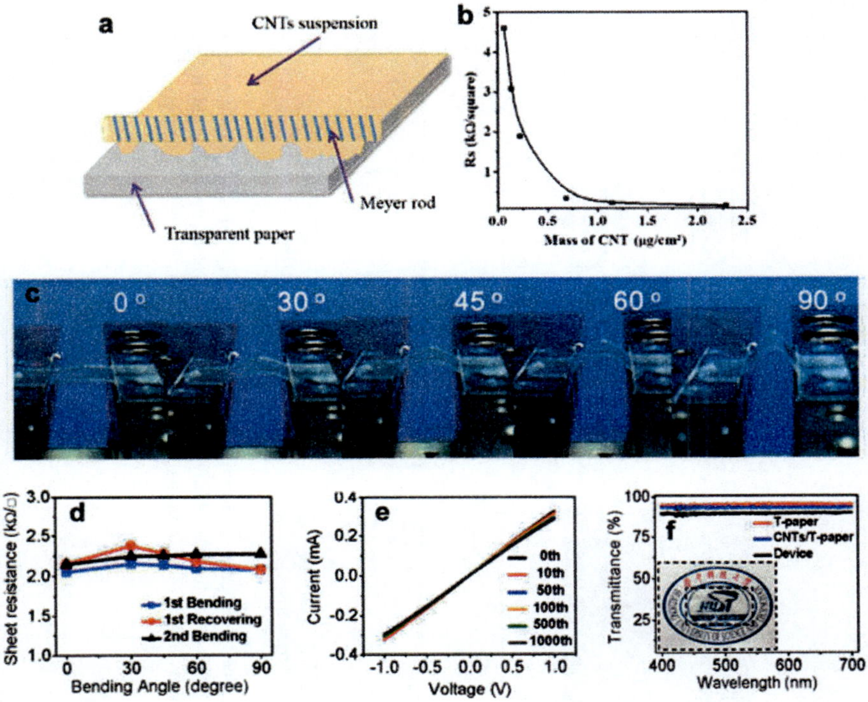

Fig. 7.8 **a** Coating procedure for CNT coating on nanocellulose paper substrate. **b** Sheet resistance of CNT electrode with various amount of CNTs [33]. Copyright 2013, Royal Society of Chemistry. **c** Pictures of CNT electrode bent at different angles. **d** Sheet resistances of CNT electrodes after first bending process, recovery process, and second bending process, respectively. **e** 1000 times bending test of CNT electrodes, with a bending angle of 60°. **f** Transmittance of the substrate with CNT electrode [34]. Copyright 2014, American Chemical Society

conduct transfers for silver nanowires of different amount in order to obtain transparent electrodes on nanocellulose paper substrate with different transmittance and conductivity (Fig. 7.9d). It is found that sample with lower sheet resistance has lower transmittance because lower sheet resistance needs a denser network. When the sheet resistance is less than 10 Ω/sq, the transparency under wavelength of 600 nm is over 80% (Fig. 7.9d).

7.3.3 Conductive Nanosheet

Two-directional conductive material like graphene is composed of one or few layers of atoms, which make it be extremely thin and transparent. Graphene has excellent mechanical strength with Young's modulus values of 2.4 ± 0.4 and 2.0 ± 0.5 TPa

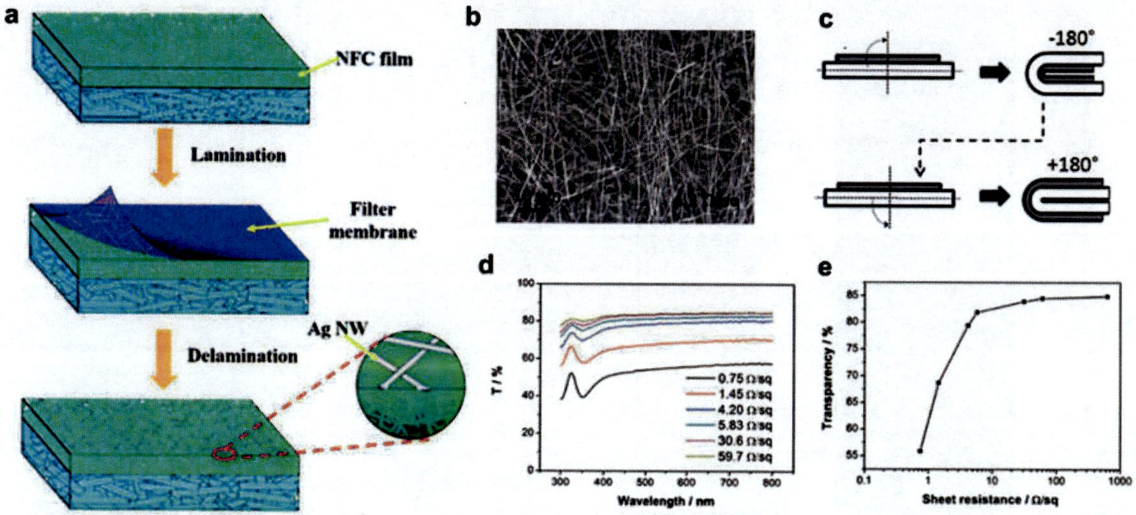

Fig. 7.9 **a** Schematic of silver nanowire network embedded on the surface of nanocellulose paper substrate. **b** Surface SEM image of silver nanowires with a sheet resistance of 4.2 Ω/sq [35]. Copyright 2014, Royal Society of Chemistry. **c** Schematics for the folding procedure of silver nanowires electrode. **d** Spectra of silver nanowires electrodes of different sheet resistance. **e** Transparency at a wavelength of 600 nm with different sheet resistances [36]. Copyright 2015, Wiley–VCH

for single- and bilayer graphene, respectively [37]. Thus, such two-directional electrodes are good candidates that can be assembled with nanocellulose paper substrate. Graphene can be covered on the substrate surface to construct the conductive network, but need to solve the connection issue. For example, Gao et al. endeavor to prepare transparent conductive nanocellulose paper through layer-by-layer assembly method with divalent copper ions as the crosslinking agent, as shown in Fig. 7.10a [38]. The thickness of the reduced graphene oxide (RGO) layer in nanocellulose paper is adjustable by changing the cycle times of the layer-by-layer assembly method. Figure 7.10b shows the picture of nanocellulose/RGO paper, demonstrating good flexibility and color retention ability. Transmittance of the nanocellulose/RGO paper at a wavelength of 550 nm is related to the number of layer-by-layer cycles (Fig. 7.10c), which is 89.2% for pure nanocellulose paper and 76% for nanocellulose/$[RGO]_{20}$ paper (see inset of Fig. 7.10c). The sheet resistance of the nanocellulose/$[RGO]_{20}$ paper reaches 2.5 kΩ/sq (20 means 20 times of layer-by-layer cycles), which keeps stable under 200 cycles of bending, as shown in Fig. 7.10d. Furthermore, nanocellulose/$[RGO]_{20}$ paper retains good mechanical property from nanocellulose paper with the final strength of ~136 MPa. Another strategy to assemble graphene electrode with nanocellulose paper substrate is to mix them. Luong et al. report a type of graphene/cellulose composite paper that has excellent performances by combining RGO and amine-modified nano-fibrillated cellulose (A-NFC) [39]. By changing graphene oxide (GO) content, graphene/cellulose nanocomposites with graphene content ranging from 0.1 wt% to 10 wt% were obtained. The synthesized RGO/A-NFC nanocomposite possesses an electrical percolation threshold of 0.3 wt% and an electrical conductivity of 4.79×10^{-4} S/m, which is higher than the antistatic value, as shown in Fig. 7.10e. It should be noted that with 5 wt% of graphene, the RGO/A-NFC composite paper had a greatly enhanced tensile strength of ~273 MPa, which is 1.4 times and 2.8 times higher than that of cellulose paper and graphene oxide paper, respectively.

7.3.4 Conductive Polymer

Conducting polymers, such as PEDOT: PSS and PANi, are a kind of highly conjugated polymers with spatially extended π-bonding that can provide unique properties [32]. Conductive polymers are easily dispersed and bonded with nanocellulose paper substrate because of the intermolecular forces generated by hydrogen bonds [32]. Gao et al. report fabricating PEDOT: PSS transparent electrode on nanocellulose paper via spinning coating method, because hydrophilic feature of nanocellulose paper and its smooth surface make it easily coated with the PEDOT: PSS aqueous ink, as shown in Fig. 7.11a [27]. Atomic force microscopy (AFM) image shows that the roughness is ~83 nm, proving the little influence of transmittance. The transmittance of nanocellulose paper and nanocellulose paper with electrode is ~90% and ~87%, respectively (Fig. 7.11b). A photo of the nanocellulose paper with electrode also shows its good transparency. Furthermore, deformation experiments with bending angles changing

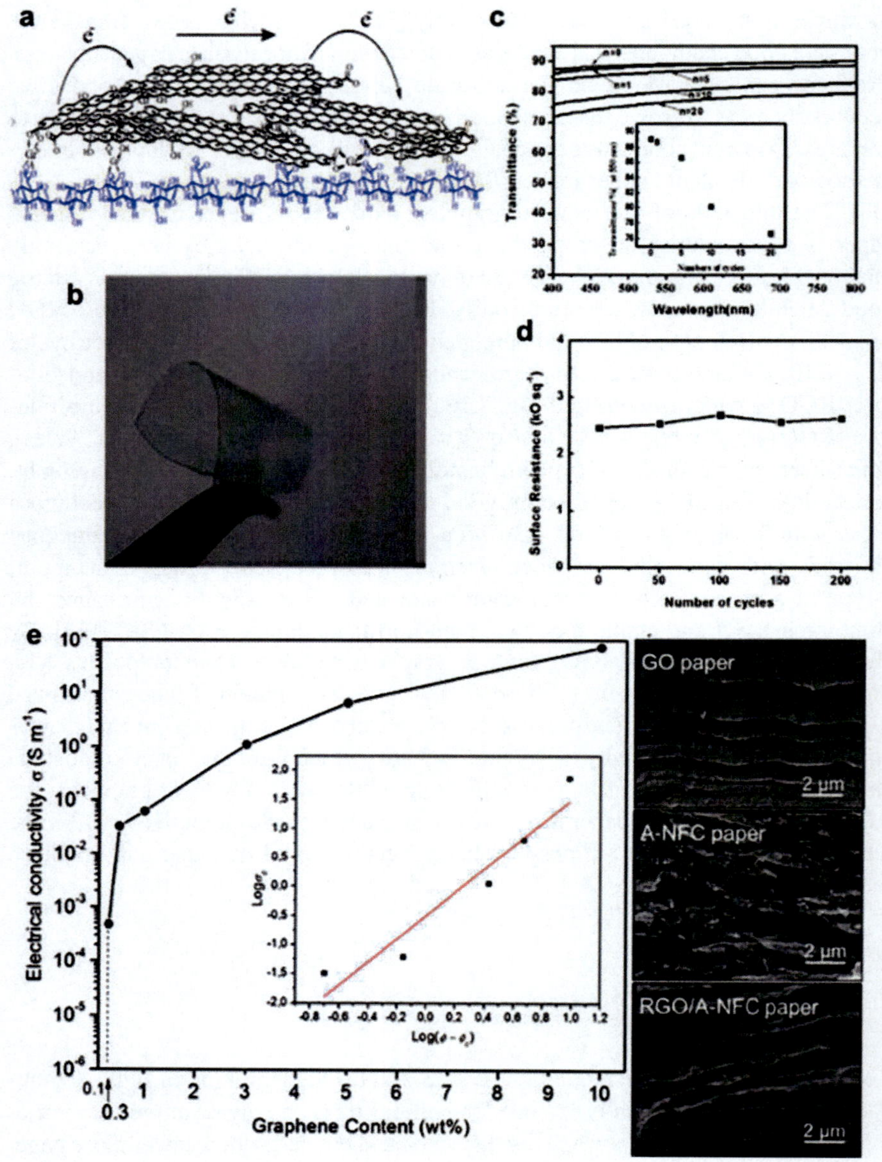

Fig. 7.10 **a** Schematic diagram of nanocellulose/RGO paper with Cu^{2+} as the crosslinking agent. **b** Optical image of nanocellulose/$[RGO]_{20}$ paper. **c** Transmittance of nanocellulose/$[RGO]_n$ papers with different layer-by-layer cycles, insets show the relationship between transmittance at a wavelength of 550 nm and the number of layer-by-layer cycles. **d** Sheet resistance of nanocellulose /$[RGO]_{20}$ paper under bending cycles [38]. Copyright 2013, Elsevier. **e** Electrical conductivity of the RGO/A-NFC nanocomposite papers, and SEM images of GO paper, A-NFC paper, and RGO/A-NFC composite paper with 0.3 wt% of grapheme [39]. Copyright 2011, Royal Society of Chemistry

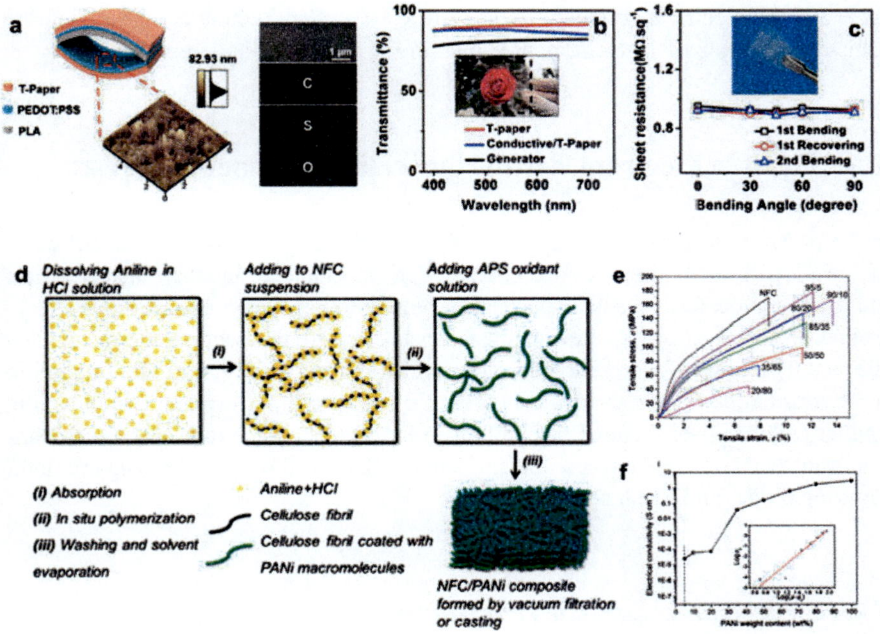

Fig. 7.11 **a** SEM image, element mapping image, and AFM image of nanocellulose paper with PEDOT-PSS electrode. **b** Transmittance of nanocellulose paper and nanocellulose paper with electrode. **c** Sheet resistance of the nanocellulose paper with electrode for the continuous three bending processes. Inset shows the optical image of bending nanocellulose paper [27]. Copyright 2016, American Chemical Society. **d** Fabrication process of NFC/PANi composite paper. **e** Stress–strain curves of pure NFC and NFC/PANi composite papers with PANi content ranging from 5 wt% to 80 wt%. **f** Electrical conductivity of the NFC/PANi composite papers at different PANi contents [40]. Copyright 2013, Elsevier

from 30° to 90° illustrate that the sheet resistance of the conductive paper substrate remains ~0.9 MΩ/sq after continuous three bending process (Fig. 7.11c). Meanwhile, the sample is continuously bent at 60° for one thousand cycles, while the I–V curves remain almost stable. Except covering conductive polymer on the substrate surface, another way is to nanocellulose conductive polymer on the nanocellulose fibers directly. Luong, et al. report NFC/PANi paper-like materials based on NFC/PANi aqueous suspensions adopting simple vacuum filtration method and solvent casting techniques, with detailed fabrication steps shown in Fig. 7.11d [40]. Tensile test is conducted to evaluate the mechanical performance of NFC and NFC/PANi composite papers, as shown in Fig. 7.11e. The tensile strength and Young's modulus of pure NFC are around 173 MPa and 5.7 GPa, respectively, with the breaking strain of ~9%. The excellent mechanical properties of the NFC paper are contributed to the high aspect ratio of cellulose nanofibrils and strong intra- and intermolecular hydrogen bonding. The NFC/PANi composite paper possesses a percolation threshold of 4.57

wt% and the electrical conductivity is ~2.6×10^{-5} S/cm, which is higher than the antistatic criterion of 10^{-8} S/cm, as shown in Fig. 7.11f.

7.4 Flexible Electronics with Nanocellulose Paper Substrate

In the past decades, various flexible electronics with nanocellulose paper as substrate are developed, attributing to the excellent optical, mechanical, and thermal properties, etc. Such nanocellulose-based electronics belong to next generation of green electronics and are the hope to alleviate the E-waste crisis. In this section, we will review the recent progress of representative nanocellulose paper-based electronics, including the construction, device structure, working mechanism, performance, application, challenge, etc. These typical nanocellulose-based devices include electromechanical transducer, energy storage devices, solar cell, thin film transistor, organic light emitting diode, and touch screen.

7.4.1 Electromechanical Transducer

So far, researchers have demonstrated various nanocellulose-based electronics. However, most of these electronic devices rely on external power sources or rigid batteries, thus it is of highly desirable to develop self-powered paper electronic systems to accumulate mechanical energy from environment and human body movement. Paper is an appropriate substrate for paper-based energy harvesters so that to ensure the flexibility and reduce the fabrication cost. Researchers try to fabricate paper-based energy harvester/sensor with working mechanism of piezoelectric and triboelectric/electrostatic effects to convert mechanical energy/signal to electrical domains [11, 41, 42]. Specifically, for nanocellulose-based energy harvester/sensor, Zhong et al. develop an active pressure sensor and energy harvester with following claimed advantages: (1) these devices are self-powered, meaning they don't need external power; (2) they are transparent; (3) they are compatible with other paper products such as artwork and packaging, etc.; (4) they have potential to be mass manufactured and can be added or removed easily [34]. Applications regarding anti-theft or anti-fake are demonstrated with such active pressure sensors, as shown in Fig. 7.12a. One type of sensor is composed of CNTs/T-paper and polyethylene (PE)/CNTs/T-paper, both are transparent and with good working stability. The other type of sensor is composed of PE/CNTs/T-paper and can be easily patterned. In these devices, PE electret is used to hold electrostatic charges and the working mechanism of these sensors is the electrostatic induction effect. In general, Pressing and releasing the device will produce alternating signals, with schematic diagram indicating the detailed working process in Fig. 7.12b. The output performance (peak current and peak voltage) of this sensor is excellent and has good stability, which is operated for 54,000 cycles with little variation. Another nanocellulose-based energy harvester

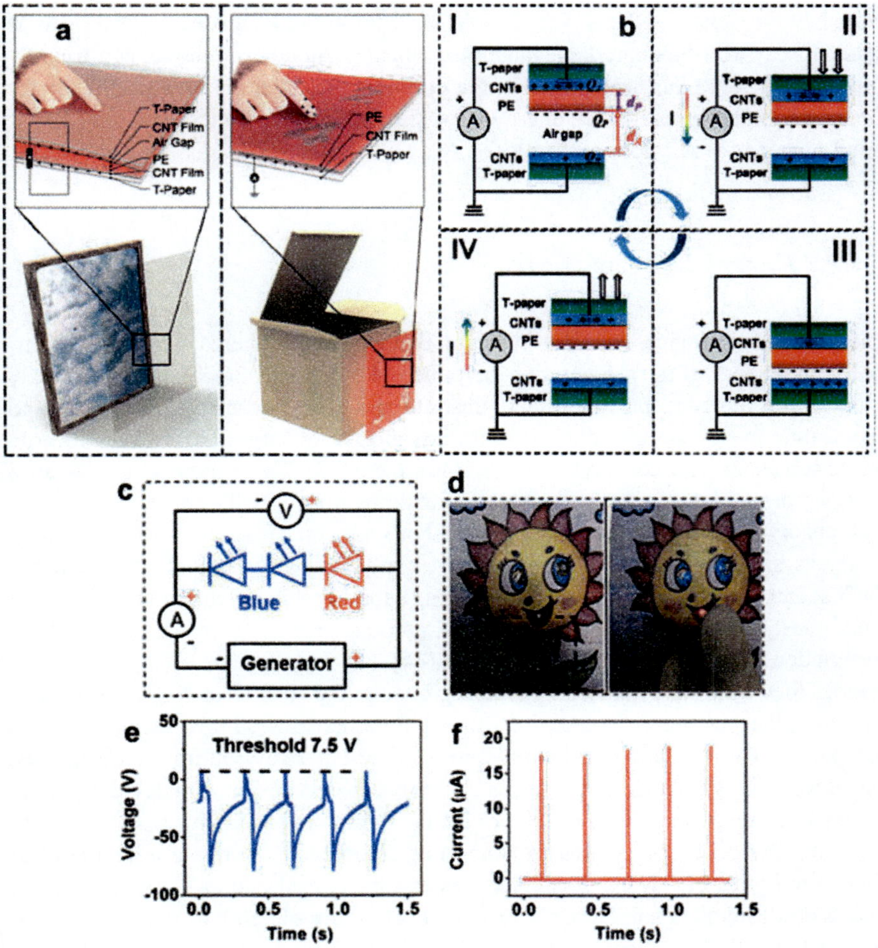

Fig. 7.12 **a** Transparent paper-based artwork anti-theft system and smart mapping anti-fake system used for artwork protection and smart packaging. **b** Working process of the device at (I) the original, (II) the pressing, (III) the equilibrium, and (IV) the releasing states [34]. Copyright 2014, American Chemical Society. **c** Equivalent circuit for the self-powered paper toy. **d** Pictures of self-powered paper toy. **e** Voltage across the LEDs when pressing the device. **f** Current through the LEDs when pressing the device [27]. Copyright 2016, American Chemical Society

with similar structure and working mechanism is demonstrated by Gao et al. [27]. The energy harvester can achieve a maximum power of ~84.4 μW and is continuously operated for 54,000 cycles stably. It is used to light one red LED in series with two blue LEDs, as shown in Fig. 7.12c. Since the energy harvester is invisible, they place it on the surface of one toy without affecting the appearance of toy (left in Fig. 7.12d). When pressing the paper toy, the device generates electricity which can light up LEDs (right in Fig. 7.12d). Once the energy harvester is pressed, the generated voltage across these LEDs is ~7.5 V (Fig. 7.12e) and the current through these LEDs reaches ∼18 μA (Fig. 7.12f). More importantly, this energy harvester is made of nearly all-biodegradable materials, which is a good example for eco-friendly concept. However, further improving the output ability is still a great challenge for the nanocellulose paper-based energy harvesters, as current devices can only power small power consumption electronics.

7.4.2 Energy Storage Device

Energy storage devices, such as lithium batteries and supercapacitors, are critical components of next-generation portable electronics. Nanocellulose is usually adopted to construct the functional substrate and electrode of the single-use and disposable energy storage devices, with advantages of large surface area to hold active substance, low cost and easy fabrication, etc [43]. These nanocellulose-based energy storage devices have potential applications in supplying power for displays and sensors in smart packages in the future. For supercapacitor demonstration, Wang et al., as shown in Fig. 7.13a, prove that surface modified nanocellulose fibers (NCFs) can serve as substrates in constructing supercapacitor electrodes, which have the highest full electrode-normalized gravimetric capacitances of ~127 F/g at high current densities (33 A/g) till reported date [44]. Before polypyrrole (PPy) polymerization, they introduce quaternary amine groups to the surface of NCFs, which can minimize the macropore volume of PPy-NCF composites without influencing the volume of the micro- and mesopores. Based on these porosity-optimized electrodes, they develop a symmetric and aqueous electrolyte-based device with device-specific volumetric energy and power density of 3.1 mWh/cm^3 and 3 W/cm^3, respectively. In addition, PPy-NCF composites are able to be combined with miniaturized flexible devices to light LEDs when the device is in extremely challenging states. Moreover, Zhang et al. report a solid-state hydroxide-conducting electrolyte that is based on nanocellulose and functionalized GO with laminated structure, and used on a flexible and rechargeable zinc–air battery, as given in Fig. 7.13b [45]. Once functionalized, the superior hydroxide conductivity can achieve 58.8 mS/cm at 70 °C. Compared to commercial A201 membrane, battery based on this membrane possesses better rechargeability and stability. In addition, the zinc-air battery has excellent output power density when it is under stress. Such new type of nanocellulose-based battery may open up an opportunity for developing flexible and solid-state electrochemical energy. However, the environmental stability of the nanocellulose-based energy storage devices like the stability against humidity and low/high temperature still need to be enhanced to ensure the practicability.

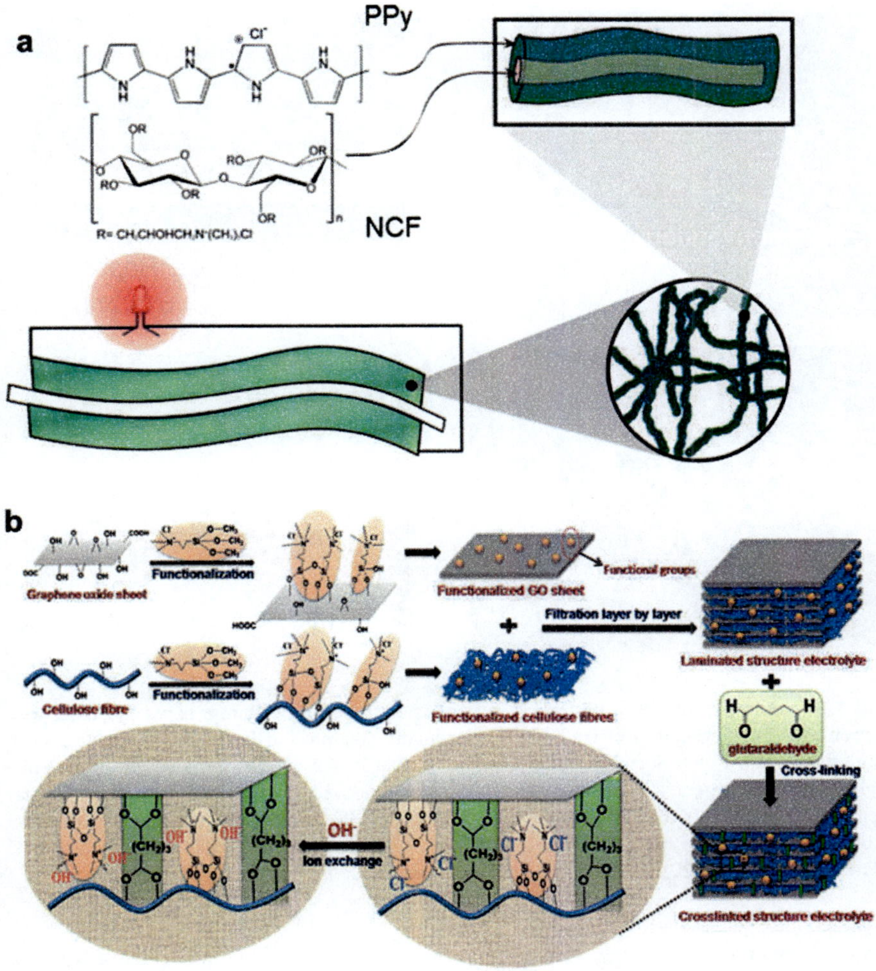

Fig. 7.13 **a** Nanocellulose-based supercapacitor, with PPy as functional electrode material and nanocellulose as substrate [44]. Copyright 2015, American Chemical Society. **b** Schematic diagram of the fabrication procedure for the nanocellulose-based zinc–air battery [45]. Copyright 2016, Wiley–VCH

7.4.3 Solar Cell

High optical performances of nanocellulose paper like high transparency and low haze make it good for the solar cell substrate, and there is great progress in the transparent paper-based solar cell recently. The nanocellulose-based solar cell is a very good example for green energy concept [46]. For example, Hu et al. firstly present an organic solar cell on the nanocellulose paper substrate with a power conversion efficiency of 0.40%, as shown in Fig. 7.14a, b [19]. It is found that the

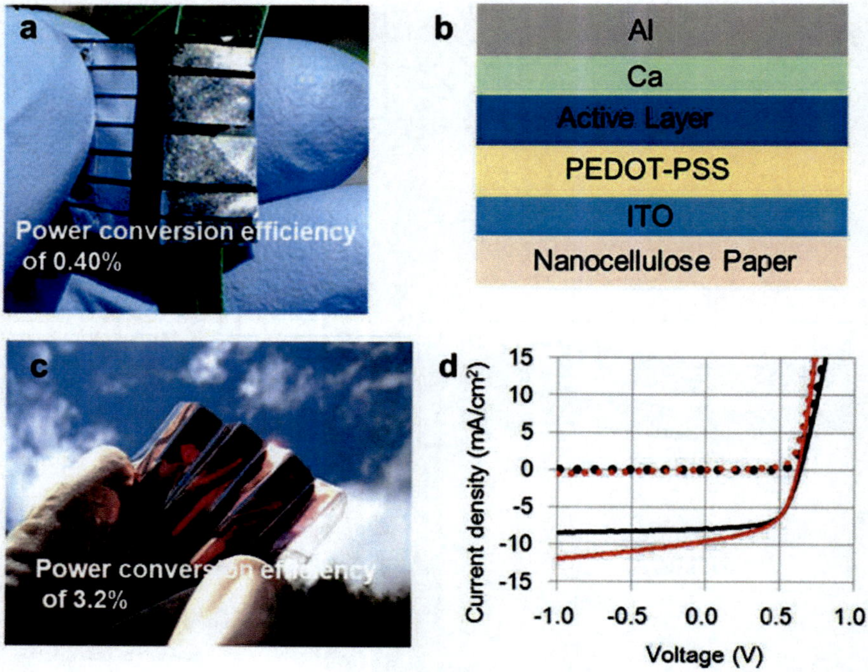

Fig. 7.14 **a** Printed solar cells on transparent and conductive nanocellulose paper substrate, with power conversion efficiency of 0.4%. **b** The schematic diagram for the solar cell structure [19]. Copyright 2013, Royal Society of Chemistry. **c** Optical image of solar cells based on foldable, lightweight, transparent, and conductive nanocellulose paper. **d** I–V characteristics of organic solar cells (P3HT/PCBM) in the dark (broken lines), and under 100 mW/cm^2 of AM 1.5 G illumination (solid lines); Red plot: solar cells based on nanofiber paper; black plot: solar cells based on ITO glass [47]. Copyright 2015, Nature Publication Group

power conversion efficiency could be further increased by maintaining the surface smoothness of substrate during fabrication. Nogi et al. report a transparent conductive paper based on cellulose nanofibers and silver nanowires, which has the comparable performance to ITO glass [47]. Solar cells based on nanocellulose are fabricated with transparent conductive nanocellulose paper, which has high power conversion efficiency of ~3.2%, as shown in Fig. 7.14c, d. Benefiting from the good affinity and entanglement between cellulose nanofibers and silver nanowires, the nanocellulose paper maintain its high conductivity, and the paper solar cells can also generate electricity after folding it. However, it's inevitable that long time of sunniness will accelerate the decay of nanocellulose, so that affect the performance of solar cells. Compared with flexible solar cells with traditional polymer substrate, the power conversion efficiency, stability, and robustness of the nanocellulose-based solar cells still need to be improved.

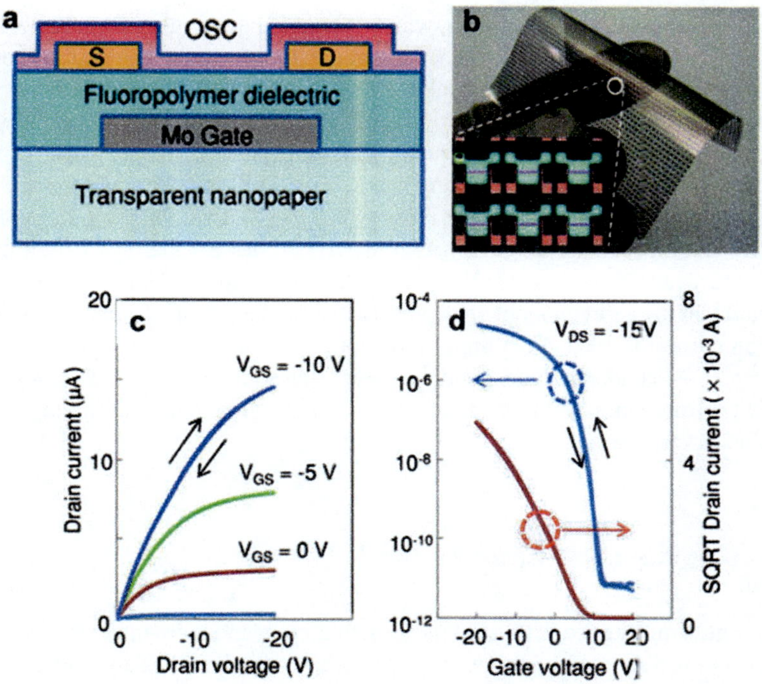

Fig. 7.15 **a** Cross-sectional structure of fabricated OTFT on nanopaper. **b** Optical image and optical microscopy image of the 20-μm-thick transparent nanopaper-based OTFT array. **c** Output and **d** transfer characteristics of OTFT under ambient air when the channel length is ~10 μm [48]. Copyright 2014, Wiley–VCH

7.4.4 *Thin Film Transistor*

Transistor is an important semiconductor device, which is used for the amplification and conversion of electronic signals or electrical power. Nanocellulose paper substrate has high mechanical strength, excellent surface property, and optical property, which make thin film transistor (TFT) constructed on such substrate have unique performance than TFT constructed on normal paper substrate or polymer substrate. For example, Fujisaki et al. report a thin and transparent organic thin-film transistor (OTFT) with high-mobility array based on nanocellulose paper substrate, which has good chemical durability, good thermal stability (>180 °C), and low coefficient of thermal expansion (CTE: 5–10 ppm/K) [48]. The advantages mentioned above ensure its capacity in building OTFT arrays on nanocellulose papers with a similar process to build OTFT arrays on normal glasses. Figure 7.15a presents the cross-sectional structure of the bottom-contact OTFT array on a nanocellulose paper. They first laminate the nanocellulose paper on a supporting glass substrate with an adhesive film. Before OTFT manufacturing, nanocellulose paper is covered with a 2 μm thick thermally crosslinked olefin polymer, which is beneficial for enhancing the mechanical flexibility and flattening the surface of nanocellulose paper. The bottom-contact OTFT array is fabricated with a soluble small-molecule and a fluoropolymer gate dielectric by lithographic and solution-based process. After integrating the whole device, the 20-μm-thick transparent nanocellulose paper is then peeled off from the supporting glass substrate without any cracks and damages. The picture of a real OTFT array on nanocellulose paper is shown in Fig. 7.15b, showing an excellent flexibility and mechanical durability. Figure 7.15c, d present the typical output and transfer characteristics of the OTFT on nanocellulose paper substrate, respectively. It is found that the short channel device has a channel length of ~10 μm and exhibits a high on-current (~10^{-5} A), a high on/off ratio (~10^6—~10^8), and a high mobility (1 cm^{-2} V^{-1} s^{-1}) under ambient air. It's believed that solution-based OTFT with low-cost and environmental friendly paper substrates have great potential for application in flexible electronics.

7.4.5 *Organic Light Emitting Diode*

Organic light emitting diode (OLED) is attracting more and more interest due to their features of thin, light-weight, and energy efficient, and has been widely applied for the display and lighting [15]. Normally, OLEDs are fabricated on rigid glass substrate or flexible plastic substrate, such as polyethylene terephthalate (PET) and polyethylene naphthalate (PEN). Glass substrate will increase the cost and bring inconvenience in device carrier due to its rigidity and density [49]. Nanocellulose paper substrate possesses good smoothness in nanoscale and tunable optical property from clear to hazy, making it suitable for flexible OLED devices [50, 51]. For example, Jin et al. demonstrate an OLED on a transparent paper based on chitin nanocellulose fibers (ChNF), with device structure given in Fig. 7.16a [50]. The OLED is fabricated on a 40 μm-thick ChNF paper, which is coated with a layer of thin (<1 μm) poly-methylmethacrylate (PMMA) as the surface planarization layer, while the reference device is developed based on a 50 μm-thick PEN substrate. Figure 7.16b presents the optical images of OLED based on ChNF paper, which operates stably in both flat and bending states. The OLED device can achieve a turn-on voltage of 2.5 V and a maximum luminance of 3.89 k cd/m^2, which is nearly the same as the reference device, as shown in Fig. 7.16c, d. Such nanocellulose-based OLED has potential application as a display unit for the smart packages. However, the long-time stability of this device still needs to be improved to ensure the practicability.

7.4.6 *Touch Screen*

Nanocellulose paper substrate has large optical haze value, making it perfect to eliminate strong light reflection when exposed to sunlight and thus a good substrate for

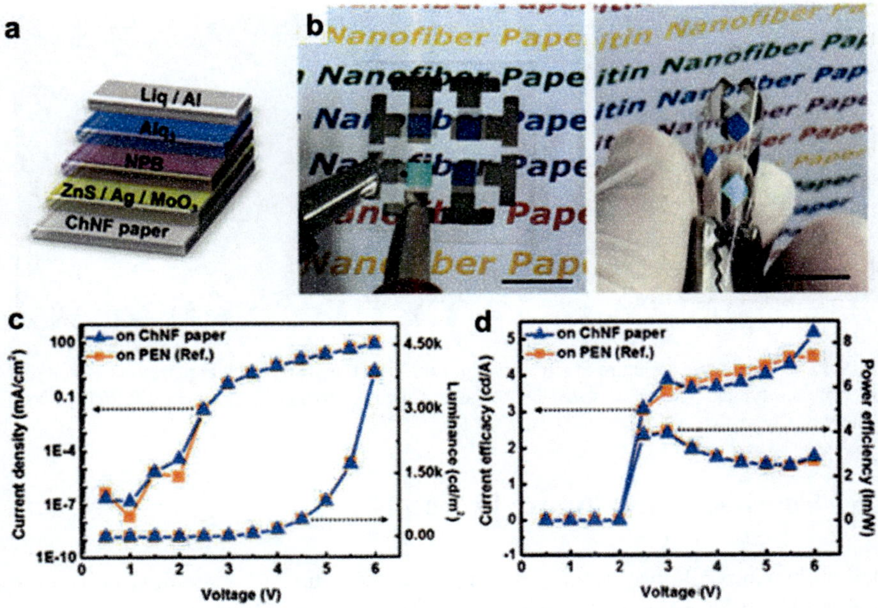

Fig. 7.16 **a** Device structure of the OLED on nanocellulose paper substrate. **b** Optical images of the flexible OLED operating in flat state (left) and bending state (right). Scale bar is 1 cm. **c** J-L-V characteristics and **d** current/power efficiency of the fabricated OLED and a reference OLED device [50]. Copyright 2016, Wiley–VCH

touch screen [15]. For touch screen application, the transparency of the nanocellulose paper and the electrode should reach more than 90% and haze smaller than 1% [52]. Normally, CNTs, graphene, and Ag nanowires are reasonable electrodes for the touch screen application. Fang et al. develop an analog resistive paper touch screen, which uses a transparent nanocellulose paper as the electrode, as shown in Fig. 7.17a [33]. Here, screen-printing method is adopted to form silver patterns on the CNT-coated transparent nanocellulose paper. The transparent nanocellulose paper with CNTs electrode is then used to fabricate paper touch screen. "eGalaxTouch" software is used to test the transparent nanocellulose paper touch screen. The word "paper" is written on the touch screen through a stylus pen and then successfully displayed on the computer through a USB port (Fig. 7.17b), which shows that the polymer substrate like PET can be replaced by transparent and conductive nanocellulose paper in electronic devices. However, the touch screen doesn't belong to single-use devices, and the durability of the nanocellulose-based touch screen still need to be improved, as nanocellulose paper can't against moisture from human hand.

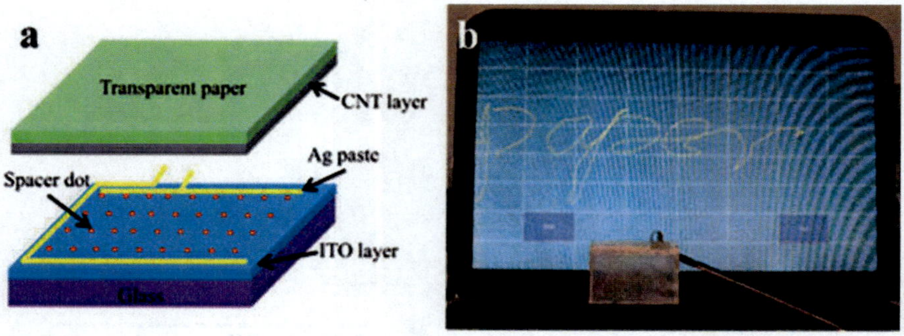

Fig. 7.17 **a** Schematic structure of a transparent four-wire touch screen. **b** Paper touch screen that can simulate typical "paper" assisted by "eGalax Touch" software [33]. Copyright 2013, Royal Society of Chemistry

7.5 Conclusion and Future Prospects

Definitely, nanocellulose paper is an appropriate candidate for the next-generation flexible electronic substrate, by virtue of the advantages of light, transparent, mechanical strong, thermal stable, biodegradable and biocompatible, etc. Based on the electrode fabrication technology that will not hamper the advantages of nanocellulose paper substrate, including conductive nanoparticles, nanowires, nanosheets, and polymers, various of flexible electronics such as energy harvester/sensor, energy storage device, solar cell, thin film transistor, and organic light emitting diode, touch screen, etc. are successfully developed. However, these novel nanocellulose-based electronics are just concepts at present. Compared with traditional glass and polymer substrate, nanocellulose paper still have the following limitations: (1) the cost for fabricating nanocellulose paper is still high; (2) humidity stability and shelf life of nanocellulose paper still need to be improved; (3) large-scale electrode fabrication on the nanocellulose paper is not mature. It's believed that more nanocellulose-based electronics will appear in the future and may become popular commercial products, if the above problems can be solved. That would depend on improving the unique performances of nanocellulose paper and developing mature large-scale manufacture technology, like roll-to-roll method. We believe that nanocellulose paper-based electronics is the one hope to solve the E-waste problem, even though there is still a long way to go.

Acknowledgements Reproduced from [15] with permission from the Royal Society of Chemistry.
Reproduced from [19] with permission from the Royal Society of Chemistry.
Reproduced from [22] with permission from the Royal Society of Chemistry.
Reproduced from [33] with permission from the Royal Society of Chemistry.
Reproduced from [35] with permission from the Royal Society of Chemistry.

References

1. R. Widmer, H. Oswald-Krapf, D. Sinha-Khetriwal, M. Schnellmann, H. Böni, Global perspectives on e-waste. Environ. Impact Assess. Rev. **25**(5), 436–458 (2005)
2. I.C. Nnorom, O. Osibanjo, Overview of electronic waste (e-waste) management practices and legislations, and their poor applications in the developing countries. Resour. Conserv. Recycl. **52**(6), 843–858 (2008)
3. C. Chen, L. Hu, Nanocellulose toward advanced energy storage devices: structure and electrochemistry. Acc. Chemical. Res. **51**(12), 3154–3165 (2018)
4. https://www.sciencedaily.com/releases/2017/12/171213143714.htm (2016)
5. B. Balakrishnan Ramesh, P. Anand Kuber, B. Chiya Ahmed, Electrical and electronic waste: a global environmental problem. Waste Manag. Res. **25**(4), 307–318 (2007)
6. D. Tobjörk, R. Österbacka, Paper electronics. Adv. Mater. **23**(17), 1935–1961 (2011)
7. S. Thiemann, S.J. Sachnov, F. Pettersson, R. Bollström, R. Österbacka, P. Wasserscheid, J. Zaumseil, Cellulose-based ionogels for paper electronics. Adv. Func. Mater. **24**(5), 625–634 (2014)
8. A. Russo, B.Y. Ahn, J.J. Adams, E.B. Duoss, J.T. Bernhard, J.A. Lewis, Pen-on-paper flexible electronics. Adv. Mater. **23**(30), 3426–3430 (2011)
9. L. Hu, J.W. Choi, Y. Yang, S. Jeong, F. La Mantia, L.-F. Cui, Y. Cui, Highly conductive paper for energy-storage devices. Proc. Natl. Acad. Sci. **106**(51), 21490–21494 (2009)
10. L. Yuan, X. Xiao, T. Ding, J. Zhong, X. Zhang, Y. Shen, B. Hu, Y. Huang, J. Zhou, Z.L. Wang, Paper-based supercapacitors for self-powered nanosystems. Angew. Chem. **124**(20), 5018–5022 (2012)
11. X. Fan, J. Chen, J. Yang, P. Bai, Z. Li, Z.L. Wang, Ultrathin, rollable, paper-based triboelectric nanogenerator for acoustic energy harvesting and self-powered sound recording. ACS Nano **9**(4), 4236–4243 (2015)
12. Q. Zhong, J. Zhong, B. Hu, Q. Hu, J. Zhou, Z.L. Wang, A paper-based nanogenerator as a power source and active sensor. Energy Environ. Sci. **6**(6), 1779–1784 (2013)
13. K. Yong-Hoon, M. Dae-Gyu, H. Jeong-In, Organic TFT array on a paper substrate. IEEE Electron Device Lett. **25**(10), 702–704 (2004)
14. R.J. Moon, A. Martini, J. Nairn, J. Simonsen, J. Youngblood, Cellulose nanomaterials review: structure, properties and nanocomposites. Chem. Soc. Rev. **40**(7), 3941–3994 (2011)
15. H. Zhu, Z. Fang, C. Preston, Y. Li, L. Hu, Transparent paper: fabrications, properties, and device applications. Energy Environ. Sci. **7**(1), 269–287 (2014)
16. Y. Habibi, L.A. Lucia, O.J. Rojas, Cellulose nanocrystals: chemistry, self-assembly, and applications. Chem. Rev. **110**(6), 3479–3500 (2010)
17. A. Isogai, T. Saito, H. Fukuzumi, TEMPO-oxidized cellulose nanofibers. Nanoscale **3**(1), 71–85 (2011)
18. M. Nogi, S. Iwamoto, A.N. Nakagaito, H. Yano, Optically transparent nanofiber paper. Adv. Mater. **21**(16), 1595–1598 (2009)
19. L. Hu, G. Zheng, J. Yao, N. Liu, B. Weil, M. Eskilsson, E. Karabulut, Z. Ruan, S. Fan, J.T. Bloking, M.D. McGehee, L. Wågberg, Y. Cui, Transparent and conductive paper from nanocellulose fibers. Energy Environ. Sci. **6**(2), 513–518 (2013)
20. H. Zhu, S. Parvinian, C. Preston, O. Vaaland, Z. Ruan, L. Hu, Transparent nanopaper with tailored optical properties. Nanoscale **5**(9), 3787–3792 (2013)
21. M. Henriksson, L.A. Berglund, P. Isaksson, T. Lindström, T. Nishino, Cellulose nanopaper structures of high toughness. Biomacromol **9**(6), 1579–1585 (2008)
22. H. Zhu, Z. Xiao, D. Liu, Y. Li, N.J. Weadock, Z. Fang, J. Huang, L. Hu, Biodegradable transparent substrates for flexible organic-light-emitting diodes. Energy Environ. Sci. **6**(7), 2105–2111 (2013)
23. M. Nogi, C. Kim, T. Sugahara, T. Inui, T. Takahashi, K. Suganuma, High thermal stability of optical transparency in cellulose nanofiber paper. Appl. Phys. Lett. **102**(18), 181911 (2013)
24. S. Soares, G. Camino, S. Levchik, Comparative study of the thermal decomposition of pure cellulose and pulp paper. Polym. Degrad. Stab. **49**(2), 275–283 (1995)

25. M.-C. Hsieh, C. Kim, M. Nogi, K. Suganuma, Electrically conductive lines on cellulose nanopaper for flexible electrical devices. Nanoscale **5**(19), 9289–9295 (2013)
26. Y.H. Jung, T.-H. Chang, H. Zhang, C. Yao, Q. Zheng, V.W. Yang, H. Mi, M. Kim, S.J. Cho, D.-W. Park, H. Jiang, J. Lee, Y. Qiu, W. Zhou, Z. Cai, S. Gong, Z. Ma, High-performance green flexible electronics based on biodegradable cellulose nanofibril paper. Nat. Commun. **6**, 7170 (2015)
27. X. Gao, L. Huang, B. Wang, D. Xu, J. Zhong, Z. Hu, L. Zhang, J. Zhou, Natural materials assembled, biodegradable, and transparent paper-based electret nanogenerator. ACS Appl. Mater. Interf. **8**(51), 35587–35592 (2016)
28. I. Siró, D. Plackett, Microfibrillated cellulose and new nanocomposite materials: a review. Cellulose **17**(3), 459–494 (2010)
29. G. Chinga-Carrasco, D. Tobjörk, R. Österbacka, Inkjet-printed silver nanoparticles on nano-engineered cellulose films for electrically conducting structures and organic transistors: concept and challenges. J. Nanopart. Res. **14**(11), 1213 (2012)
30. H. Koga, T. Saito, T. Kitaoka, M. Nogi, K. Suganuma, A. Isogai, Transparent, conductive, and printable composites consisting of TEMPO-oxidized nanocellulose and carbon nanotube. Biomacromolecules **14**(4), 1160–1165 (2013)
31. https://www.cheaptubes.com/resources/graphene-battery-users-guide/ (2019)
32. Z. Shi, G.O. Phillips, G. Yang, Nanocellulose electroconductive composites. Nanoscale **5**(8), 3194–3201 (2013)
33. Z. Fang, H. Zhu, C. Preston, X. Han, Y. Li, S. Lee, X. Chai, G. Chen, L. Hu, Highly transparent and writable wood all-cellulose hybrid nanostructured paper. J. Mater. Chem. C **1**(39), 6191–6197 (2013)
34. J. Zhong, H. Zhu, Q. Zhong, J. Dai, W. Li, S.-H. Jang, Y. Yao, D. Henderson, Q. Hu, L. Hu, J. Zhou, Self-powered human-interactive transparent nanopaper systems. ACS Nano **9**(7), 7399–7406 (2015)
35. C. Preston, Z. Fang, J. Murray, H. Zhu, J. Dai, J.N. Munday, L. Hu, Silver nanowire transparent conducting paper-based electrode with high optical haze. J. Mater. Chem. C **2**(7), 1248–1254 (2014)
36. W. Kang, C. Yan, C.Y. Foo, P.S. Lee, Foldable electrochromics enabled by nanopaper transfer method. Adv. Func. Mater. **25**(27), 4203–4210 (2015)
37. J.-U. Lee, D. Yoon, H. Cheong, Estimation of young's modulus of graphene by Raman spectroscopy. Nano Lett. **12**(9), 4444–4448 (2012)
38. K. Gao, Z. Shao, X. Wu, X. Wang, J. Li, Y. Zhang, W. Wang, F. Wang, Cellulose nanofibers/reduced graphene oxide flexible transparent conductive paper. Carbohyd. Polym. **97**(1), 243–251 (2013)
39. N.D. Luong, N. Pahimanolis, U. Hippi, J.T. Korhonen, J. Ruokolainen, L.-S. Johansson, J.-D. Nam, J. Seppälä, Graphene/cellulose nanocomposite paper with high electrical and mechanical performances. J. Mater. Chem. **21**(36), 13991–13998 (2011)
40. N.D. Luong, J.T. Korhonen, A.J. Soininen, J. Ruokolainen, L.-S. Johansson, J. Seppälä, Processable polyaniline suspensions through in situ polymerization onto nanocellulose. Eur. Polymer J. **49**(2), 335–344 (2013)
41. K.-H. Kim, K.Y. Lee, J.-S. Seo, B. Kumar, S.-W. Kim, Paper-based piezoelectric nanogenerators with high thermal stability. Small **7**(18), 2577–2580 (2011)
42. C. Wu, T.W. Kima, S. Sung, J.H. Park, F. Li, Ultrasoft and cuttable paper-based triboelectric nanogenerators for mechanical energy harvesting. Nano Energy **44**, 279–287 (2018)
43. M. Jiao, T. Liu, C. Chen, M. Yue, G. Pastel, Y. Yao, H. Xie, W. Gan, A. Gong, X. Li, L. Hu, Holey three-dimensional wood-based electrode for vanadium flow batteries. Energy Stor. Mater. **27**, 327–332 (2020)
44. Z. Wang, D.O. Carlsson, P. Tammela, K. Hua, P. Zhang, L. Nyholm, M. Strømme, Surface modified nanocellulose fibers yield conducting polymer-based flexible supercapacitors with enhanced capacitances. ACS Nano **9**(7), 7563–7571 (2015)
45. J. Zhang, J. Fu, X. Song, G. Jiang, H. Zarrin, P. Xu, K. Li, A. Yu, Z. Chen, Laminated cross-linked nanocellulose/graphene oxide electrolyte for flexible rechargeable zinc–air batteries. Adv. Energy Mater. **6**(14), 1600476 (2016)

46. L. Gao, L. Chao, M. Hou, J. Liang, Y. Chen, H.-D. Yu, W. Huang, Flexible, transparent nanocellulose paper-based perovskite solar cells. npj Flex. Electron.**3**(1), 4 (2019)
47. M. Nogi, M. Karakawa, N. Komoda, H. Yagyu, T.T. Nge, Transparent conductive nanofiber paper for foldable solar cells. Sci. Rep. **5**, 17254 (2015)
48. Y. Fujisaki, H. Koga, Y. Nakajima, M. Nakata, H. Tsuji, T. Yamamoto, T. Kurita, M. Nogi, N. Shimidzu, Transparent nanopaper-based flexible organic thin-film transistor array. Adv. Func. Mater. **24**(12), 1657–1663 (2014)
49. Z. Fang, H. Zhu, C. Preston, L. Hu, Development, application and commercialization of transparent paper. Transl. Mater. Res. **1**(1), 015004 (2014)
50. J. Jin, D. Lee, H.-G. Im, Y.C. Han, E.G. Jeong, M. Rolandi, K.C. Choi, B.-S. Bae, Chitin nanofiber transparent paper for flexible green electronics. Adv. Mater. **28**(26), 5169–5175 (2016)
51. J. Tao, R. Wang, H. Yu, L. Chen, D. Fang, Y. Tian, J. Xie, D. Jia, H. Liu, J. Wang, F. Tang, L. Song, H. Li, Highly transparent, highly thermally stable nanocellulose/polymer hybrid substrates for flexible OLED devices. ACS Appl. Mater. Inter. **12**(8), 9701–9709 (2020)
52. H. Zhu, Z. Fang, Z. Wang, J. Dai, Y. Yao, F. Shen, C. Preston, W. Wu, P. Peng, N. Jang, Q. Yu, Z. Yu, L. Hu, Extreme light management in mesoporous wood cellulose paper for optoelectronics. ACS Nano **10**(1), 1369–1377 (2016)

Chapter 8
Energy Storage Applications

Zhaohui Wang and Leif Nyholm

Abstract This chapter briefly explains the function of electrochemical energy storage devices, such as batteries and supercapacitors, and how nanocellulose can be used in the manufacturing of such devices. It is demonstrated that the use of nanocellulose facilitates the development of sustainable, versatile, and inexpensive electrochemical energy storage devices since nanocellulose can be used in the manufacturing of electrode materials, separators, and current collectors. Such nanocellulose-based energy storage devices can most likely be manufactured at low costs and may therefore serve as an interesting complement to conventional batteries and supercapacitors.

8.1 Fundamentals of Energy Storage Devices

Electrochemical energy storage is typically based on the utilization of batteries or supercapacitors. As seen in the schematic illustration in Fig. 8.1, such devices contain two electrodes separated by an electrolyte [1, 2]. The latter can be an aqueous or non-aqueous solution containing at least one salt that ideally should be electrochemically inert. The electrolyte is often situated within the pores of a separator that physically separates the electrodes preventing short-circuit of the battery [2]. In a charged battery, the two electrodes (which are electrically contacted using current collectors) have different potentials, which cause electrochemical reactions to take place at the electrodes, i.e., reduction at the cathode and oxidation at the anode when using the battery. The equilibrium potential of each electrode, E, can be calculated using Nernst equation $E = E_0 - RT/nF \ln Red/Ox$, where E_0 is the standard potential, R denotes the gas constant, T is the temperature in Kelvin, n is the number of electrons involved in the process, F is the Faraday constant while Red and Ox denote the activities (often

Z. Wang
College of Materials Science and Engineering, Hunan University, Changsha 410082, China
e-mail: sunnywang@hnu.edu.cn

L. Nyholm (✉)
Department of Chemistry-Ångström Laboratory, Uppsala University, Uppsala, Sweden
e-mail: leif.nyholm@kemi.uu.se

L. Hu et al. (eds.), *Emerging Nanotechnologies in Nanocellulose*, NanoScience and Technology, https://doi.org/10.1007/978-3-031-14043-3_8

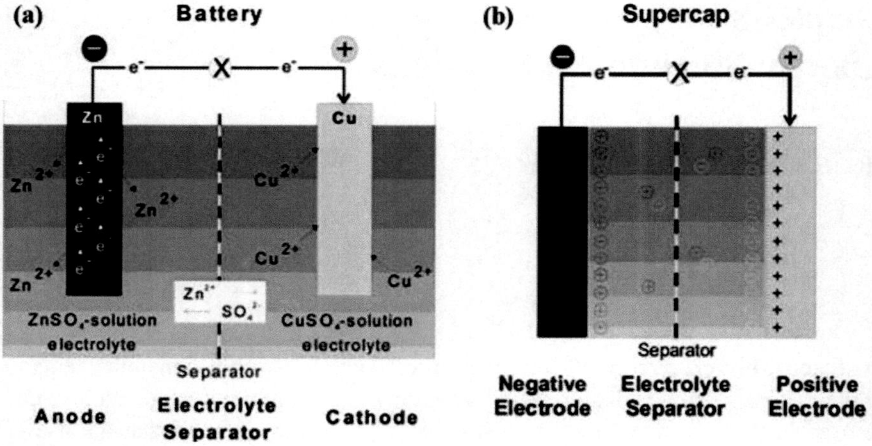

Fig. 8.1 Schematic illustrations of (**a**) a battery and (**b**) a supercapacitor. The operation of a battery is based on electrochemical reactions at the electrodes while the charge in an electrochemical capacitor (supercapacitor) comprising porous carbon electrodes mainly is stored in the electric double layers at the electrode-electrolyte interfaces. Reprinted from [2]. Copyright 2004 American Chemical Society (ACS)

approximated with the concentrations) of the reduced and oxidized form of the redox couple, respectively. The electrochemical reactions (or redox reactions) are consequently responsible for the current delivered by the battery. The transfer of electrons between the electrodes in the outer circuit gives rise to a simultaneous movement (i.e., migration) of ions in the electrolyte between the electrodes, which ensures that the charge balance is maintained in the battery. Since the electrode reactions alter the concentrations of species at the electrode surfaces there is also always mass transport involving diffusion to and from the electrodes. For many battery materials, the charge time is limited by the mass transport (diffusion) within the electrode material, particularly when thick electrode materials are used to obtain high capacities. During discharge, the cell voltage is always lower than the equilibrium cell voltage (i.e., the open circuit cell voltage) due to the presence of electron transfer overpotentials, mass transport effects and an ohmic drop (i.e., iR drop) due to the cell resistance. As the battery is discharged, the cell voltage decreases with time. This effect can be more or less pronounced depending on the electrode materials employed as a result of the shapes of their current–voltage curves.

The energy, W, stored in the battery depends on the cell voltage, E_{cell}, i.e., the difference between the cathode and anode potentials, and the amount of available charge, Q, in the capacity-limiting electrode according to $W = E_{cell} \cdot Q$. Non-aqueous electrolytes are often used to enable an increase in E_{cell} due to the use of redox reactions which cannot be exploited in aqueous electrolytes due to the reduction and/or oxidation of water. To increase Q, electrode reactions involving many electrons are generally desirable although the rates of such reactions often are relatively low due to their inherent complexity (e.g., due to significant structural changes). The capacity of a battery can naturally also be increased by increasing the amount of the electrode materials while balancing the electrode capacities properly. In practice, the success of the latter approach may, however, be limited by the longer mass transport distances associated with the use of thicker electrodes. The risk that the full capacity may not be accessible on the time scale of the experiments consequently generally increases when increasing the amount of electrode material.

While the energy of a battery readily can be obtained once E_{cell} and Q are known, the current that can be delivered by the battery is more difficult to predict as it cannot be calculated based on thermodynamic data. This is a consequence of the fact that a charged battery does not constitute an equilibrium system. The current, and hence the power, $P = W/t$ (where t denotes the discharge time) that can be delivered by a battery is hence determined by the cell resistance, or the electron transfer or mass transport rate associated with the rate-limiting electrode. Since the cell resistance of a well-functioning battery should be relatively low, the performance of a battery during discharge or charge is generally governed by the rate-limiting electrode reaction.

The energy stored in a supercapacitor, which is given by $W = C_{cell} \cdot (E_{cell})^2/2$ where C_{cell} denotes the cell capacitance, typically stems from both double layer charging and pseudo-capacitive effects (Fig. 8.2a, b). Double layer charging involves charging of the double layer capacitance, C_d, associated with the interface between the electrode and the electrolyte involving a redistribution of the ions in the double layer as a result of a change in the potential (and hence the charge) of the electrode. The thickness of the double layer is typically of the order of a few nanometers (assuming an electrolyte concentration of about 1 M). As the double layer charging process does not involve any redox reactions it is analogous to the charging of a conventional capacitor. The double charging current, i_c, is given by $i_c = C_d \cdot dE/dt$, where dE/dt denotes the rate by which the potential of the electrode is changed. As C_d depends on the electroactive area of the electrode, supercapacitor electrodes are typically based on materials with high surface areas, e.g., 2000 $m^2\ g^{-1}$. If the supercapacitor is charged using a potential step, the charge time is determined by the $R_{cell} \cdot C_d$ time constant of the cell where R_{cell} denotes the cell resistance. Due to the surface confined charge storage process, supercapacitors can generally be charged and discharged very rapidly which is why supercapacitors generally exhibit high power densities. Their charge storage capacities and hence energy densities are, on the other hand, generally much lower than those for batteries since the redox reactions in a battery ideally involve, not only the electrode surfaces, but the entire volume of the battery material. The self-discharge rates are normally also significantly higher for supercapacitors than for batteries since the rapid surface-based mechanisms are active both during the charge and discharge steps. The cell voltage for a supercapacitor also decreases continuously as the supercapacitor is discharged, which is why the influence of a self-discharge is more clearly seen for a supercapacitor than for a battery.

Although the distinction between batteries and supercapacitors may appear straightforward based on the description above, the situation is commonly complicated by the fact that surface-confined redox reactions often are found to be present also for supercapacitor electrode materials. These reactions, which typically involve oxidation or reduction of surface groups, give rise to a so-called pseudo-capacitance

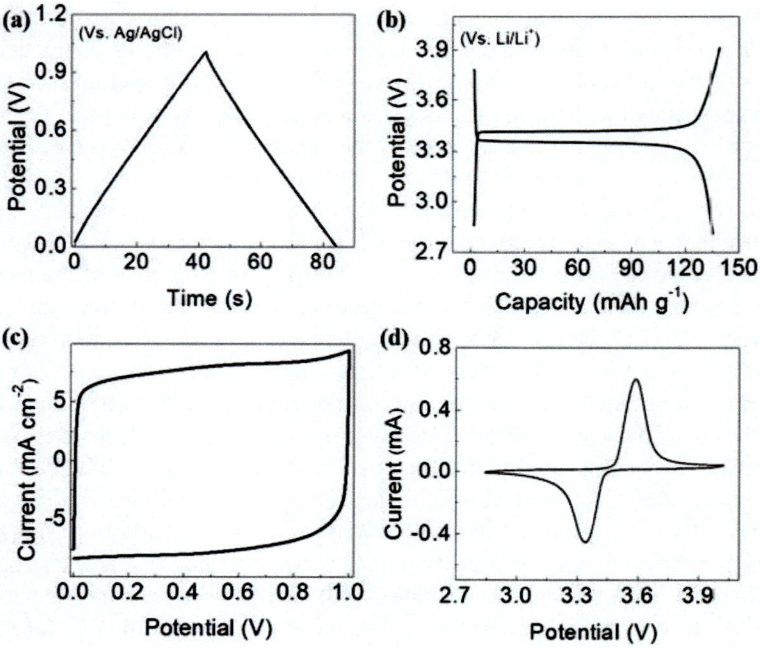

Fig. 8.2 Electrochemical performance of (**a**) a supercapacitor and (**b**) a battery material. Under constant current conditions the potential depends linearly on time for a supercapacitor material (**a**) while the battery material (e.g., $LiFePO_4$ (LFP)) exhibits a potential plateau (**b**). Cyclic voltammograms for a supercapacitor material (**c**) and a battery material (e.g., LFP). Note that the latter voltammogram features faradaic redox peaks (**d**)

that is not straightforwardly distinguished from the double layer capacitance. The latter stems from the fact that the rates of these surface-confined redox reactions (unlike those of the redox reactions employed in battery electrodes) are not limited by mass transport effects since only the electrode surface is electroactive. The capacitive-like behavior of the surface confined redox reaction can often be explained based on a distribution of redox potentials (rather than a single redox potential) since the oxidation or reduction of one site changes the potential needed to oxidize or reduce the remaining sites on the electrode surface. Due to the presence of surface confined redox reactions, the capacitances of many electrode materials with high surface areas consequently often exceed the values expected based on the double layer capacitances.

The performances of batteries and supercapacitors are generally studied using electrochemical methods such as voltammetry, chronopotentiometry (i.e., controlled current techniques), and electrochemical impedance spectroscopy (EIS) using two- or three-electrode set-ups. In voltammetry, the current is measured as a function of the electrode potential and oxidation and reduction processes are typically seen as peaks at different potentials (Fig. 8.2c, d). In chronopotentiometry, the potential of the electrode is often measured while applying a constant current to the electrode to induce oxidations or reductions while the impedance of the electrochemical cell is studied as a function of the ac frequency in EIS. As these techniques have different advantages and disadvantages, it is generally wise to employ several techniques when studying the performance of an energy storage device. The use of a three-electrode cell (containing a reference electrode, a counter electrode, and the working electrode, i.e., the electrode under study), allows the electrochemical performance of the working electrode be directly studied. In a two-electrode cell the electrochemical behavior will, on the other hand, be determined by the limiting electrode which may not always be the electrode of interest.

The characterization of battery materials mainly includes the determination of the specific capacity of the electrode, the capacity retention during cycling, the dependence of the capacity on the rate of the cycling and the Coulombic efficiency (i.e., the ratio between the discharge and charge capacities). The capacity of the electrode is commonly normalized with respect to the mass of the electroactive components in the electrode, e.g., in mAh/g. It should, however, be noted that this value can be misleading as high values readily can be obtained for monolayers of electroactive compounds even though the amount of the electroactive species clearly would be too low to be of interest in any practical application. Capacities normalized with respect to the surface area are often used for porous or thin film electrodes, especially when it is difficult to estimate the mass of the electroactive compounds. Since the area-based capacity depends on the thickness of the electrode, the latter should always be specified. The capacities for porous electrodes are hence best given in the form of volumetric capacities (e.g., in mAh/cm^3). In addition, the mass of all inactive components (e.g., binders and conducting additives) in the electrode should always be stated, including the estimated weight of the electrolyte present in the pores of a porous electrode. The capacity retention and rate capability tests are normally carried out under constant current conditions and the cycling rate is then typically reported in the form of a C-rate, where 1C corresponds to discharge and charge steps, each with a duration of one hour. Analogously, rates of C/2 and 2C indicate that the charge and discharge steps were 2 h and 30 min long, respectively. When using constant current experiments, a higher and higher current is used to increase the rate of the experiments. This gives rise to an increased iR drop, which, unless compensated for, results in a decreasing capacity with increasing rate. The Coulombic efficiency (i.e., $Q_{discharge}/Q_{charge}$) is used to evaluate the reversibility of the overall charge storage process where 100% implies that the stored charge could be completely recovered during the discharge. Given that the expected capacity retention after 100 cycles merely is 37% assuming a Coulombic efficiency of 99% (as $0.99^{100} = 0.37$), it is immediately clear that the Coulombic efficiency must remain very close to 100% if the battery or a supercapacitor is to remain operational during many cycles. It should also be noted that the presence of irreversible reactions during the initial charging steps often decreases the Coulombic efficiency considerably during the initial cycles. This effect requires proper balancing of the cells as the irreversible reactions may decrease the cell capacity depending on which electrode is capacity limiting.

The characterizations of electrodes for supercapacitors generally involve the same type of measurements as for battery electrodes apart from the fact that capacitances (rather than capacities), are reported. In analogy with the specific capacities for battery electrodes, specific electrode capacitances based on the mass, area or volume of the electrode are commonly used. As in the battery electrode case, a normalization with respect to the mass of the electrode can yield very high values if very small amounts of the electrode material are used as this increases the likelihood of accessing the entire surface of the material. The mass of the complete electrode (including the mass of the electrolyte in the pores of a porous electrode) should always be given in addition to the mass of the active components. Reports of area-based capacitances should be accompanied by the thickness of the electrodes whereas volumetric capacitances should be employed for porous electrodes. It should be clearly stated if the reported capacitance corresponds to a cell capacitance or an electrode capacitance, note that $C_{cell} = C_{electrode}/4$ for symmetrical supercapacitors (i.e., devices containing two identical electrodes). As the cell capacitance is limited by the lowest capacitance it important to balance the electrode capacitances properly. Electrode material exhibiting oxidation and reduction peaks in the voltammograms or plateaus in the chronopotentiograms should not be classified as supercapacitor materials since a supercapacitor electrode should exhibit a capacitance that is practically constant within its operating voltage window.

8.2 Key Challenges Addressed Using Nanocellulose

The increasing use of batteries and supercapacitors in our society generates a strong need for sustainable and inexpensive devices with improved energy and power densities as well as enhanced safety and lifespan [2–5]. As will be explained below, many of these issues can be addressed using nanocellulose-based energy storage devices. This is a consequence of the physical and chemical properties of nanocellulose and the fact that cellulose is a very common, versatile, sustainable, and relatively inexpensive material that straightforwardly can be produced in large quantities [6, 7].

One very important contemporary challenge is to develop energy storage devices that combine the advantages of batteries and supercapacitors in devices exhibiting both high energy and power densities. This typically requires the use of porous electrodes with high mass loadings in which the electrode material can be completely utilized. As will be described in the next section, such electrodes can be made in the form of nanocellulose composites in which the nanocellulose serves as a large-surface area and porous scaffold while the electroactive material is present as a thin film on the nanocellulose fibers.

Another important issue involves the realization of up-scalable energy storage devices. While there are many reports on electrode materials with very high energy and/or power densities, the majority of these results were in fact obtained with electrodes having very low mass loadings or which also contained large amounts of inactive components [5, 8]. This results in significantly lower values when the normalization is carried out with respect to the electrode mass rather than the mass of the electroactive material. There is hence a large demand for the development of inexpensive electrode with high mass loadings and electrode materials that readily can undergo straightforward up-scaling. Using nanocellulose this can be accomplished employing the nanocellulose as a combined host and binder for many electroactive materials. As the nanocellulose is lightweight and as the electrode manufacturing process can be based on, e.g., filtration steps this provides new possibilities for the manufacturing of inexpensive cellulose-based energy storage devices based on conventional paper-making approaches. This can pave the way for the realization of inexpensive paper-based batteries and supercapacitors in which nanocellulose is used not only in the electrodes but also in the separator and current collectors.

Currently, there is a large interest in various types of flexible electronic devices, which in turn has generated significant research aimed at the development of flexible batteries and supercapacitors [9]. In this work, nanocellulose and composites containing nanocellulose have been found to be very useful due to their mechanical properties. As a result, there has been a development of a range of flexible energy storage devices, some of which are described in more detail in the next section.

Using nanocellulose, it is also possible to manufacture carbon fibers employing carbonization of the nanocellulose. The obtained carbon materials (the structure of which mimicking that of the nanocellulose) can either be used as electrode materials or as conductive materials in the manufacturing of different types of composites used as electrodes or current collectors [10].

As already indicated, nanocellulose can in fact be used in the manufacturing of essentially all parts used in a battery or supercapacitor, i.e., in the current collectors, electrodes and separator. The manufacturing processes (involving e.g., filtration) developed so far also allow the thicknesses of these parts to be readily controlled. The porosity and pore size distributions, which are very important to the electrochemical performances of electrodes as well as that of the separator, can be controlled using several approaches.

The properties of the nanocellulose can also be altered via surface modifications of the cellulose fibers in order to control the interaction between the nanocellulose and other components present in different composites, e.g., via modifications of the nanocellulose surface charge [11, 12].

At present, there is a growing interest in the use of nanocellulose-based separators for batteries and supercapacitors. One reason for this interest is the desire to replace the polyolefin-based separators commonly used in batteries containing non-aqueous electrolytes with more sustainable separators exhibiting better electrolyte wettabilities and thermal stabilities [13]. The attainment of a homogeneous current distribution at the electrode, which is very important for the electrochemical performance of the cell, can likewise be facilitated since the pore structure of nanocellulose-based separators can be made more homogeneous [14, 15]. The surfaces of the nanocellulose separators can also readily be functionalized to transform the separator from an electroinactive to an electroactive component in the battery [16, 17]. The latter approach indicates that it is possible to increase the energy and power density of a battery merely by replacing the conventional separator with a nanocellulose-based separator. It should be noted that the approach does not require any modification of the other parts of the battery.

In batteries, it is generally desirable to decrease the thickness of the separator in order to improve the energy and power densities of the battery. The problem here is that it is difficult to manufacture and handle very thin separators and that the safety of the battery may be jeopardized in the presence of a very thin separator. With nanocellulose-based separators, it has been shown that these problems can be solved employing separators composed of a thin nanocellulose layer (serving as the separator) and a thicker layer of a porous electroactive nanocellulose composite [16]. In this case, the latter layer serves as a mechanical support layer as well as an extra source of capacity.

The crystallinity of the nanocellulose is another factor that needs to be considered for batteries containing non-aqueous electrolytes as nanocellulose with a high degree of crystallinity should be used to the minimize the introduction of water in the battery [13, 18]. The application of nanocellulose in electrodes, separators and in different energy storage devices will be discussed in more detail in the next section.

8.3 Nanocellulose for Energy Storage Applications

8.3.1 Nanocellulose-Structured Electrodes

Cellulose-based electrodes hold great promise compared to traditional material-based electrodes regarding environmental friendliness, low-cost, sustainability, feasible chemical modification and structure [5, 10, 19, 20]. As a relatively new member of the cellulose family, nanocellulose is composed of fibers with nanoscale diameters and lengths ranging from nanometers to micrometers. The properties of the nanocellulose entities, including their capacity for extensive chemical modification and the ability to form versatile semi-crystalline nanofibril morphologies, are combined with specific features such as very large surface areas [21]. The high surface area of nanocellulose and the reactive hydroxyl groups on the nanocellulose surface are anticipated to create interactions with other materials and the possibility of forming uniform coatings. This has resulted in the development of nanocellulose composite electrodes with tunable structures and properties. Nanocellulose substrates are also flexible and are believed to be able to accommodate large volumetric expansions. This may improve the cycling stability of the electrode materials, which is an important aspect when employing different electroactive materials [5]. Some recent progress in the development of nanocellulose-based electrodes within this rapidly growing field is reviewed below.

8.3.1.1 Nanocellulose as Electrode Building Blocks

Nanocellulose fibers (NCFs) have been widely explored as 1D building blocks in the development of sustainable freestanding electrodes with high electrochemical performances and mechanical flexibility. In this approach, the electroactive materials are embedded in the interconnected nanocellulose fiber matrix by either adding the pristine active components to the fibers, or by coating the individual nanofibrils using chemical reactions, to obtain a cellulose-based electrode. The electrode can be prepared in the form of microfibers, paper-like membranes, and foams and aerogels, depending on the solvent removal method (i.e., vacuum filtration, supercritical drying, solvent exchange) used together with the nanocellulose-active component mixtures.

Electronically conducting polymers (ECPs), such as polypyrrole (PPy), polyaniline (PANI), and poly(3,4-ethylenedioxythiophene) (PEDOT), are particularly attractive materials in the making of novel electrode materials because of their inherent high conductivities, fast redox switching, lightweight and straightforward manufacturing processes. Nanocellulose has a high affinity for the ECPs since the abundant hydroxyl groups on the surface of nanocellulose give rise to strong interactions with the latter. Upon polymerization of the conducting polymer monomer, uniform ECPs coatings on the individual nanocellulose can be formed which can give mechanically flexible and robust electrode structures [9]. All-polymer-electrodes consisting of nanocellulose and ECPs may hence feature high surface areas, high active mass loadings as well as high charge capacities with respect to the total electrode weight and volume. Several approaches towards the manufacturing of such electrodes have, therefore, been described recently [22–25].

The idea of fabricating paper-based electrodes from nanocellulose and ECPs was initially demonstrated in 2008; [26, 27] when pyrrole was chemically polymerized by iron (III) chloride to obtain a PPy@NCF paper composite with a typical specific surface area of ~80 $m^2\ g^{-1}$. In Fig. 8.3a, it can be seen that the PPy@NCF composite had an interconnected nanofibrous and porous structure and that the thickness of the PPy layer formed on the nanocellulose was ~50 nm (Fig. 8.3b) [12, 27]. This paper electrode exhibited a charge storage capacity of 370 C g^{-1} in an aqueous 2 M NaCl electrolyte (Fig. 8.3c) [25]. Using the same approach, paper electrodes composed of PANI@NCFs and PEDOT@NCFs were also prepared which demonstrated good capacities as well as good cycling and rate capabilities in aqueous electrolytes [28–30]. It is worth noting that the ECP paper electrodes also work well as organic cathodes in lithium-ion batteries (LIBs) containing non-aqueous electrolytes, as capacities of about 100 mAh g^{-1} have been reported [31]. The excellent electrochemical performance of these paper electrodes is due to the fast ion transport and the complete utilization of the active material arising from their porous structure, the thin ECP layers and the continuous conductive and flexible nanofiber network. This is further supported by their good rate capabilities and cycling performances obtained with high (up to 30 mg cm^{-2}) active mass loadings and thick (e.g. 580 μm) paper electrodes [23, 32, 33]. It has been demonstrated that a 290 μm-thick PPy/NCF

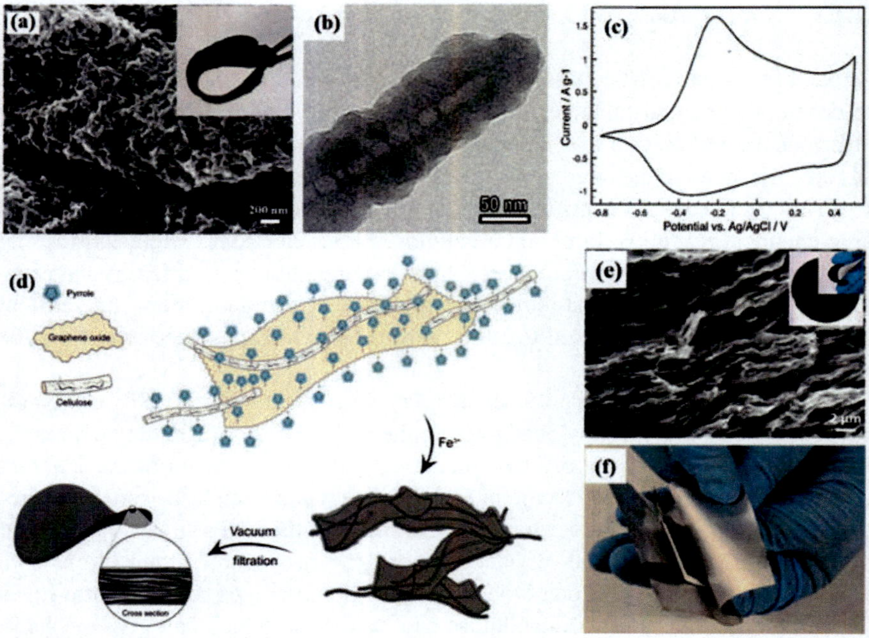

Fig. 8.3 **a** SEM image of a flexible PPy-NCF composite paper. **b** TEM image of a PPy/NCF composite. **c** Cyclic voltammogram obtained with a PPy/NCF paper electrode using a three-electrode set-up. **d** Schematic illustration of the preparation and structure of PPy/NCF/GO (PNG) paper electrodes. **e** Cross-section SEM images of a PNG paper. The inset shows a digital image of a flexible PNG paper electrode. **f** Photograph showing the assembly of an ECP paper electrode-based cell with graphite current collectors. Reproduced with permission from [12] (a), 27 (b), 25 (c), 40 (d, e), 9 (f). Copyright 2015 ACS, 2011 ACS, 2012 Elsevier, 2015 The Royal Society of Chemistry (RSC) and 2011 WILEY–VCH, respectively

electrode with a weight of 20 mg cm^{-2} yielded a gravimetric capacitance of 192 F g^{-1} based on the full electrode weight at a current density of 1 mA cm^{-2} [23].

Recently, the ECP-NCF hybrid composites have been combined with different carbon materials to develop paper electrodes with enhanced flexibilities and improved electrochemical performances [34–39]. Wang et al. [40] thus introduced small amounts of graphene oxide (GO) into a pyrrole-NCF solution prior to the polymerization yielding PPy. In this design, the high conductivity and capacitance of PPy are combined with those of the highly flexible GO, which combined with the strong chemical bonding between the NCF and GO as well as the pyrrole yield a robust PPy/NCF/GO paper electrode with a compact layered structure (Fig. 8.3d, e). It should be noted that the abovementioned paper-based electrodes are sustainable, and metal-free as carbon foil current collectors generally were employed (Fig. 8.3f).

In addition to coating electroactive materials on individual NCFs, cellulose-based electrodes can also be made by adding the active materials to an aqueous NCF containing solution, followed by vacuum-filtration or freeze drying. In this design, the NCFs act as a dispersion agent and 1D nanofibrous binders in the making of the paper electrodes. CNTs and rGO are commonly chosen as active materials due to their high conductivities and capacitances originated mainly from the charging of the electrical double layer [41–44]. Inorganic electroactive material can likewise be included to obtain ternary paper composites [45]. While the carbon-based paper electrodes generally show lower capacitances than the ECP/nanocellulose paper electrodes, their cycling stability is often excellent as well as their capacity retention under mechanical stress. The latter can be attributed to the non-faradic reactions and the flexibility of the CNTs and graphene [41, 43, 44].

8.3.1.2 Nanocellulose as Structural Substrate

Films/aerogels consisting of nanocellulose often exhibit excellent mechanical integrity and strength owing to strong interactions between the cellulose chains and have hence been used as versatile structural substrates in the development of free-standing cellulose electrodes composed of electrically conductive and/or electroactive materials. Lightweight and mechanically strong nanocrystal cellulose (CNC) aerogels were recently manufactured via the generation of cross-links between aldehyde-modified CNC and hydrazide modified CNCs. The chemically cross-linked CNC aerogel exhibited a fundamentally different mechanical properties compared to those found in other CNC materials [45–47]. A schematic illustration of the synthesis of CNC aerogel is shown in Fig. 8.4a, the resulting free-standing aerogel which had a density of 5.6 mg cm^{-3}, was so lightweight that it could stand on a feather (Fig. 8.4b) [46]. Such CNC aerogels can serve as universal lightweight substrates in the preparation of aerogel electrodes containing various nanosized active materials, including PPy nanofibers, PPy-coated CNTs, and MnO_2 nanoparticles [45]. The aqueous sol–gel process involved mixing, freezing, and freeze-drying to entangle the chosen nanoparticles within the cross-linked CNC network. The nanoparticles were maintained within the cross-linked CNC as a result of hydrogen bonding and/or nonpolar interactions to yield a gel-like structure consisting of hybrid cross-linked clusters which eventually was transformed into a hybrid aerogel by freeze-drying [45]. More importantly, the CNC-nanoparticle aerogels were strong and remained intact after compression in either air or aqueous electrolyte, indicating that it should be well-suited for use in compressible storage devices. The manufactured hybrid aerogel-based symmetrical supercapacitors featured a capacity retention of 94% after 2000 charge/discharge cycles at a current density of 0.1 mA cm^{-2}.

In addition to CNC, cross-linked NCF networks have also been fabricated and employed as a platform to make an electroactive NCF aerogel [48]. In this approach, a robust and rapid method involving on the layer-by-layer (LbL) assembly of polymers and nanoparticles on strong and elastic aerogels has been developed and used to manufacture supercapacitors (Fig. 8.4c). The LbL method successfully provided an ultrathin nanoporous coating that provided good connectivity throughout the whole NCF network structure while thin layers of biomolecules, ECPs, and carbon nanotubes were embedded in the NCF aerogel network. The latter resulted in aerogels with interesting properties, including a high charge storage capacity.

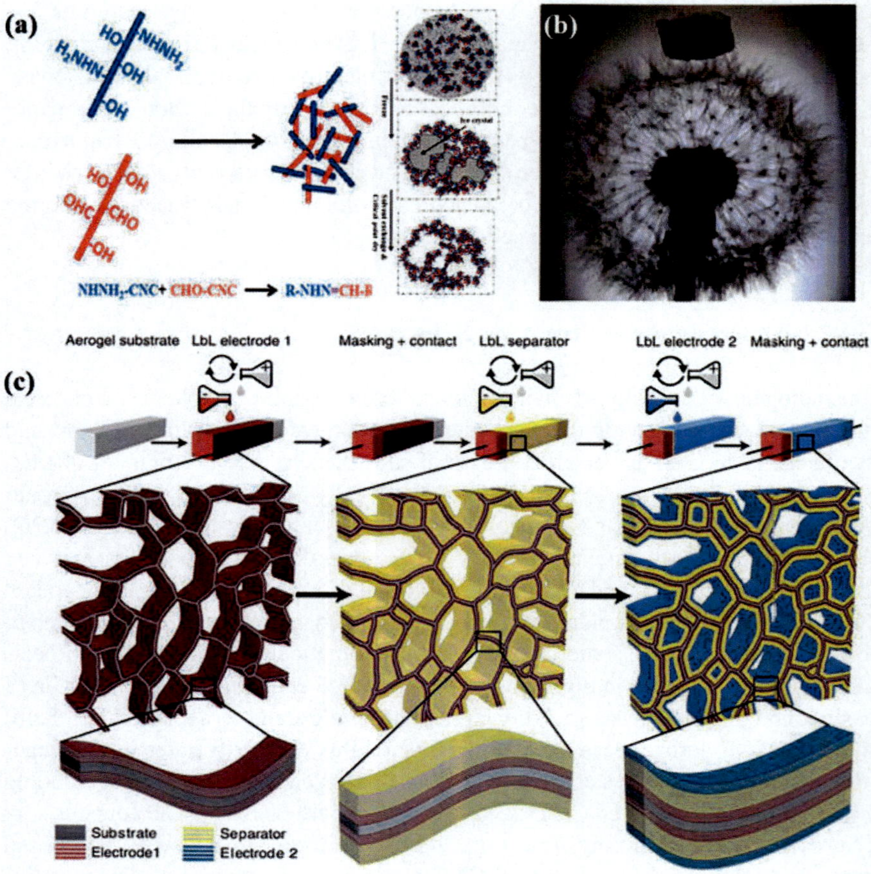

Fig. 8.4 **a** Schematic illustration of the manufacturing of a CNC aerogel using a suspension of cross-linked clusters. **b** Photograph showing the lightweight nature of the aerogel. **c** Schematics of fabrication of 3D devices in an NCF aerogel using LbL approach. Reproduced with permission from [46] (a, b), 48 (c). Copyright 2014 ACS and 2015 Creative Commons license, respectively

Relatively little attention has so far been paid to the manufacturing of NCF composite electrodes, containing inorganic materials, particularly for supercapacitor applications. This is most likely due to the problems associated with the coating of such material on the NCFs. Almost all the described inorganic materials/NCF electrodes were hence are made by embedding pre-synthesized inorganic materials within highly conductive NCFs networks. In this case, the optimization of the conductivity of the electrode is clearly very important to the realization of a high-capacity NCF-based electrode. The excellent film-forming and dispersion ability of nanocellulose do, however, enable the design of a greener fabrication method for electrodes in comparison with the conventional electrode preparation processes requiring the use of synthetic polymer binders and toxic organic solvents [49, 50].

In a typical NCF-based battery electrode preparation process, the electroactive materials, carbon, and NCFs are first mixed in an aqueous solution. After sonication and high-speed stirring, nanocellulose paper electrodes can then be formed by filtrating the mixture through a filter membrane. After drying, the composite electrode can then be peeled off and used as a battery electrode. This paper-making type of electrode manufacturing approach can be expected to have several advantages with respect to current state-of-the art electrode-making methods since [14, 51]: (i) the NCF matrix, acting as the composite backbone material, provides the electrode with sufficient mechanical integrity and paper-like properties; (ii) the NCFs act both as a dispersion agent and binder facilitating the attainment of a uniform distribution of carbon and active materials as well as a good contact between the components; (iii) the high surface area NCF scaffold yields porous paper-electrodes with interconnected channels allowing fast mass transport; (iv) the flexible NCF network may accommodate the volume changes associated with the cycling; and finally v) the process can, most likely, undergo up-scaling based on the methods currently employed within the paper-making industry.

Cornell and co-workers [52] thus prepared flexible LFP paper electrodes by vacuum filtration of an electrode dispersion containing TEMPO-oxidized NCF (TONCF), super-P carbon and LFP nanoparticles (Fig. 8.5a). The finely dispersed TONCF demonstrated strong dispersing and binding properties even at concentration as low as 4 wt.%, and the LFP particles were found to be relatively uniformly distributed (Fig. 8.5b), in good agreement with previous findings [53]. More recently, Hu et al. [54] produced thick and compact nanocellulose paper electrodes using freeze-drying of LFP:CB:TONCF aqueous suspensions. The as-obtained freestanding aerogel electrode was compressed under pressure and transformed into a compact paper electrode with maintained integrity and porosity (Fig. 8.5c, d). It was found that an array of CB particles was present on the NCF backbone, forming an interconnected conductive network and a tight wrapping of the LFP particles (Fig. 8.5e). The formation of the conformal conductive nanofiber was likely due to a spontaneous electrostatic self-assembly mechanism involving the NCF and the CB nanoparticles (Fig. 8.5f, g). The conductive cellulose nanofibers with large aspect ratios provide electron transfer pathways and enable facile fabrication of flexible nanopaper electrodes with a compact structure and high mass loading (up to 60 mg cm^{-2}). These results demonstrate that the nanocellulose paper electrode design constitutes a promising strategy towards the realization of sustainable high-performance battery electrodes.

A conductive NCF network can likewise be straightforwardly made by introducing 1D conductive nanofibers (i.e., CNT or Ag nanowires) within a NCF network to obtain e.g. an interpenetrated robust CNF-NCF network [14, 55–57]. The good compatibility of various electroactive materials present within the conductive NCFs makes the latter a versatile material that can be employed in connection with many other material systems. Cui and co-workers [58] hence described a lightweight, conductive silicon nanopaper electrode suitable for use as an LIB anode. A thin-layer Si was deposited on a CNT-NCF aerogel substrate by plasma-enhanced chemical vapor deposition method (Fig. 8.5h, i). The as-obtained Si nanopaper electrode

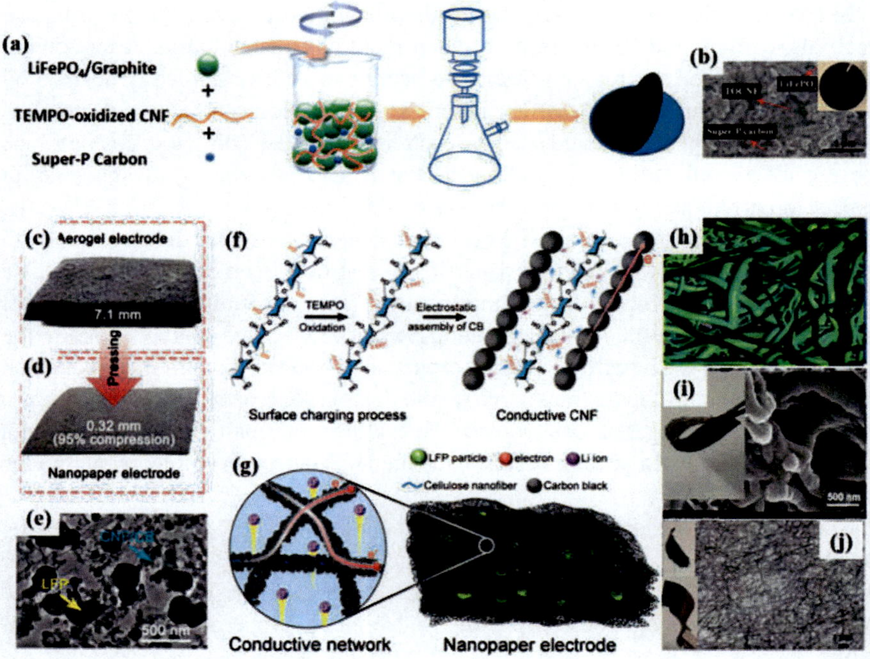

Fig. 8.5 **a** Schematic illustration of an aqueous paper-making process for the making of flexible NCF-based paper electrodes. **b** SEM image of an LFP:CB:TONCF paper electrode. The inset shows a digital image of the paper electrode. **c, d** Photographs illustrating the highly compact nanopaper electrode with a layer-by-layer structure obtained by compressing the aerogel electrode. **e** TEM image of the compact LFP:CB:TONCF nanopaper composite. **f** Schematic illustration of the preparation of conductive NCF materials by electrostatic self-assembly of NCF and CB. **g** The structure of the resulting porous LFP:CB:TONCF nanopaper electrode with its ion and electron transfer pathways. **h, i** Conductive NFC:CNT nanocomposites, where the black lines are the CNTs while the green lines are the NFCs. SEM image of a silicon nanoparticle coated conductive and flexible nanopaper. **j** SEM image and a photograph (inset) of a flexible Ag nanowire-NCF nanopaper. Reproduced with permission from [52] (a, b), 54 (c–g), 58 (h, i), 57 (j). Copyright 2016 ACS, 2018 WILEY–VCH, 2013 Elsevier and 2018 WILEY–VCH, respectively

delivered a capacity of 1200 mAh g^{-1} based on the Si mass after 100 cycles. Recently, Wang and co-workers [57] described a highly conducting (0.6 Ω sq^{-1}, 8200 S cm^{-1}), lightweight and thin silver nanopaper composed of 1D Ag nanowires and nanocellulose (Fig. 8.5j). This flexible Ag nanopaper, which had a low silver mass loading of 0.88 mg cm^{-2} and a thickness of only 2 μm, was used as current collector for a stable sodium metal anode. Due to the low mass and volume of the electrode, high electrode-mass-normalized gravimetric and volumetric capacities could be obtained.

8.3.1.3 Integrated Nanocellulose Paper Electrodes

Beside the use of NCFs for the manufacturing of various paper electrodes, the NCF platform can also be utilized to make integrated electrode/separator and electrode/current collector structures in the form of single sheets of paper [14, 59]. A single structure comprising multiple battery components could be advantageous owing to a straightforward manufacturing procedure and an improved mechanical performance due to the nanocellulose containing layers. Wang et al. [51] thus described a flexible Si paper electrode for use as a high-capacity LIB anode. As shown in Fig. 8.6a, the lightweight Si paper anode was composed of a Si-CNT-NCF active layer and a CNT-NCF current collector layer. Benefitting from the uniform distribution and strong adhesion of the Si nanoparticles to the NCF and CNT (see Fig. 8.6b), as well as the enhanced mechanical integrity and the integrated current collector, this flexible Si-based paper anode exhibited an electrode capacity of 800 mAh g^{-1} based on the weight of the electrode. In a similar study, Cornell and coworkers [60] prepared bendable $Li_4Ti_5O_{12}$ (LTO) electrodes in which carbon fiber (CF) layers were integrated as current collectors to enhance the mechanical properties and to increase the specific energy density of the electrodes (Fig. 8.6c, d). An increase in the weight fraction of the active material (LTO) from around 19% to 71% of the combined weight of the electrode and current collector was demonstrated using the CFs rather than conventional Cu current collector. Kim et al. [61] reported an all-solid-state flexible supercapacitor with promising electrochemical performance at different bending states. This performance was realized with bacteria nanocellulose integrated CNT paper electrodes (Fig. 8.6e), which have high mechanical integrity even during bending (Fig. 8.6f).

An all-nanocellulose structured single paper flexible battery was prepared by Leijonmarck et al. [59] The cross-section SEM indicates a structure consisting of three discrete and tightly adhering layers with a total thickness of 250 μm, where the layers serve as the negative electrode, nanocellulose separator and positive electrode, respectively (Fig. 8.6g). The resulting battery was mechanically robust and flexible and exhibited good capacity retention. More recently, Wang and coworkers [14] reported the first NCF paper-based LMB, composed of a NCF separator, NCF/CNT/LFP cathode, and Li-NCF/CNF anode. The latter LMB demonstrated excellent cycling stability, exhibiting a capacity retention of 91% after 800 cycles at a current density of 1.27 mA cm^{-2}.

During the last few years, Lee et al. [62–64] have described hetero-nanonet (HN) batteries based on a structured separator/electrode architecture containing 1D NCF and CNT building blocks. As shown in Fig. 8.6i, the separator/electrode layer was composed of a nanoporous NCF separator and a current collector and binder-free electrode containing only CNT-netted electrode active materials. The 1D nanobuilding block-based structural/physicochemical features of the separator/electrode design resulted in significant improvements with respect to the electron conduction pathways, electrolyte accessibility, dimensional stability, and the thermal properties [62]. The binder- and metallic current collector-free, NCF/CNT heteronet architecture allows repeated stacking of the electroactive materials layer.

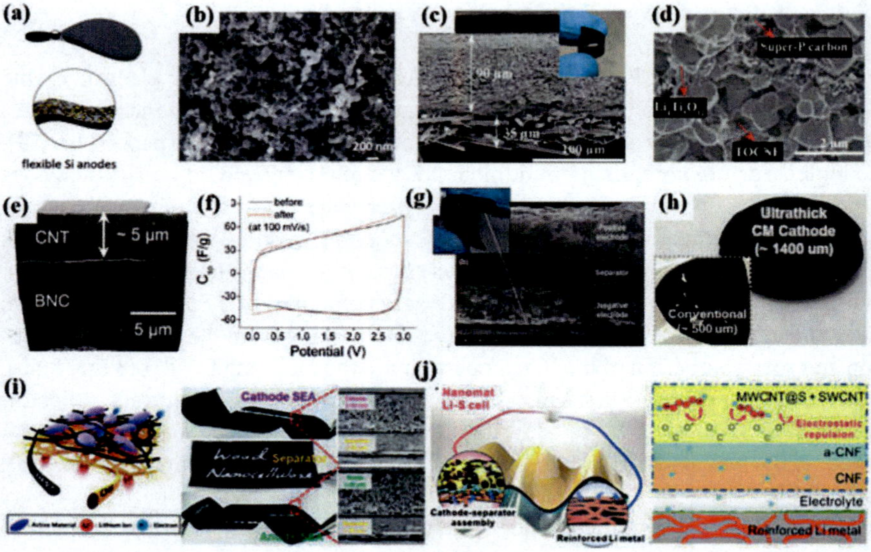

Fig. 8.6 **a** Schematic illustration of a flexible Si anode integrated with a CNT/NCF current collector layer. **b** SEM image showing that the Si nanoparticles were uniformly distributed within the CNT/NCF matrix. **c** and **d** SEM images of a flexible LTO anode integrated with a CF/NCF current collector. **e** Cross-section SEM image of a CNT electrode supported by a BC layer, **f** Cyclic voltammograms for a CNT-BC electrode before and after bending. **g** Cross-section SEM image of a paper battery. The inset shows its mechanical flexibility. **h** Photographs of an ultra-thick CM cathode with a thickness of about 1.4 mm. The inset shows the poor structural stability of a 0.5 mm thick conventional cathode. **i** A schematic illustration, photographs, and cross-sectional SEM images of a flexible LFP cathode and LTO anode using a separator integrated electrode design. **j** Conceptual illustrations of a nanomat Li–S battery architecture and its electrochemical functions. Reproduced with permission from [51] (a, b), 60 (c, d), 61 (e, f), 59 (j), 65 (h), 62 (i), 63 (j). Copyright 2015 RSC, 2017 Elsevier, 2012 ACS, 2013 RSC, 2015 WILEY–VCH, 2014 ACS and 2019 RSC, respectively

The latter can be used to make tailored, very thick (e.g. 1400 μm) LFP electrodes with high-mass loadings, see Fig. 8.6h, [65] which are difficult to manufacture using conventional battery electrode technologies. Driven by the structural advantages of the combined separator/electrode design and the chemical functionalities of the NCF, a new class of nanomat Li–S batteries with all-fibrous cathode-separator assemblies and conductive nonwoven-reinforced Li metal anodes has likewise been developed [63]. These nanomat Li–S cells provide significant advantages with respect to the electrochemical performance and mechanical properties (Fig. 8.6j). It should be noted that the fibrous sulfur cathodes made by depositing sulfur on MWCNT and SWCNT were integrated with bi-layered NCF separators consisting of a pristine NCF layer and an anionic NCF layer, respectively. The anionic CNFs were used to suppress the polysulfide shuttling effect in the Li–S battery based on electrostatic repulsion while Li foils were embedded into nickel-/copper-plated conductive poly(ethylene terephthalate) nonwovens to yield Li metal anodes with high electrochemical and dimensional stabilities.

8.3.2 *Nanocellulose-Derived Carbon*

Biomass-derived carbonaceous materials have received considerable attention for use in energy storage applications due to their charge storage capacities, cost, conductivity, versatile form, availability, ease of processing, and eco-friendly nature [66, 67]. Among many natural biopolymers, nanocellulose is a particularly promising precursor in the manufacturing of high-performance carbon electrodes and hybrid carbon electrodes. The latter have several advantages stemming from their 1D fibrous structures, free-standing film-forming properties, and versatility in combination with various functional materials [5, 10, 19].

8.3.2.1 Nanocellulose-Derived Carbon Electrodes

As shown in Figure 8.7a, b, the manufacturing of NCF-based carbon materials is straightforward. Typically, porous carbon materials can be prepared by heat-treatment of the nanocellulose precursor in an inert atmosphere and depending on the shape and morphology of the nanocellulose, carbon powder, [68] carbon nanofiber paper [69, 70] or carbon aerogels [71–73] can be prepared. The surface areas of these carbon materials, which generally are in the range of 100 to 400 m^2/g, [68, 73–76] can be further increased while still maintaining the initial morphology of the nanocellulose using additions of an activation agent such as KOH during the carbonization process. Carbon nanofiber webs with hierarchically porous structures and surface areas of up to 970 m^2/g were thus obtained by KOH activation of the pyrolyzed bacterial cellulose (pBC) [69]. This activated-pBC was composed of a 3D homogeneous conductive carbon nanofiber network, with a hierarchical structure comprising micro- and mesopores and a large surface area. The latter gave rise to a large electrode-electrolyte contact area decreasing the mass transfer resistance as well as increasing the electrode capacitance [69, 77, 78].

Hybrid electrodes composted of electroactive inorganic materials and carbon can be prepared in-situ by carbonization of a mixture containing the inorganic materials precursors and nanocellulose. For example, the inorganic Fe_xO_y nanomaterials are generally coated on NCF-derived carbon frameworks based on adsorption, hydrothermal reactions, and sequential post-treatment. Using bacteria cellulose as a starting material, flexible inorganic nanocomposite-based paper electrodes can be successfully prepared employing, e.g., Fe_2O_3-carbon paper, [79] Fe_3O_4-carbon paper, [80] and Si-carbon paper [81].

In addition to the making of NCF-derived pure carbon materials and carbon/inorganic hybrid materials, is also possible to employ nanocellulose to manufacture heteroatom-doped carbon electrodes. This is an interesting approach as heteroatoms including nitrogen (N), phosphorus (P), sulfur (S), and boron (B) have been shown to increase the rate capability and capacity of carbon electrodes [82, 83]. While many N-doped carbon materials have been produced via post-treatments in ammonia gas, involving hazardous operating conditions or yielding uncontrollable N-doping levels, an alternative approach could be to utilize NCFs functionalized with various dopant-containing materials. The latter hence constitutes a versatile and promising approach for the manufacturing of dopant enriched precursors. Recently, Ruan and coworkers [68] described the use of a one-pot reductive amination process to functionalize 2,3-dialdehyde cellulose beads (DAC) with chitosan and L-cysteine to provide single (N)- and dual (N/S)-doped precursors. With chitosan and L-cysteine as the nitrogen and sulfur sources, respectively, the latter approach gave rise to N-doped or N/S co-doped carbon beads after the pyrolysis (Fig. 8.7c), which was verified using EDX. This type of functionalization enables the physicochemical properties of the cellulose materials to be tailored and can result in carbon precursors with heteroatom doping levels facilitating energy storage applications. Heteroatom-doped carbon electrodes can, however, also be prepared by carbonization of ECP/NCF composite materials, where the use of PPy and PANI would give N-doping [75, 84, 85], while that of PEDOT would yield S-doping [86].

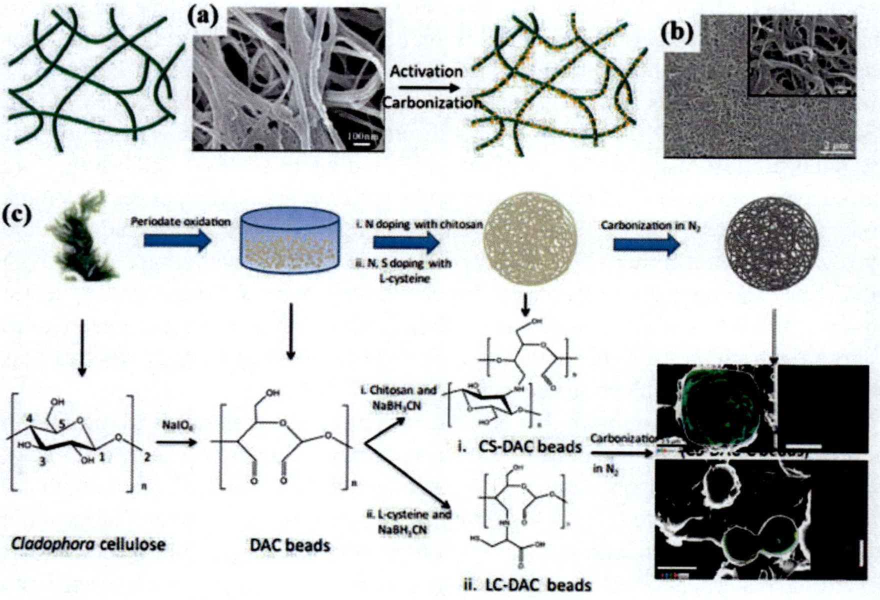

Fig. 8.7 **a** and **b** A schematic illustration of the NCF-derived carbon-making process, and SEM images of the NCF precursor before and after the calcination. **c** The preparation procedure and SEM and EDX mapping images of doped carbon derived from 2,3-dialdehyde cellulose beads. Reproduced with permission from [69] (a, b), 68 (c). Copyright 2015 Elsevier and 2018 Springer, respectively

8.3.2.2 Nanocellulose-Derived Carbon Substrates

NCF-derived carbon materials can also be directly employed as porous conductive carbon substrates to produce electrodes for tailorable energy storage applications. BC pellicles were, for example, first cut into the designed shape to fit the requirements of different types of LIBs. After freeze-drying and pyrolysis, a black, lightweight, and conductive membrane was obtained. The latter was composed of numerous intertwined ultrathin nanofibrils giving rise to a large surface area that could be used to support electroactive nanomaterials. The pore abundance is clearly a key point with respect to the attainment of a good distribution of the active nanomaterials, and to obtain fast mass transport of ions, when the material is to be used as a LIB electrode [87]. Electroactive materials such as SnO_2 and Ge nanoparticles were introduced into the BC-derived carbon nanofibril network through an in-situ assembly process to form flexible paper-based LIB anodes.

Inspired by the anisotropically aligned channel structure of natural wood, a series of hybrid carbon electrodes were recently produced based on wood-derived carbon substrates with high porosities and low tortuosities [88–90]. Benefiting from the efficient ion and gas transport system present in wood, a high-performance flexible Li-CO_2 battery was realized with a CNT-Ru modified wood-carbon based cathode [88]. The wood-derived carbon was also employed as thick, mechanical strong current collectors to produce thick battery cathodes with high mass loadings, [90] in addition to all-wood-based supercapacitors [91]. Efforts have likewise been made to utilize pyrolyzed nanocellulose in the manufacturing of complex structures. Hu and coworkers, [92] for example, employed BC as a template to produce hybrid electrolyte membranes with a garnet-type $Li_7La_3Zr_2O_{12}$ (LLZO) solid electrolyte particle interconnected network and a poly(ethylene oxide) (PEO) polymer containing bis(trifluoromethane)sulfonimide lithium (LiTFSI). As shown in Fig. 8.8, the fabrication process comprised three key steps: (1) preparation of a BC/LLZO precursor membrane employing a BC aerogel membrane as a porous and hydrophilic template incorporating the LLZO precursor; (2) production of a freestanding LLZO membrane by removal of BC nanofibers via calcination in air; (3) infusion of a PEO-LiTFSI mixture into the LLZO nanofiber network to improve its flexibility and to form a hybrid electrolyte. It is important to note that the BC-templated LLZO membrane contained interconnected channels that yielded a continuous ion-conducting network of LLZO-PEO/LiTFSI. Based on the latter and the high aspect ratio of the LLZO network, the hybrid electrolyte exhibited an ionic conductivity of 1.12×10^{-4} S cm^{-1} at room temperature. This biopolymer nanofiber templating approach can most likely also be extended to prepare other types of hybrid electrolytes suitable for use in various electrochemical energy storage systems.

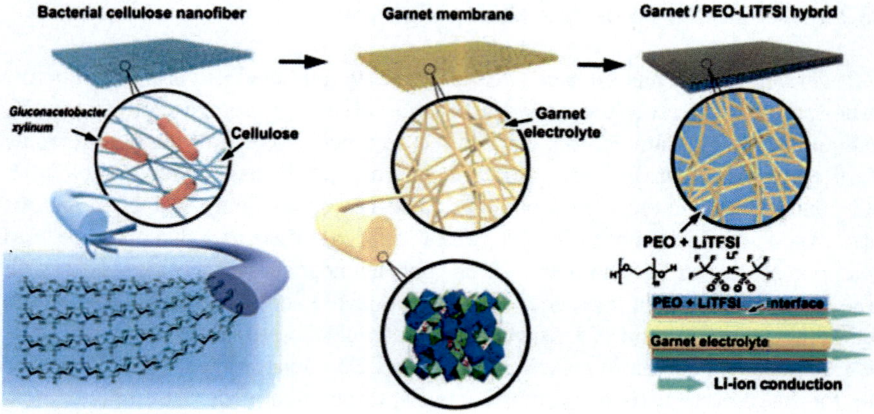

Fig. 8.8 The procedure used for the synthesis of a BC-templated hybrid electrolyte. The microstructures intended for the Li-ion transport are also schematically indicated. Reproduced with permission from [92]. Copyright 2018 WILEY–VCH, respectively

8.3.3 Nanocellulose-Based Separators

While much effort has been made to develop NCF-based electrodes and carbonized nanocellulose composites, very little attention has so far been paid to the use of NCFs as a battery separator material. The separator, which is an indispensable component of a battery, is typically composed of an insulating porous membrane located between the electrodes to prevent a short-circuit of the battery and to provide the ionic pathways required in the battery [93, 94]. Synthetic polyolefin separators dominate the battery separator market owing to their good mechanical properties, high chemical stabilities and low cost [95]. Polyolefin separators, however, exhibit shortcomings with respect to their relatively low electrolyte wettabilities and thermal stabilities, which may compromise the performance and safety of the battery. It was, however, recently demonstrated that separators made of cellulose can exhibit higher electrolyte wettabilities, better thermal stabilities and higher degrees of sustainability [96, 97]. The latter cellulose separators were, however, generally much thicker and less porous than the conventional polyolefin-based separators due to the large diameter of the cellulose microfibrils and their low surface area. LIBs equipped with such cellulose separators, therefore, exhibited larger cell resistances and decreased performances. With NCFs, the high surface area and structure of the nanofibrils, on the other hand, provide new possibilities for the manufacturing of separators with tailor-made thicknesses, adjustable pore structures and advanced functionalities that can boost the battery performance [98–101]. It should, however, also be pointed out that the type of NCF used also is very important with respect to the water content in the separator. A high-water content would give rise to a large irreversible capacity in a LIB or LMB which is why a nanocellulose with a high degree of crystallinity should be used [13, 18].

8.3.3.1 Separator morphology, Structure, and Basic Functions

NCF separators can be made employing a straightforward paper-making method involving vacuum filtration of nanocellulose dispersion [13]. The obtained NCF separator exhibited a uniform porous structure with randomly distributed and intertwined cellulose nanofibers, whereas the commercial polyolefin (PE) separator displayed a non-uniform pore distribution with fibers of thicknesses ranging from 20 nm to 1 μm. The pore structures of the separators were evaluated by N_2 adsorption (using the BJH method). As shown in Fig. 8.9, the nanocellulose separator exhibited a mesoporous structure with an average pore size of 20 nm, while the PE separator was less mesoporous with an average pore diameter of 80 nm in good agreement with previous findings [94]. The NCFs separator exhibited a surface area of 100 $m^2 \cdot g^{-1}$, while the corresponding value for the PE separator was 23 $m^2 \cdot g^{-1}$. In addition, the porosity of the nanocellulose separator was 46%, which was significantly higher than the 20% found for the polyolefin separator [13]. As a result of this, the nanocellulose separator exhibited a higher ionic conductivity, i.e., 0.4 mS cm^{-1} compared to 0.13 mS cm^{-1}, than a conventional separator. The higher porosity and ionic conductivity of the nanocellulose separator are favorable when using the nanocellulose separator in a LIB separator due to the resulting decreased cell resistance which enhances the battery performance [13]. It should also be noted that the pore structure of the NCF separator can be optimized by controlling the thickness of the NCF separator. The preparation of a thin NCF separator results in a larger average pore size and hence a higher porosity, which enhances the mass transport and lowers the electrolyte resistance in the pores of the separator (Fig. 8.9c) [101].

Due to their hydrophilic nature, NCF separators exhibit high electrolyte wettabilities, as can be seen in Fig. 8.9d, where the LIB electrolyte can be seen to have wetted the whole NCF separator while the electrolyte droplet remained intact on the polyolefin separator. Since the electrolyte wettability is related to the rate of electrolyte penetration in the separator, a rapid access to the pores within the NCF separator is expected to facilitate the ion transport and the attainment of a uniform ion flux, which should improve the capacity and stability of the battery. Owing to the high melting point of cellulose (around 300 °C), NCF separators should also be more thermally stable than the polyolefin separators which melt at 180 °C. The latter was experimentally confirmed by comparing the thermal stabilities of nanocellulose and polyolefin separators (Fig. 8.9e). While the NCF separator maintained its original shape at the studied temperatures, the polyolefin separator was totally deformed after the high-temperature test. It is therefore reasonable to assume that the higher thermal stability of the NCFs separator would assist in preventing battery short-circuits if the temperature in the LIBs were to be significantly increased for some reasons.

Studies involving different types of NCFs [102] have further demonstrated that NCFs separators can have good mechanical performances, mesoporous structures, uniform pore size distributions, large porosities, high electrolyte wettabilities, and good thermal stabilities, Based on these findings, it is immediately clear that NCF-based separators provide new possibilities for the realization of new generations

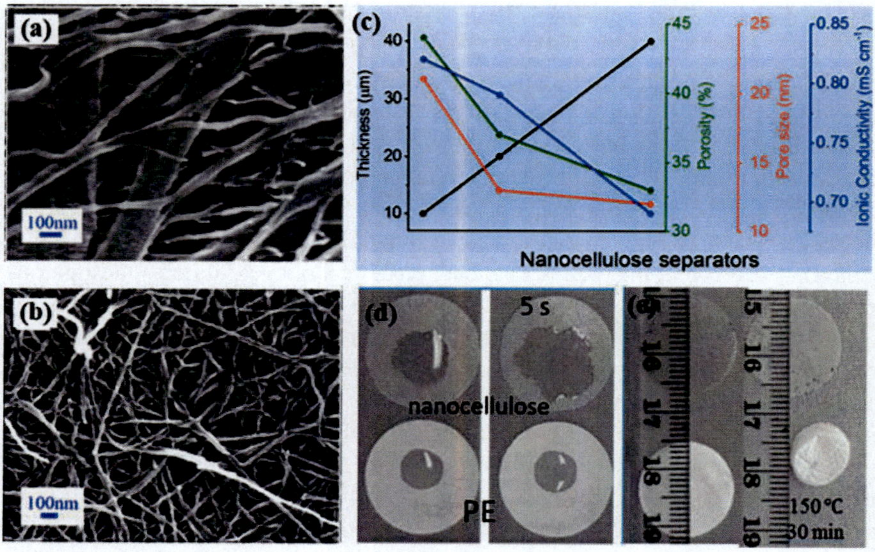

Fig. 8.9 Morphology and structural comparison of NCF and PE separators. SEM image of a PE separator (**a**) and NCF separator (**b**). **c** Comparison of the thickness, pore size, porosity, and ionic conductivity for three CC separators in a LIBs electrolyte. **d** Electrolyte wettabilities. **e** Thermal stabilities. Reproduced with permission from [13] (a, b, d, e), 101 (c). Copyright 2016 Elsevier, 2017 Springer, respectively

of separators, which can be used to address safety and stability issues, not only in different types of batteries but also in energy storage devices in general.

8.3.3.2 Advanced Separator Functions

The increased energy density demand of rechargeable batteries has revived the interest in lithium-metal batteries (LMBs) as an alternative to LIBs, based on the high-capacity Li-metal anodes (3860 mAh g^{-1} or 2060 mAh cm^{-3}) compared to those of commercial graphite anodes (372 mAh g^{-1} or 450 mAh cm^{-3}). The commercialization of LMBs is, however, hindered by the formation of dendrites and mossy lithium on the lithium electrodes, which give rise to significant safety and stability problems. Recently, it was demonstrated that the influence of the latter problems can be decreased by using a more homogeneous current distribution at the electrodes via the use of separators with more homogeneous pore distributions [103]. A more uniform current distribution on the lithium electrode surface can hence give rise to better lithium deposition and stripping performances and hence stabilize the Li metal anode. Several attempts have also been made to develop dendrite-proof separators with improved ion diffusion properties. The use of a separator with high wettability should enable a rapid filling of the separator pores which should facilitate the attainment of uniform mass transport through the entire separator [104]. A narrow pore size distribution as well as a homogeneous pore structure are also desirable since this combination should favor the attainment of a homogeneous current distribution [105]. The homogeneous ion transport through the separator favors uniform and compact deposition of Li metal. Due to the high electrolyte wettability and uniform pore structure, nanocellulose-based separator can provide a relatively uniform Li^+ mass transport throughout the separator which can be expected to extend the lifespan of LMBs. As is schematically illustrated in Fig. 8.10a, a homogeneous current distribution should result in the formation of numerous homogeneously distributed Li nuclei on the electrode surface, which facilitate the formation of dense Li deposits.

Conventional polyolefin separators have a thickness of about 20 μm, which, in fact, may constitute a significant fraction of the volume of a LIB, e.g., nearly 20% for cells containing a graphite anode and a $LiCoO_2$ cathode [106]. It should, therefore, be possible to increase the gravimetric and volumetric energy densities for LIBs by transforming a part of the separator into a porous electronically and electrochemically active layer without adversely affecting the overall properties of the separator [16]. While the use of a very thin separator usually is associated with mechanical limitations and safety concerns, a redox-active bi-layered separator would then appear to be a promising alternative.

By combining a thin insulating cellulose layer with a porous supporting layer composed of conductive and electrochemically-active materials, a flexible redox-active separator with a thickness similar to that of a conventional separator can be obtained, which also can be used to increase the capacity of the batteries (Fig. 8.10b) [16]. The redox-active separator, which can be manufactured using a straightforward filtration-based approach, consists of an insulating nanocellulose layer (3 μm thick) that is integrated with a redox-active PPy-nanocellulose supporting layer which also serves as a current collector layer. The extra electrochemical capacity hence stems from the oxidation and reduction of the PPy (see Fig. 8.10c). By replacing a commercial separator with the present PPy-based redox-active separator, the capacity of a LIB comprising an LFP cathode, and a Li anode could be increased from 0.16 to 0.276 mAh (see Fig. 8.10d). When normalized with respect to the total volume (or weight) of the separator and cathode, the capacity increase was 300% (Fig. 8.10e), i.e., from 18 $\mu Ah\ cm^{-3}$ (or 26 $mAh\ g^{-1}$) to 67 $\mu Ah\ cm^{-3}$ (or 81 $mAh\ g^{-1}$). The redox-active separator approach is particularly promising in cases where the capacity of a LIB needs to be increased without changing the electrode materials or the volume of the cell [16]. Based on the same idea, polydopamine can be used as anode materials to prepare a PDA-based redox-active (PRA) separator, which can provide extra capacity within the potential region of a typic anode material. Pan et al. [107] thus prepared a PDA-based redox-active separator composed of a 2 μm thick insulating nanocellulose layer and an 18 μm thick PDA-CNTs containing redox-active layer. A proof-of-concept LTO/Li cell was employed to demonstrate that the use of the PRA separator resulted in an enhanced capacity and rate capability compared to the corresponding PE separator-based cell.

The generation of species such as manganese ions (Mn^{2+}) due to the dissolution of manganese and hydrofluoric acid (HF) due to the degradation of, e.g., PF_6^-

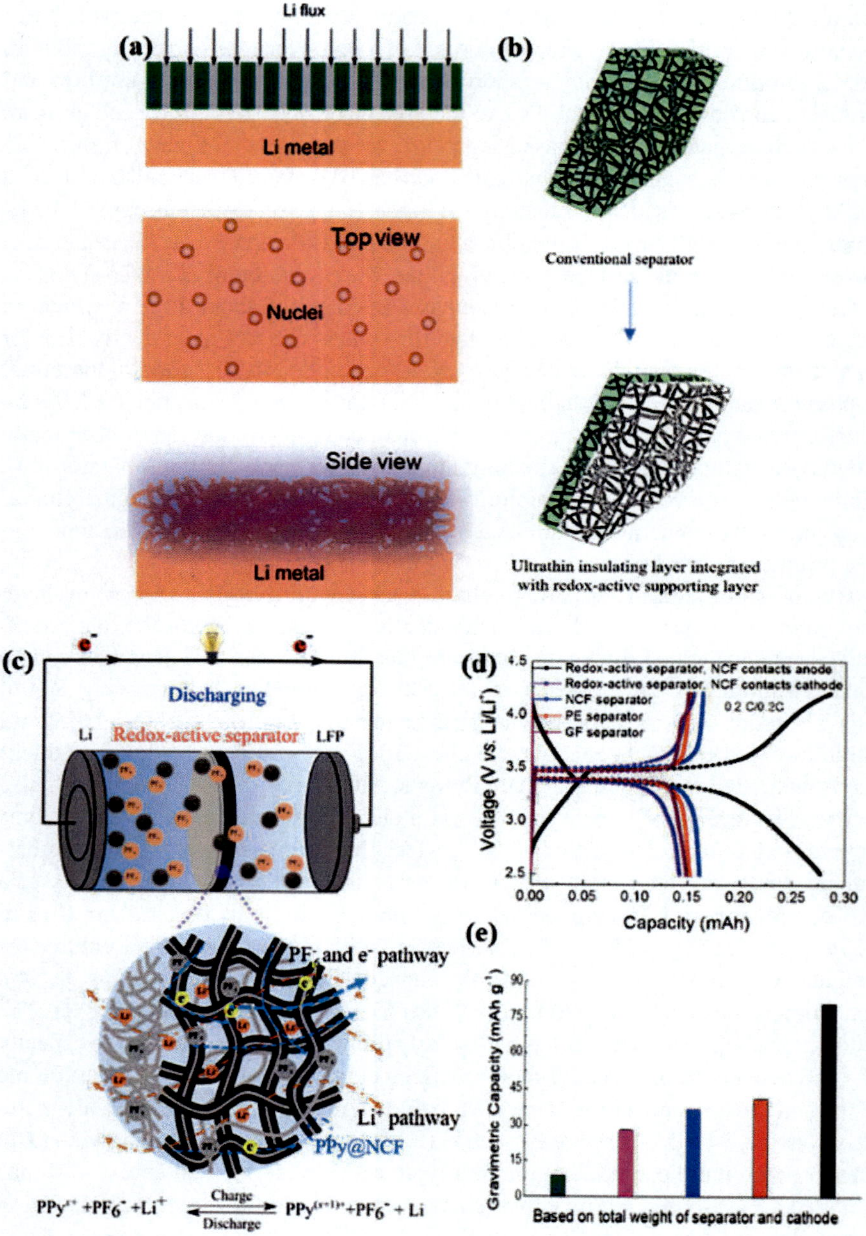

Fig. 8.10 **a** Schematic illustration of the impact of the separator pore distribution on the morphology of the deposited Li. **b** Schematic side-view illustrations of a conventional and a redox-active separator, respectively. The light green regions represent the separator layer, while the light gray region represents the redox-active supporting layer. **c** Schematic description of an LFP/Li battery equipped with a redox-active separator with the PPy layer facing the LFP electrode. **d** Charge/discharge profiles for LFP/Li cells comprising different separators. **e** Gravimetric capacity as a function of the combined separator and cathode weight for different LFP/Li cells. Reproduced with permission from [103] (a), 16 (b–e). Copyright 2018 WILEY–VCH, 2018 WILEY–VCH, respectively

ions are often assumed to give rise to capacity losses for transition metal oxide-based cathode materials. These interfering species can, however, be scavenged using functionalized separators equipped with suitable chemical functionalities before the species can reach the other electrode [108]. Lee and co-workers [109] have developed a novel layered nanomat-based hierarchical and asymmetric porous membrane with coupled chemical functionalities as a nanocellulose-mediated green material strategy to develop Mn^{2+} and HF scavenging separator membranes. The bi-layered nanocellulose-based membrane was composed of a TPY-functionalized NCF nanoporous thin mat as the top layer and an electrospun PVP/PAN macroporous thick mat as the support layer (Fig. 8.11a). The nanoporous TPY-NCF top layer played a critical role in suppressing the leakage current between electrodes, while the TPY (used to chelate the Mn^{2+} ions) and PVP (used to capture HF)-mediated chemical functionalities gave rise to a substantial improvement in the cycling performance of cells operated at elevated temperature (Fig. 8.11b).

It is hence clear that nanocellulose-based separators have gained significant attention and that the results have demonstrated their efficiency and feasibility to improve the performance of batteries. Additional studies are, nevertheless, needed before this type of new separators is suitable for inclusion in commercial batteries. These studies should, e.g., focus on the mechanical properties of nanocellulose separators in their wet state, the influence of the separator pore structure on the battery performance and straightforward ways of eliminating the water that can be found as an impurity in many types of nanocellulose fibers.

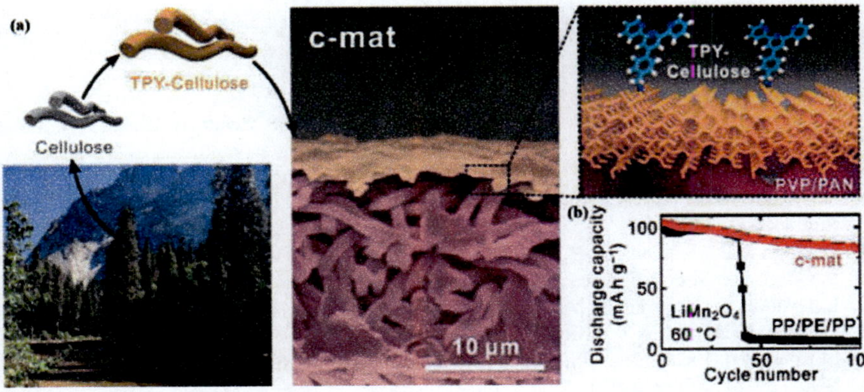

Fig. 8.11 **a** A schematic illustration of a functionalized NCF-integrated nanomat separator (c-mat) consisting of a terpyridine (TPY)-NCF thin layer and an electrospun polyvinylpyrrolidone (PVP)/polyacrylonitrile (PAN) support layer. **b** High-temperature (60 °C) cycling performance of $LiMnO_2$ containing cells equipped with c-mat separators. Reproduced with permission from [109]. Copyright 2016 ACS

8.4 Future Perspectives

The results of the research carried out during the last few years indicate that nanocellulose will play an important role in the development of new inexpensive, flexible, and versatile energy storage devices including both supercapacitors and different types of batteries (e.g., lithium-ion, lithium-metal, and sodium-based batteries). As nanocellulose can serve as an inexpensive and sustainable scaffold in the manufacturing of porous and flexible electrodes with high surface areas and thin layers of electroactive materials it is possible to manufacture energy storage devices exhibiting high energy as well as power densities. High mass loadings of electroactive species can be obtained merely using filtration-based manufacturing approaches which most likely also can be scaled-up with the aid of the manufacturing techniques currently employed within the paper-making industry. This paves the way for a very exciting development of new types of energy storage devices in which nanocellulose or nanocellulose-based composites are used in the current collectors, electrodes as well as in functionalized separators. Such all-cellulose energy storage devices can most likely be manufactured at low costs and may therefore serve as an interesting complement to conventional batteries and supercapacitors, particularly when the cost of the device is more important than the size of the device.

References

1. P. Simon, Y. Gogotsi, B. Dunn, Science **343**, 1210 (2014)
2. M. Winter, R.J. Brodd, Chem. Rev. **104**, 4245 (2004)
3. M. Winter, B. Barnett, K. Xu, Chem. Rev. **118**, 11433 (2018)
4. S. Muench, A. Wild, C. Friebe, B. Haupler, T. Janoschka, U.S. Schubert, Chem. Rev. **116**, 9438 (2016)
5. Z.H. Wang, P. Tammela, M. Strømme, L. Nyholm, Adv. Energy Mater. **7**, 1700130 (2017)
6. B. Thomas, M.C. Raj, A.K. B, R. M. H, J. Joy, A. Moores, G.L. Drisko, C. Sanchez, Chem. Rev. **118**, 11575 (2018)
7. D. Klemm, B. Heublein, H.-P. Fink, A. Bohn, Angew. Chem. Int. Ed. **44**, 3358 (2005)
8. Y. Gogotsi, P. Simon, Science **334**, 917 (2011)
9. L. Nyholm, G. Nyström, A. Mihranyan, M. Strømme, Adv. Mater. **23**, 3751 (2011)
10. C. Chen, L. Hu, Acc. Chem. Res. **51**, 3154 (2018)
11. E. Kontturi, P. Laaksonen, M.B. Linder, Nonappa, A.H. Groschel, O.J. Rojas, O. Ikkala, Adv. Mater. **30**, e1703779 (2018)
12. Z. Wang, D.O. Carlsson, P. Tammela, K. Hua, P. Zhang, L. Nyholm, M. Strømme, ACS Nano **9**, 7563 (2015)
13. R. Pan, O. Cheung, Z. Wang, P. Tammela, J. Huo, J. Lindh, K. Edström, M. Strømme, L. Nyholm, J. Power Sources **321**, 185 (2016)
14. Z. Wang, R. Pan, R. Sun, K. Edström, M. Strømme, L. Nyholm, ACS Appl. Energy Mater. **1**, 4341 (2018)
15. R. Pan, R. Sun, Z. Wang, J. Lindh, K. Edström, M. Strømme, L. Nyholm, Nano Energy **55**, 316 (2019)
16. Z. Wang, R. Pan, C. Ruan, K. Edström, M. Strømme, L. Nyholm, Adv. Sci. **5**, 1700663 (2018)
17. Z. Wang, R. Pan, C. Xu, C. Ruan, K. Edström, M. Strømme, L. Nyholm, Energy Storage Mater. **13**, 283 (2018)

18. H. Kim, V. Guccini, H.R. Lu, G. Salazar-Alvarez, G. Lindbergh, A. Cornell, ACS Appl. Energy Mater. **2**, 1241 (2019)
19. J.H. Kim, D. Lee, Y.H. Lee, W. Chen, S.Y. Lee, Adv. Mater. e1804826 (2018)
20. W. Chen, H. Yu, S.Y. Lee, T. Wei, J. Li, Z. Fan, Chem. Soc. Rev. **47**, 2837 (2018)
21. D. Klemm, F. Kramer, S. Moritz, T. Lindström, M. Ankerfors, D. Gray, A. Dorris, Angew. Chem. Int. Ed. **50**, 5438 (2011)
22. M. Paakko, J. Vapaavuori, R. Silvennoinen, H. Kosonen, M. Ankerfors, T. Lindström, L.A. Berglund, O. Ikkala, Soft Matter **4**, 2492 (2008)
23. Z. Wang, P. Tammela, P. Zhang, M. Strømme, L. Nyholm, J. Mater. Chem. A **2**, 16761 (2014)
24. S. Liu, T. Yu, Y. Wu, W. Li, B. Li, RSC Adv. **4**, 34134 (2014)
25. G. Nyström, M. Strømme, M. Sjödin, L. Nyholm, Electrochim. Acta **70**, 91 (2012)
26. A. Razaq, A. Mihranyan, K. Welch, L. Nyholm, M. Strømme, J. Phys. Chem. B **113**, 426 (2008)
27. G. Nyström, A. Razaq, M. Strømme, L. Nyholm, A. Mihranyan, Nano Lett. **9**, 3635 (2009)
28. Z. Wang, P. Tammela, J. Huo, P. Zhang, M. Strømme, L. Nyholm, J. Mater. Chem. A **4**, 1714 (2016)
29. H. Wang, E. Zhu, J. Yang, P. Zhou, D. Sun, W. Tang, J. Phys. Chem. C **116**, 13013 (2012)
30. J. Xu, L. Zhu, Z. Bai, G. Liang, L. Liu, D. Fang, W. Xu, Org. Electron. **14**, 3331 (2013)
31. Z. Wang, C. Xu, P. Tammela, P. Zhang, K. Edström, T. Gustafsson, M. Strømme, L. Nyholm, Energy Technol. **3**, 563 (2015)
32. Z. Wang, P. Tammela, P. Zhang, M. Strømme, L. Nyholm, J. Mater. Chem. A **2**, 7711 (2014)
33. Z. Wang, P. Tammela, P. Zhang, J. Huo, F. Ericson, M. Strømme, L. Nyholm, Nanoscale **6**, 13068 (2014)
34. Q. Zheng, Z. Cai, Z. Ma, S. Gong, ACS Appl. Mater. Interf. **7**, 3263 (2015)
35. C. Yang, C. Chen, Y. Pan, S. Li, F. Wang, J. Li, N. Li, X. Li, Y. Zhang, D. Li, Electrochim. Acta **182**, 264 (2015)
36. C. Yang, D. Li, Mater. Lett. **155**, 78 (2015)
37. S. Li, D. Huang, J. Yang, B. Zhang, X. Zhang, G. Yang, M. Wang, Y. Shen, Nano Energy **9**, 309 (2014)
38. Q. Liu, S. Jing, S. Wang, H. Zhuo, L. Zhong, X. Peng, R. Sun, J. Mater. Chem. A **4**, 13352 (2016)
39. Y. Liu, J. Zhou, J. Tang, W. Tang, Chem. Mater. **27**, 7034 (2015)
40. Z. Wang, P. Tammela, M. Strømme, L. Nyholm, Nanoscale **7**, 3418 (2015)
41. B. Andres, C. Dahlström, N. Blomquist, M. Norgren, H. Olin, Mater. Des. **141**, 342 (2018)
42. K. Gao, Z. Shao, J. Li, X. Wang, X. Peng, W. Wang, F. Wang, J. Mater. Chem. A **1**, 63 (2013)
43. K. Gao, Z. Shao, X. Wang, Y. Zhang, W. Wang, F. Wang, RSC Adv. **3**, 15058 (2013)
44. Q. Niu, K. Gao, Z. Shao, Nanoscale **6**, 4083 (2014)
45. X. Yang, K. Shi, I. Zhitomirsky, E.D. Cranston, Adv. Mater. **27**, 6104 (2015)
46. X. Yang, E.D. Cranston, Chem. Mater. **26**, 6016 (2014)
47. K. Shi, X. Yang, E.D. Cranston, I. Zhitomirsky, Adv. Func. Mater. **26**, 6437 (2016)
48. G. Nystrom, A. Marais, E. Karabulut, L. Wågberg, Y. Cui, M.M. Hamedi, Nat. Commun. **6**, 7259 (2015)
49. L. Jabbour, C. Gerbaldi, D. Chaussy, E. Zeno, S. Bodoardo, D. Beneventi, J. Mater. Chem. **20**, 7344 (2010)
50. L. Jabbour, R. Bongiovanni, D. Chaussy, C. Gerbaldi, D. Beneventi, Cellulose **20**, 1523 (2013)
51. Z. Wang, C. Xu, P. Tammela, J. Huo, M. Strømme, K. Edström, T. Gustafsson, L. Nyholm, J. Mater. Chem. A **3**, 14109 (2015)
52. H. Lu, M. Behm, S. Leijonmarck, G. Lindbergh, A. Cornell, ACS Appl. Mater. Interf. **8**, 18097 (2016)
53. S. Leijonmarck, A. Cornell, G. Lindbergh, L. Wågberg, Nano Energy **2**, 794 (2013)
54. Y. Kuang, C. Chen, G. Pastel, Y. Li, J. Song, R. Mi, W. Kong, B. Liu, Y. Jiang, K. Yang, L. Hu, Adv. Energy Mater. **8**, 1802398 (2018)
55. A. Hajian, S.B. Lindström, T. Pettersson, M.M. Hamedi, L. Wågberg, Nano Lett. **17**, 1439 (2017)

56. W. Luo, J. Hayden, S.-H. Jang, Y. Wang, Y. Zhang, Y. Kuang, Y. Wang, Y. Zhou, G.W. Rubloff, C.-F. Lin, L. Hu, Adv. Energy Mater. **8**, 1702615 (2018)
57. Z. Wang, X. Zhang, S. Zhou, K. Edström, M. Strømme, L. Nyholm, Adv. Funct. Mater. **28**, 1804038 (2018)
58. L. Hu, N. Liu, M. Eskilsson, G. Zheng, J. McDonough, L. Wågberg, Y. Cui, Nano Energy **2**, 138 (2013)
59. S. Leijonmarck, A. Cornell, G. Lindbergh, L. Wågberg, J. Mater. Chem. A **1**, 4671 (2013)
60. H. Lu, J. Hagberg, G. Lindbergh, A. Cornell, Nano Energy **39**, 140 (2017)
61. Y.J. Kang, S.-J. Chun, S.-S. Lee, B.-Y. Kim, J.H. Kim, H. Chung, S.-Y. Lee, W. Kim, ACS Nano **6**, 6400 (2012)
62. K.-H. Choi, S.-J. Cho, S.-J. Chun, J.T. Yoo, C.K. Lee, W. Kim, Q. Wu, S.-B. Park, D.-H. Choi, S.-Y. Lee, S.-Y. Lee, Nano Lett. **14**, 5677 (2014)
63. J.-H. Kim, Y.-H. Lee, S.-J. Cho, J.-G. Gwon, H.-J. Cho, M. Jang, S.-Y. Lee, S.-Y. Lee, Energy Environ. Sci. **12**, 177 (2019)
64. J.-M. Kim, J.A. Kim, S.-H. Kim, I.S. Uhm, S.J. Kang, G. Kim, S.-Y. Lee, S.-H. Yeon, S.-Y. Lee, Adv. Energy Mater. **7**, 1701099 (2017)
65. S.-J. Cho, K.-H. Choi, J.-T. Yoo, J.-H. Kim, Y.-H. Lee, S.-J. Chun, S.-B. Park, D.-H. Choi, Q. Wu, S.-Y. Lee, S.-Y. Lee, Adv. Func. Mater. **25**, 6029 (2015)
66. K. Fic, A. Platek, J. Piwek, E. Frackowiak, Mater. Today **21**, 437 (2018)
67. L. Zhang, Z. Liu, G. Cui, L. Chen, Prog. Polym. Sci. **43**, 136 (2015)
68. C.Q. Ruan, Z. Wang, J. Lindh, M. Strømme, Cellulose **25**, 3545 (2018)
69. W. Wang, Y. Sun, B. Liu, S. Wang, M. Cao, Carbon **91**, 56 (2015)
70. P. Tammela, Z. Wang, S. Frykstrand, P. Zhang, I.-M. Sintorn, L. Nyholm, M. Strømme, RSC Adv. **5**, 16405 (2015)
71. H.W. Liang, Z.Y. Wu, L.F. Chen, C. Li, S.H. Yu, Nano Energy **11**, 366 (2015)
72. L. Zuo, W. Fan, Y. Zhang, Y. Huang, W. Gao, T. Liu, Nanoscale **9**, 4445 (2017)
73. X.Z. Xu, J. Zhou, D.H. Nagaraju, L. Jiang, V.R. Marinov, G. Lubineau, H.N. Alshareef, M. Oh, Adv. Func. Mater. **25**, 3193 (2015)
74. W. Luo, J. Schardt, C. Bommier, B. Wang, J. Razink, J. Simonsen, X. Ji, J. Mater. Chem. A **1**, 10662 (2013)
75. J. Liu, Z. Wang, J. Zhu, J. Power Sources **306**, 559 (2016)
76. L.F. Chen, Z.H. Huang, H.W. Liang, H.L. Gao, S.H. Yu, Adv. Func. Mater. **24**, 5104 (2014)
77. K.-Y. Lee, H. Qian, F.H. Tay, J.J. Blaker, S.G. Kazarian, A. Bismarck, J. Mater. Sci. **48**, 367 (2012)
78. C. Long, D. Qi, T. Wei, J. Yan, L. Jiang, Z. Fan, Adv. Func. Mater. **24**, 3953 (2014)
79. Y. Huang, Z. Lin, M. Zheng, T. Wang, J. Yang, F. Yuan, X. Lu, L. Liu, D. Sun, J. Power Sources **307**, 649 (2016)
80. Y. Wan, Z. Yang, G. Xiong, H. Luo, J. Mater. Chem. A **3**, 15386 (2015)
81. Y. Chang, M. Zhou, X. Li, Y. Zhang, L. Zhi, ChemElectroChem **2**, 1238 (2015)
82. Z. Wang, X. Xiong, L. Qie, Y. Huang, Electrochim. Acta **106**, 320 (2013)
83. L. Qie, W.-M. Chen, Z. Wang, Q.-G. Shao, X. Li, L.-X. Yuan, X.-L. Hu, W.-X. Zhang, Y.-H. Huang, Adv. Mater. **24**, 2047 (2012)
84. D. Xu, C. Chen, J. Xie, B. Zhang, L. Miao, J. Cai, Y. Huang, L. Zhang, Adv. Energy Mater. **6**, 1501929 (2016)
85. Z. Wang, M. Li, C. Ruan, C. Liu, C. Zhang, C. Xu, K. Edström, M. Strømme, L. Nyholm, J. Phys. Chem. C **122**, 23352 (2018)
86. L. Qie, W. Chen, X. Xiong, C. Hu, F. Zou, P. Hu, Y. Huang, Adv. Sci. **2**, 1500195 (2015)
87. B. Wang, X. Li, B. Luo, J. Yang, X. Wang, Q. Song, S. Chen, L. Zhi, Small **9**, 2399 (2013)
88. S. Xu, C. Chen, Y. Kuang, J. Song, W. Gan, B. Liu, E.M. Hitz, J.W. Connell, Y. Lin, L. Hu, Energy Environ. Sci. **11**, 3231 (2018)
89. C. Chen, Y. Zhang, Y. Li, Y. Kuang, J. Song, W. Luo, Y. Wang, Y. Yao, G. Pastel, J. Xie, Adv. Energy Mater. **7**, 1700595 (2017)
90. L.-L. Lu, Y.-Y. Lu, Z.-J. Xiao, T.-W. Zhang, F. Zhou, T. Ma, Y. Ni, H.-B. Yao, S.-H. Yu, Y. Cui, Adv. Mater. **30**, 1706745 (2018)

91. C. Chen, Y. Zhang, Y. Li, J. Dai, J. Song, Y. Yao, Y. Gong, I. Kierzewski, J. Xie, L. Hu, Energy Environ. Sci. **10**, 538 (2017)
92. H. Xie, C. Yang, K.K. Fu, Y. Yao, F. Jiang, E. Hitz, B. Liu, S. Wang, L. Hu, Adv. Energy Mater. **8**, 1703474 (2018)
93. P. Arora, Z. Zhang, Chem. Rev. **104**, 4419 (2004)
94. X. Huang, J. Solid State Electrochem. **15**, 649 (2011)
95. Y.D. Pan, S.L. Chou, H.K. Liu, S.X. Dou, Natl. Sci. Rev. **4**, 917 (2017)
96. I. Kuribayashi, J. Power Sources **63**, 87 (1996)
97. L.C. Zhang, X. Sun, Z. Hu, C.C. Yuan, C.H. Chen, J. Power Sources **204**, 149 (2012)
98. S.-J. Chun, E.-S. Choi, E.-H. Lee, J.H. Kim, S.-Y. Lee, S.-Y. Lee, J. Mater. Chem. **22**, 16618 (2012)
99. J.-H. Kim, J.-H. Kim, E.-S. Choi, H.K. Yu, J.H. Kim, Q. Wu, S.-J. Chun, S.-Y. Lee, S.-Y. Lee, J. Power Sources **242**, 533 (2013)
100. F. Jiang, L. Yin, Q. Yu, C. Zhong, J. Zhang, J. Power Sources **279**, 21 (2015)
101. R. Pan, Z. Wang, R. Sun, J. Lindh, K. Edström, M. Strømme, L. Nyholm, Cellulose **24**, 2903 (2017)
102. J. Sheng, S. Tong, Z. He, R. Yang, Cellulose **24**, 4103 (2017)
103. R. Pan, X. Xu, R. Sun, Z. Wang, J. Lindh, K. Edström, M. Strømme, L. Nyholm, Small **14**, 1704371 (2018)
104. M.-H. Ryou, Y.M. Lee, J.-K. Park, J.W. Choi, Adv. Mater. **23**, 3066 (2011)
105. W. Liu, D. Lin, A. Pei, Y. Cui, J. Am. Chem. Soc. **138**, 15443 (2016)
106. M.N. Obrovac, V.L. Chevrier, Chem. Rev. **114**, 11444 (2014)
107. R. Pan, Z. Wang, R. Sun, J. Lindh, K. Edström, M. Strømme, L. Nyholm, J. Materiomics **5**, 204 (2018)
108. A. Banerjee, B. Ziv, Y. Shilina, S. Luski, I.C. Halalay, D. Aurbach, Adv. Energy Mater. **7**, 1601556 (2016)
109. J.H. Kim, M. Gu, H. Lee do, J.H. Kim, Y.S. Oh, S.H. Min, B.S. Kim, S.Y. Lee, Nano Lett. **16**, 5533 (2016)

Chapter 9
Cellulose-Based Printed Power Sources

Sang-Young Lee, Jung-Hwan Kim, and Donggue Lee

Abstract The emerging demand of ubiquitous smart electronics and the Internet of Things (IoT) calls for high-energy–density power sources with aesthetic versatility. Printed power sources have recently been in the spotlight as a promising system because of their various form factors and performance compatibility with electronics. Hence, extensive research on advanced materials has been conducted to achieve both facile processing and high performance of printed power sources. Cellulose, due to its natural abundance, environmental friendliness, chemical versatility, and dimensional stability, has been widely used as a reliable and sustainable basic constituent for a vast variety of materials/devices in our daily lives. The characteristic one-dimensional (1D) structure and chemical functionalities of cellulose bring unprecedented advantages to the fabrication and performance of energy storage materials and systems, which overcome the limitations of conventional synthetic materials. Notably, to obtain both form factor diversity and electrochemical reliability of printed power sources, cellulose has been extensively investigated as a promising building element because of its compatibility with the printing-based fabrications, mechanical stability, and electrochemically favorable morphology for printed power source components. In this chapter, we overview recent studies of cellulose as a building block for advanced printed energy storage systems, with a particular focus on application to printed electrodes, electrolytes/separator membranes, and substrates of Li-ion batteries/supercapacitors.

S.-Y. Lee (✉)
Department of Chemical and Biomolecular Engineering, Yonsei University, 50 Yonsei-ro, Seodaemun-gu, Seoul 03722, Republic of Korea
e-mail: syleek@yonsei.ac.kr

J.-H. Kim · D. Lee
School of Energy and Chemical Engineering, Ulsan National Institute of Science and Technology (UNIST), 50 UNIST-gil, Eonyang-eup, Ulju-gun, Ulsan 44919, Republic of Korea

L. Hu et al. (eds.), *Emerging Nanotechnologies in Nanocellulose*, NanoScience and Technology, https://doi.org/10.1007/978-3-031-14043-3_9

9.1 Introduction

The ever-increasing needs for wearable electronics, the Internet of Things (IoT), and electric vehicles have promoted the development of functional power sources with outstanding electrochemical performance and mechanical compliance. Battery form factors and flexibility are the key factors to fulfill this challenging objective, with on-going efforts in the development of high-performance energy storage materials as well.

Conventional batteries of fixed form factors and sizes are assembled with cell components (such as anodes, cathodes, and separator membranes) by winding or stacking and then inserting these components into cylindrical, prismatic, or pouch cases, followed by liquid electrolyte injection. The use of liquid electrolytes limits packaging materials because of its strict packaging conditions to avoid leakage issues and requires separator membranes to prevent short-circuit issues. Thus, the conventional battery fabrication process and components limit the design diversity of the power sources, and their integration with versatile-shaped electronic devices [1, 2]. For the past few years, printed power sources have been extensively investigated because of their outstanding form factor diversity and flexibility. Because of the simple, scalable, and elaborate printing techniques, printed batteries show significant characteristics such as form factor versatility and monolithic integration with devices [3, 4].

The design and preparation of battery materials are of crucial strategic importance for accomplishing practical form factor-free, rechargeable power sources [5] so various candidate materials have been explored as structural components [6–8]. Cellulose, due to its remarkable mechanical properties, chemical functionalities, and natural abundance have shown great attention in various energy storage devices [9]. The remarkable mechanical property of cellulose (i.e., Young's modulus of 138 GPa and tensile strength of 2–3 GPa) [10] and the one-dimensional (1D) fiber morphology (length: > 2 μm, width: 2–30 nm) of cellulose allow the fabrication of flexible self-standing films or porous foams with interwoven web-like structures. This unique morphology of cellulose is beneficial for facilitating ion and electron transport during battery operations [8]. Another major feature of cellulose is the numerous functional groups and its various substitutes that can be easily modified to achieve the desired chemical properties [11]. Benefiting from such characteristics, cellulose has been applied to various energy storage systems as an alternative to outperform its conventional synthetic counterparts. The research activities on cellulose in various energy storage systems have been described in several review articles and progress reports [8, 10, 12] including applications to permselective separator membranes, electrode binders, electrode substrates, and conductive host materials.

Intrigued by the uniqueness of printed batteries (process viewpoint) and cellulose (material viewpoint) described above, cellulose-based printed batteries have been investigated as an unprecedented concept of form factor-free power sources with extraordinary conformability and design diversity (Fig. 9.1). In this chapter, we investigate the current progress of cellulose-based printed power sources, with

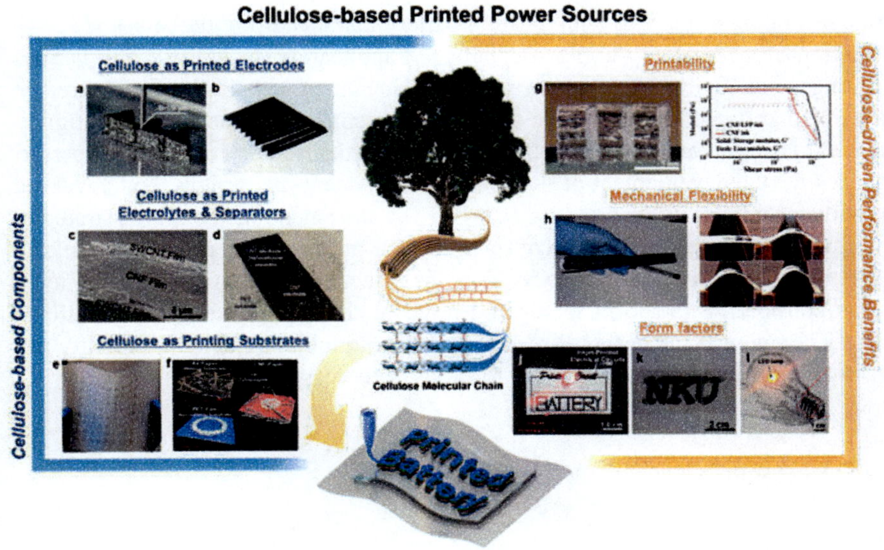

Fig. 9.1 Schematic illustration of cellulose-based printed batteries and their physiochemical/electrochemical performance. An illustration of the hierarchical structure of wood (Reprinted with permission from [112]. Copyright 2018 Nature Publishing Group.) Photographs of "Cellulose as printed electrodes". **a** Reprinted with permission from [111]. Copyright 2019 Wiley. **b** Reprinted with permission from [79]. Copyright 2017 MDPI. An SEM image and a photograph of "Cellulose as printed electrolytes & separators". **c** Reprinted with permission from [81]. Copyright 2017 American Chemical Society. **d** Reprinted with permission from [82]. Copyright 2014 IEEE. A photograph and illustrations of "Cellulose as printing substrates". **e** Reprinted with permission from [87]. Copyright 2010 AIP Publishing. **f** Reprinted with permission from [33]. Copyright 2016 Royal Society of Chemistry. A photograph and rheological analysis of "Printability". **g** Reprinted with permission from [111]. Copyright 2019 Wiley. Photographs of "Mechanical flexibility". **h** Reprinted with permission from [23]. Copyright 2009 National Academy of Sciences. **i** Reprinted with permission from [101]. Copyright 2016 Royal Society of Chemistry. Photographs of "Form factors". **j** Reprinted with permission from [33]. Copyright 2016 Royal Society of Chemistry. **k** Reprinted with permission from [81]. Copyright 2017 American Chemical Society. **l** Reprinted with permission from [110]. Copyright 2018 Wiley

a focus on their physicochemical/electrochemical superiority compared to those of conventional rechargeable power sources. And, future development direction will be discussed to address the technical challenges facing the commercial application of cellulose-based printed power sources.

9.2 Principles and Uniqueness of Printed Batteries

Recently, the, noteworthy developments of printed power sources are leading conventional energy storage devices to a new regime that power sources can be shape customized, miniaturized, and monolithically integrated with other devices, which are challenging to acquire with existing battery manufacturing processes [13, 14]. The simple processability, high throughput, and application versatility enable various printing technologies to be extensively utilized in numerous fields [3, 15]. Currently, research on printed energy storage devices is concentrated on the battery configuration and form factors, tuning of rheological properties and electrochemical performances for the development of new battery component ink designs, and advanced printing techniques utilization (Fig. 9.2) [4]. The aforementioned research strategies eventually aim for the realization of so-called "all-printed batteries". The successful development of printed batteries requires the selection of a suitable printing technique considering the form factor and scale of the final system, while rational battery component ink designs are elucidated regarding both facile printability and electrochemical performances of the resulting batteries.

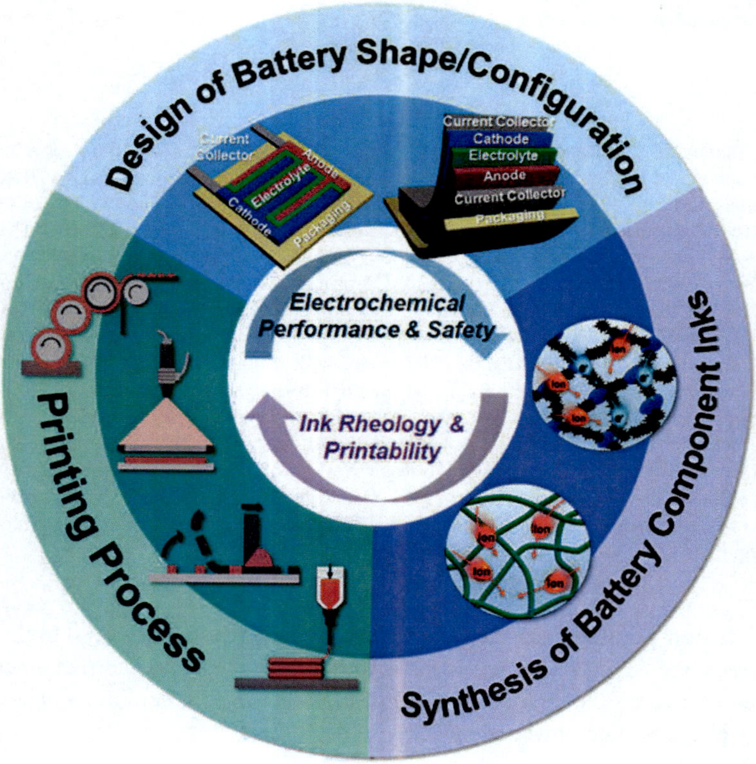

Fig. 9.2 A research strategy for the development of printed batteries: (i) Design of battery shape/configuration, (ii) Synthesis of battery component inks, and (iii) Application of suitable printing process. The key fundamentals are based on the rheology/printability of the battery component inks and the electrochemical performance/safety tolerance of the resulting printed batteries. Reprinted with permission from [4]. Copyright 2018 American Chemical Society

9.2.1 *Printing Techniques*

Rheological properties of the battery component inks are the key parameter in fabricating printed batteries with various printing techniques. Because of the simple and low-cost properties, mask-based printing techniques, including stencil, screen, flexographic, and spray printing, have been widely used in various commercial fields and a vast variety of inks with different rheological properties.

9.2.1.1 Mask-Based Printing Techniques

Screen/stencil printing [13, 14] is a printing technique in which the ink is deposited through pre-designed masks (e.g., screen or stencil) (Fig. 9.3a). The technique is advantageous for large-scale, high-throughput fabrication. As the pre-patterned mask is the key element of this technique, the design and fabrication of the mask is highly dependent on the diversified and sophisticated patterning of battery components. Flexographic printing is a technique that utilizes flexible printing plates wrapped on multiple cylinders to transfer and pattern inks on the targeted substrate (Fig. 9.3b) [14]. Flexographic printing can apply various kinds of inks such as water- and oil-based inks and a broad range of substrates such as metallic films, plastic, cellophane, and paper. Although flexographic printing is a fast, adaptable technique suitable for continuous manufacturing processes, the relatively low printing thickness (1–3 μm) is not suitable for typical battery component thickness (>20 μm) and requires multiple printing to acquire the desirable thickness. Spray printing [13, 14] deposits nano- to micro-sized ink droplets onto a substrate through pre-patterned masks via air (Fig. 9.3c). The atomized ink droplets induce immediate solvent drying which facilitate ink deposition on curved or rough surfaces preventing ink spreading and result in thin, even layers. Thus, the drying behavior of the ink should be precisely controlled to achieve uniform printed layers by considering the temperature, humidity of the environment, and surface properties of the substrate.

9.2.1.2 Direct-Ink-Writing (DIW) Printing Techniques

Direct-ink-writing (DIW) printing is an on-demand deposition technique which can construct arbitrary-shaped objects with low material waste due to the exclusion of pre-designed masks [14]. The unique on-demand deposition trait enables the manufacturing of sophisticated battery arrays or monolithically integrating batteries with devices of various form factors. As inspired by everyday handwriting, pen/pencil drawing technique is the simplest DIW process in which the pen ink or graphitic components of the pencil are deposited onto the substrate through hand pressure applied to pen or pencil tip (Fig. 9.3d) [16–18]. This DIW method is an easily accessible and highly cost-effective fabrication process for energy storage devices unlike other printing techniques which need complicated and expensive printing

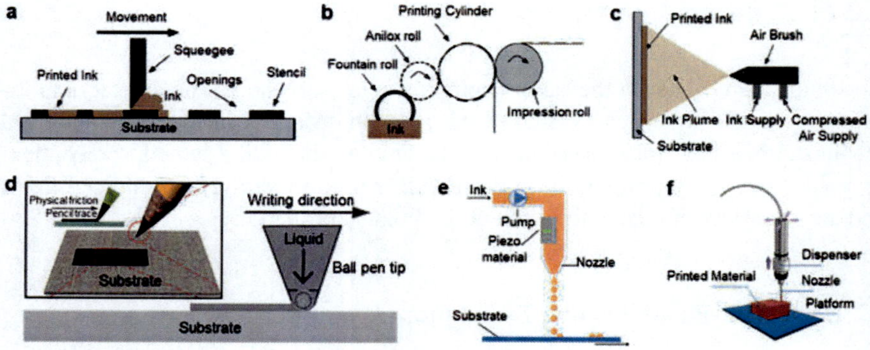

Fig. 9.3 Printed batteries fabricated through various printing techniques. Schematic illustrations of: **a** stencil, **b** flexographic, **c** spray, (Reprinted with permission from [14]. Copyright 2015 Wiley.) **d** pen drawing technique using a ball pen, (Reprinted with permission from [16]. Copyright 2017 Nature Publishing Group.) inset showing the working principle of the pencil drawing, (Reprinted with permission from [18]. Copyright 2016 Wiley.) **e** inkjet, (Reprinted with permission from [14]. Copyright 2015 Wiley.) and **f** 3D printing techniques. Reprinted with permission from [21]. Copyright 2015 Royal Society of Chemistry

equipment which require longer process time. Pen/pencil-drawn energy storage devices show user-customized form factors and facile integration with arbitrary-shaped electronic devices. However, the manual process of pen/pencil drawing techniques results in low throughput and low resolution which makes the process to be suitable for customized prototyping. Inkjet printing is a non-contact printing technique that can create various shapes in high precision through pre-programmed digital patterns. Inkjet printing favors ink designs of low viscosity and low surface tension to promote facile jetting (Fig. 9.3e) [13, 14, 19, 20]. Although inkjet printing can produce high-resolution patterns, the fine inkjet nozzles are vulnerable to clogging in which securing the stability of the ink is a major challenge. Moreover, limitation in printing speed is a disadvantage of inkjet printing which hinder the fabrication of thick, high-loading battery components. On the contrary, inkjet printing is effective in printing microscale, thin-film energy storage devices because of its high-precision fabrication capability. 3D printing is an advanced DIW technology which is capable of shaping 3D structures in a bottom-up manner (Fig. 9.3f) [21]. Typical 3D printing extrudes filaments or drops of the ink which requires stringent requirements of viscoelastic rheological properties and mechanical strength after deposition [22]. Therefore, rational considerations of battery electrode ink design are required to formulate the composition of electrode active materials, conductive additives, binders, and solvents. The large particle size distribution and the different densities between the solid particles and solvents complicate dispersion stability of the inks leading to poor printability. The solid content of the ink should be maximized to physically prevent sedimentation at an extent in which the particles cannot aggregate. Also the active material and conductive additive particles are synthesized or mechanically milled to reduce the average particle size while modifying the surface properties to promote facile dispersion. Moreover, optimizing the rheological properties is crucial to secure smooth flowing of the ink through the nozzles and recover structural integrity after deposition. An additional advantage of 3D printing is the capability of printing on nonplanar, uneven surfaces owing to the non-contact nature, which is difficult to achieve with aforementioned on-contact printing techniques.

9.2.2 *Preparation and Characterization of Battery Component Inks*

Battery component inks should be designed to ensure facile printability, while achieving the desired electrochemical performance. The top priorities of battery component ink design include solid particle dispersion, optimized rheological properties for the selected printing technique, structural integrity after printing (e.g., mechanical tolerance upon external stress, adhesion with the substrate), and electrochemical performance of the resulting printed battery) [4]. For thin-film printed batteries, low viscosity and low surface tension of battery component inks is required. In particular, aggregation of electrode particles (e.g., active materials, conductive additives) needs to be avoided to improve nozzle clogging issues [23]. Addition of surfactants or surface modifications on electrode particles is effective to improve particle dispersion state and to control the ink surface tension [24]. In the case of thick-film printed batteries, fine-tuned viscoelastic properties promote component ink flow under shear stress and ensure structural integrity sufficient to support the printed dimension.

9.2.2.1 Printed Electrodes and Current Collectors

To develop printed electrodes with various form factors and reliable physical/electrochemical properties, fine-tuning of rheological properties is an important prerequisite. Theoretical understanding of the rheological properties and their correlation with printing techniques of the electrode inks have been reported in previous reviews [14, 22, 25]. Note that the materials (i.e., electrochemically active materials, electrically conductive additives, binders), formulation, and rheological properties of electrode inks should be controlled to assure stable printability for the selected printing technique.

Conductive additives play a pivotal role in facilitating electron transport in the electrode. Among various conductive additives, carbon materials (e.g., carbon nanotubes (CNTs) and graphenes) have been widely applied because of their high electrical conductivity and sufficient capacitance [21]. However, carbonaceous substances suffer from low dispersion in water and polar solvents due to severe particle aggregation which arise from strong van der Waals interactions and intrinsic hydrophobicity [15, 24]. Various strategies for CNT dispersion have been reported by using surface functionalization of CNTs, or by using dispersants such as sodium dodecylbenzene sulfonate (SDBS) (Fig. 9.4a) [15, 24, 26, 27]. Note that the key is to minimize the deterioration of the intrinsic electrical conductivity of CNTs while achieving high dispersibility.

To induce facile electrochemical reactions of the active materials in the printed electrodes, both ion and electron conductions within the electrodes must be secured.

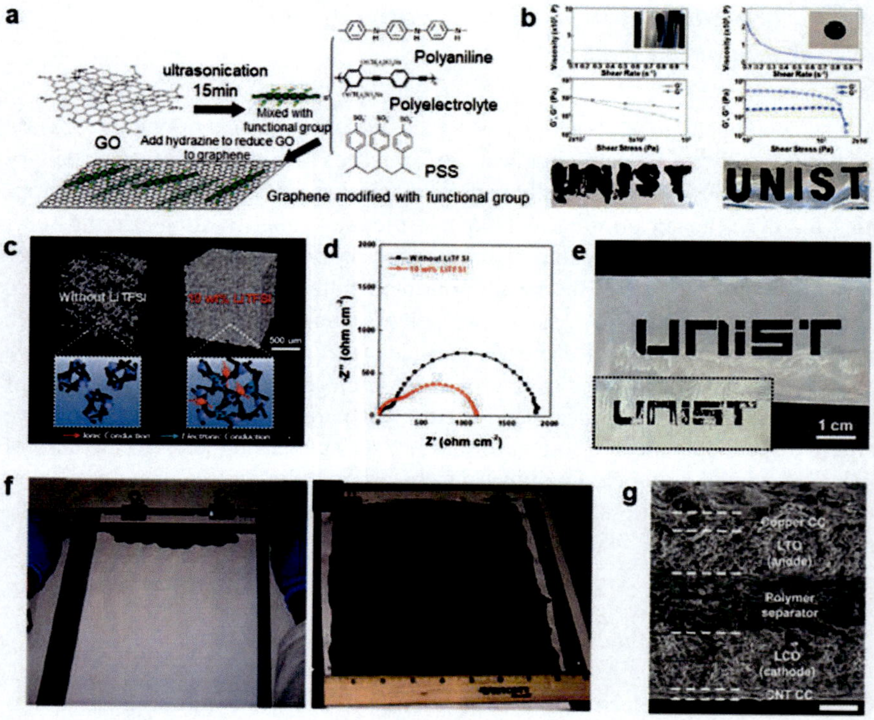

Fig. 9.4 Printed electrode and current collector inks with a focus on dispersibility and rheological control and electrochemical properties. **a** Different chemical structures of modified graphene inks. Reprinted with permission from [26]. Copyright 2011 IOP Publishing. **b** Rheological properties of printable $LiFePO_4$ (LFP) electrode inks with an ionically conductive matrix and photographs of stencil-printed, letter-shaped LFP electrode inks. Reprinted with permission from [1]. Copyright 2015 American Chemical Society. **c** Micro-CT images of the T-shirt-integrated supercapacitor electrodes (without LiTFSI salt versus with LiTFSI salt). Insets show the conceptual representation of the colloidal network structure. **d** EIS plots of the printed supercapacitor electrodes (without LiTFSI salt vs. with LiTFSI salt). **e** A photograph of a letter ("UNIST")-shaped electrode printed on an aluminum (Al) substrate. Inset is a photograph of the control sample (without LiTFSI). Reprinted with permission from [2]. Copyright 2018 Wiley. **f** Meyer-rod coating of carbon nanotubes (CNTs) or silver nanowire (Ag NW) ink on commercial Xerox paper. Reprinted with permission from [23]. Copyright 2009 National Academy of Sciences. **g** Cross-sectional scanning electron microscopy (SEM) image of a spray-printed Li-ion battery full cell containing the spray-printed CNT and copper (Cu) current collectors (scale bar is 100 μm). Reprinted with permission from [29]. Copyright 2012 Nature Publishing Group

Interestingly, a majority of the research on printed electrodes focus on the electrically conductive components without sufficient consideration of the ionically conductive element. Accordingly, Kim et al. reported a new concept of electrode formulation of a bicontinuous electron/ion conducting system [1]. The printed cathode and anode inks were comprised of active materials, carbon black conductive additives, and an ionically conductive matrix (i.e., UV-crosslinkable triacrylate monomer and liquid electrolyte). Unlike previous electrodes, these printed electrodes included the electrolyte in the slurry formulation, which enabled the exclusion of processing solvents such as N-methyl-2-pyrrolidone and water. This results in elimination of the solvent drying and liquid-electrolyte injection process which simplifies the electrode fabrication process. For facile stencil printing, the thixotropic rheological properties of the electrode inks were developed (Fig. 9.4b) and solidified by UV-crosslinking. Furthermore, to achieve high electrical conductivities of mixed electron/ion conducting printed electrodes, an increase in solid content (active, conductive materials) of the electrode ink is favorable. However, such an approach causes the electrode ink to show a glass-like response which hinders proper printing [28]. To address this issue, Lee et al. exploited the colloidal network concept [2]. The colloidal particles of electrode components were designed to facilitate a network structure induced by surface charge shielding and van der Waals interactions based on particle flocculation. To promote well-controlled particle flocculation, lithium bis(trifluoromethylsulfonyl)imide (LiTFSI) was added to the electrode ink which results in the formation of the percolation of activated carbon (AC) and multi-walled carbon nanotubes (MWCNTs) compared to the electrode ink without the LiTFSI salt (Fig. 9.4c). Based on this concept, well-interconnected electrically conducting pathways were formed within the ionically conducting pathway (i.e., ionic liquid electrolyte matrix). Owing to this bicontinuous ion/electron conducting structure of the printed electrodes, low resistance of the electrodes was proven (Fig. 9.4d). The ionic strength-induced electrode ink with LiTFSI also showed optimized rheological properties which enabled facile printability compared to the electrode ink without LiTFSI (Fig. 9.4e).

Current collectors are also an essential component of the electrode which enable stable electrochemical reactions in battery electrodes by supporting facile electron conduction. However, the mostly used current collectors in conventional electrodes are metallic foils which their fixed form factors and rigid properties limit the form factor diversification of printed batteries. Therefore, to develop all-printed batteries, current collectors should also be printable while maintaining their main requirements of high electrical conductivity and electrochemical stability. Unfortunately, printed current collectors have only been reported in a few works. Owing to its high electrical conductivity and electrochemical stability, carbon-based printed current collectors have been developed such as CNT-based conductive current collectors (Fig. 9.4f) [23] or spray-printed CNT current collectors [29]. Metal-based printed current collectors have also been reported using spray-printed copper (Cu) nanowire (NW) inks. This study showed the potential of metallic inks as printed current collectors (Fig. 9.4g).

9.2.2.2 Printed Electrolytes and Separator Membranes

Liquid electrolyte of conventional batteries requires the rigid packaging substance and separator membranes to prevent the leakage of liquid electrolyte and electrical short circuit, respectively, which ultimately limits the design diversity and form factors of conventional batteries. Unfortunately, most reported printed batteries have focused on the printed electrode rather than liquid electrolyte and separator membrane [29, 30]. Blake et al. demonstrated a phase-inversion-based 3D-printed separator membrane (Fig. 9.5a) [31]. In addition, Kil et al. demonstrated a shape-conformable and imprintable solid-state gel polymer electrolyte [32]. The printed solid-state electrolyte comprised a high-boiling point electrolyte, a UV-crosslinked polymer network as a mechanical skeleton, and alumina nanoparticles as a filler/rheology tuning agent (Fig. 9.5b). Based on the UV-assisted nanoimprint lithography (UV-NIL), the printed solid-state electrolyte enabled a high-fidelity, shape-conformable 3D structure through the imprinting of electrolyte pastes onto a maze-patterned PDMS stamp and UV-irradiation-based solidification (Fig. 9.5c). Ionic-liquid-based solid-state polymer electrolyte also has been introduced based on similar approaches. A UV-curable acrylate monomer incorporated ionic-liquid-based electrolyte (Fig. 9.5d) [33] and sol–gel reaction utilized mesoporous silica matrix-based printed solid-state electrolytes which were reported [34]. Moreover, thiol-ene chemistry also have been investigated as polymer network skeleton of ionic liquid-based printed solid-state electrolytes (Fig. 9.5e) [2]. The printability and rheology of the printable electrolyte paste was tuned through the control of colloidal network structure of SiO_2 nanoparticles. The mitigation of the electrostatic repulsions between SiO_2 leads to fluid-to-gel transition of the nanoparticle/ionic liquid mixtures, which was caused by the formation of a highly developed colloidal network of electrolyte ink. Lee et al. elucidated such phenomena by introducing SiO_2 nanoparticles in terms of different surface polarities. The hydrophilic SiO_2 nanoparticles in the electrolyte ink showed a well-interconnected SiO_2 colloidal network, which enabled a drastic increase in modulus compared to hydrophobic SiO_2 resulting in enhanced printability (Fig. 9.5f).

9.2.3 Recent Advances in Printed Batteries and Their Applications

Printing-based fabrication process enables the monolithic integration of batteries with various electronic devices such as smart cards/toys, radio frequency identification (RFID) tags, sensors, and biomedical devices. Generally, these electronics devices have important development guidelines such as device form factor, mechanical flexibility, thin films, and sometimes disposability. In particular, printed batteries should endure unexpected mechanical deformations like bending, folding, winding and twisting for their practical application to wearable devices. Furthermore, in other

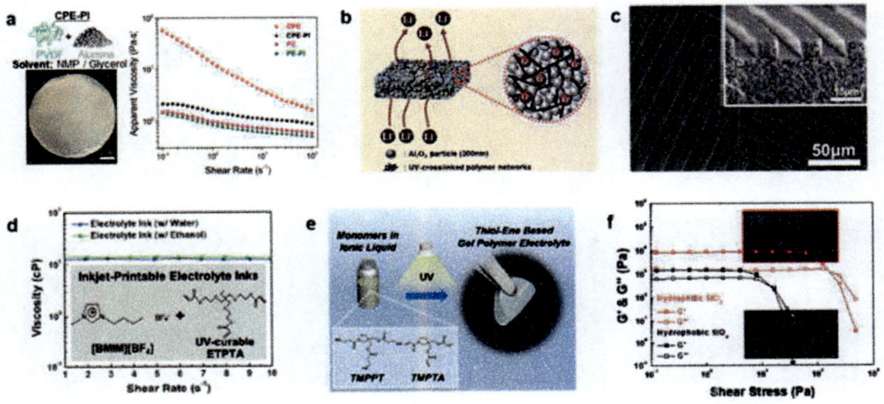

Fig. 9.5 Printed electrolytes and separator membranes. **a** Components and viscosity of the inks for the fabrication of 3D-printed separator membranes (photograph) through a phase inversion technique. Reprinted with permission from [31]. Copyright 2017 Wiley. **b** Conceptual illustration of an imprintable and shape-conformable composite gel polymer electrolyte. **c** SEM images of the maze-patterned composite gel polymer electrolyte fabricated via a UV-NIL technique (inset cross-sectional image). Reprinted with permission from [32]. Copyright 2013 Wiley. **d** Viscosity of the inkjet-printable electrolyte ink (composed of ionic liquid and UV-curable acrylate monomer) as a function of the shear rate. Reprinted with permission from [67]. Copyright 2016 Royal Society of Chemistry. **e** Schematic illustration depicting the synthesis of thiol-ene polymer networks via a click reaction of UV-crosslinkable TMPPT (as a thiol monomer) and TMPTA (as an ene monomer) in the presence of ionic liquid [EMIM][TFSI]. **f** Viscoelastic properties (G′ and G″) of the electrolyte pastes (hydrophilic SiO_2 versus hydrophobic SiO_2) as a function of the shear stress. Insets show the photographs of the letter ("UNIST")-shaped, printed electrolyte pastes after they were vertically tilted. Reprinted with permission from [2]. Copyright 2018 Wiley

to accomplish both the high performance and the aesthetic versatility of printed batteries, the cell design and configuration are also important factors as well as the component materials and associated electrochemistry. Meanwhile, Zn-based electrochemistry was widely adapted to many previously reported printed batteries [30, 35–45]. Recently, other electrochemical systems for printed batteries have been demonstrated such as supercapacitors [2, 33, 46–59], Li-based batteries (such as Li-ion [1, 29, 60–67], Li–S [68], and Li-O_2 [69] batteries) and polymer batteries [70].

In the following section, an overview of printed energy storage devices is discussed based on the electrochemical systems and form factor diversification:

(i) Supercapacitors. Supercapacitors (i.e., electrical double-layer capacitors (EDLCs)), functioned by capacitive desorption and adsorption of charged species at the surface of electrode. Supercapacitors feature long cycle life, high power density, and reliable electrochemical performance. However, supercapacitors suffer from self-discharge problem, short operation time, and low cell potential [71]. Several

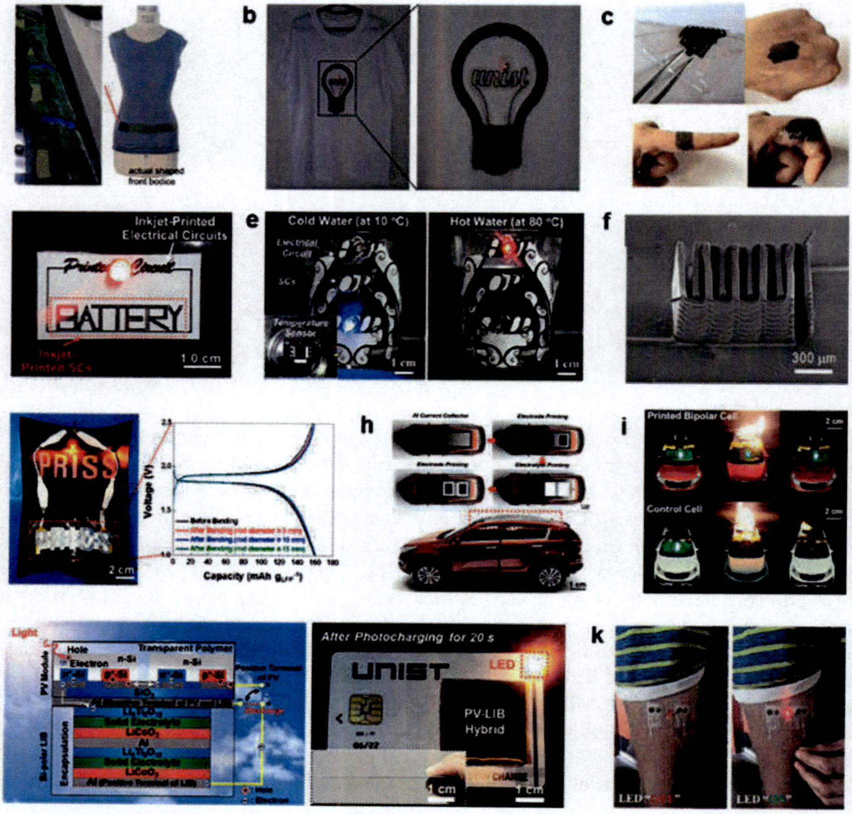

printing techniques have been applied to various printed supercapacitors such as screen printing [2, 46–48], gravure printing [49], inkjet printing [33, 50–53], spraying [54–57], and dispenser printing [58, 59]. Figure 9.6a shows the knitted carbon fibers and screen-printed electrode-based textile supercapacitors [46]. The textile supercapacitor showed reliable electrochemical performance while subjected to mechanical deformation. Lee et al. presented screen-printed wearable supercapacitors on T-shirts showing potential of integrating power sources on garments (Fig. 9.6b) [2]. The wearable supercapacitor T-shirt showed exceptional mechanical flexibility and thermal tolerance under the abusive conditions of the wearable tests such as walking, running, laundering, ironing, and folding, which demonstrate versatile applicability as a wearable power source.

Conductive polymers (e.g., PEDOT:PSS, polypyrrole (PPy), and polyaniline (PANI)) and low-dimensional carbon materials (e.g., graphenes, reduced graphene oxides, and CNTs,) were introduced as printed supercapacitor active materials due to their low electrical resistance and solution processability [49, 51–53, 56–58]. An inkjet-printed flexible micro-supercapacitor with PEDOT:PSS conductive polymer

◀**Fig. 9.6** Recent advances in printed batteries and their applications. **a** (left) Knitted and screen-printed carbon-fiber supercapacitor electrodes. (right) A shaped front bodice knitted in one piece with the carbon-fiber-based electrode integrated as a part of the textile. Reprinted with permission from [46]. Copyright 2013 Royal Society of Chemistry. **b** Application of a SC-printed T-shirt as a potential electronic garment. Photographs of the bulb symbol-shaped supercapacitor that powered an LED lamp. Reprinted with permission from [2]. Copyright 2018 Wiley. **c** Photographs of ultra-thin and flexible micro-supercapacitors. The micro-supercapacitors are thinner than a human hair, wearable on human hands, and conformable to bent fingers. Reprinted with permission from [51]. Copyright 2016 Wiley. **d** Photographs of the inkjet-printed, letter ("BATTERY")-shaped supercapacitors (marked by a red box) that were seamlessly connected with the inkjet-printed electrical circuits and an LED lamp. **e** Photograph depicting the operation of a blue LED lamp in the smart cup (for cold water (~10 °C)), wherein the inset is a photograph of a temperature sensor. (right) Photograph depicting the operation of a red LED lamp in the smart cup (for hot water (~80 °C)). Reprinted with permission from [33]. Copyright 2016 Royal Society of Chemistry. **f** SEM image of printed and annealed 16-layer 3D-interdigitated LTO-LFP electrode architectures. Reprinted with permission from [60]. Copyright 2013 Wiley. **g** (left) Photograph of a "PRISS" letter-shaped solid-state Li-ion battery and (right) its charge/discharge profiles while being completely wound along rods with different diameters (5, 10, and 15 mm). Reprinted with permission from [1]. Copyright 2015 American Chemical Society. **h** Photographs of the stepwise fabrication of the printed bipolar two-stacked cell on the curved roof of a miniature toy car. **i** Nonflammability test of the printed bipolar two-stacked cell and control cell; the Al pouch-based packaging substances were removed prior to this test. Reprinted with permission from [66]. Copyright 2018 American Chemical Society. **j** (left) Schematic representation of the internal structure and operating principle of the SiPV-LIB device under sunlight illumination and (right) photograph showing the operation of the SiPV-LIB-embedded smartcard. Inset shows the overall thickness of the resulting smartcard. Reprinted with permission from [67]. Copyright 2017 Royal Society of Chemistry. **k** Demonstration of the printed tattoo battery for lighting a red LED after transfer onto skin. Reprinted with permission from [41]. Copyright 2014 Royal Society of Chemistry

and exfoliated graphene electrodes hybrid inks was reported (Fig. 9.6c) [51]. The inkjet-printed micro-supercapacitor presented reliable capacitance retention under 1000 cycles of mechanical deformation (bending radius = 5 mm) and was applied to various body parts (e.g., human finger), which implied excellent mechanical flexibility and wearable capability. 3D-printed graphene composite aerogels (3D-GCA) were presented for supercapacitor electrodes, which consist of graphene nanoplatelets (GNP) and graphene oxide (GO) [59]. The 3D-printed GO-GNP structure was freeze-dried for highly porous structure and electrically conductive supercapacitor electrode. Inkjet-printed solid-state flexible supercapacitors were demonstrated on the conventional A4 paper by Choi et al. [33]. The inkjet-printed supercapacitors showed high aesthetic diversity by presenting letter shapes or various patterns (Fig. 9.6d). The inkjet printing enabled direct connection of supercapacitors in series or in parallel configuration to customize the voltage or capacitance without introducing extra metal interconnects. Furthermore, inkjet-printed supercapacitors were easily integrated with smart glass cups with artistic patterns (Fig. 9.6e).

(ii) *Li-ion batteries*. Lightweight, high cell voltage (~3.7 V), and high energy densities (~260 Wh kg^{-1}, ~780 Wh L^{-1}) have contributed to the widespread applications of Li-ion batteries [72]. However, the flammable organic electrolytes impose severe safety and swelling problems. Furthermore, poor moisture stability restricted the implementation of various printing techniques. Wang et al. reported a painted Li-ion battery, in which all battery components were prepared in the form of spray-printable inks including the electrodes, separator membranes, and current collectors, and subsequently printed through pre-patterned masks [29]. 3D structured batteries based on the micro- and nanostructured electrodes could be advantageous for microscale power sources. 3D-printed Li-ion microbatteries with interdigitated high-aspect-ratio electrode arrays were demonstrated (Fig. 9.6f) [60, 61]. However, the large inter-electrode area hampers the energy/power density, which remains a challenge to practical use. Also, a harsh post-treatment induces serious limitations for the material selection for other cell components.

Many of the abovementioned previous studies have used liquid electrolytes as an ionic media and mask-based printing techniques that require supplementary spatial alignments. Particularly, liquid electrolytes require an additional electrolyte injection step thereby imposing unwanted problems, such as leakage, packaging limitations, and safety failure. To address these long-standing issues, screen-printed solid-state Li-ion batteries were reported by Kim et al. [1]. A UV-curable solid-state composite polymer electrolyte was introduced as the printed electrolyte/ separator membrane which excluded the electrolyte injection process and energy-/time-consuming solvent drying processes during cell fabrication. The letter-shaped printed cells presented stable charge/discharge behavior under repeated bending deformation (bending radius = 5, 10, and 15 mm) (Fig. 9.6g). Further research of all-solid-state printed Li-ion batteries was focused to address the safety issues of Li-ion batteries. Kim et al. developed a new class of mechanically and thermally tolerant flexible bipolar Li-ion batteries [66]. The all-solid-state Li-ion batteries were fabricated via multistage printing of a UV-curable component precursor ink by introducing a novel flexible and sebaconitrile (SBN)-based nonflammable gel electrolyte and a semi-interpenetrating (semi-IPN) polymer network (Fig. 9.6h). Compared with conventional inorganic electrolyte-based solid-state Li-ion batteries, the as-fabricated Li-ion batteries demonstrated excellent thermal/mechanical stability when exposed to flame or even after they were cut in half (Fig. 9.6i). Due to the UV-assisted multistage printing process, in-series and in-plane bipolar configurations of the Li-ion batteries were realized, showing the electrochemical viability of the printed batteries with versatile form factors. Additionally, photorechargeable portable power sources were demonstrated through the monolithic integration of printed Li-ion batteries onto miniaturized silicon (Si) photovoltaic (PV) cells (Fig. 9.6j) [67]. The Li-ion battery and Si PV-integrated power sources showed notably improved electrochemical performances such as rapid photocharging (charging within 2 min), high discharge rate, and cycling performance (photocharge/galvanostatic discharge) compared to conventional PVs or Li-ion batteries alone.

(iii) *Zn-based batteries*. High safety and low cost are the main features of Zn-based batteries because of aqueous electrolytes and naturally abundant Zn metal [73]. In particular, aqueous electrolytes enable the ambient environment fabrication of Zn-based battery without a costly dehumidification facility (i.e., dry room) which is an essential prerequisite for the fabrication of Li-ion batteries. Unfortunately, the relatively low cell voltage of Zn-based batteries (~1.5 V) limits their potential application for high-voltage operation electronic devices [74, 75]. In addition, their charge/discharge cyclability needs to be improved to a satisfactory level, as compared to other rechargeable batteries. Nylon mesh-containing electrode-based printed, flexible Zn/MnO_2 batteries were reported by Gaikwad et al. [42]. The mesh provides a crucial role in the flexibility and mechanical stability of the Zn/MnO_2 battery electrode which minimized the loss of electrochemical performance after repeated bending deformation. A direct ink writing (DIW) technique-based printed Zn/MnO_2 microbattery was reported by Ho et al. [43]. Given that in the Zn electrode undergoes shape change during stripping/plating reaction, dendritic growth of Zn metal is the main limitation in achieving extensive rechargeability. To alleviate the Zn dendrite formation issue, ionic liquid-based new electrolytes were introduced by G.G. Kumar et al. [44]. In addition, Zn/Ag batteries have attracted much attention due to its high energy density and air stability [37]. However, the expensive material cost of silver hinders the practical applications of Zn/Ag-printable batteries. Moreover, printing-based fabrication was suggested as an efficient way to reduce material waste, thereby accelerating the development of practical Zn/Ag batteries as promising electrochemical systems. A skin-patchable, epidermal Zn/Ag rechargeable battery consisted of an alkaline gel electrolyte, screen-printed electrodes, a silicon packaging, and temporary tattoo paper was reported by Berchmans et al. (Fig. 9.6k) [41]. The shape-conformable tattoo battery was successfully applied to human skin. Kumar et al. reported a highly elastic, all-printed Zn/Ag rechargeable battery using screen-printable conductive inks [45]. Owing to the polystyrene-block-polyisoprene-block-polystyrene (SIS) elastic binder, the Zn/Ag stretchable battery presented the hyperelastic properties (1300% elongation).

Printed batteries have garnered much attention as a game-changer for form factor-free architecture, which breaks conventional perspectives on power sources. All battery components (including electrodes, electrolyte, current collectors, and packaging material) are highly required to be form factor-free and flexible to truly achieve the design diversity and facile integration capability of printed batteries. Unfortunately, the electrochemically inactive component (i.e., current collector or packaging) has been paid little attention compared to electrochemically active component (i.e., electrodes, electrolytes, and separator membranes). To accomplish the ultimate goal of so-called "all-printed batteries", thorough consideration should be taken to the printing substrates. The synthesis and rational design of new substrate materials such as cellulose could contribute to further advances in printed batteries in terms of form factors and aesthetic versatility.

9.3 Cellulose in Printed Batteries

Recently, cellulose has been exploited as a promising eco-friendly material compared to conventional materials for printed batteries. Notably, the structural uniqueness, mechanical robustness, and chemical functionalities of cellulose enabled substantial improvements in flexibility, printability, and performance which are hard to achieve with conventional synthetic materials, of printed batteries. Benefiting from the unique characteristics described above, cellulose can be used in printed batteries: (i) as battery components including electrodes, electrolytes, and separator membranes and (ii) as printing substrates in the form of paper.

9.3.1 Cellulose as Battery Components

As described above, inks should be rationally formulated to ensure process compatibility with printing techniques. Cellulose can be used in ink formulations owing to the combination of natural abundance and the unique properties of cellulose in aqueous-based systems. High water retention capability due to the rich hydroxyl groups of cellulose allows the inks to possess unique viscoelastic properties [76]. Cellulose acts as a dispersing agent to assist dispersion state of various materials in aqueous solution due to its extremely high negative zeta potential ($\approx$ 60 mV) [77]. The 1D fibrous feature and chemical functionalities of cellulose are beneficial for maintaining structural integrity (e.g., adhesion between the electrode and the substrate, cohesion between component particles, and mechanical compliance) [8]. Cellulose and its derivatives have been reported for component inks of printed batteries (i) as an electrode binder or (ii) as a gelling agent and polymer matrix for printed electrolytes/separator membranes.

(i) *Printed Electrodes*. In the printed batteries, the binders play the role of holding the electrode components together and enhancing the adhesion between the components and the substrate [3]. Cellulose can be suggested as an alternative binder material that replaces conventional synthetic polymer-based ones. Among cellulose and its derivatives, carboxymethyl cellulose (CMC), which is a derivative of linear polymeric cellulose with substitutional anionic carboxymethyl groups in various degrees, is frequently used as electrode binders. Although CMC shows high stiffness and a small strain at fracture, it shows better bonding with the electrode materials and the electrode substrates due to the large number of carboxymethyl groups of cellulose. Park et al. fabricated the dispenser-printed $LiFePO_4$ (LFP) or $Li_4Ti_5O_{12}$ (LTO)/Ag NW/CMC composites, in which the CMC acted as a polymeric stabilizer and a viscosity thickening agent [78]. The Ag NW/CMC conductor was prepared to construct electrical percolation networks in the electrode composites. Subsequently, electroactive materials (LFP or LTO) were added to the Ag NW/CMC conductor. The prepared electrodes showed well-developed percolation networks (Fig. 9.7a) and shear thinning behavior which is suitable for dispenser printing (Fig. 9.7b). Liu et al. also demonstrated highly porous LFP electrodes using CMC binders through dispenser-printed electrode and freeze-drying [79]. The rheological properties of electrode inks, which consisted of LFP and CMC in a water and 1,4 dioxane (freezing agent) solvent mixture, were tailored to be suitable for dispenser printing (Fig. 9.7c). The electrode ink was dispenser-printed in a grid pattern (Fig. 9.7d), followed by freeze-drying to form a porous structure in the electrodes. Figure 9.7e shows the morphology of CMC-based printed electrodes. LFP or graphite electrodes containing

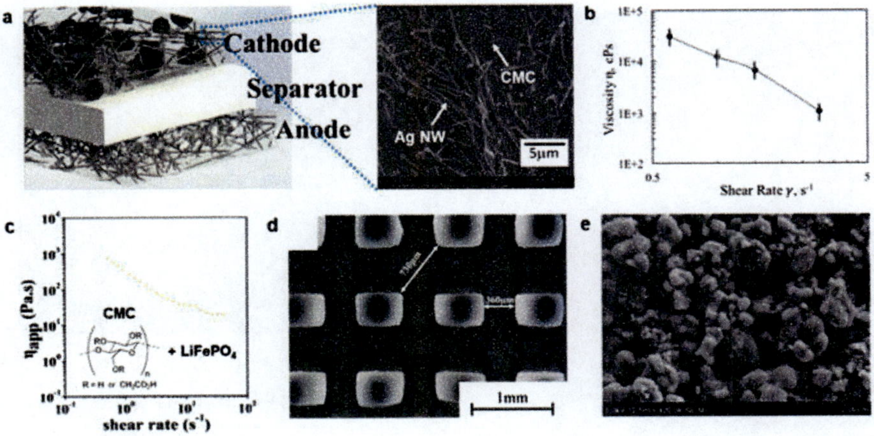

Fig. 9.7 Cellulose-based printed electrodes. **a** (left) Schematic illustration of dispenser-printed battery with electroactive materials/Ag NW/CMC composites, in which black particles represent LFP (cathode) particles while white particles represent LTO (anode) particles. (right) SEM images of the Ag NW/CMC conductor in the electrodes. **b** Viscosity of anode ink as a function of shear rate. Reprinted with permission from [78]. Copyright 2017 Nature Publishing Group. **c** Rheological properties of LFP inks. Inset shows the chemical structure of CMC. **d** Dispenser-printed LFP electrode in grid pattern. **e** SEM images of the dispenser-printed LFP electrode. Reprinted with permission from [79]. Copyright 2017 MDPI

cellulose were screen-printed on both sides of a paper separator to fabricate an LIB [80]. The inks were composed of active materials (LFP or graphite), carbon black as a conducting agent, cellulose nanofiber (CNF) as a binder, and CMC as a dispersant. CMCs also can act as CNFs flocculation inhibitor during the ink preparation and printing process. The printed electrodes showed high adhesion with metallic current collectors owing to the presence of CNFs.

(ii) *Printed Electrolytes and Separator Membranes*. Most of the previous studies on the printed batteries tend to focus on electrodes, however, much attention should be paid to separator membranes and electrolytes to realize all-printed-batteries with design versatility and flexibility. Only a few studies have been reported on cellulose-based printed separator membranes and electrolytes. Wang et al. reported thin CNF films as separator membranes, which were fabricated onto electrodes by a spray printing process [81]. The homogeneous CNF aqueous suspension can be easily printed on an appropriate substrate. The resulting separator membrane was seamlessly integrated with electrodes as shown in Fig. 9.8a. Tukkanen et al. fabricated nanocellulose-based separator membranes through spray and stencil printing process [82]. Cellulose nanocrystal (CNC) ink and CNF ink were sequentially printed onto CNT electrodes (Fig. 9.8b, c). The CNC layer was applied to promote adhesion between thick electrodes and separator films. Due to the nanoscale dimensions and the capability to form an entangled nanoporous network, nanocellulose can be suggested as a promising material for separator membranes. Areir et al. developed dispenser-printed gel polymer electrolytes (Fig. 9.8d), which consisted of CMC,

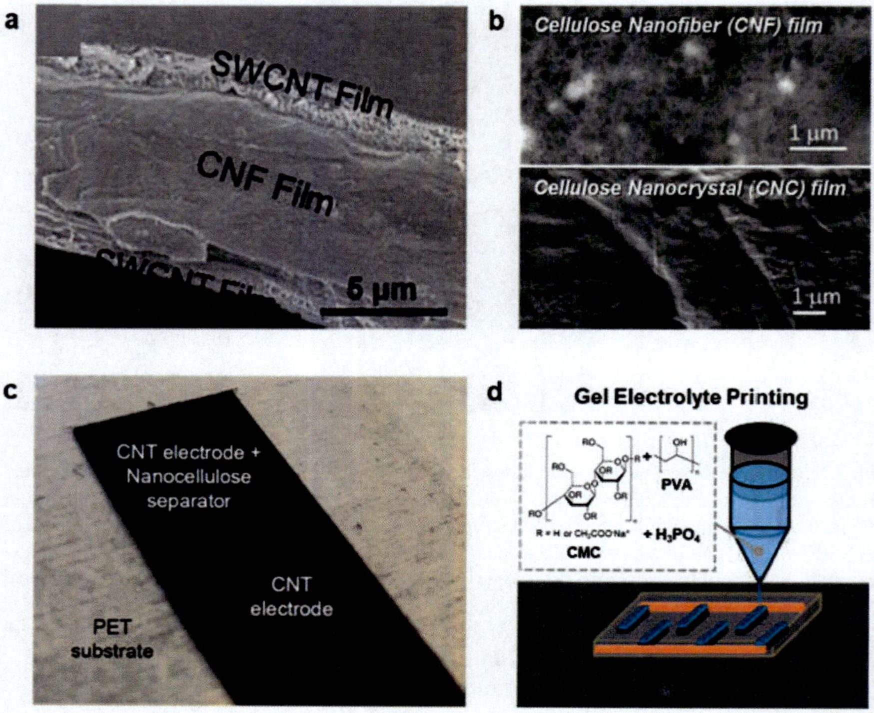

Fig. 9.8 Printed electrolytes and separator membranes based on cellulose. **a** SEM image of the spray-printed CNF separator membrane with SWCNT electrodes. Reprinted with permission from [81]. Copyright 2017 American Chemical Society. **b** SEM images of the spray-printed CNF (top) and CNC films (bottom). **c** Optical image of CNT electrodes and nanocellulose separator layers on PET substrate. Reprinted with permission from [82]. Copyright 2014 IEEE. **d** Scheme of the dispenser printing process of the CMC-based gel polymer electrolyte. Reprinted with permission from [83]. Copyright 2017 Springer

poly(vinyl alcohol) (PVA), and H_3PO_4 electrolyte [83]. CMCs were used to tailor the rheological properties of the gel polymer electrolyte ink for enabling the dispenser printing. Braam et al. demonstrated methyl cellulose-based gel polymer electrolytes fabricated through the dispenser printing process [37]. The gel polymer electrolyte consisted of methyl cellulose and KOH electrolyte. The degree of polymerization of methyl cellulose should be rationally designed to control the crystallinity of the cellulose structure that crucially affects the dispersion state of the resulting inks.

9.3.2 *Cellulose as Printing Substrates*

Printing substrates, along with the ink chemistry/formulation, affect the fabrication of printed batteries. Cellulose paper is an attractive printing substrate compared to currently widespread polymeric materials such as polyethylene terephthalate (PET) [84] due to its superior mechanical flexibility and its recyclability. In addition, cellulose paper has good thermal stability up to 250 °C without structural/chemical disruption. Furthermore, the roughness of cellulose promotes the adhesion with electrode materials and its porous structure facilitates the transport of ions [85]. Hu et al. demonstrated a CNT-coated conductive paper for printed supercapacitors [23]. CNT inks were printed on commercial paper by Meyer-bar coating which resulted in a conductive paper of low sheet resistance of ~10 Ω sq^{-1}. The strong binding of CNTs with paper substrates led to high structural stability against external stresses such as peeling-off and bending (Fig. 9.9a). Due to the large surface area and high electric conductivity, the conductive paper was exploited as electrodes and current collectors of supercapacitors (Fig. 9.9b). Gaikwad et al. reported the Zn-MnO_2 batteries directly printed on polyvinylalcohol/cellulose (PAC) membranes which serve as the separator and mechanical support for the printed Zn and MnO_2 electrodes [36]. Due to the fibrous structure of the PAC membrane, the printed Zn, MnO_2 electrodes, and silver current collectors showed great adhesion on the membrane, resulting in great contact between the membrane, electrodes, and consecutively printed silver current collectors (Fig. 9.9c, d).

Traditional paper is not a suitable substrate for high-precision printed batteries due to its surface roughness and highly porous structure [86]. On the paper, inks tend to be randomly spread out due to the unexpected capillary force generated by the inhomogeneously distributed pores. To mitigate these problems of traditional paper,

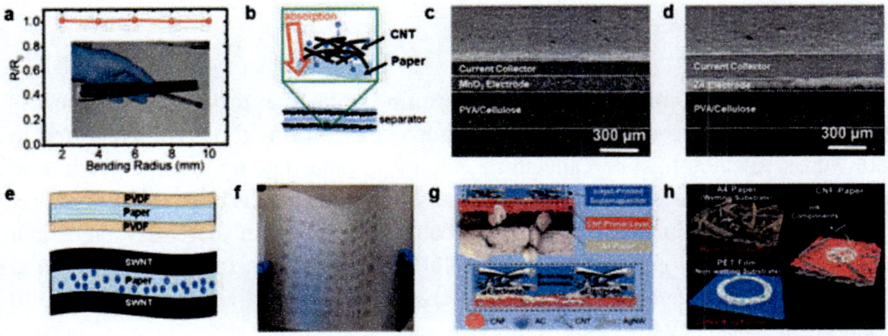

Fig. 9.9 Cellulose as printing substrates. **a** Sheet resistance changes after bending conductive paper into different radii. **b** Schematic illustration of all-paper SCs based on CNT conductive paper. Zoomed-in schematic illustrates that ion accessibility is enhanced by the strong solvent absorption. Reprinted with permission from [23]. Copyright 2009 National Academy of Sciences. Cross-sectional SEM image of (**c**) the MnO_2 electrodes, **d** the Zn electrodes and silver current collectors printed on the PAC fibrous substrate. Reprinted with permission from [36]. Copyright 2013 AIP Publishing. **e** Schematic of paper treatment with PVDF (top). Paper supercapacitor structure with SWCNTs film printed on both sides of the treated paper (bottom). **f** Optical image of inkjet-printed supercapacitor on paper. Reprinted with permission from [87]. Copyright 2010 AIP Publishing. **g** Schematic illustration of inkjet-printed supercapacitors with CNF layers on commercial A4 paper. **h** Effect of the substrate on the printability of the inkjet printing process. Reprinted with permission from [33]. Copyright 2016 Royal Society of Chemistry

the surface roughness has been reduced by applying polymer layers on the top of the paper [84]. Hu et al. demonstrated paper substrates coated with a thin layer of PVdF to fabricate fully integrated printed supercapacitors [87]. Printing single-walled carbon nanotube (SWCNT) inks directly onto paper results in penetration of the inks through the pores, giving rise to internal short-circuit failure. On the other hand, the SWCNT ink was easily printed on both sides of the PVDF-treated paper (Fig. 9.9e). Figure 9.9f shows the printed supercapacitor, in which the electrodes and separators were integrated into a single sheet of paper substrate. Meanwhile, non-wetting substrates often provoke a ring-like deposition (i.e., coffee-ring effect) of ink particles, in addition to poor adhesion between the printed layers and substrates [88]. To address the substrate-related issues described above, most of the previous studies have focused on ink materials and formulations which, however, often limits the widening of ink chemistry. Choi et al. demonstrated an inkjet-printed cellulose nanomat primer layer on commercial A4 paper (Fig. 9.9g) [33]. The cellulose nanomat primer layers provided nanoporous surface structure, eventually showing the combined advantageous attributes of both wetting and non-wetting substrates (Fig. 9.9h). As a result, the cellulose nanomat primer layer suppresses the ink spreading or coffee-formation, which contributed to achieving high-resolution printed images on conventional A4 paper and also to broadening the material design flexibility of electrode/electrolyte inks.

9.4 Recent Advances in Cellulose-Based Printed Battery Systems

Cellulose has been combined with various printing techniques to develop form factor-free, flexible power sources. Many of cellulose-based printed batteries reported to date were dedicated to supercapacitors. Exploring other battery systems, including Li-ion batteries, Zn-based batteries, and metal-air batteries, has been recently undertaken. However, cellulose-based printed batteries are still in the early stages and require further development for commercial use. The application of cellulose and its derivatives to printed batteries is summarized in Table 9.1, with special attention to cellulose types, battery systems, printing techniques, and electrochemical performance of printed power sources. More details are systematically described in the subsequent sections.

9.4.1 Cellulose-Based Printed Supercapacitors (SCs)

Cellulose-based supercapacitors have been demonstrated through various printing techniques, such as screen/stencil printing [89, 92, 97, 99], spray printing [81, 82], pen/pencil drawing [93–95, 98, 100], inkjet printing [33, 87, 90, 96, 101], and

Table 9.1 Battery systems, types/roles of cellulose, printing techniques, components, and electrochemical performance of cellulose-based printed power sources

Battery system	Cellulose	Purpose	Printing method	Cathode	Anode	Electrolyte	Capacitance or capacity/voltage	References
SC	Algae cellulose	Binder	Stencil	AC/Graphite	Ppy/Graphite	PVP/PEG/$NaNO_3$	120 mF cm^{-2}/0–0.8 V	[89]
	Ethyl cellulose	Binder, Dispersant	Inkjet	Exfoliated graphene	Exfoliated graphene	1 M Na_2SO_4	0.59 mF cm^{-2} (100 mV s^{-1})/−0.35–0.35 V	[90]
	CMC	Gelling agent	Dispenser	AC	AC	CMC/PVA/H_3PO_4	11.2 mF (100 mV s^{-1})/0–0.8 V	[83]
	CMC	Gelling agent	Dispenser	AC	AC	CMC/PVA/H_3PO_4	38.5 mF (15 mA)/0–1.0 V)	[91]
	CNF	Separator	Spray	SWCNT/PANI	SWCNT/PANI	PVA/H_2SO_4	322 F g^{-1} (20 mV s^{-1})/0–0.8 V	[81]
	CNF	Substrate, Separator	Screen	Graphene	Graphene	1 M NaCl	4.9 F g^{-1}	[92]
	Paper, CNF	Substrate	Inkjet	AC	AC	ETPTA/[BMIM][BF_4]	100 mF cm^{-2} (0.2 mA cm^{-2})/0–2.0 V	[33]
	Paper	Substrate	Pen drawing	MnO_2/Ni/Graphite	MnO_2/Ni/Graphite	CMC/Na_2SO_4	1020 mF cm^{-3}, 16.3 F g^{-1} (5 mV s^{-1})/0–0.8 V	[93]
	Paper	Substrate	Pen drawing	Graphite/PANI	Graphite/PANI	PVA/H_2SO_4	77.8 mF cm^{-2} (0.1 mA cm^{-2})/0.-0.8 V	[94]
	Paper	Substrate, Separator	Pen drawing	Graphite	Graphite	1 M H_2SO_4	2.3 mF cm^{-2} (200 mA g^{-1})/0–0.9 V	[95]
	Paper	Substrate	Inkjet	rGO-MnO_2	AC	PVA/LiCl	1.586 F cm^{-2}, 1023 F g^{-1} (4 mA cm^{-2})/0–2.0 V	[96]
	Paper	Substrate	Screen	AC/Graphite	AC/Graphite	[EMIM][TFSI]	7 Wh kg^{-1}/0.2–1.5 V	[97]
	Paper	Substrate	Pen drawing	MXene	MXene	PVA/H_2SO_4	5 mF cm^{-2}/0–0.6 V	[98]

(continued)

Table 9.1 (continued)

Battery system	Cellulose	Purpose	Printing method	Cathode	Anode	Electrolyte	Capacitance or capacity/voltage	References
	Filter paper	Substrate	Screen	MnO_2	MnO_2	CMC/Na_2SO_4	25.8 mF cm^{-2} (0.305 mA cm^{-2})/0–1.5 V	[99]
	Notebook paper	Substrate, Separator	Pen drawing	Graphite/MWCNT	Graphite/MWCNT	PVA/Na_2SO_4	9.16 mF cm^{-2} (50 mV s^{-1})/0–0.8 V	[100]
	Photo paper	Substrate	Inkjet	PEDOT:PSS-CNT	PEDOT:PSS-CNT	PVA/H_3PO_4	23.6 F cm^{-3} (10 mV s^{-1})/0–0.9 V	[101]
	Newspaper	Substrate, Separator	Inkjet	SWCNT	SWCNT	1 M $LiPF_6$ in EC/DEC	3 F g^{-1}/0–3.0 V	[87]
LIB	Ethyl cellulose	Binder	Screen	LCO	Li	$Li_7La_3Zr_2O_{12}$	100 mAh g^{-1}(0.05C)/3.0–4.05 V	[102]
	Microfibrillated cellulose, CMC	Binder	Spray	Graphite	Li	1 M $LiClO_4$ in EC/DEC	350 mAh g^{-1}(0.1C)/0.02–1.2 V	[103]
	Microfibrillated cellulose, CMC	Binder	Screen	Graphite	Li	1 M $LiPF_6$ in EC/PC/DMC	100 mAh g^{-1}(0.1C)/0.02–0.15 V	[104]
	CMC	Binder	Inkjet	LFP	Li	1 M $LiPF_6$ in EC/DMC	150 mAh g^{-1}(0.1C)/2.0–4.0 V	[105]
	CMC	Binder	Inkjet	LFP	LTO	PYR_{13}-Li-TFSI, SiO_2	60 mAh g^{-1}(0.1C)/1.0–2.5 V	[34]
	CMC	Binder	Dispenser	LFP	Li	–	60 mAh g^{-1}(10C)/2.5–4.2 V	[79]
	CMC	Binder	Spray	LFP	LTO	1 M $LiPF_6$ in EC/DMC	300 Wh kg^{-1}(0.1C)/1.0–2.5 V	[106]
	CMC	Binder	Spray	TiO_2(B)	Li	1 M $LiPF_6$ in EC/DMC	140 mAh g^{-1}(30C)/1.0–3.0 V	[107]
	CMC	Binder	Screen	LFP	Graphite	1 M $LiPF_6$ in EC/DMC	80 mAh g^{-1}(0.1C)/2.0–4.0 V	[80]
	CMC	Binder	Spray	LFP	Carbon nanofiber	1 M $LiPF_6$ in EC/DMC	1.3 mWh cm^{-2}(40 mA g^{-1})/2.0–3.8 V	[108]
$Li-O_2$	Paper	Substrate, GDL	Brush painting	Carbon	Li	–	6500 mAh g^{-1} (at 200 mA g $^{-1}$)/0–2.75 V	[109]

(continued)

Table 9.1 (continued)

Battery system	Cellulose	Purpose	Printing method	Cathode	Anode	Electrolyte	Capacitance or capacity/voltage	References
Zn-MnO_2	PVA/cellulose	Substrate	Stencil	MnO_2	Zn nanoparticle	5.6 M KOH (0.37 M ZnO)	0.8 mA h/0–14 V	[36]
Zn-air	Paper	Substrate, GDL	Pen drawing	Graphite	Zn nanoparticle	PAA/6 M KOH	600 mAh g^{-1}(Zn), 5 mAh cm^{-2}/1.2 V	[110]
	Paper	Substrate, GDL	Air-brush, inkjet	PEDOT	Zn/Carbon/PC	LiCl/LiOH	0.5 mAh cm^{-2}/0.8 V	[40]

dispenser printing [83, 91]. Wang et al. reported all-spray-printed supercapacitors, in which the SC components (including the CNF-based separator) were spray-printed on various substrates [81]. Commercial separator membranes pose a formidable challenge in diversifying the form factors of power sources. To address this issue, they introduced the CNFs in the separator membranes by spray-printing of the aqueous CNF suspension. The stepwise spray-printing of CNF-based separators and SWCNT electrodes led to the fabrication of SWCNT/CNF/SWCNT multilayers. With the use of pre-patterned masks, the spray-printed SCs with the design versatility (e.g., letter ("NKU")-shaped SCs) were fabricated (Fig. 9.10a). Furthermore, the all-spray-printed SCs were fabricated on the substrates with uniform surfaces (such as PET and glass) and also on the substrates with non-uniform surfaces (such as cloth and paper). Such versatile adaptability with substrates allowed the integration of the spray-printed SCs with a cloth (Fig. 9.10b) and a glass bottle on a curved surface.

The cellulose paper was combined with direct writing techniques to fabricate flexible SCs. Quain et al. developed MXene-based micro-supercapacitors (MSCs) on cellulose paper substrate using pen writing techniques [98]. The Ti_3C_2 MXene-in-water inks were loaded in a rollerball pen to directly deposit MXene material on the cellulose paper without ink clogging (Fig. 9.10c). The high electrical conductivity of MXenes enables single-step fabrication of the MSCs without the use of additional current collectors. To further demonstrate the versatility of the pen writing technique, four series-connected MXene-based MSCs with in-plane configuration were integrated on curved surfaces of a paper cup (Fig. 9.10d). Liu et al. reported paper-based all-solid-state flexible MSCs fabricated by sequential inkjet-printing [101]. The PEDOT:PSS-CNT ink and Ag nanoparticle ink were sequentially printed on cellulose paper substrate followed by drop-casting of PVA/H_3PO_4 gel electrolytes (Fig. 9.10e). The PEDOT:PSS-CNT/Ag electrodes show good adhesion to the cellulose paper substrate without additional binders. As a consequence, the MSC showed stable capacitive performance under mechanical deformation at different bending angles (Fig. 9.10f). Guo et al. developed MnO_2-based MSCs on highly conductive Ni patterns that were fabricated on paper substrates through screen printing of catalytic ink and electroless deposition [99]. The MSCs were fabricated through the formation of Ni patterns, deposition of MnO_2, and dropping of CMC-based gel electrolyte (Fig. 9.10g). Owing to paper-assisted strong interface adhesion between MnO_2 and the Ni film, the MSCs showed superior flexibility with negligible capacitance degradation after 10,000 bending cycles (Fig. 9.10h). Additionally, the MSCs were integrated with UV and gas sensors on a one-piece paper (Fig. 9.10i). The integrated device on paper opens a new route for portable and wearable electronics.

9.4.2 Cellulose-Based Printed Lithium-Ion Batteries (LIBs)

Lithium-ion batteries (LIBs) have been explored as a promising electrochemical system for printed power sources due to their high energy density and high operating

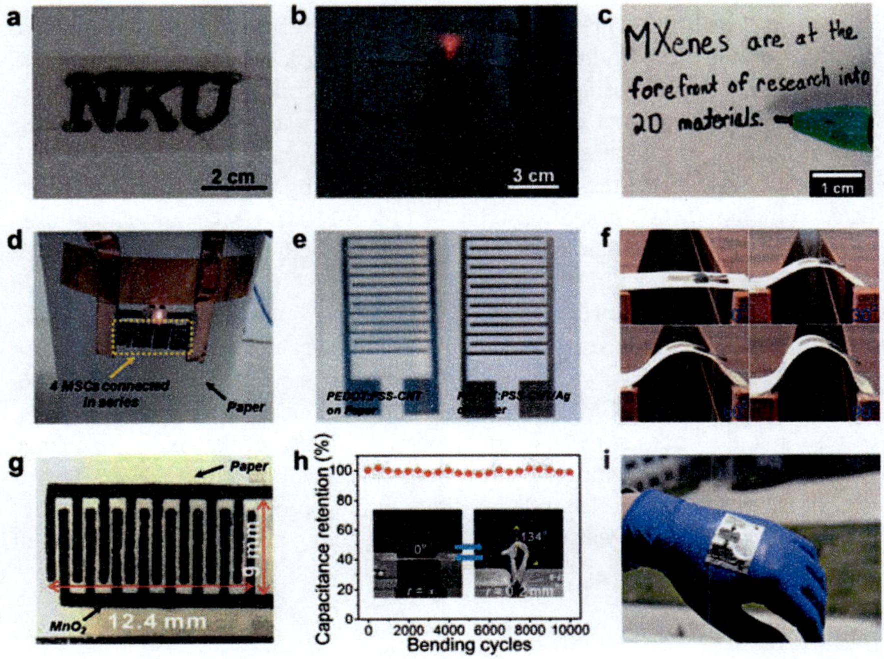

Fig. 9.10 Cellulose-based printed supercapacitors. **a** Optical image of a letter ("NKU")-shaped supercapacitor. **b** SWCNT and CNF film patterns on a coarse surface for assembling four supercapacitors into one unit in series. Optical images of an LED illuminated using four supercapacitors in series on a white coat. Reprinted with permission from [81]. Copyright 2017 American Chemical Society. **c** MXene-in-water inks for writing of Ti_3C_2 on paper. **d** Four MXene micro-supercapacitors (MSCs) connected in series on a curved paper cup. Reprinted with permission from [98]. Copyright 2019 Wiley. **e** Optical images of MSCs with the interdigitated PEDOT:PSS-CNT film before (left) and after (right) overprinting of the Ag film. **f** Optical images of paper-based MSCs bent at different angles. Reprinted with permission from [101]. Copyright 2016 Royal Society of Chemistry. **g** Optical image of interdigital electrodes electrochemically depositing MnO_2. **h** Capacitance retention of MSCs during 10,000 bending. The inset shows the MSC bending at different states (flat: $r = \infty$ and $\theta = 0°$, bent: $r = 0.2$ mm, and $\theta = 134°$), **i** Optical image of integrated circuit attached on the back of the hand with tandem MSC and UV sensor. Reprinted with permission from [99]. Copyright 2017 Wiley

voltage. Most of the previous studies have focused on the material design for electrode ink. In the formulation of electrode inks, binders play a crucial role as dispersing agents as well as their basic role as electrode binders. In specific, they contribute to the dispersion of electrode materials including lithium cobalt oxide (LCO) [102], lithium iron phosphate (LFP) [34, 79, 80, 105], lithium titanate (LTO) [106], TiO_2(B) [107], graphite [103, 104, 106], and silicon [108]. Beneventi et al. fabricated the LIB electrodes based on CNF binders and paper separators through screen printing [80]. The aqueous electrode inks were prepared using CNFs as a binder to print LIB electrodes directly on a paper separator. Cao et al. demonstrated high-performance lithium metal batteries (LFP cathode and infused-Li anode) through a 3D printing technique using

CNF binder (Fig. 9.11a–c) [111]. The porous structure of the printed CNF scaffold improved the ion accessibility and also decreased the local current density of Li anode, eventually suppressing the non-uniform/random dendrite formation during Li plating/stripping. Ohta et al. presented the all-solid-state LIBs by fabricating the screen-printed LCO cathodes, in which ethyl cellulose was used as a binder [102]. Baradai et al. fabricated flexible graphite LIB anodes by screen-printing microfibrillated cellulose (MFC)-based inks (Fig. 9.11d–f) [104]. CMCs are widely used in printed electrode inks due to their capabilities as both a binder and viscosifier. Gu et al. presented printed LFP cathodes containing CMC binder via water-based inkjet printing [105]. Bideau et al. demonstrated lithium-ion microbatteries by inkjet printing of CMC binder-based LFP cathodes and ionogel electrolyte [34]. Cellulose binders were also combined with spray printing to fabricate LIB electrodes [103, 106, 107]. Lee et al. fabricated CMC binder-containing LIB electrodes by spray printing. The resulting electrodes allowed a low fraction ($\leq$ 5wt%) of inactive components, thereby improving the gravimetric and volumetric capacities [108]. Liu et al. fabricated 3D-strtuctured LFP electrode by direct writing process (dispenser-printing) (Fig. 9.11g–i) [79]. CMC was used as a viscosity-tuning agent of LFP inks as well as an electrode binder.

9.4.3 *Cellulose-Based Printed Batteries Based on Other Electrochemical Systems*

Cellulose-based printed batteries based on other electrochemical systems such as Zn-based batteries and metal-air batteries have been reported. Similar to the previously mentioned approaches, cellulose (usually in the form of paper) is mainly used as a physical support for printed electrodes and also as a separator membrane. Due to its microporous structure, cellulose paper shows strong adhesion to the printed battery components and allows diffusion of liquid electrolytes through its pores. Gaikwad et al. reported a high-potential printed Zn-MnO_2 battery comprised of ten cells by fabricating series-connected electrode arrays on single sheets of cellulose-based membranes which serve as both the separator and mechanical support (Fig. 9.12a) [36]. The Zn and MnO_2 electrode inks were stencil-printed on separate sheets of cellulose-based membranes. In order to develop ten cells on a single sheet of substrate, a hydrophobic-amorphous-fluoropolymer-based ionic barrier was printed between the electrodes to prevent ionic contact between adjacent unit cell on the same sheet. Sequentially, Ag current collectors are printed on top of each cathode and anode array, resulting in the series-connected electrodes (Fig. 9.12b). Finally, electrolyte-deposited Zn and MnO_2-printed sheets are overlaid to match the cathode and anodes and develop a sandwich configuration for each unit cell. Such in-plane cell array enables a thin thickness and high-potential battery compared to vertically stacked multiple cells, which allows facile integration with other thin electronic devices. The high-potential Zn-MnO_2 battery showed an open-circuit voltage

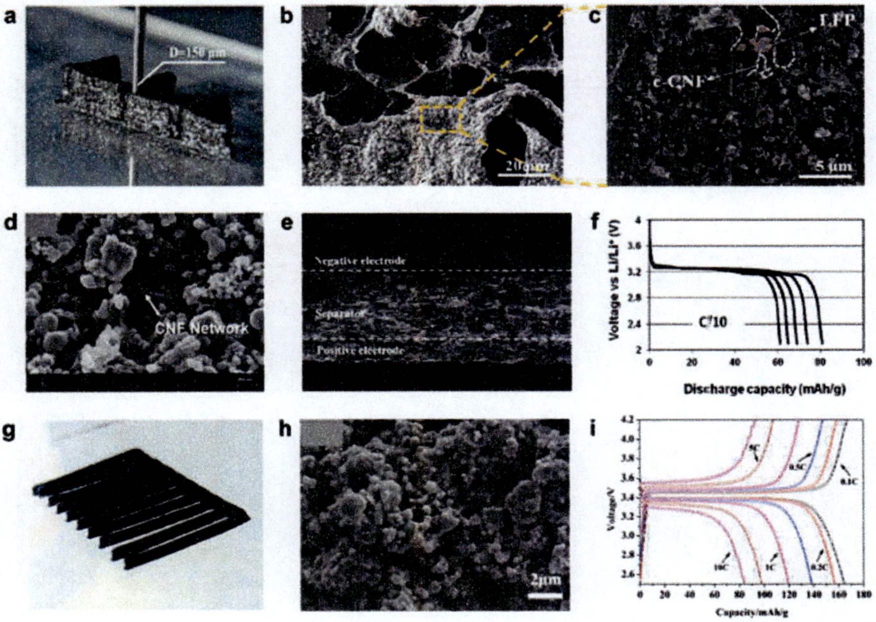

Fig. 9.11 Printed Li-ion batteries with cellulose. **a** CNF/LFP electrode printed layer by layer. **b** SEM images of the boundary between the adjacent filaments. **c** SEM image of the electrode. The carbon fiber linking with the LFP particles is highlighted. Reprinted with permission from [111]. Copyright 2019 Wiley. **d** Network structure between the graphite (GP) platelets and CNF in printed graphite electrode by screen-printing. **e** The cross section of the full printed battery is shown with a three-layer structure corresponding to the separator and the positive (LFP) and negative (graphite) electrodes using CNF binder, and delamination between the separator and the positive electrode is displayed. **f** Discharge capacity profiles of LFP/graphite full cell at C/10. Reprinted with permission from [80]. Copyright 2018 MDPI. **g** Printed comb-shaped 3D LFP electrode (3 layers) using CMC binder by direct writing process (dispenser-printing). **h** SEM images of the cross section of printed 3D LFP electrode. **i** Charge/discharge curves of LFP electrode. Reprinted with permission from [79]. Copyright 2017 MDPI

(OCV) of 14 V and a capacity of 0.8 mAh which was integrated with a printed complementary five-stage ring oscillator.

In the metal-air batteries, cellulose paper is mostly used as substrates for air electrodes. The micropores of the cellulose paper-based substrates can be utilized as air pathways forming a facile three-phase boundary between air, liquid electrolyte, and solid electrode materials (catalyst/conductive materials). As a result, effective oxygen reduction reactions (ORR) and oxygen evolution reactions (OER) can occur within the air electrode of metal-air batteries. Thus, by printing air cathode materials directly on cellulose paper substrates, the substrates can function as both the mechanical support and gas diffusion layer of the printed cathode. Hilder et al. reported paper-based printed Zn-air batteries using photo-quality paper as both the electrode substrate and as the separator membrane by developing a two-faced cell configuration [40]. The zinc/carbon/polycarbonate composite anode was fabricated

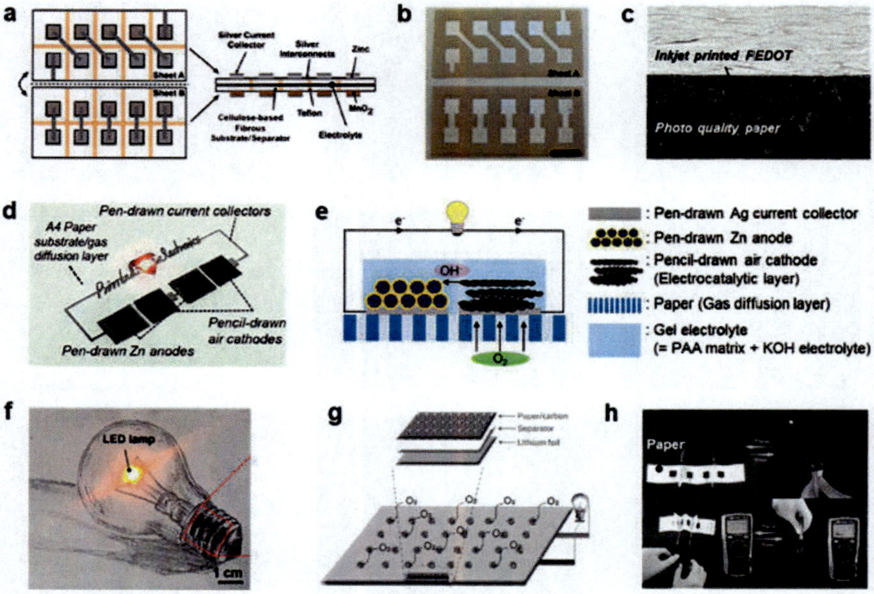

Fig. 9.12 Cellulose-based printed batteries based on other electrochemical systems (Zn-based and metal-air batteries). **a** Schematic diagram of the top and cross section of the ten Zn-MnO_2 cells, printed on a cellulose-based fibrous substrate. Printed silver electrodes serve as the current collectors and printed silver lines connect the individual cells to form a high-potential battery. Amorphous fluoropolymer solution printed in between the Zn and MnO_2 electrodes segregates the electrolyte in individual cells. After the printing steps, the membrane is cut into two sheets (sheet A and sheet B). Each electrode in sheets A and B is soaked with KOH/ZnO electrolyte and stacked together to complete the battery. **b** Optical image of sheets A and B after printing the Zn and MnO_2 electrodes, the silver current collectors, and interconnects (scale bar = 2 cm). Reprinted with permission from [36]. Copyright 2013 AIP Publishing LLC. **c** SEM image of cross section of inkjet-printed PEDOT on photo-quality paper. PEDOT is absorbed into the top 25 μm, apparent from change in texture. Reprinted with permission from [40]. Copyright 2009 Elsevier. **d** A photograph of the all-hand-drawn Zn-air battery (in series connection of two cells) that powered an LED lamp, wherein the cell components were consecutively drawn on the A4 paper substrate. **e** Schematic representation depicting the major components and discharge (i.e., spontaneous Faradaic) reaction of the Zn-air unit cell with in-plane electrode configuration. **f** A photograph of the e-sketch of a light bulb, in which the all-hand-drawn Zn-air battery (in series connection of two cells, marked by a red box) powered an LED lamp in the e-sketch. Reprinted with permission from [110]. Copyright 2018 Wiley. **g** Schematic illustration of a Li-O_2 battery assembly composed of brush-painted paper-based cathode, glass fiber (separator), and lithium foil (anode). **h** Photograph of a foldable Li-O_2 battery pack (top) turning on a red LED and (bottom) the open-circuit voltage of this battery packs before and after folding. Reprinted with permission from [109]. Copyright 2015 Wiley

by air-brushing the electrode ink on one side of the photo-quality paper, and sequentially the air cathode was fabricated by inkjet printing electrocatalytic PEDOT on the other side of the paper followed by vapor-phase polymerization (Fig. 9.12c). Aqueous LiCl/LiOH electrolyte ink was inkjet printed on the electrode printed paper and dried to eliminate pre-added processing solvents which were needed to obtain a suitable viscosity for inkjet printing. The resulting paper-based printed Zn-air battery showed a discharge voltage of 0.8 V and a capacity of 0.5 mAh cm^{-2}. Such two-faced electrode configuration enabled thin Zn-air cell fabrication; however, the author stated that the poor performance arises from the low ionic mobility of the paper-electrolyte system which requires further optimization. Recently, Choi et al. reported all-hand-drawn Zn-air batteries fabricated directly on commercial A4 paper which serves as a mechanical support and gas diffusion layer of the air cathodes (Fig. 9.12d) [110]. The current collector, anode, and cathode components were hand-drawn on A4 paper simply using a silver ink pen, oil painting-inspired Zn ink pen, and 8B pencil. Consequently, the hand-drawn components were activated with the pen-drawn PAA/KOH-based gel electrolytes (Fig. 9.12e). The pencil-drawn electrocatalytic air cathode consisted of graphitic carbon which showed stable ORR catalytic behavior. As air was produced through the porous paper GDL, the hand-drawn Zn-air battery showed stable discharge voltage plateau of 1.2 V. Due to the fabrication simplicity derived from the pen/pencil-based hand drawing techniques, the all-hand-drawn Zn-air batteries showed high aesthetic versatility. The two Zn-air cells connected in series were seamlessly integrated into a pencil-drawn electronic sketch of a light bulb made on A4 paper and successfully powered an LED lamp (Fig. 9.12f). The hand-drawn Zn-air battery showed high potential as an on-the-fly power source due to their facile customizing capability with easy accessibility to end-users.

Other metal-air battery systems such as Li-O_2 batteries which utilize cellulose paper-based GDL were also reported. Liu et al. presented a flexible Li-O_2 battery based on brush-painted paper-based cathodes (Fig. 9.12g) [109]. The microporous structure of the paper enabled the carbon-based cathode ink to be effectively absorbed by the substrate and the carbon nanoparticles to be homogeneously adsorbed to the paper skeleton without additional binders. The resulting paper cathode demonstrated high electrochemical and mechanical stability. Moreover, a foldable Li-O_2 battery pack was assembled using paper as the cathode substrate, separator, and the outer package enabling the whole battery to be bendable and foldable (Fig. 9.12h). The assembled battery achieved a capacity of 6500 mAh g^{-1} and maintained its performance even after 1000 bending cycles.

9.5 Summary and Future Outlook

Cellulose has garnered considerable attention as an eco-friendly green material for energy storage systems due to the natural abundance, structural uniqueness, mechanical compliance, and chemical diversity that lie far beyond achievable with conventional synthetic materials. Here, we reviewed the recent progress of cellulose, with a particular focus on application to printed batteries. The unique 1D fibrous structure and mechanical flexibility, and chemical functionalities of cellulose, through the combination with printing technologies, contribute to developing form factor-free printed batteries with aesthetic versatility and reliable electrochemical performance.

Cellulose has also been explored as a promising material for printed battery component ink, which has to be elaborately designed to secure both process compatibility and electrochemical performance. For the printed electrodes, various electroactive materials and conducting agents were combined with cellulose and its derivatives that were used as a viscosity thickening agent, a dispersant, and a binder. The resulting electrodes showed excellent printability, well-developed percolation networks, and strong cohesion between particles. In addition, cellulose and its derivatives were used as a key material for printed electrolytes and separator membranes to improve printability and form porous ion conducting channels. Cellulose has also been extensively investigated as printing substrates, onto which the battery component inks (e.g., electrodes, current collectors, electrolytes, and separator membranes) are deposited. Cellulose-mediated paper, driven by its porous microstructure, allowed the formation of bicontinuous ion/electron pathways. However, the capillary force-induced random ink spreading issue should be resolved to acquire high-resolution printed patterns.

Through the elaborate combination of cellulose with various printing technologies, cellulose-based printed batteries enabled considerable progress in the form factors, aesthetic versatility with flexibility. Representative examples include the letter-shaped batteries and monolithically integrated batteries with other objects and electronic devices. Furthermore, the printed batteries can be readily connected in series or parallel to tailor the voltage and current of the resultant power sources.

Despite the research activities described above, the commercial application of the printed power sources containing cellulose and its derivatives still remains sluggish due to several unresolved challenges. For example, the hydrophilic nature of cellulose due to its abundant hydroxyl groups hampers its use to the aqueous ink system and integration with hydrophobic active components. Future studies on cellulose-based printed batteries will be directed to designing chemistry and structures of cellulose, which enables fabrication of functional substrates and new battery component inks with various active materials and solvents. Furthermore, the newly emerging printing techniques such as high-fidelity inkjet, electrohydrodynamic, and 3D printing will be combined with the cellulose-based component inks. We envision that cellulose will play a viable role as a key material in boosting up the development of high-performance printed power sources for upcoming IoT and flexible/wearable electronics era.

References

1. S.-H. Kim, K.-H. Choi, S.-J. Cho, S. Choi, S. Park, S.-Y. Lee, Nano Lett. **15**(8), 5168–5177 (2015)
2. S.-S. Lee, K.-H. Choi, S.-H. Kim, S.-Y. Lee, Adv. Funct. Mater. **28**(11), 1705571 (2018)
3. J. Kim, R. Kumar, A.J. Bandodkar, J. Wang. Adv. Electron. Mater. **3**(1), 1600260 (2017)
4. K.-H. Choi, D.B. Ahn, S.-Y. Lee, ACS Energy Lett. **3**(1), 220–236 (2018)
5. D. Chen, Z. Lou, K. Jiang, G. Shen, Adv. Funct. Mater. **28**(51), 1805596 (2018)
6. S.C. Dhanabalan, B. Dhanabalan, X. Chen, J.S. Ponraj, H. Zhang, Nanoscale **11**(7), 3046–3101 (2019)

7. L. He, K. Wen, Z. Zhang, L. Ye, W. Lv, J. Fei, S. Zhang, W. He, J. Mater. Res. **33**(16), 2281–2296 (2018)
8. J.-H. Kim, D. Lee, Y.-H. Lee, W. Chen, S.-Y. Lee. Adv. Mater. 1804826 (2018)
9. C.-Y. Liao, H.-H. Chien, Y.-C. Hao, C.-W. Chen, I.-S. Yu, J.-Z. Chen, J. Electron. Mater. **47**(7), 3861–3868 (2018)
10. W. Chen, H. Yu, S.-Y. Lee, T. Wei, J. Li, Z. Fan, Chem. Soc. Rev. **47**(8), 2837–2872 (2018)
11. Y. Habibi, Chem. Soc. Rev. **43**(5), 1519–1542 (2014)
12. Z. Wang, P. Tammela, M. Strømme, L. Nyholm, Adv. Energy Mater. **7**(18), 1700130 (2017)
13. R.E. Sousa, C.M. Costa, S. Lanceros-Méndez, Chemsuschem **8**(21), 3539–3555 (2015)
14. A.M. Gaikwad, A.C. Arias, D.A. Steingart, Energy Technol. **3**(4), 305–328 (2015)
15. Y. Aleeva, B. Pignataro, J. Mater. Chem. C. **2**(32), 6436–6453 (2014)
16. Y.L. Han, J. Hu, G.M. Genin, T.J. Lu, F. Xu, Sci. Rep. **4**, 4872 (2014)
17. X. Liao, Q. Liao, X. Yan, Q. Liang, H. Si, M. Li, H. Wu, S. Cao, Y. Zhang, Adv. Funct. Mater. **25**(16), 2395–2401 (2015)
18. Z. Li, H. Liu, C. Ouyang, W. Hong Wee, X. Cui, T. Jian Lu, B. Pingguan-Murphy, F. Li, F. Xu. Adv. Funct. Mater. **26**(2), 165–180 (2016)
19. L.J. Deiner, T.L. Reitz, Adv. Eng. Mater. **19**(7), 1600878 (2017)
20. M. Gao, L. Li, Y. Song, J. Mater. Chem. C. **5**(12), 2971–2993 (2017)
21. A. Ambrosi, M. Pumera, Chem. Soc. Rev. **45**(10), 2740–2755 (2016)
22. R.D. Farahani, M. Dubé, D. Therriault, Adv. Mater. **28**(28), 5794–5821 (2016)
23. L. Hu, J.W. Choi, Y. Yang, S. Jeong, F. La Mantia, L.-F. Cui, Y. Cui, Proc. Natl. Acad. Sci. **106**(51), 21490 (2009)
24. S. Lawes, A. Riese, Q. Sun, N. Cheng, X. Sun, Carbon **92**, 150–176 (2015)
25. F. Bonaccorso, A. Bartolotta, J.N. Coleman, C. Backes, Adv. Mater. **28**(29), 6136–6166 (2016)
26. D. Wei, H. Li, D. Han, Q. Zhang, L. Niu, H. Yang, C. Bower, P. Andrew, T. Ryhänen, Nanotechnology **22**(24), 245702 (2011)
27. K. Kordás, T. Mustonen, G. Tóth, H. Jantunen, M. Lajunen, C. Soldano, S. Talapatra, S. Kar, R. Vajtai, P.M. Ajayan, Small **2**(8–9), 1021–1025 (2006)
28. J.C. Conrad, S.R. Ferreira, J. Yoshikawa, R.F. Shepherd, B.Y. Ahn, J.A. Lewis, Curr. Opin. Colloid Interf. Sci. **16**(1), 71–79 (2011)
29. N. Singh, C. Galande, A. Miranda, A. Mathkar, W. Gao, A.L.M. Reddy, A. Vlad, P.M. Ajayan, Sci. Rep. **2**, 481 (2012)
30. K. Braam, V. Subramanian, Adv. Mater. **27**(4), 689–694 (2015)
31. A.J. Blake, R.R. Kohlmeyer, J.O. Hardin, E.A. Carmona, B. Maruyama, J.D. Berrigan, H. Huang, M.F. Durstock, Adv. Energy Mater. **7**(14), 1602920 (2017)
32. E.-H. Kil, K.-H. Choi, H.-J. Ha, S. Xu, J.A. Rogers, M.R. Kim, Y.-G. Lee, K.M. Kim, K.Y. Cho, S.-Y. Lee, Adv. Mater. **25**(10), 1395–1400 (2013)
33. K.-H. Choi, J. Yoo, C.K. Lee, S.-Y. Lee, Energy Environ. Sci. **9**(9), 2812–2821 (2016)
34. P.E. Delannoy, B. Riou, B. Lestriez, D. Guyomard, T. Brousse, J. Le Bideau, J. Power Sources **274**, 1085–1090 (2015)
35. Z. Wang, R. Winslow, D. Madan, P.K. Wright, J.W. Evans, M. Keif, X. Rong, J. Power Sources **268**, 246–254 (2014)
36. A.M. Gaikwad, D.A. Steingart, T. Nga Ng, D.E. Schwartz, G.L. Whiting. Appl. Phys. Lett. **102**(23), 233302 (2013)
37. K.T. Braam, S.K. Volkman, V. Subramanian, J. Power Sources **199**, 367–372 (2012)
38. C.C. Ho, K. Murata, D.A. Steingart, J.W. Evans, P.K. Wright, J. Micromech. Microeng. **19**(9), 094013 (2009)
39. J. Fu, D.U. Lee, F.M. Hassan, L. Yang, Z. Bai, M.G. Park, Z. Chen, Adv. Mater. **27**(37), 5617–5622 (2015)
40. M. Hilder, B. Winther-Jensen, N.B. Clark, J. Power Sources **194**(2), 1135–1141 (2009)
41. S. Berchmans, A.J. Bandodkar, W. Jia, J. Ramírez, Y.S. Meng, J. Wang, J. Mater. Chem. A. **2**(38), 15788–15795 (2014)
42. A.M. Gaikwad, G.L. Whiting, D.A. Steingart, A.C. Arias, Adv. Mater. **23**(29), 3251–3255 (2011)

43. C.C. Ho, J.W. Evans, P.K. Wright, J. Micromech. Microeng. **20**(10), 104009 (2010)
44. G.G. Kumar, S. Sampath, J. Electrochem. Soc. **150**(5), A608–A615 (2003)
45. R. Kumar, J. Shin, L. Yin, J.-M. You, Y.S. Meng, J. Wang. Adv. Energy Mater. **7**(8), 1602096 (2017)
46. K. Jost, D. Stenger, C.R. Perez, J.K. McDonough, K. Lian, Y. Gogotsi, G. Dion, Energy Environ. Sci. **6**(9), 2698–2705 (2013)
47. Y. Xu, M.G. Schwab, A.J. Strudwick, I. Hennig, X. Feng, Z. Wu, K. Müllen, Adv. Energy Mater. **3**(8), 1035–1040 (2013)
48. H. Zhang, Y. Qiao, Z. Lu, ACS Appl. Mater. Interf. **8**(47), 32317–32323 (2016)
49. Y. Xiao, L. Huang, Q. Zhang, S. Xu, Q. Chen, W. Shi, Appl. Phys. Lett. **107**(1), 013906 (2015)
50. D. Pech, M. Brunet, P.-L. Taberna, P. Simon, N. Fabre, F. Mesnilgrente, V. Conédéra, H. Durou, J. Power Sources **195**(4), 1266–1269 (2010)
51. Z. Liu, Z.-S. Wu, S. Yang, R. Dong, X. Feng, K. Müllen, Adv. Mater. **28**(11), 2217–2222 (2016)
52. W.J. Hyun, E.B. Secor, C.-H. Kim, M.C. Hersam, L.F. Francis, C.D. Frisbie, Adv. Energy Mater. **7**(17), 1700285 (2017)
53. S. Sollami Delekta, A.D. Smith, J. Li, M. Östling. Nanoscale. **9**(21), 6998–7005 (2017)
54. M. Kaempgen, C.K. Chan, J. Ma, Y. Cui, G. Gruner, Nano Lett. **9**(5), 1872–1876 (2009)
55. C. Huang, N.P. Young, P.S. Grant, J. Mater. Chem. A. **2**(29), 11022–11028 (2014)
56. Z.-S. Wu, Z. Liu, K. Parvez, X. Feng, K. Müllen, Adv. Mater. **27**(24), 3669–3675 (2015)
57. X. Shi, Z.-S. Wu, J. Qin, S. Zheng, S. Wang, F. Zhou, C. Sun, X. Bao, Adv. Mater. **29**(44), 1703034 (2017)
58. G. Sun, J. An, C.K. Chua, H. Pang, J. Zhang, P. Chen, Electrochem. Commun. **51**, 33–36 (2015)
59. C. Zhu, T. Liu, F. Qian, T.Y.-J. Han, E.B. Duoss, J.D. Kuntz, C.M. Spadaccini, M.A. Worsley, Y. Li, Nano Lett. **16**(6), 3448–3456 (2016)
60. K. Sun, T.-S. Wei, B.Y. Ahn, J.Y. Seo, S.J. Dillon, J.A. Lewis, Adv. Mater. **25**(33), 4539–4543 (2013)
61. K. Fu, Y. Wang, C. Yan, Y. Yao, Y. Chen, J. Dai, S. Lacey, Y. Wang, J. Wan, T. Li, Z. Wang, Y. Xu, L. Hu, Adv. Mater. **28**(13), 2587–2594 (2016)
62. R.E. Sousa, J. Oliveira, A. Gören, D. Miranda, M.M. Silva, L. Hilliou, C.M. Costa, S. Lanceros-Mendez, Electrochim. Acta. **196**, 92–100 (2016)
63. A. Izumi, M. Sanada, K. Furuichi, K. Teraki, T. Matsuda, K. Hiramatsu, H. Munakata, K. Kanamura, Electrochim. Acta. **79**, 218–222 (2012)
64. K.-Y. Kang, Y.-G. Lee, D.O. Shin, J.-C. Kim, K.M. Kim, Electrochim. Acta. **138**, 294–301 (2014)
65. L. Hu, H. Wu, F. La Mantia, Y. Yang, Y. Cui, ACS Nano **4**(10), 5843–5848 (2010)
66. S.-H. Kim, K.-H. Choi, S.-J. Cho, J. Yoo, S.-S. Lee, S.-Y. Lee, Energy Environ. Sci. **11**(2), 321–330 (2018)
67. H.-D. Um, K.-H. Choi, I. Hwang, S.-H. Kim, K. Seo, S.-Y. Lee, Energy Environ. Sci. **10**(4), 931–940 (2017)
68. C. Milroy, A. Manthiram, Chem. Commun. **52**(23), 4282–4285 (2016)
69. C.Y. Jung, T.S. Zhao, L. An, L. Zeng, Z.H. Wei. J. Power Sources **297**(174–180 (2015)
70. Z. Tehrani, T. Korochkina, S. Govindarajan, D.J. Thomas, J. O'Mahony, J. Kettle, T.C. Claypole, D.T. Gethin, Org. Electron. **26**, 386–394 (2015)
71. G. Wang, L. Zhang, J. Zhang. Chem. Soc. Rev. **41**(2), 797–828 (2012)
72. W. Li, B. Song, A. Manthiram, Chem. Soc. Rev. **46**(10), 3006–3059 (2017)
73. Y. Li, H. Dai, Chem. Soc. Rev. **43**(15), 5257–5275 (2014)
74. T.N. Ng, S. Sambandan, R. Lujan, A.C. Arias, C.R. Newman, H. Yan, A. Facchetti, Appl. Phys. Lett. **94**(23), 233307 (2009)
75. T.N. Ng, D.E. Schwartz, L.L. Lavery, G.L. Whiting, B. Russo, B. Krusor, J. Veres, P. Bröms, L. Herlogsson, N. Alam, O. Hagel, J. Nilsson, C. Karlsson, Sci. Rep. **2**, 585 (2012)
76. H. Zhu, W. Luo, P.N. Ciesielski, Z. Fang, J.Y. Zhu, G. Henriksson, M.E. Himmel, L. Hu, Chem. Rev. **116**(16), 9305–9374 (2016)

77. Y. Li, H. Zhu, F. Shen, J. Wan, S. Lacey, Z. Fang, H. Dai, L. Hu, Nano Energy **13**, 346–354 (2015)
78. J.S. Park, T. Kim, W.S. Kim, Sci. Rep. **7**(1), 3246 (2017)
79. C. Liu, X. Cheng, B. Li, Z. Chen, S. Mi, C. Lao, Materials **10**(8), 934 (2017)
80. O. El Baradai, D. Beneventi, F. Alloin, Y. Bultel, D. Chaussy, Nanomaterials **8**(12), 982 (2018)
81. X. Wang, Q. Lu, C. Chen, M. Han, Q. Wang, H. Li, Z. Niu, J. Chen, ACS Appl. Mater. Interf. **9**(34), 28612–28619 (2017)
82. S. Tuukkanen, S. Lehtimäki, F. Jahangir, A. Eskelinen, D. Lupo, S. Franssila. Proceedings of the 5th Electronics System-integration Technology Conference (ESTC), pp. 1–6 (2014)
83. M. Areir, Y. Xu, D. Harrison, J. Fyson, J. Mater. Sci. Mater. Electron. **28**(23), 18254–18261 (2017)
84. F. Brunetti, A. Operamolla, S. Castro-Hermosa, G. Lucarelli, V. Manca, G.M. Farinola, T.M. Brown. Adv. Funct. Mater. 1806798 (2019)
85. M.M. Pérez-Madrigal, M.G. Edo, C. Alemán, Green Chem. **18**(22), 5930–5956 (2016)
86. D. Tobjörk, R. Österbacka, Adv. Mater. **23**(17), 1935–1961 (2011)
87. L. Hu, H. Wu, Y. Cui, Appl. Phys. Lett. **96**(18), 183502 (2010)
88. M. Kuang, L. Wang, Y. Song, Adv. Mater. **26**(40), 6950–6958 (2014)
89. J. Keskinen, S. Tuurala, M. Sjödin, K. Kiri, L. Nyholm, T. Flyktman, M. Strømme, M. Smolander, Synth. Met. **203**, 192–199 (2015)
90. J. Li, F. Ye, S. Vaziri, M. Muhammed, M.C. Lemme, M. Östling, Adv. Mater. **25**(29), 3985–3992 (2013)
91. M. Areir, Y. Xu, D. Harrison, J. Fyson, R. Zhang, Mater. Manuf. Process. **33**(8), 905–911 (2018)
92. K. Torvinen, S. Lehtimäki, J.T. Keränen, J. Sievänen, J. Vartiainen, E. Hellén, D. Lupo, S. Tuukkanen, Electron. Mater. Lett. **11**(6), 1040–1047 (2015)
93. J.-X. Feng, Q. Li, X.-F. Lu, Y.-X. Tong, G.-R. Li, J. Mater. Chem. A. **2**(9), 2985–2992 (2014)
94. B. Yao, L. Yuan, X. Xiao, J. Zhang, Y. Qi, J. Zhou, J. Zhou, B. Hu, W. Chen, Nano Energy **2**(6), 1071–1078 (2013)
95. G. Zheng, L. Hu, H. Wu, X. Xie, Y. Cui, Energy Environ. Sci. **4**(9), 3368–3373 (2011)
96. P. Sundriyal, S. Bhattacharya, ACS Appl. Mater. Interf. **9**(44), 38507–38521 (2017)
97. F. Pettersson, J. Keskinen, T. Remonen, L. von Hertzen, E. Jansson, K. Tappura, Y. Zhang, C.E. Wilén, R. Österbacka, J. Power Sources **271**, 298–304 (2014)
98. E. Quain, T.S. Mathis, N. Kurra, K. Maleski, K.L. Van Aken, M. Alhabeb, H.N. Alshareef, Y. Gogotsi, Adv. Mater. Technol. **4**(1), 1800256 (2019)
99. R. Guo, J. Chen, B. Yang, L. Liu, L. Su, B. Shen, X. Yan, Adv. Funct. Mater. **27**(43), 1702394 (2017)
100. Y. Yu, J. Zhang, J. Mater. Chem. A. **5**(9), 4719–4725 (2017)
101. W. Liu, C. Lu, H. Li, R.Y. Tay, L. Sun, X. Wang, W.L. Chow, X. Wang, B.K. Tay, Z. Chen, J. Yan, K. Feng, G. Lui, R. Tjandra, L. Rasenthiram, G. Chiu, A. Yu, J. Mater. Chem. A. **4**(10), 3754–3764 (2016)
102. S. Ohta, S. Komagata, J. Seki, T. Saeki, S. Morishita, T. Asaoka, J. Power Sources **238**, 53–56 (2013)
103. D. Beneventi, D. Chaussy, D. Curtil, L. Zolin, E. Bruno, R. Bongiovanni, M. Destro, C. Gerbaldi, N. Penazzi, S. Tapin-Lingua, Chem. Eng. J. **243**, 372–379 (2014)
104. O. El Baradai, D. Beneventi, F. Alloin, R. Bongiovanni, N. Bruas-Reverdy, Y. Bultel, D. Chaussy, J. Mater. Sci. Technol. **32**(6), 566–572 (2016)
105. Y. Gu, A. Wu, H. Sohn, C. Nicoletti, Z. Iqbal, J.F. Federici, J. Manuf. Process. **20**, 198–205 (2015)
106. S.H. Lee, C. Huang, C. Johnston, P.S. Grant, Electrochim. Acta. **292**, 546–557 (2018)
107. S.H. Lee, A. Mahadevegowda, C. Huang, J.D. Evans, P.S. Grant, J. Mater. Chem. A. **6**(27), 13133–13141 (2018)
108. S.H. Lee, K. Li, C. Huang, J.D. Evans, P.S. Grant, ACS Appl. Mater. Interf. **11**(1), 603–612 (2019)

109. Q.-C. Liu, L. Li, J.-J. Xu, Z.-W. Chang, D. Xu, Y.-B. Yin, X.-Y. Yang, T. Liu, Y.-S. Jiang, J.-M. Yan, X.-B. Zhang, Adv. Mater. **27**(48), 8095–8101 (2015)
110. K.-H. Choi, H.-W. Kim, S.-S. Lee, J. Yoo, D.-G. Lee, S.-Y. Lee, Adv. Sustain. Syst. **2**(5), 1700132 (2018)
111. D. Cao, Y. Xing, K. Tantratian, X. Wang, Y. Ma, A. Mukhopadhyay, Z. Cheng, Q. Zhang, Y. Jiao, L. Chen, H. Zhu. Adv. Mater. 1807313 (2019)
112. J. Song, C. Chen, S. Zhu, M. Zhu, J. Dai, U. Ray, Y. Li, Y. Kuang, Y. Li, N. Quispe, Y. Yao, A. Gong, U.H. Leiste, H.A. Bruck, J.Y. Zhu, A. Vellore, H. Li, M.L. Minus, Z. Jia, A. Martini, T. Li, L. Hu, Nature **554**, 224 (2018)

Chapter 10
Nanocellulose for Water Treatment Applications

Yudi Kuang, Penghui Zhu, Yeling Zhu, and Feng Jiang

Abstract Water treatment, referring to the process that unwanted impurities (particles, oils, pathogens, organics, inorganics, etc.) are removed for improved water quality, is not only a cornerstone for human society and almost all industrial sectors but also crucial for sustainable development. Compared to energy-intensive thermal processing or petrochemical-derived materials that are currently deployed for water treatment, recent technological development has demonstrated the great potential of nanocellulose, a sustainable and the most abundant biopolymer on Earth, as the platform material for next-generation water treatment. This chapter provides a comprehensive overview of valorizing nanocellulose for water treatment through membrane filtration, absorption/adsorption, and solar-assisted water generation, with an emphasis on the surface modification, structural engineering, and/or composite techniques associated with these approaches. Opportunities and challenges are elaborated at the finishing part of the chapter to pave the road for future research.

Y. Kuang and P. Zhu contribute to this work equally.

Y. Kuang (✉)
School of Biomedical Sciences and Engineering, South China University of Technology, Guangzhou International Campus, Guangzhou 511442, P. R. China
e-mail: kuangyd@scut.edu.cn

P. Zhu · Y. Zhu · F. Jiang (✉)
Sustainable Functional Biomaterial Lab, Department of Wood Science, the University of British Columbia, 2424 Main Mall, Vancouver, BC V6T1Z4, Canada
e-mail: feng.jiang@ubc.ca

Y. Kuang
National Engineering Research Center for Tissue Restoration and Reconstruction, South China University of Technology, Guangzhou 510006, P. R. China

L. Hu et al. (eds.), *Emerging Nanotechnologies in Nanocellulose*, NanoScience and Technology, https://doi.org/10.1007/978-3-031-14043-3_10

10.1 Introduction

Freshwater scarcity is perceived as a global systematic challenge that significantly influences the sustainable development of human society. According to Mekonnen's report in 2016, at least two-thirds of the global population (over 4 billion people) has experienced severe water scarcity at least 1 month of the year, which is worse than the previous estimated number of 1.7–3.1 billion [1]. It is believed that this situation will become more severe in the near future because of the intensified water consumption and pollution accompanied with increasing human activities. Other than its serious effect on the development of economy, water scarcity and pollution crisis is also highly related to the human health. The World Health Organization (WHO) reported that as much as 24% of all deaths worldwide were attributable to an unhealthy environment, representing approximately 12.6 million people [2, 3]. Emergent and prolonged environmental remediation actions are urgently needed to ensure a sustainable future for all. From a technical point of view, cost-effective materials and practical approaches are the keys to tackle these crises.

Nanocellulose, as an emerging bio-based nanomaterial of the twenty-first century, is regarded as a promising solution for environmental remediation due to its abundant availability, environmental friendliness, and unique physicochemical properties. Diverse cellulose-based materials in forms of slurry, coating, membrane, aerogel, hydrogel, or monolith have been widely reported for filtration, pollutant absorption and adsorption, and drinkable water generation (Fig. 10.1). In this chapter, we will provide a state-of-the-art overview of the applications of these cellulose-based materials, as well as the fundamental principles living behind.

10.2 Nanocellulose-Based Membranes or Filters

Compared to distillation, disinfection, sublimation, or crystallization, filtration or membrane separation technology is considered as the most energy efficient technology for water purification, which selectively removes the insoluble substances or ions via pressure-driven physical processes. In general, the separation processes can be categorized into four types according to the pore size or molecular weight cut-off (MWCO) of the filters: microfiltration (MF), ultrafiltration (UF), nanofiltration (NF), and reverse osmosis (RO). The typical sieving effect of these processes on different components is shown in Fig. 10.2. Given the unique hierarchical fibrous morphologies and intra/inter-fiber hydrogen bonding capabilities of cellulose, it is one of the ideal raw materials for the preparation of filtration membranes.

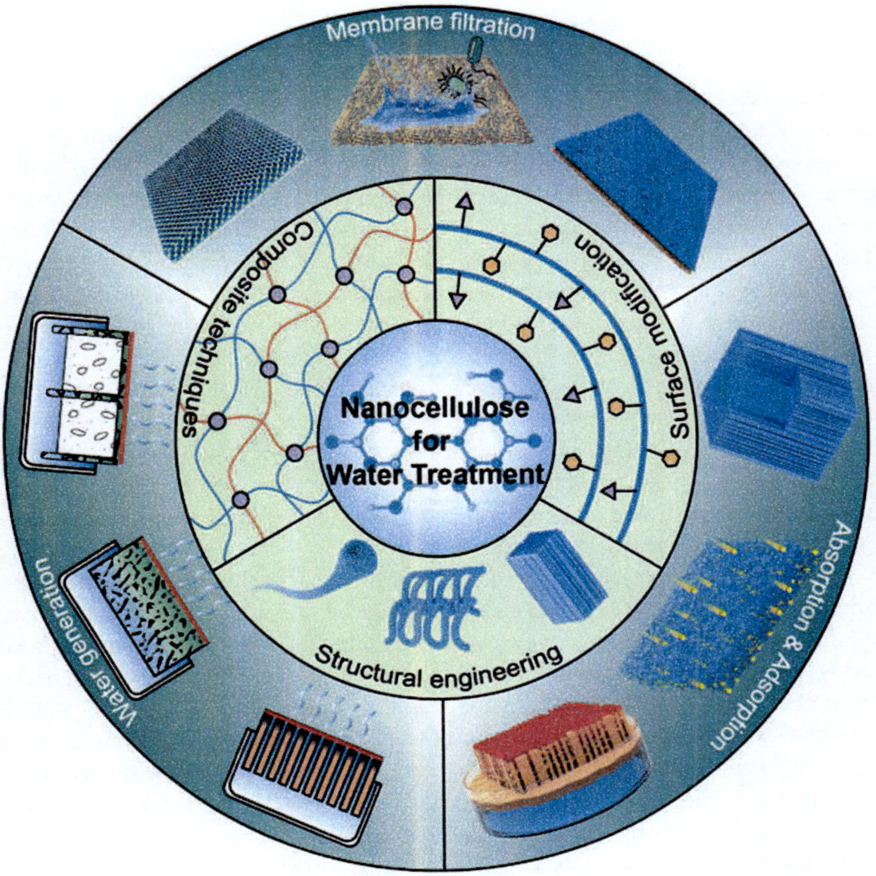

Fig. 10.1 Scheme of nanocellulose-based materials for water treatment

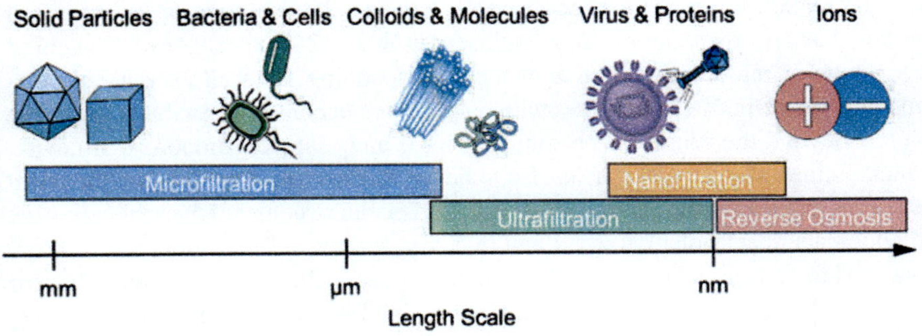

Fig. 10.2 The typical sieving range of diverse membrane separation processes

10.2.1 All-Cellulose Membranes

Ascribed to the superb hydrophilicity and fibrous morphology, cellulose fibers can be easily processed into free-standing membranes via scalable and water-based paper-making processes, which is much more cost-effective and environmentally friendly compared to petroleum-based or ceramic-based membranes. The ubiquitous coffee or tea filters are two of the most typical products that utilize the size exclusion effect of cellulose network to remove the suspended particulates in water by using gravity as the driving force. However, due to size limitation of natural wood fibers (tens of micrometers in width), conventional cellulose-based membranes (or so-called filter papers) are generally working in the range of MF. In these regions, size exclusion effect (or the so-called mechanical interception mechanism) is the predominant filtration mechanism, which selectively removes pollutants bigger than the pore size of the filters, leading to a high contaminant removal efficiency with a relatively low driving pressure [4]. As for the removal of bacterium, pollen, yeast, viruses, or even ions, rationally tuned size and stacking patterns of the cellulose fibers are needed, together with an increased driving force.

With the rise of nanotechnology, the application of nanocellulose for water purification has been widely explored in the past decade. Compared to conventional cellulose fiber, nanocellulose possesses distinguishing features including nano-sized diameter, high mechanical strength (up to 3 GPa in tensile strength for single nanocellulose fiber and 130 GPa in modulus for crystal regions), large aspect ratio (>300), and significantly enlarged specific surface area (>1000 $m^2 \cdot g^{-1}$) [5–8]. These properties allow nanocelluloses to form a highly compact fibrous network with randomly distributed nanopores and robust mechanical strength. But an interesting thing is that despite the molecular structure of cellulose (i.e., a linear polymer consisting of β-1,4-linked anhydro-D-glucose units as repeating units) has already been confirmed for over 100 years, [9, 10] the basic cellulose macromolecules needed for the formation of elementary fibril are still under dispute, making the definition of minimum nanocellulose size remain unclear. However, according to the modern observation techniques, it is agreed that the diameter of nanocellulose is as low as 2–5 nm, suggesting a similar magnitude of minimum pore size of the corresponding nanocellulose membranes (regenerated cellulose membranes with no pores are excluded given the scope of the topic). Beyond the superb mechanial properties and tunable nanoporous structure, nanocellulose-based membrane is also believed to have excellent anti-fouling property. This function is correlated to the unique molecular structure of the nanocellulose, in which abundant polar carboxyl and hydroxyl groups are uniformly and densely arranged to form a polar corona that wraps the nanocellulose. These polar groups with excellent hydration property can strongly inhibit hydrophobes from adsorbing on the surface and also effectively remove hydrophobes, if they adsorb, upon contact with water [11]. All these properties make nanocellulose-based membranes highly desirable for UF and NF applications.

Exciting developments have been made in recent years by utilizing nanocellulose as the building block for the preparation of next generation virus filter. The first trial of using pure nanocellulose membranes for virus removal was reported in 2014 by Metreveli et al. [12]. In their work, they prepared a pure nanocellulose paper (also known as nanopaper) as the nanofiltration membrane via a straightforward vacuum filtration process followed with hot-pressing treatment. The average pore size, porosity, and thickness of the resultant membranes are 19 nm, 35%, and 70 μm, respectively. Size exclusion properties of the membranes are verified by using polystyrene latex beads (30, 100, and 500 nm in diameter) and swine influenza virus (SIV, a typical particle size of 80–120 nm in diameter) as the evaluation models. An excellent virus removal performance (with $\log_{10}$ reduction value (LRV) $\geq$ 6.3) is obtained in their experiments, which is comparable to the best industrial synthetic polymer virus filters. After that, a series of follow-up works were reported showing the universal ability of nanocellulose membrane for pathogen removal [13–17].

Despite these achievements, there is still a significant downside that hinders the practical application of pure nanocellulose membranes: relative low maximum operating pressure and permeation flux (several to tens of L m^{-2} h^{-1} bar^{-1}) compared to the commercialized counterpart (tens to hundreds of m^{-2} h^{-1} bar^{-1}). To conquer this problem, great efforts have been made in recent years, including fiber-size modification, hot pressing, solvent exchange, chemical treatment and so on [18–21]. However, most of the attempts to increase the membrane permeation flux generally resulted in a sacrifice of the minimum sieve size. For example, by controlling the grammage and drying condition of the wet laid nanocelluloses, the average pore size of the nanocellulose membranes can be rationally tuned between 10 to 25 nm, which leads to a tunable filtrate flux of 10–80 L m^{-2} h^{-1} bar^{-1}, while the virus removal efficiency LRV dramatically decreased from 3.82 to 1.97 for the nanocellulose membrane with the highest filtrate flux. A potentially long latency time for the practical application of pure nanocellulose membranes is expected.

10.2.2 Nanocellulose-Based Composite Filtration Membranes

Compared to all cellulose membrane, previous works indicate that the inclusion of nanocellulose within conventional polymeric membranes can significantly improve the mechanical strength, operating pressure, permeability, and anti-fouling property of the membranes. Despite a bit sacrifice of degradability, the nanocellulose-based composite membranes are found to be more applicable for practical application. The typical configuration of nanocellulose-based composite membranes includes three types: percolation network type, co-network type, and heterolayered type (designated as P-membrane, C-membrane, and H-membrane, respectively) (Fig. 10.3).

The P-membranes are generally prepared by directly impregnating nanocellulose slurry into another preformed scaffold to form an intertwined composite network containing both nanocellulose and scaffold networks (Fig. 10.3a). In such a design, the scaffold network is serving as the primary network for hosting the

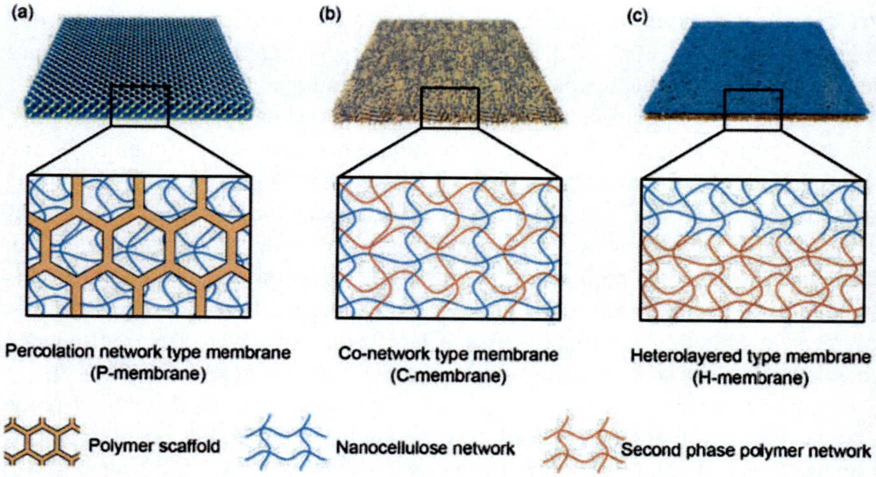

Fig. 10.3 Schemes of different nanocellulose-based composite membranes: **a** percolation network type, **b** co-network type, and **c** heterolayered type

nanocellulose network, while the post-formed nanocellulose network with fine-tuned pore size functions as the pollutant sieve. One of the most obvious advantage for the P-membranes is that the primary scaffold network can be chosen from a variety of conventional materials, including polyethylene terephthalate (PET) [22], poly(ether)sulfones (PES) [23], polyacrylonitrile (PAN) [24], polyvinylidene fluoride (PVDF) [25], and so on. This allows the P-membranes to be easily prepared in a large scale without significant technical bottlenecks. However, despite these advantages, P-membranes also face the risk of fragile interfacial bonding due to the swelling of the hydrophilic nanocellulose network in aqueous environment. Even a microscale separation of nanocellulose network from the scaffold network will lead to a significantly deteriorated pollutant separation efficiency. Therefore, more efforts are needed to address this problem, such as interlocking interfacial structure design and surface chemical modification of the nanocellulose.

In comparison, C-membranes possess a relatively homogeneous composite network with inseparable nanocellulose phase and another polymer phase (Fig. 10.3b). In a typical fabrication workflow, nanocellulose in forms of powder or slurry firstly needs to be homogeneously dispersed in the second polymer phase. The mixture will then be processed into a porous membrane via phase conversion or electrospinning methods. This kind of co-forming process thus allows the nanocellulose and polymeric molecules to cling together at a molecular or supramolecular level with improved underwater stability. However, despite the straightforward fabrication process, the production cost of the blending type membrane is generally high due to the time-/energy-consuming electrospinning processes and drying processes. Moreover, the utilization of organic solvent during the preparation of the mixture is also troublesome in perspective of both environment and economy.

The typical configuration of H-membrane is an asymmetric multilayer structure with nanocellulose barrier layer laminated on top of another polymer scaffold (Fig. 10.3c). Owing to the mechanical supporting effect of the scaffold substrate, the layered architecture can effectively increase the pressure tolerance and decrease the thickness of the nanocellulose barrier layer, which in turn leads to a significantly improved permeation flux of the H-membranes. Besides, the application of nanocellulose barrier layer also brings the resultant H-membrane a superior antifouling property. Recently, Hsiao's group reported that the antifouling property of the nanocellulose-based heterolayered membrane can be further improved by regulating the oxidation degree (OD, mmol/g) of the nanocellulose [26]. A permeation flux of 25.3 $\pm$ 1.7 L m^{-2} h^{-1} at 0.5 psi and flux recovery ratio of 98 $\pm$ 2% wer achieved for the membrane with NCs layer of ~270 nm thick and 1.80 OD. Noting that such permeation and flux recovery performances are about two times and three times higher than that of the commercial PVDF membrane. They attribute these positive results to the synergistic effect of hydrophilic groups and negative charges of the nanocellulose layer, which minimizes the water passage resistance and maximizes the repellence to negatively charged nonpolar foulants. So far, H-membranes are considered as the most promising solution for the industrialization of nanocellulose-based composite membrane due to its competitive performance and roll-to-roll processability.

10.2.3 Wood-Based Nanocellulose Monolith Filters

Except for the membranes obtained through "bottom-up" assembly of nanocellulose, great efforts have also been devoted to the fabrication of monolith filters via "top-down" processing of natural wood. Compared to the "bottom-up" strategy, "top-down" approaches generally demonstrate time and energy-saving process with a preservation of the microstructure of wood. By rationally utilizing these microstructure characteristics, superior water purification performance can be achieved. Charcoal derived from the incomplete combustion of natural wood is one of the most representative monolith materials that have been used for water purification by ancient Hindus as early as 450 BC [27]. The straightforward fabrication method, unique microporous structure, as well as physicochemical robustness make charcoal still an ideal filter for pollutant absorption or adsorption till now [28–30]. Recently, with the deep looking of structure–function relationship of nanocellulose in the composition of wood, significant attentions have been devoted to the fabrication of these wood-derived nanocellulose-based monolith filters, opening a new paradigm for bulk filter design.

Typically, wood possesses a complex hierarchical pore structure with low tortuosity derived from diverse plant cells, including the vessel channels with several hundred micrometers, the tracheids channels with tens of micrometers, and the cellulose fibril spacing with several nanometers, all in diameters. Between the fibers, the channels are interconnected through pits on the cell walls. Owing to the unique pore structure, wood is born with the talent for water pumping and filtration. By decorating with catalysts, diverse wood filters have been reported with excellent catalytic contaminant removal. In 2017, Hu group reported a mesoporous 3D wood filter decorated with palladium (Pd) nanoparticles (NPs) for efficient wastewater treatment [31]. Four major advantages of using wood as filter are summarized: (1) the lignin component in natural wood allows the *in-situ* formation of Pd NPs via reduction reaction of Pd(II) ions; (2) the abundant hydroxyl groups of cellulose can help the immobilization of the formed Pd NPs; (3) the mesoporous structure of wood with enlarged surface area can increase the likelihood of wastewater contacting Pd NPs; (4) the long, low-tortuosity, Pd NPs-decorated channels can significantly facilitate the bulk flows of filtration. Owing to these advantages, the Pd NPs decorated wood membrane demonstrated a high water treatment rate of 1×10^5 $L{\cdot}m^{-2}{\cdot}h^{-1}$ and superb turnover frequency (TOF) of ~2.02 $mol_{MB}{\cdot}mol_{Pd}{}^{-1}{\cdot}min^{-1}$ with a methylene blue (MB) removal efficiency of >99.8%, which is much higher than the previous reported values. Interestingly, soon afterwards the same group further found that the TOF performance of the wood filter can be improved to 2.2 $mol_{MB}{\cdot}mol_{Pd}{}^{-1}{\cdot}min^{-1}$ by just fabricating the membrane in a reverse direction (cross-flow design) [32], suggesting a great structural modification potential of the wood filter design as well. Except for MB catalytic decomposition, wood-based filters that are capable of killing germs, removing metal ions, absorbing oil have also been prepared by utilizing a similar functional strategy. For example, Che et al. have reported a wood-based mesoporous filter decorated with silver nanoparticles (designated as Ag/wood filter) for water purification [33]. Benefitting from the Ag NPs, the Ag/wood filter not only demonstrates high MB removal efficiency (>98.5%) but also gives superior bacterial removal capabilities (6.0 and 5.2 orders of magnitude for *E. coli* and *S. aureus*, given as the log removal of the viability: log (A/B), where A and B are the number of bacteria before and after filtering).

These works show a sustainable future and a great deal of potential of utilizing wood to cope with water contamination. However, despite great achievements that have been made, it is understandable that studies of nanocellulose-based monolith filters are still in the early stages. More efforts are still needed to better understand the fundamental mechanisms and structure–function relationships of these monolith filters for diverse water contaminants treatment.

10.3 Nanocellulose-Based Adsorbents/Absorbents

Adsorbent is a solid substance capable of effectively adsorbing certain components from gas or liquid to its surface and further achieves the removal of pollutants from air, water, or soil [35–37], which typically features large specific surface area (SSA), suitable pore structure, and excellent chemical stability. Depending on the nature of feedstock used, adsorbents can be divided into three categories: inorganic adsorbents, synthetic polymers adsorbents, and bio-based adsorbents. Inorganic adsorbents mainly include activated carbon and inorganic minerals [38]. Owing to their porous structure and large SSA, these materials demonstrate excellent adsorption performance, but they are mostly used in the form of powder or granules, bringing problems such as difficult separation and recovery after adsorption [39–41]. In contrast, synthetic polymer adsorbents show good mechanical properties and can be reused many times after elution. However, most of them are nonbiodegradable and pose negative impacts on environment [42, 43]. Apart from inorganic and synthetic ones, bio-based adsorbents obtained from natural polymers have attracted special attention due to their abundant availability and environmental friendliness [44–47]. Among various bio-based options, nanocellulose derived from cellulose by hydrolysis or mechanical nanofibrillation stands out in this application primarily due to its nanoscale size, high SSA, good biodegradability, and ease of chemical functionalization for desirable purposes [48–51]. These characteristics endow nanocellulose a great potential as adsorbents for environmental remediation applications, such as the removal of dyes, heavy metal ions, and oil.

10.3.1 Adsorption Theory of Nanocellulose-Based Adsorbents

In general, solution-based adsorption is a spontaneous mass transfer process where the adsorbates in the solution/suspension are accumulated at the surface and interior of the solid adsorbent materials [52]. The adsorption process basically consists of three consecutive steps (Fig. 10.4): (1) attraction of the adsorbates by the boundary of the adsorbents (external diffusion); (2) diffusion of the adsorbates to the surface or interior of the adsorbents (intraparticle diffusion); and (3) adsorption reaction between the adsorbates and the adsorbents (intraparticle adsorption). Depending on the nature of the adsorbates and adsorbents, one or more adsorption mechanisms could be involved in an adsorption process.

10.3.1.1 Adsorption Mechanisms

The phenomenon of adsorbate-adsorbent interaction may originate from three mechanisms: (1) physisorption; (2) chemisorption; (3) ion exchange. Physisorption is an exothermic process that occurs based on the van der Waals forces between the adsorbate and the adsorbent (Fig. 10.5a) [53, 54]; the adsorbate may accumulate in either monolayer or multilayer at the adsorbent's surface. Since no chemical bonds are formed during adsorption and low activation energy between the reversible adsorption/desorption, the adsorption interaction in physisorption is weak and unstable with dynamic equilibrium between adsorption and desorption. In contrast, chemisorption involves the transfer or pairing of electron between the adsorbate and the adsorbent for the formation of chemical bonds (Fig. 10.5b) [53, 55]. Due to the strong interaction, this kind of adsorption has comparatively higher activation energy, higher

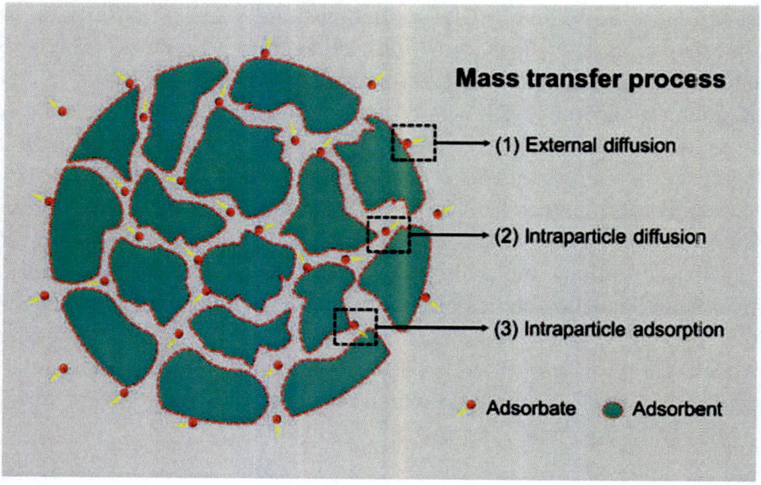

Fig. 10.4 Scheme of mass transfer process of adsorption

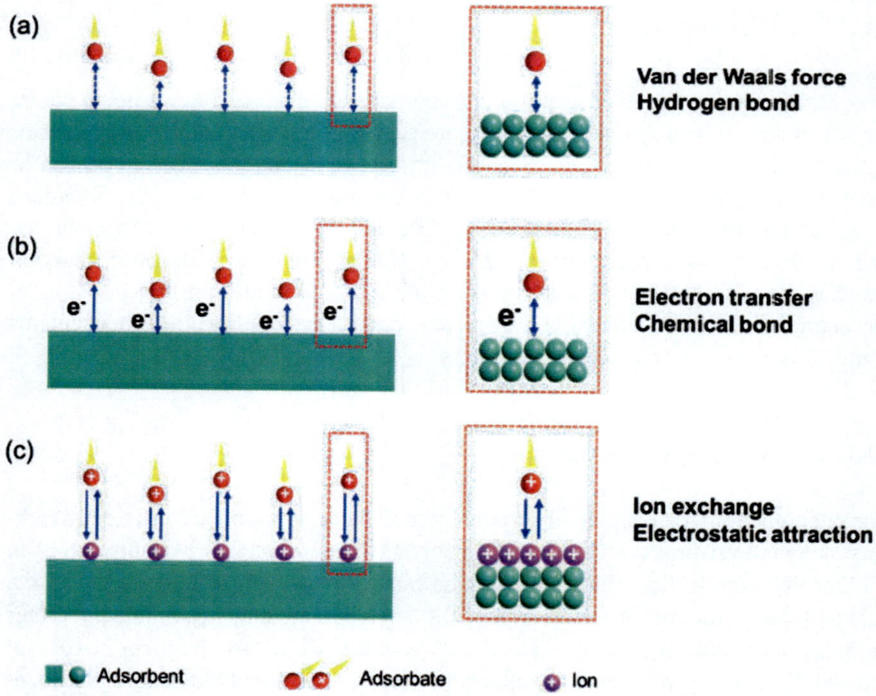

Fig. 10.5 Scheme of adsorption mechanisms: **a** Physisorption, **b** Chemisorption, and **c** Ion exchange

selectivity towards adsorbate, as well as the feature of monolayer and irreversible adsorption. As for ion exchange, it is to replace other ions which originally fix on the active adsorption site of the adsorbent with adsorbate ions (Fig. 10.5c) [56]. This adsorption is generally driven by electrostatic attraction, which is a reversible adsorption process. Overall, the adsorption of nanocellulose-based adsorbents is a complicated process, which usually involves a combination of multiple adsorption mechanisms. To further study the mechanisms of the adsorption process, adsorption isotherms and kinetic models are generally used.

10.3.1.2 Adsorption Isotherms

Adsorption isotherm can macroscopically reflect the adsorption process. By analyzing the variation trend of the adsorption isotherm, the strength of the interaction between adsorbent and adsorbate can be well judged, which is an important indicator for investigating the adsorption efficiency of the adsorbent. Currently, there are many adsorption isotherm models to study adsorption equilibrium, while the most useful and commonly employed are Langmuir and Freundlich models.

a. **Langmuir model**

Langmuir model is established based on the following assumptions: (1) the adsorbent has a homogeneous surface; (2) the active adsorption sites on the surface of the adsorbent are identical and energetically equivalent; (3) one active adsorption site on the adsorbent surface can only adsorb one adsorbate molecule, and there is no interaction between the neighboring adsorbed adsorbate molecules; (4) the adsorption process is a monolayer adsorption [57]. Langmuir adsorption isotherm equation is expressed as follows:

$$\frac{C_e}{q_e} = \frac{1}{K_L q_{max}} + \frac{C_e}{q_{max}} \tag{10.1}$$

where q_e is the equilibrium adsorption capacity, q_{max} is the maximum adsorption capacity, C_e is the equilibrium concentration of adsorbate, and K_L is the Langmuir constant. Also, a dimensionless constant R_L is defined to determine the favorability and feasibility of adsorption process.

$$R_L = \frac{1}{1 + K_L C_0} \tag{10.2}$$

where C_0 is the initial concentration of adsorbate and K_L is the Langmuir constant. Adsorption process is irreversible if $R_L = 0$ and linear if $R_L = 1$. If $0 < R_L < 1$, it is favorable for adsorption, and it is unfavorable for adsorption if $R_L > 1$.

b. **Freundlich model**

Freundlich model is a semi-empirical equation assuming that adsorption process is a multilayer adsorption with non-uniform distribution of affinity on a heterogeneous adsorbent surface; meanwhile, the adsorption capacity is dependent on the concentration of adsorbate in liquid phase [58]. Freundlich model is represented by the following equation:

$$ln\, q_e = lnK_f + \frac{1}{n} ln\, C_e \tag{10.3}$$

where q_e is the equilibrium adsorption capacity, C_e is the equilibrium concentration of adsorbate, K_f is the Freundlich constant, and n is the constant related to the interaction intensity between adsorbent and adsorbate. The larger value of n stands for stronger interaction between these two. Specifically, when $1/n$ is between 0.1 and 0.5, it is favorable for adsorption. In contrast, it is unfavorable for adsorption when $1/n$ is between 0.5 and 1.

10.3.1.3 Adsorption Kinetics

Adsorption kinetics are governed by both adsorption and desorption behaviors. When the rates of adsorption and desorption are equal, the macroscopic adsorption behavior reaches a dynamic equilibrium state. Adsorption kinetic studies can be used to quantitatively describe the adsorption and desorption rate, which are important to investigate the adsorption mechanism. For this reason, many adsorption kinetic models have been developed and studied in various adsorption systems, with the pseudo-first-order and pseudo-second-order kinetic models being the most commonly used.

a. **Pseudo-first-order kinetic model**

Pseudo-first-order kinetic model is developed based on the fundamental principles of interfacial kinetics [59]. Same as Langmuir model, it is also created based on the assumption that one adsorbate molecule can only react with one active adsorption site on adsorbent. External diffusion in such model is considered as the rate-limiting step and the mechanism of the adsorption process is usually physisorption. The equation can be expressed as follows:

$$log(q_e - q_t) = logq_e - \frac{k_1 t}{2.303} \tag{10.4}$$

where q_e and q_t are the adsorption capacity at equilibrium state and at time t, respectively, k_1 is the rate constant of the pseudo-first-order adsorption.

b. **Pseudo-second-order kinetic model**

In contrast, pseudo-second-order kinetic model is identified as chemisorption that involves high binding energy due to the chemical bonding formed between adsorbent and adsorbate [60]. The rate-limiting step in this model is intra-particle diffusion. The equation is represented as follows:

$$\frac{t}{q_t} = \frac{1}{k_2 {q_e}^2} + \frac{t}{q_e} \tag{10.5}$$

where q_e is the adsorption capacity at equilibrium state, q_t is the adsorption capacity at time t, and k_2 is the rate constant of the pseudo-second-order adsorption.

10.3.2 Nanocellulose-Based Adsorbents for the Removal of Dyes and Heavy Metal Ions

Adsorption capacity of cellulose materials has been identified to be determined by both the materials' SSA and surface functional groups [61, 62]. For pristine nanocellulose, though it has a higher SSA with high-aspect-ratio structure compared to cellulose fibers, no significant improvement has been seen in its adsorption capacity. To enhance its adsorption capacity, the surface of nanocellulose needs to be tuned to favor the interaction with adsorbates, which can be achieved through the following strategies (Fig. 10.6).

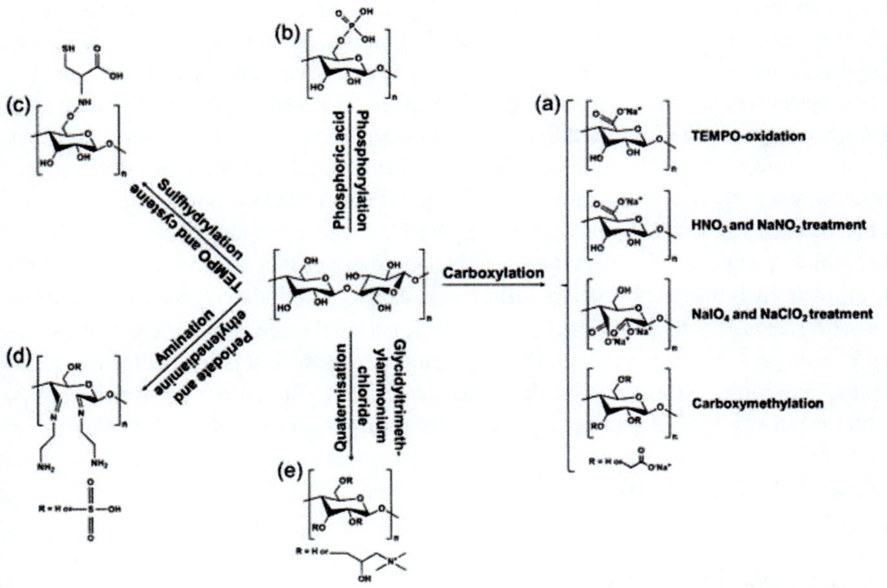

Fig. 10.6 Modification strategies of nanocellulose: **a** Carboxylation, **b** Phosphorylation, **c** Sulfhydrylation, **d** Amination, and **e** Quaternization

10.3.2.1 Carboxylation

Carboxylation is a functionalization process in which the hydroxyl groups of nanocellulose convert into carboxylic acid groups, as shown in Fig. 10.6a. Common strategy for carboxylation of nanocellulose includes: (1) 2,2,6,6-Tetramethylpiperidinyloxy (TEMPO)-oxidation; (2) carboxylation using nitric acid and sodium nitrite (HNO_3 and $NaNO_2$); (3) periodate-chlorite ($NaIO_4$ and $NaClO_2$) oxidation, and (4) carboxymethylation. The most popular one among these methods is TEMPO-oxidation, in which the C6-hydroxyls on the cellulose chains are selectively oxidized to negatively charged carboxyl groups. Therefore, the surface carboxylated nanocellulose can be applied to adsorb cationic dyes and positively charged heavy metal ions. For example, activation of sulfuric acid (H_2SO_4) hydrolyzed CNC by TEMPO oxidation can introduce 2.1 mmol/g of surface carboxyl groups, which was significantly higher than the unmodified CNC with 0.2 mmol/g of sulfate groups [63]. As a result, the adsorption capacity towards MB showed a dramatic improvement from 118 mg/g to 769 mg/g after surface modification. Similar to CNC, TEMPO-CNF can also act as an effective adsorbent. The pioneering work for applying TEMPO-CNF as adsorbent was conducted by Saito et al. [64]. In their research, the ion exchange behavior of CNF was explored by soaking it in solutions of different metal ions. The results demonstrated that lead (Pb^{2+}), calcium (Ca^{2+}), and silver (Ag^{+}) ions can be preferably adsorbed onto CNF at a metal ion/carboxylate molar ratio of up to 1:1. Moreover, CNF also showed high adsorption selectivity towards various metal ions. Sehaqui et al. systematically investigated the effect of environmental pH and the carboxyl group content of CNF on its adsorption efficiency [65]. The results suggested that the adsorption capacity of CNF followed a linear trend with the carboxyl groups content when pH was in the range of 3–7. Apart from TEMPO-oxidation, CNC with a high carboxyl groups content can be prepared through a two-step periodate-chlorite oxidation. Yang et al. developed a novel biopolymer aerogel adsorbent based on the composite of bifunctional nanocellulose and carboxymethylated chitosan (CMCT) through freeze-drying [66]. The CNC with bifunctional moieties (carboxyl and aldehyde groups) was prepared from the sequential periodate and partial chlorite oxidation of cellulose, followed by a hot water treatment. Benefiting from the porous structure of the aerogel and the two functional groups on its surface, the adsorption capacity of MB using the prepared aerogel adsorbent reached up to 785 mg/g, as calculated by Langmuir formula.

10.3.2.2 Phosphorylation

Phosphorylated nanocellulose can be prepared by reacting nanocellulose with phosphoric acid, adenosine-5′-triphosphate (ATP), or sodium alendronate trihydrate. In the process of phosphorylation, the hydroxyl groups on nanocellulose would be partially or totally replaced by phosphate groups (Fig. 10.6b). Similar to the carboxylated one, phosphorylated nanocellulose can be used to remove cationic dyes and positively charged heavy metal ions. For example, Liu et al. demonstrated the synthesis of phosphorylated CNC and CNF using ATP [67]. The results showed greatly improved adsorption capacity of the modified nanocellulose, as reflected by up to 99% removal efficiency of copper ion (Cu^{2+}) and iron ion (Fe^{3+}) when treating effluent from mirror making industry. In another study, Sirviö et al. developed a phosphorylated CNC using sodium periodate oxidation, followed by the reaction with sodium alendronate trihydrate [68]. The results indicated that the amount of phosphate groups grafted on nanocellulose was highly determined by the preceding oxidation. With 0.32 mmol/g phosphate groups content, the modified CNC enabled a vanadium ion (V^{5+}) adsorption capacity of 1.98 mmol/g.

10.3.2.3 Sulfhydrylation

Sulfhydrylation is another way to introduce negative charge on the surface of nanocellulose. Previous studies have confirmed that the thiol group has a strong affinity with mercury ion (Hg^{2+}), and thus the thiol-modified nanocellulose could be specialized for Hg^{2+} adsorption (Fig. 10.6c) [69]. For example, Geng et al. proposed a thiol-functionalized CNF for the removal of Hg^{2+}, which was prepared in two steps [70]. In the first step, CNF was isolated from bamboo pulp by TEMPO oxidization followed by mechanical disintegration. After that, TEMPO-CNF was treated with 3-mercaptopropyl-trimethoxysilane (MPTs) to obtain thiol-modified CNF. The removal rate of Hg^{2+} by the modified CNF was 92%, which were contributed by the synergistic electrostatic interaction of both thiol groups and carboxy groups with Hg^{2+}.

10.3.2.4 Amination

Different from the aforementioned modification method, the introduction of amino groups on the surface of nanocellulose, due to its positively charged nature, can significantly increase the adsorption capacity towards anionic dye and negatively charged metallic species. Such adsorption is more effective over a low pH range from 2.5 to 5.0, where amine groups are sufficiently protonated. For example, Singh et al. presented modified CNC using a combined succination and amination, and further investigated its adsorption efficiency for chromium ions (Cr^{3+} and Cr^{6+}) [71]. Benefiting from the increased surface area-to-volume ratio and the rich functional groups on the obtained CNC, the CNC adsorbent can remove as high as 98% of Cr^{6+} in a 25 mg/mL solution at $pH = 2.5$. Jin et al. reported using amino-functionalized CNC to remove anionic dyes [72]. In their study, CNC was firstly extracted from bleached hardwood kraft pulp via H_2SO_4 hydrolysis, followed by sodium periodate oxidation to prepare dialdehyde nanocellulose. Then, the nanocellulose was further grafted with ethylenediamine (EDA) and the amino-functionalized CNC was obtained, as shown in Fig. 10.6d. The modified CNC acted as an excellent adsorbent for the removal of anionic dyes such as Acid Red GR (AR), Congo Red 4BS (CR), and Reactive Light

Yellow K-4G (RLY). The results showed that the maximum adsorption capacity of the aminated CNC for AR was approximately 555.6 mg/g.

10.3.2.5 Quaternization

In addition to amination, quaternization modification is also an effective method to introduce positive charges onto CNF, making it suitable to adsorb anionic dyes. Pei et al. demonstrated a kind of surface quaternized CNF which was obtained by a combined glycidyltrimethylammonium chloride pretreatment and the followed high-pressure homogenization (Fig. 10.6e) [73]. The resulting quaternized CNF with trimethylammonium chloride content of 0.59–2.31 mmol/g was then able to remove anionic dyes. It was observed that the adsorption capacity of the quaternized CNF for CR and Acid Green 25 increased with the trimethylammonium chloride content from 0.59 mmol/g to 1.32 mmol/g, and the peak values were 664 mg/g and 683 mg/g, respectively. The quaternized CNF can be further processed into CNF nanopaper for CR removal from its bulky solution (0.5 mg/mL). The results showed that the nanopaper adsorbent can reach final equilibrium with 95% dye removal rate after 72 h.

10.3.3 Regeneration of Nanocellulose-Based Adsorbents

Used adsorbents can be regenerated by properly separating the contained adsorbates from their structure. In view of the following aspects of (1) retaining the original adsorption capacity of the adsorbents, (2) prolonging or enlarging the adsorbents' service capability, especially in consideration of the expensive adsorbents, and (3) reducing environmental pollution resulting from the disposal of the used adsorbents, regeneration of adsorbents to their maximum capacity is crucial for practical application in industrial wastewater treatment. As discussed above, the adsorption mechanism towards adsorbates usually involves electrostatic interactions, van der Waals forces, complexation, and so on [55]. Therefore, the basic principle for the selection of eluents is to break the adsorbent and adsorbate interactions. Accordingly, the eluents typically cover three types: (1) strong acids or bases, (2) organic solvents, and (3) complexing agents.

10.3.3.1 Strong Acids and Bases

In some cases, adsorption efficiency is highly dependent on pH condition, especially in the system that electrostatic interaction dominates the adsorption process. For instance, the adsorption of positively charged metal ions and cationic dyes onto carboxylated nanocellulose is favored at a high pH condition. If strong acids such as HCl, H_2SO_4, or HNO_3 are added into the solution, the adsorbed metal ions or dyes will desorb from nanocellulose due to the protonation of surface carboxyl groups. Similarly, aminated nanocellulose acts as an effective adsorbent for the removal of negatively charged metal ions and anionic dyes at a low pH condition, and desorption can then be carried out by introducing strong bases (such as NaOH). Therefore, strong acids or bases can be used as good eluents in such circumstances. Geng et al. developed a thiol-functionalized TEMPO-CNF aerogel for highly efficient removal of Hg^{2+} [70]. After adsorption, 1 M HCl solution with 5 wt% thiourea was used to desorb Hg^{2+} from the CNF aerogel. The results showed that the adsorption efficiency of the regenerated adsorbent barely displayed a change from 97.5% to 93.2% even after four adsorption/desorption cycles. In addition, Song et al. demonstrated a CNC-reinforced keratin adsorbent can be regenerated by 0.1 g/L NaOH solution [74]. The adsorbent after fifth regeneration demonstrated high removal efficiency of over 80% for both Reactive Black 5 and Direct Red 80.

10.3.3.2 Organic Solvents

Due to the presence of hydrophobic groups in dyes, they can also be eluted from nanocellulose adsorbent by organic solvents. He et al. demonstrated a kind of carboxylated CNC adsorbent with a MB adsorption capacity of 101 mg/g [75]. Using ethanol as eluent demonstrated satisfactory adsorbent regeneration efficiency, with more than 90% of the adsorbates successfully removed after seven cycles.

10.3.3.3 Complexing Agents

Ethylenediaminetetraacetic acid (EDTA) can also be used as an eluent for metal cations. For example, Yang et al. reported a thiol-functionalized TEMPO-CNF membrane adsorbent [69]. Desorption of the adsorbed Pb^{2+} from the adsorbent was accomplished using EDTA, and the TEMPO-CNF-based adsorbent could retain high adsorption efficiency (>93%) after three recycles.

10.3.4 Nanocellulose Absorbents for Oil Remediation

Oil spills and the discharge of oily pollutants to water body have caused great challenges to the aquatic and seashore ecosystems. Effective removal of these oily waste from the contaminated water has become one of the exigent and arduous tasks regarding environmental remediation [76, 77]. In recent years, nanocellulose-based materials are considered as promising candidates for oily wastewater treatment, while the hydrophilic characteristic of nanocellulose due to the existence of massive hydroxyl groups on its molecular impedes its use as an oil absorbent with high selectivity [78]. Therefore, certain surface modification is crucial to endow nanocellulose with hydrophobic property. There are two common ways to achieve such target: (1) introducing surface hydrophobic agents such as silane coupling agents; (2) carbonization of nanocellulose-based materials in high-temperature condition.

10.3.4.1 Hydrophobically Functionalized Nanocellulose-Based Oil Absorbents

Hydrophobically functionalized nanocellulose-based absorbents can be prepared by reacting nanocellulose with methyltrimethoxysilane (MTMS), trimethylchlorosilane (TMCS), triethoxyl (octyl) silane, or other hydrophobic agents. According to the target of hydrophobic treatment, the modification approaches can be divided into two types: (1) hydrophobic modification of nanocellulose followed by preparation of aerogel/sponge; (2) hydrophobic modification of as-prepared nanocellulose-based aerogel/sponge. Following the above two strategies, many works have been conducted. Zhang et al. reported a flexible, porous, and ultralightweight CNF sponge by freeze-drying CNF dispersion modified with MTMS [79]. The modified CNF sponge with hydrophobic and oleophilic properties showed good removal efficiency of dodecane, a model of oily waste, from wastewater, and the absorption capacity can reach 52 g/g. The same CNF sponge was also able to collect other organic solvents and oils, with the absorption capacity increasing with the density of the liquids. This facile functionalization method provided a new way for the construction of nanocellulose-based absorbents with good performance. Besides, Gu et al. reported an efficient CNF-based absorbent through oleic acid and Fe_3O_4 functionalization combined with freeze-drying method [80]. In this design, the oleic acid served as a hydrophobic agent, while Fe_3O_4 was used as a magnetic agent. Experimental results showed the absorption capacities of this absorbent to ethyl acetate, cyclohexane, and vacuum pump oil were 56.32 g/g, 68.06 g/g, and 33.24 g/g, respectively. Moreover, with the help of the excellent magnetic responsivity, the absorbent can be easily recycled by a permanent magnet after use. Jiang et al. developed an ultralightweight and ultraporous CNF aerogel followed by vapor depositing triethoxyl (octyl) silane onto its surface [81]. The aerogel absorbent demonstrated superior absorption capacity ranging from 139 g/g to 356 g/g towards organic solvents and oil, outperforming most absorbents obtained from natural or synthetic polymers.

10.3.4.2 Nanocellulose-Derived Carbonized Oil Absorbents

Another strategy to provide nanocellulose-based aerogel/sponge with hydrophobicity is through high-temperature carbonization. Based on carbonization treatment, Meng and coworkers synthesized a porous and ultralightweight sponge-like nanocellulose-based carbon aerogel with excellent hydrophobic property [82]. Taking advantage of the 3D network structure formed by the entanglement of carbonized cellulose fibers and the low surface energy of carbonized fibers, this absorbent showed high absorption capacity for a range of oils such as canola oil, diesel oil, paraffin, and pump oil. The maximum paraffin absorption capacity reached 86 g/g. Furthermore, after thoroughly cleaning in ethanol and drying of the used carbon aerogel, it also retained the same oil absorption capacity after ten absorption/desorption cycles. To further enhance the mechanical strength of the carbonized nanocellulose-based materials, Long et al. developed a hierarchical wave-shaped CNF-graphene oxide (GO)/glucose-kaolin carbon aerogel (C-NGGK) via unidirectional freeze-drying and carbonization [34]. Herein, carbonized CNF functioned as a supporting skeleton to provide mechanical strength for the whole system; carbonized glucose is used to interact with the rGO layer, and kaolin endowed the aerogel with hydrophobicity. Based on such structural design, the aerogel could retain its original height even after 100 compressive cycles, indicating outstanding fatigue resistance and mechanical robustness. More importantly, C-NGGK showed good absorption performance, and the absorption capacity for organic contaminants and oils can achieve 75–225 g/g. In addition, the carbon aerogel absorbent also demonstrated excellent reusability.

In summary, nanocellulose shows great potential to be used as effective adsorbents/absorbents for wastewater treatment. However, there are still challenges remained to be addressed, such as complex fabrication process, high cost, insufficient reusability, and lack of scaling-up trials. Great efforts are needed to further optimize the preparation process and the product performance for the industrial application of nanocellulose-based adsorbents/absorbents.

10.4 Nanocellulose-Based Solar Conversion Devices Toward the Water–Energy Nexus

Beyond regional water shortage (so called physical water scarcity), approximate 0.66 billion population have limited access to drinkable water due to affordability (so-called economic water scarcity) [83]. Seeking for a cost-effective and sustainable way for people who are struggling with economic water scarcity becomes an urgent challenge. Solar still, which directly utilizes solar energy to meet the enthalpy required for water vaporization, is the simplest device that has been used by humans to generate clean water since ancient times [84]. To date, the fundamental mechanism of solar-driven steam generation processes has not changed. But the application of solar stills is becoming increasingly scarce due to the low solar-to-steam conversion efficiency and high capital cost associated with the facility. Recently, with the emerging of interfacial evaporation theory [85], this conventional water generation technology has regained popularity in response to economic water scarcity. Among all reported interfacial solar evaporation devices, nanocellulose-based one with affordable cost, environmental friendliness, and competitive performance is considered as one of the most promising solutions for off-grid water supplying.

10.4.1 Mechanism and Evolution of Solar-Driven Steam Generators

Solar-driven steam generation process is essentially a phase transition process of water from liquid phase to vapor phase driven by the solar converted thermal energy. The thermal-to-steam efficiency η_{sg} of this process can be assessed as following

$$\eta_{sg} = \frac{\tilde{m}h_{LV}}{C_{opt}q} \tag{10.6}$$

where $\tilde{m}$ is the steam generation flux, h_{LV} is the phase change enthalpy of liquid water to vapor (both sensible heat and latent heat), q is the nominal solar irradiation under 1 sun (1 kW m^{-2}), C_{opt} is the optical concentration. Generally, the phase change enthalpy and solar irradiation of an actual solar evaporation system is certain and fixed. Therefore, the solar-to-steam efficiency of the evaporation system can be calculated by measuring the steam generation flux.

According to the structural design, there are three typical designs of solar evaporation system, as shown in Fig. 10.7. The most primitive form is the bottom-heating design, of which the absorber is placed at the bottom of the container (Fig. 10.7a). Because thermal generated at the surface of solar absorber will soon dissipate to the bulk water and ambient via thermal conduction and convection, the heating temperature and vapor generation efficiency for such a solar still device is usually very low (30–45%) [86]. To increase the efficiency, solar evaporation system using optical nanofluids has been invented (Fig. 10.7b). The application of nanofluids leads to a volumetric solar absorption and brings more vapor generation sites, which contributes to an improved vapor generation efficiency. However, since the heat losses to the bulk water is inevitable, the efficiency improvement of such design is limited. Moreover, the long-term dispersion and pumping of nanofluids under solar radiation also remain challenging [87, 88]. To reduce the surface heat loss of absorber, a novel interfacial solar evaporation system of which the solar absorber with water pumping capability floated on water was reported by Chen et al. in 2014 (Fig. 10.7c). Since the solar absorber is placed at the water-air interface, the solar thermal energy can be effectively localized at the liquid surface and mainly used to heat the evaporative portion of water rather than the bulk water. Owing to the improved energy efficiency, such interfacial evaporator design could easily reach a solar-to-steam efficiency of over 70%. Moreover, the self-floating evaporator design also allows to greatly reduce the volume of solar still system, which is promising for portable applications.

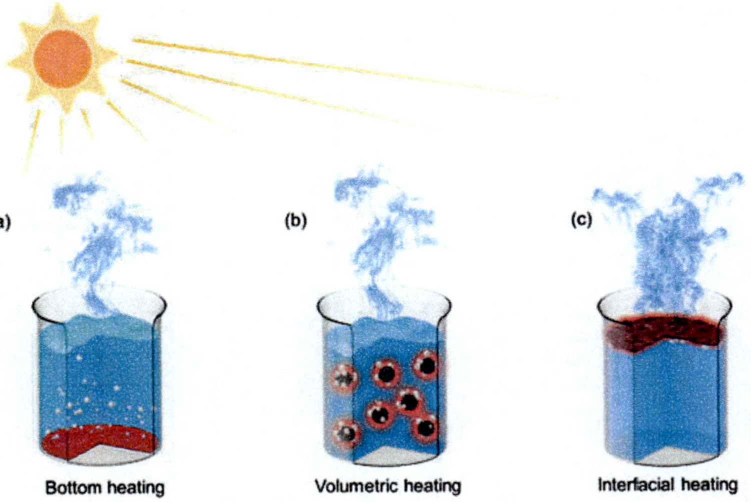

Fig. 10.7 Schemes of typical solar evaporation system designs: **a** bottom heating design, **b** volumetric heating design, and **c** interfacial heating design

10.4.2 Design Principles for Nanocellulose-Based Solar Evaporators

10.4.2.1 Photothermal Materials

Photothermal material, which responses for solar absorption and solar-to-heat conversion, is one of the key components for solar evaporator. Since the spectrum of sunlight reaching the Earth distributes in a wide wavelength range of 0.3–2.5 μm, the absorber is required to have a broad absorption spectrum across the full solar spectrum to utilize the solar energy. In general, there are three types of photothermal material that can be chosen for the fabrication of nanocellulose based evaporator, which corresponds to three categories of solar-heating mechanisms (Fig. 10.8) [89].

Metal nanoparticle with strong light-to-heat capability, also known as plasmonic nanoparticle, is one of the most representative photothermal materials. When the photon frequency of the incident light matches the natural frequency of conduction electrons on the surface of nanoparticles, the illumination will cause the resonant oscillation of these electrons (so-called surface plasmon resonance). The plasmon-excited electrons will then redistribute their energy via electron-electron and electron-phonon scattering processes to generate heat with a nearly 100% solar-to-heat conversion efficiency (Fig. 10.8a) [90, 91].

Metal oxide-based semiconductors have also been widely used for absorber fabrication. When the photon energy of the incident light is higher than the bandgap of the semiconductor, the valence electrons will be excited to the conduction band with higher energy level and generate above-bandgap electron-hole pairs. Accompanied

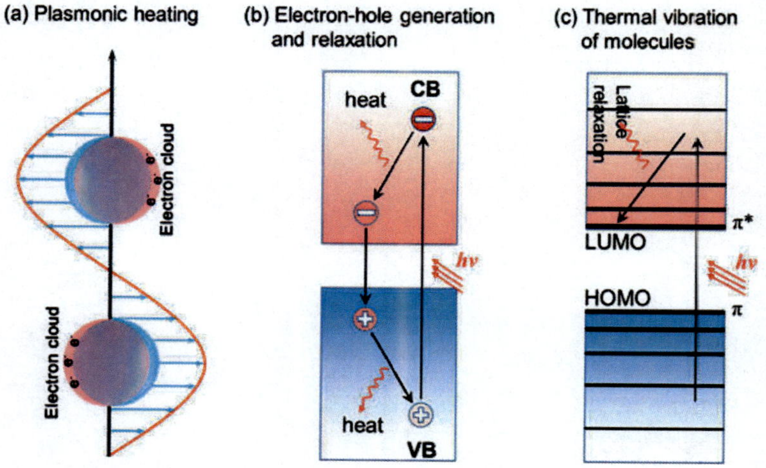

Fig. 10.8 Photothermal mechanisms of different types of photothermal materials: **a** plasmonic heating, **b** electron-hole generation and relaxation, and **c** thermal vibration of molecules

by the relaxation of these above-bandgap carriers, the absorbed optical energy will then be released in forms of heat or light (Fig. 10.8b). As for photothermal application, narrow bandgap semiconductors that facilitate higher photothermal efficiency are more preferred compared to the opposite. Some representative narrow bandgap semiconductors that have been investigated in solar evaporation including black titania [92], Ti_2O_3 nanoparticles [93], Fe_3O_4, [94] $CoFe_2O_4$, [95] and so on.

Except for the plasmonic nanoparticles and metal oxide-based semiconductors, recently, carbonaceous materials and some conjugated polymers have gained great attentions in fabricating solar absorbers due to their exceptional processability and wide choices of source materials. Because of the complex sp, sp^2, and sp^3 hybrid orbitals formed by the carbon atoms, these materials demonstrate high optical absorption across a wide range of wavelength, resulting in a black appearance. Under solar illumination, the electrons at ground state (generally refers to the highest occupied molecular orbital, HOMO) will be first excited to a higher molecular orbital and subsequently release heat by relaxing to the lower orbitals via electron-electron or electron-phonon scattering (Fig. 10.8c).

10.4.2.2 Thermal Management and Structure Design

Compared to the photothermal materials, how these materials are constructed into solar evaporator is equally important in determining the steam generation efficiency. Typically, nanocellulose-based solar evaporators possess highly integrated structure with the solar absorber layer piling up on a supporting layer. It is crucial that the evaporator should not only offer enough buoyance and water pumping capacity but also possess excellent heat management property (Fig. 10.9). Particularly, a favorable

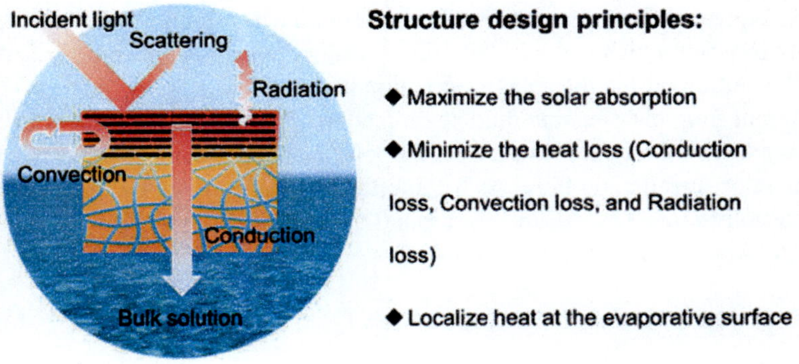

Fig. 10.9 Thermal management scheme of solar evaporator

heat management for minimum dissipation can be achieved from perspective of containing the three basic heat transfer means: radiation, conduction, and convection.

Thermal radiation is the emission of a spectrum of electromagnetic radiation due to an object's temperature. According to the Stefan-Boltzmann equation, radiation loss of the solar-heated evaporator can be calculated as following,

$$\Phi = \varepsilon A\sigma(T_e^4 - T_i^4) \tag{10.7}$$

$$\text{Radiationloss} = \frac{\Phi}{AC_{opt}q_i} = \frac{\varepsilon\sigma(T_e^4 - T_i^4)}{C_{opt}q_i} \tag{10.8}$$

where Φ is the heat flux, ε is the emissivity of the evaporator, A represents the surface area of the irradiator, σ represents the Stefan-Boltzmann constant which is 5.67×10^{-8} J m^{-2} s^{-1} K^{-4}, T_i and T_e are the equilibrium temperature of the evaporator surface before and after solar irradiation, respectively, C_{opt} and q_i are the optical concentration of solar irradiation and the nominal solar illumination as defined in (6).

Conduction loss is one of the major thermal dissipation pathways in interfacial solar evaporation. The solar converted heat at the irradiated surface spontaneously migrates from the hotter surface to the colder body and solution nearby. As for a relatively enclosed solar evaporation system, the lost conduction heat will finally turn to the internal energy of the whole system. Since the internal energy loss of the evaporation system to the ambient environment is comparatively negligible, the conduction loss can be approximately calculated based on the bulk water absorption heat Q as following

$$Q = mC\Delta T_1 \tag{10.9}$$

$$\text{Conductionloss} = \frac{Q}{AC_{opt}q_i\Delta t} = \frac{mC\Delta T_1}{AC_{opt}q_i\Delta t} \tag{10.10}$$

where m represents the mass of the bulk water solution, C represents the specific heat capacity of water which is 4.2 kJ °C^{-1} kg^{-1}, ΔT_1 represents temperature change of the bulk solution during the solar evaporation period of Δt.

Different from the dynamic fluid phenomenon of natural convection, convection loss here means the transfer of heat from the solar evaporator to the ambient due to the movement of surrounding fluid. As for interfacial solar evaporation, the convection loss is usually caused by the air flow around the evaporator, which can be analyzed by the Newton's Law of Cooling as below,

$$\Psi = \xi A \Delta T_2 \tag{10.11}$$

$$\text{Convectionloss} = \frac{\Psi}{AC_{opt}q_i} = \frac{\xi \Delta T_2}{C_{opt}q_i} \tag{10.12}$$

where Ψ is the thermal energy, ξ represents the convection heat transfer coefficient, ΔT_2 represents the temperature difference between the solar heating surface and the ambient temperature. Theoretically, the convection loss issue can be effectively alleviated by simply placing the evaporator in a steady environment without strong air flow; unfortunately, it is often impractical in real scenarios.

Evaporators with well-designed architecture can not only improve the solar absorption efficiency but also minimize the heat losses during solar evaporation. With the revolution of the evaporator structure from simply 2D structure to an elaborately designed 3D structure, significantly improved steam generation efficiency (close to and even beyond 100% efficiency) and environmental adaptability have been achieved in the past years. Currently, there are three commonly used structure designs for the fabrication of nanocellulose based solar evaporator, as shown in Fig. 10.10. Air-supported structure is one of the most common reported architecture in nanocellulose-based solar evaporator design, of which a hydrophilic cellulose/photothermal materials composite layer is placed at the top of an air-trapped insulation panel (Fig. 10.10a). During solar evaporation, part of the extended composite photothermal layer would connect to the bulk solution and continuously pump water to the solar-heated surface via capillary effect for vapor generation. At the meantime, the self-floating air-trapped insulation layer will support the localized solar-heating layer from directly contacting with the bulk solution. Since the thermal conductivity of the trapped air (0.0267 Wm^{-1} K^{-1} at 20 °C) is much lower than water and other evaporator components, such an air-supported structure design can significantly improve the solar steam efficiency by minimizing the conduction loss.

Nanocellulose-based aerogels have also been used in the fabrication of solar evaporator because of their lightweight and low thermal conductivity at dry conditions [96]. However, unlike the air-supported design, the empty space in the nanocellulose-based aerogel is not able to trap air due to its open-pore structure (Fig. 10.10b). When the aerogel is placed in water, it will turn to its hydrogel form with dramatically increased thermal conductivity, leading to a diminished solar-to-vapor efficiency. A recent strategic trial to resolve this problem is to introduce a hydrophobic coating

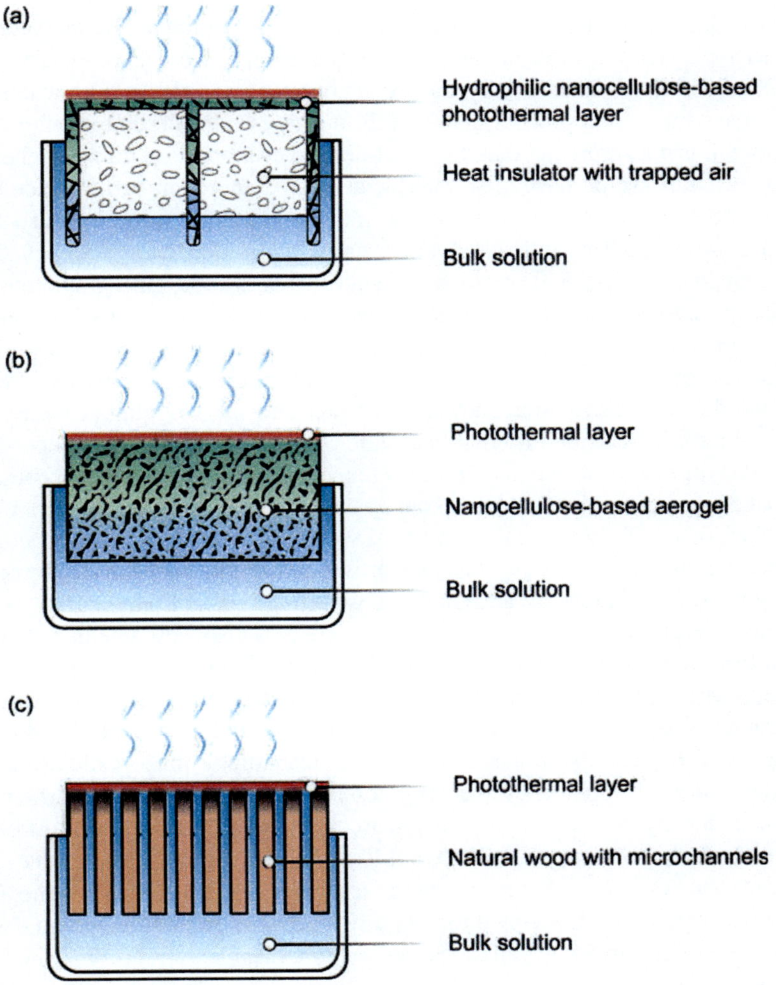

Fig. 10.10 Schematics of diverse nanocellulose-based solar evaporator designs: **a** air-supported evaporator design, **b** nanocellulose-based aerogel evaporator design, and **c** wood-derived evaporator design

at the surface of the aerogel backbone [97]. The modified aerogel would able to trap air in the open pores, while still allows water to transfer along the hydrophilic nanocellulose backbone. Moreover, the hydrophobic treatment further endows the aerogel evaporator an excellent anti-salt clogging property that is highly desired in seawater treatment.

Except for the air-supported aerogel evaporator designs, recently, wood-derived cellular structures have also been proved to be suitable for steam generation [98, 99]. Through in-depth study of the microstructure of wood using modern observation methods, scientists found that the tracheids and vessels in the wood form a bunch of micro-channels with low tortuosity (Fig. 10.10c). This unique microstructure is believed to play a critical role in the efficient sap transfer of wood via capillary water transfer effect. Most coincidentally, the growth of trees also accompanies with interfacial solar evaporation process that is well known as transpiration phenomenon. Therefore, it is not surprising that the wood-derived evaporators with well-inherited cellular structure would have great interfacial solar evaporation performance [99]. Besides the water transfer advantage, the microstructure of wood has also been confirmed with excellent anti-salt clogging property. By decorating fine palladium (Pd) nanoparticles into the 3D mesoporous matrix of basswood, Hu's group reported a kind of plasmonic wood that could stably work in seawater without salt formation under one-sun illumination [100]. They attributed this unique property to the low-tortuous and open pore structure of wood. Soon after, they found that balsa wood with bimodal pore distribution shows even better anti-salt clogging property. The reason was ascribed to the different hydraulic conductivities (diameter-dependent) of the two type channels, which drives ions to move out from the high concentration area to the low counterpart [101]. Inspired by this discovery, the scientists from Hu's group then successfully fabricated universal wood evaporators with excellent self-regeneration property that could stably work even in a highly concentrated brine (up to 20 wt% NaCl solution) by artificially constructing bimodal pores [102]. Moreover, this structure-related self-regeneration property is also workable in non-wood evaporators, indicating a promising practical value of bionic engineering.

In summary, cellulose demonstrates great opportunities in the generation of drinkable water, especially in interfacial solar steam. The ubiquitous plant resources together with the energy-cost-free solar steam technology make cellulose-based evaporator widely affordable even at the poorest part of the world, and particularly promise in dealing with economic water scarcity. However, noting that most of the current interfacial solar steam studies are still at laboratory level of where the evaporators are tested in a simplified environment, leading to an inflating water generation capability. There is an urgent need for a highly integrated evaporator system design that is portable and working well in diverse water bodies.

10.5 Perspective

This chapter overviews the emerging technologies of using nanocellulose-based materials for water and wastewater treatment, with subcategories including membrane filtration, absorption/adsorption, and solar-assisted water generation. Despite much progress achieved in proof-of-concept studies, very few technologies have fulfilled industrial expectations for contaminants removal and/or water generation, and hence their commercialization is still not practical. To push forward technology development, more attention should be paid to the following aspects.

- **Identification of unique benefits among various water treatment strategies**

The urgent demand to deal with water scarcity and pollution crisis has been raising a great deal of attempts in incorporating different materials and/or strategies to develop water treatment technologies. Apart from cellulose, a variety of reagents and materials for water and wastewater treatment have been recently developed from other biomass feedstocks, including alginate [103], lignin, [104, 105] and protein [106, 107]. Understanding the unique characteristics and benefits of the products from each category is thus vital in identifying the most appropriate scenario for their practical application. To achieve this, factors taken into consideration include at least local feedstock availability, cost evaluation, and product features aligned with target performance.

- **Concerns regarding materials safety**
  Albeit nanocellulose is identified "generally regarded as safe", limited research has been conducted on its impact on human/animal health and safety. As elaborated by Jiang et al. [108], there are few studies on in vivo impact of nanocellulose; for studies on in vitro impact, though they are relatively more reported, their experiments often lead to conflicting results. More efforts are thus required to clarify these unknowns. This is particularly important when using certain types of nanocellulose-based materials that could involve the interaction with surface/groundwater quality and affect the relevant ecosystems. Specific circumstances require further study include chemical release during industrial effluent cleanup and seawater desalinization, which is important to minimize any secondary contamination that is induced by the use of nanocellulose-based materials.
- **Consideration of sustainability and environmental footprint**
  Though nanocellulose is generally considered green for water treatment application due to its feedstock's sustainable nature, using nanocellulose-based materials may not necessarily meet the criteria of sustainability unless an overall life cycle assessment (LCA) is implemented. As elaborated by Jiang et al. [108] a comprehensive LCA covers chemical/water use, energy consumption, and treatment of process waste that are involved in materials synthesis and fabrication, as well as the disposal of the spent materials after use. All this information would help provide an unbiased assessment of the environmental footprint of the developed materials.

References

1. M.M. Mekonnen, A.Y. Hoekstra, Four billion people facing severe water scarcity. Sci. Adv. **2**(2), e1500323 (2016)
2. A. Prüss-Üstün, J. Wolf, C. Corvalán, R. Bos, M. Neira, Preventing disease through healthy environments: a global assessment of the burden of disease from environmental risks. World Health Organization (2016)
3. X. Xu, S. Nie, H. Ding, F.F. Hou, Environmental pollution and kidney diseases. Nat. Rev. Nephrol. **14**(5), 313 (2018)
4. P.R. Sharma, S.K. Sharma, T. Lindström, B.S. Hsiao, Nanocellulose-enabled membranes for water purification: perspectives. Adv. Sustain. Syst. **4**(5), 1900114 (2020)
5. T. Saito, R. Kuramae, J. Wohlert, L.A. Berglund, A. Isogai, An ultrastrong nanofibrillar biomaterial: the strength of single cellulose nanofibrils revealed via sonication-induced fragmentation. Biomacromol **14**(1), 248–253 (2013)
6. S. Iwamoto, W. Kai, A. Isogai, T. Iwata, Elastic modulus of single cellulose microfibrils from tunicate measured by atomic force microscopy. Biomacromol **10**(9), 2571–2576 (2009)
7. H. Yano, J. Sugiyama, A.N. Nakagaito, M. Nogi, T. Matsuura, M. Hikita, K. Handa, Optically transparent composites reinforced with networks of bacterial nanofibers. Adv. Mater. **17**(2), 153–155 (2005)
8. A. Šturcová, G.R. Davies, S.J. Eichhorn, Elastic modulus and stress-transfer properties of tunicate cellulose whiskers. Biomacromol **6**(2), 1055–1061 (2005)
9. W. Haworth, E. Hirst, H. Thomas, The existence of the cellobiose residue in cellulose. Nature **126**(3177), 438–438 (1930)
10. O. Sponsler, W. Dore, The structure of ramie cellulose as derived from X-ray data, in *Fourth Colloid Symposium Monograph* (1926), pp. 174–202
11. B. Thomas, M.C. Raj, J. Joy, A. Moores, G.L. Drisko, C.M. Sanchez, Nanocellulose, a versatile green platform: from biosources to materials and their applications. Chem. Rev. **118**(24), 11575–11625 (2018)
12. G. Metreveli, L. Wågberg, E. Emmoth, S. Belák, M. Strømme, A. Mihranyan, A size-exclusion nanocellulose filter paper for virus removal. Adv. Healthcare Mater. **3**(10), 1546–1550 (2014)
13. L. Manukyan, J. Padova, A. Mihranyan, Virus removal filtration of chemically defined Chinese Hamster Ovary cells medium with nanocellulose-based size exclusion filter. Biologicals **59**, 62–67 (2019)
14. O. Gustafsson, L. Manukyan, S. Gustafsson, G.K. Tummala, S. Zaman, A. Begum, M.A. Alfasane, K. Siddique-e-Rabbani, A. Mihranyan, Scalable and sustainable total pathogen removal filter paper for point-of-Use drinking water purification in Bangladesh. ACS Sustain. Chem. Eng. **7**(17), 14373–14383 (2019)
15. O. Gustafsson, L. Manukyan, A. Mihranyan, High-performance virus removal filter paper for drinking water purification. Global Chall. **2**(7), 1800031 (2018)
16. S. Gustafsson, P. Lordat, T. Hanrieder, M. Asper, O. Schaefer, A. Mihranyan, Mille-Feuille paper: a novel type of filter architecture for advanced virus separation applications. Mater. Horiz. **3**(4), 320–327 (2016)
17. M. Asper, T. Hanrieder, A. Quellmalz, A. Mihranyan, Removal of xenotropic murine leukemia virus by nanocellulose based filter paper. Biologicals **43**(6), 452–456 (2015)
18. S. Gustafsson, A. Mihranyan, Strategies for tailoring the pore-size distribution of virus retention filter papers. ACS Appl. Mater. Interface **8**(22), 13759–13767 (2016)
19. H. Sehaqui, B. Michen, E. Marty, L. Schaufelberger, T. Zimmermann, Functional cellulose nanofiber filters with enhanced flux for the removal of humic acid by adsorption. ACS Sustain. Chem. Eng. **4**(9), 4582–4590 (2016)
20. P. Orsolini, T. Marchesi D'Alvise, C. Boi, T. Geiger, W.R. Caseri, T. Zimmermann, Nanofibrillated cellulose templated membranes with high permeance. ACS Appl. Mater. Interface **8**(49), 33943–33954 (2016)

21. Z. Karim, S. Claudpierre, M. Grahn, K. Oksman, A.P. Mathew, Nanocellulose based functional membranes for water cleaning: tailoring of mechanical properties, porosity and metal ion capture. J. Membr. Sci. **514**, 418–428 (2016)
22. R. Wang, S. Guan, A. Sato, X. Wang, Z. Wang, R. Yang, B.S. Hsiao, B. Chu, Nanofibrous microfiltration membranes capable of removing bacteria, viruses and heavy metal ions. J. Membr. Sci. **446**, 376–382 (2013)
23. D. Zhang, A. Karkooti, L. Liu, M. Sadrzadeh, T. Thundat, Y. Liu, R. Narain, Fabrication of antifouling and antibacterial polyethersulfone (PES)/cellulose nanocrystals (CNC) nanocomposite membranes. J. Membr. Sci. **549**, 350–356 (2018)
24. X. Cao, M. Huang, B. Ding, J. Yu, G. Sun, Robust polyacrylonitrile nanofibrous membrane reinforced with jute cellulose nanowhiskers for water purification. Desalination **316**, 120–126 (2013)
25. J. Lv, G. Zhang, H. Zhang, C. Zhao, F. Yang, Improvement of antifouling performances for modified PVDF ultrafiltration membrane with hydrophilic cellulose nanocrystal. Appl. Surf. Sci. **440**, 1091–1100 (2018)
26. M. Yang, P. Hadi, X. Yin, J. Yu, X. Huang, H. Ma, H. Walker, B.S. Hsiao, Antifouling nanocellulose membranes: how subtle adjustment of surface charge lead to self-cleaning property. J. Membr. Sci. **618**, 118739 (2021)
27. D. Mohan, K.P. Singh, Granular activated carbon. Water Encyclopedia **1**, 92–107 (2005)
28. M. Jiao, Y. Yao, C. Chen, B. Jiang, G. Pastel, Z. Lin, Q. Wu, M. Cui, S. He, L. Hu, Highly efficient water treatment via a wood-based and reusable filter. ACS Mater. Lett. **2**(4), 430–437 (2020)
29. H. Xia, Z. Zhang, J. Liu, Y. Deng, D. Zhang, P. Du, S. Zhang, X. Lu, Novel Fe-Mn-O nanosheets/wood carbon hybrid with tunable surface properties as a superior catalyst for Fenton-like oxidation. Appl. Catal. B **259**, 118058 (2019)
30. G. Liu, D. Chen, R. Liu, Z. Yu, J. Jiang, Y. Liu, J. Hu, S. Chang, Antifouling wood matrix with natural water transfer and microreaction channels for water treatment. ACS Sustain. Chem. Eng. **7**(7), 6782–6791 (2019)
31. F. Chen, A.S. Gong, M. Zhu, G. Chen, S.D. Lacey, F. Jiang, Y. Li, Y. Wang, J. Dai, Y. Yao, Mesoporous, three-dimensional wood membrane decorated with nanoparticles for highly efficient water treatment. ACS Nano **11**(4), 4275–4282 (2017)
32. S. He, C. Chen, G. Chen, F. Chen, J. Dai, J. Song, F. Jiang, C. Jia, H. Xie, Y. Yao, High-performance, scalable wood-based filtration device with a reversed-tree design. Chem. Mater. **32**(5), 1887–1895 (2020)
33. W. Che, Z. Xiao, Z. Wang, J. Li, H. Wang, Y. Wang, Y. Xie, Wood-based mesoporous filter decorated with silver nanoparticles for water purification. ACS Sustain. Chem. Eng. **7**(5), 5134–5141 (2019)
34. G. Crini, E. Lichtfouse, L. Wilson, N. Morin-Crini, Green adsorbents for pollutant removal. Environ. Chem. Sustain. World 18, 23–71 (2018)
35. R.E. Abouzeid, R. Khiari, N. El-Wakil, A. Dufresne, Current state and new trends in the use of cellulose nanomaterials for wastewater treatment. Biomacromol **20**(2), 573–597 (2018)
36. N. Mahfoudhi, S. Boufi, Nanocellulose as a novel nanostructured adsorbent for environmental remediation: a review. Cellulose **24**(3), 1171–1197 (2017)
37. A.J. Cramer, J.M. Cole, Removal or storage of environmental pollutants and alternative fuel sources with inorganic adsorbents via host–guest encapsulation. J. Mater. Chem. A **5**(22), 10746–10771 (2017)
38. W. Maatar, S. Boufi, Poly (methacylic acid-co-maleic acid) grafted nanofibrillated cellulose as a reusable novel heavy metal ions adsorbent. Carbohyd. Polym. **126**, 199–207 (2015)
39. K. Köse, M. Mavlan, J.P. Youngblood, Applications and impact of nanocellulose based adsorbents. Cellulose **27**(6), 2967–2990 (2020)
40. C.Y. Yin, M.K. Aroua, W.M.A.W. Daud, Review of modifications of activated carbon for enhancing contaminant uptakes from aqueous solutions. Sep. Purif. Technol. **52**(3), 403–415 (2007)

41. G. Reshma, C. Reshmi, S.V. Nair, D. Menon, Superabsorbent sodium carboxymethyl cellulose membranes based on a new cross-linker combination for female sanitary napkin applications. Carbohyd. Polym. **248**, 116763 (2020)
42. H. Wang, Y. Wu, C. Cui, J. Yang, W. Liu, Antifouling super water absorbent supramolecular polymer hydrogel as an artificial vitreous body. Adv. Sci. **5**(11), 1800711 (2018)
43. M.-Y. Chang, R.-S. Juang, Adsorption of tannic acid, humic acid, and dyes from water using the composite of chitosan and activated clay. J. Colloid Interface Sci. **278**(1), 18–25 (2004)
44. A. Bhatnagar, M. Sillanpää, A. Witek-Krowiak, Agricultural waste peels as versatile biomass for water purification—a review. Chem. Eng. J. **270**, 244–271 (2015)
45. H. Dai, H. Huang, Enhanced swelling and responsive properties of pineapple peel carboxymethyl cellulose-g-poly (acrylic acid-co-acrylamide) superabsorbent hydrogel by the introduction of carclazyte. J. Agric. Food Chem. **65**(3), 565–574 (2017)
46. X. Qi, W. Wei, T. Su, J. Zhang, W. Dong, Fabrication of a new polysaccharide-based adsorbent for water purification. Carbohyd. Polym. **195**, 368–377 (2018)
47. R. Reshmy, D. Thomas, E. Philip, S.A. Paul, A. Madhavan, R. Sindhu, P. Binod, A. Pugazhendhi, R. Sirohi, A. Tarafdar, Potential of nanocellulose for wastewater treatment. Chemosphere **281**, 130738 (2021)
48. A.W. Carpenter, C.-F. de Lannoy, M.R. Wiesner, Cellulose nanomaterials in water treatment technologies. Environ. Sci. Technol. **49**(9), 5277–5287 (2015)
49. R. Reshmy, E. Philip, S.A. Paul, A. Madhavan, R. Sindhu, P. Binod, A. Pandey, R. Sirohi, Nanocellulose-based products for sustainable applications-recent trends and possibilities. Rev. Environ. Sci. Bio/Technol. **2020**, 1–28
50. H. Voisin, L. Bergström, P. Liu, A.P. Mathew, Nanocellulose-based materials for water purification. Nanomaterials **7**(3), 57 (2017)
51. R. Kecili, C.M. Hussain, *Mechanism of Adsorption on Nanomaterials* (Nanomater, Chromatogr, Elsevier, 2018), pp. 89–115
52. D.M. Ruthven, *Principles of Adsorption and Adsorption Processes* (Wiley, 1984)
53. N. Srivastava, I. Eames, A review of adsorbents and adsorbates in solid–vapour adsorption heat pump systems. Appl. Therm. Eng. **18**(9-10), 707–714 (1998)
54. J.N. Putro, A. Kurniawan, S. Ismadji, Y.-H. Ju, Nanocellulose based biosorbents for wastewater treatment: study of isotherm, kinetic, thermodynamic and reusability. Environ. Nanotechnol., Monit. Manage. **8**, 134–149 (2017)
55. S. Hokkanen, A. Bhatnagar, M. Sillanpää, A review on modification methods to cellulose-based adsorbents to improve adsorption capacity. Water Res. **91**, 156–173 (2016)
56. I. Langmuir, The adsorption of gases on plane surfaces of glass, mica and platinum. J. Am. Chem. Soc. **40**(9), 1361–1403 (1918)
57. A. Proctor, J. Toro-Vazquez, The Freundlich isotherm in studying adsorption in oil processing. J. Am. Oil Chem. Soc. **73**(12), 1627–1633 (1996)
58. S. Lagergren, Zur theorie der sogenannten adsorption geloster stoffe (1898)
59. G. Blanchard, M. Maunaye, G. Martin, Removal of heavy metals from waters by means of natural zeolites. Water Res. **18**(12), 1501–1507 (1984)
60. A. Qiao, M. Cui, R. Huang, G. Ding, W. Qi, Z. He, J.J. Klemeš, R. Su, Advances in nanocellulose-based materials as adsorbents of heavy metals and dyes. Carbohydr. Polym. **272**, 118471 (2021)
61. V. Gupta, P. Carrott, R. Singh, M. Chaudhary, S. Kushwaha, Cellulose: a review as natural, modified and activated carbon adsorbent. Bioresour. Technol. **216**, 1066–1076 (2016)
62. R. Batmaz, N. Mohammed, M. Zaman, G. Minhas, R.M. Berry, K.C. Tam, Cellulose nanocrystals as promising adsorbents for the removal of cationic dyes. Cellulose **21**(3), 1655–1665 (2014)
63. T. Saito, A. Isogai, Ion-exchange behavior of carboxylate groups in fibrous cellulose oxidized by the TEMPO-mediated system. Carbohydr. Polym. **61**(2), 183–190 (2005)
64. H. Sehaqui, U.P. de Larraya, P. Liu, N. Pfenninger, A.P. Mathew, T. Zimmermann, P. Tingaut, Enhancing adsorption of heavy metal ions onto biobased nanofibers from waste pulp residues for application in wastewater treatment. Cellulose **21**(4), 2831–2844 (2014)

65. H. Yang, A. Sheikhi, T.G. Van De Ven, Reusable green aerogels from cross-linked hairy nanocrystalline cellulose and modified chitosan for dye removal. Langmuir **32**(45), 11771–11779 (2016)
66. P. Liu, P.F. Borrell, M. Božič, V. Kokol, K. Oksman, A.P. Mathew, Nanocelluloses and their phosphorylated derivatives for selective adsorption of Ag+, Cu2+ and Fe3+ from industrial effluents. J. Hazard. Mater. **294**, 177–185 (2015)
67. J.A. Sirviö, T. Hasa, T. Leiviskä, H. Liimatainen, O. Hormi, Bisphosphonate nanocellulose in the removal of vanadium (V) from water. Cellulose **23**(1), 689–697 (2016)
68. R. Yang, K.B. Aubrecht, H. Ma, R. Wang, R.B. Grubbs, B.S. Hsiao, B. Chu, Thiol-modified cellulose nanofibrous composite membranes for chromium (VI) and lead (II) adsorption. Polymer **55**(5), 1167–1176 (2014)
69. B. Geng, H. Wang, S. Wu, J. Ru, C. Tong, Y. Chen, H. Liu, S. Wu, X. Liu, Surface-tailored nanocellulose aerogels with thiol-functional moieties for highly efficient and selective removal of Hg (II) ions from water. ACS Sustainable Chem. Eng. **5**(12), 11715–11726 (2017)
70. K. Singh, J.K. Arora, T.J.M. Sinha, S. Srivastava, Functionalization of nanocrystalline cellulose for decontamination of Cr (III) and Cr (VI) from aqueous system: computational modeling approach. Clean Technol. Environ. Policy **16**(6), 1179–1191 (2014)
71. L. Jin, W. Li, Q. Xu, Q. Sun, Amino-functionalized nanocrystalline cellulose as an adsorbent for anionic dyes. Cellulose **22**(4), 2443–2456 (2015)
72. A. Pei, N. Butchosa, L.A. Berglund, Q. Zhou, Surface quaternized cellulose nanofibrils with high water absorbency and adsorption capacity for anionic dyes. Soft Matter **9**(6), 2047–2055 (2013)
73. K. Song, H. Xu, L. Xu, K. Xie, Y. Yang, Cellulose nanocrystal-reinforced keratin bioadsorbent for effective removal of dyes from aqueous solution. Bioresour. Technol. **232**, 254–262 (2017)
74. X. He, K.B. Male, P.N. Nesterenko, D. Brabazon, B. Paull, J.H. Luong, Adsorption and desorption of methylene blue on porous carbon monoliths and nanocrystalline cellulose. ACS Appl. Mater. Interface **5**(17), 8796–8804 (2013)
75. B. Liu, L. Zhang, H. Wang, Z. Bian, Preparation of MCC/MC silica sponge and its oil/water separation apparatus application. Ind. Eng. Chem. Res. **56**(20), 5795–5801 (2017)
76. W.-J. Yang, A.C.Y. Yuen, A. Li, B. Lin, T.B.Y. Chen, W. Yang, H.-D. Lu, G.H. Yeoh, Recent progress in bio-based aerogel absorbents for oil/water separation. Cellulose **26**(11), 6449–6476 (2019)
77. H. Jin, M. Kettunen, A. Laiho, H. Pynnönen, J. Paltakari, A. Marmur, O. Ikkala, R.H. Ras, Superhydrophobic and superoleophobic nanocellulose aerogel membranes as bioinspired cargo carriers on water and oil. Langmuir **27**(5), 1930–1934 (2011)
78. Z. Zhang, G. Sèbe, D. Rentsch, T. Zimmermann, P. Tingaut, Ultralightweight and flexible silylated nanocellulose sponges for the selective removal of oil from water. Chem. Mater. **26**(8), 2659–2668 (2014)
79. H. Gu, X. Zhou, S. Lyu, D. Pan, M. Dong, S. Wu, T. Ding, X. Wei, I. Seok, S. Wei, Magnetic nanocellulose-magnetite aerogel for easy oil adsorption. J. Colloid Interface Sci. **560**, 849–856 (2020)
80. F. Jiang, Y.-L. Hsieh, Amphiphilic superabsorbent cellulose nanofibril aerogels. J. Mater. Chem. A **2**(18), 6337–6342 (2014)
81. Y. Meng, T.M. Young, P. Liu, C.I. Contescu, B. Huang, S. Wang, Ultralight carbon aerogel from nanocellulose as a highly selective oil absorption material. Cellulose **22**(1), 435–447 (2015)
82. S. Long, Y. Feng, F. He, S. He, H. Hong, X. Yang, L. Zheng, J. Liu, L. Gan, M. Long, An ultralight, supercompressible, superhydrophobic and multifunctional carbon aerogel with a specially designed structure. Carbon **158**, 137–145 (2020)
83. W.U.J.W. Supply, S.M. Programme, W.H. Organization, Progress on sanitation and drinking water: 2015 update and MDG assessment. World Health Organization (2015)
84. H.M. Qiblawey, F. Banat, Solar thermal desalination technologies. Desalination **220**(1–3), 633–644 (2008)

85. P. Tao, G. Ni, C. Song, W. Shang, J. Wu, J. Zhu, G. Chen, T. Deng, Solar-driven interfacial evaporation. Nat. Energy **3**(12), 1031–1041 (2018)
86. A. Kabeel, S. El-Agouz, Review of researches and developments on solar stills. Desalination **276**(1–3), 1–12 (2011)
87. R. Prasher, P.E. Phelan, P. Bhattacharya, Effect of aggregation kinetics on the thermal conductivity of nanoscale colloidal solutions (nanofluid). Nano Lett. **6**(7), 1529–1534 (2006)
88. F. Yu, Y. Chen, X. Liang, J. Xu, C. Lee, Q. Liang, P. Tao, T. Deng, Dispersion stability of thermal nanofluids. Progr. Nat. Sci. Mater. Int. **27**(5), 531–542 (2017)
89. C. Chen, Y. Kuang, L. Hu, Challenges and opportunities for solar evaporation. Joule **3**(3), 683–718 (2019)
90. M.L. Brongersma, N.J. Halas, P. Nordlander, Plasmon-induced hot carrier science and technology. Nat. Nanotechnol. **10**(1), 25–34 (2015)
91. H.H. Richardson, M.T. Carlson, P.J. Tandler, P. Hernandez, A.O. Govorov, Experimental and theoretical studies of light-to-heat conversion and collective heating effects in metal nanoparticle solutions. Nano Lett. **9**(3), 1139–1146 (2009)
92. G. Zhu, J. Xu, W. Zhao, F. Huang, Constructing black titania with unique nanocage structure for solar desalination. ACS Appl. Mater. Interface **8**(46), 31716–31721 (2016)
93. J. Wang, Y. Li, L. Deng, N. Wei, Y. Weng, S. Dong, D. Qi, J. Qiu, X. Chen, T. Wu, High-performance photothermal conversion of narrow-bandgap Ti2O3 nanoparticles. Adv. Mater. **29**(3), 1603730 (2017)
94. R. Chen, Z. Wu, T. Zhang, T. Yu, M. Ye, Magnetically recyclable self-assembled thin films for highly efficient water evaporation by interfacial solar heating. RSC Adv. **7**(32), 19849–19855 (2017)
95. Y. Shi, R. Li, Y. Jin, S. Zhuo, L. Shi, J. Chang, S. Hong, K.-C. Ng, P. Wang, A 3D photothermal structure toward improved energy efficiency in solar steam generation. Joule **2**(6), 1171–1186 (2018)
96. F. Jiang, H. Liu, Y. Li, Y. Kuang, X. Xu, C. Chen, H. Huang, C. Jia, X. Zhao, E. Hitz, Lightweight, mesoporous, and highly absorptive all-nanofiber aerogel for efficient solar steam generation. ACS Appl. Mater. Interface **10**(1), 1104–1112 (2018)
97. R. Hu, J. Zhang, Y. Kuang, K. Wang, X. Cai, Z. Fang, W. Huang, G. Chen, Z. Wang, A Janus evaporator with low tortuosity for long-term solar desalination. J. Mater. Chem. A **7**(25), 15333–15340 (2019)
98. M. Zhu, Y. Li, G. Chen, F. Jiang, Z. Yang, X. Luo, Y. Wang, S.D. Lacey, J. Dai, C. Wang, Tree-inspired design for high-efficiency water extraction. Adv. Mater. **29**(44), 1704107 (2017)
99. C. Jia, Y. Li, Z. Yang, G. Chen, Y. Yao, F. Jiang, Y. Kuang, G. Pastel, H. Xie, B. Yang, Rich mesostructures derived from natural woods for solar steam generation. Joule **1**(3), 588–599 (2017)
100. M. Zhu, Y. Li, F. Chen, X. Zhu, J. Dai, Y. Li, Z. Yang, X. Yan, J. Song, Y. Wang, Plasmonic wood for high-efficiency solar steam generation. Adv. Energy Mater. **8**(4), 1701028 (2018)
101. S. He, C. Chen, Y. Kuang, R. Mi, Y. Liu, Y. Pei, W. Kong, W. Gan, H. Xie, E. Hitz, Nature-inspired salt resistant bimodal porous solar evaporator for efficient and stable water desalination. Energy Environ. Sci. **12**(5), 1558–1567 (2019)
102. Y. Kuang, C. Chen, S. He, E.M. Hitz, Y. Wang, W. Gan, R. Mi, L. Hu, A high-performance self-regenerating solar evaporator for continuous water desalination. Adv. Mater. **31**(23), 1900498 (2019)
103. M. Nasrollahzadeh, M. Sajjadi, S. Iravani, R.S. Varma, Starch, cellulose, pectin, gum, alginate, chitin and chitosan derived (nano) materials for sustainable water treatment: A review. Carbohydr. Polym. **251**, 116986 (2021)
104. P. Carrott, M.R. Carrott, Lignin–from natural adsorbent to activated carbon: a review. Bioresour. Technol. **98**(12), 2301–2312 (2007)
105. S. Zhu, J. Xu, Y. Kuang, Z. Cheng, Q. Wu, J. Xie, B. Wang, W. Gao, J. Zeng, J. Li, Lignin-derived sulfonated porous carbon from cornstalk for efficient and selective removal of cationic dyes. Ind. Crops Prod. **159**, 113071 (2021)

106. Y. Zhu, M. Chae, B. Adhikari, V. Khatri, H. Kaminsky, P. Mussone, D.C. Bressler, Valorizing Biowaste for Wastewater Treatment: Dewatering Sludges Using Specified Risk Material-Based Flocculants for Industrial Sustainability. ACS Sustainable Chem. Eng. **9**(5), 2037–2046 (2021)
107. Y. Zhu, M. Chae, J. Wang, B. Adhikari, P. Mussone, D.C. Bressler, Biowaste-based biodegradable flocculants for clean and sustainable tailings management in industrial mining and mineral processing. J. Cleaner Prod. **323**, 129195 (2021)
108. J. Jiang, Y. Zhu, F. Jiang, Sustainable isolation of nanocellulose from cellulose and lignocellulosic feedstocks: recent progress and perspectives. Carbohydr. Polym. **267**, 118188 (2021)

Chapter 11
Engineering of Nanocellulose Thin Films for Triboelectric Nanogenerator Development

Xudong Wang, Long Gu, and Chunhua Yao

Abstract Cellulose is the most abundant natural polymer on earth, providing a sustainable green resource that is renewable, degradable, biocompatible and cost-effective. The nanometer-sized cellulose, e.g. cellulose nanofibrils (CNFs) provides wide processing possibilities and greater application potential as a functional soft material building block. This chapter reviews a series of recent research from our group regarding the functionalization, fabrication, and assembly of CNFs aiming at triboelectric nanogenerator (TENG) development for mechanical energy harvesting. Specifically, TENG made from a flexible and compact CNF film with nanoscale surface roughness is introduced as a promising strategy for harvesting energy from footsteps. Chemical reaction approaches can be employed to introduce different functionalities thus different triboelectric polarities to CNFs, and thus significantly improve the TENG's output. The output quantification and Figure of Merit (FOM) characterization are discussed in detail. Converting CNF films into conductive transparent films by aluminum-doped zinc oxide (AZO) coating is then reviewed. The mesoporous structure of the CNF film renders strong adhesion of the AZO layer and exhibits excellent mechanical integrity and electrical conductivity within a wide range of tensile and compressive strains, which provides a new transparent and degradable TENG material with decent output. At last, we briefly discuss the perspectives and opportunities of using CNFs as a new and potential functional soft material component in energy-related fields.

11.1 Introduction

Cellulose is the most abundant natural polymer on Earth, representing ~ 10^{12} tons of total annual biomass production [1, 2]. It is used extensively in various applications, commonly including paper, textiles, chemical raw material, and pharmaceutical filler [3]. More recently, cellulose is also applied in enzymatic hydrolysis for biofuels [4],

X. Wang (✉) · L. Gu · C. Yao
Department of Materials Science and Engineering, University of Wisconsin-Madison, Madison, WI 53706, USA
e-mail: xudong.wang@wisc.edu

L. Hu et al. (eds.), *Emerging Nanotechnologies in Nanocellulose*, NanoScience and Technology, https://doi.org/10.1007/978-3-031-14043-3_11

and biomedical implant and microsurgery [5–7]. Cellulose embodies the versatile properties such as biodegradability [8–10], biocompatibility [5], hydrophilicity [5, 11], and chirality [12–14]. Many of its applications and properties are dependent on its specific molecular structure, supramolecular structure, and morphology.

As various cellulose structures obtained from conventional micrometer scale cellulose find wide applications, we can expect greater potential with structures fabricated from nanometer scale cellulose. Extracting cellulose at a nanometer scale and then building structures from these nanosized building blocks allows for better control of material properties. Recently, researchers have successfully disintegrated nanoscale cellulose from cellulose source materials, such as wood pulp and microcrystalline cellulose. Nanosized fibrils and crystallites were obtained, and termed as cellulose nanofibrils (CNF) and cellulose nanocrystals (CNCs), respectively [15–18]. Chemical treatment and mechanical grinding were performed for disintegration. During these processes, the micrometer scale cellulose fiber aggregates were broken down to nanoscale, and charges were introduced to the surface of cellulose. These combined effects resulted in the uniform dispersion of nanocellulose in water as a stable colloidal suspension.

Compared to conventional cellulose, CNF possesses a higher aspect ratio, larger specific surface area, less hierarchical structure-induced material defects, and better dispersity in water thus better processibility for various structures [17–20]. Therefore, CNF are more and more applied in new areas that require novel and improved material properties, such as fiber reinforced composites, [21, 22] flexible and transparent electronics, [23, 24] strain sensing [25], and oxygen and water vapor barrier [25, 26].

Triboelectric nanogenerator (TENG) is a new technology for harvesting mechanical energy by the combined effects of triboelectrification and electrostatic induction [27, 28]. The TENGs developed so far are mostly built on synthetic, non-recyclable materials which demonstrate a strong tendency of either attracting or repelling electrons. Cellulose, as a very attractive natural polymer with multiple functions, can be a good TENG material candidate as well. As it contains abundant oxygen atoms, cellulose exhibits a tendency of losing electrons and becomes triboelectric positive [21]. In this chapter, we will discuss the application of CNF as an active layer to develop flexible and transparent TENGs. First, triboelectric positive CNF films can be paired with triboelectric negative films, such as FEP to fabricate an efficient TENG. An all-cellulose TENG was then developed by chemically functionalizing pristine nanocellulose towards opposite triboelectric polarity. The metal charge collector on CNF films could also be replaced by a water-dissolvable aluminum-doped zinc oxide (AZO) thin film. Thanks to the mesoporous structure of the CNF film, the AZO layer was strongly adhered on the CNF film and presented excellent electrical conductivity and mechanical integrity within a wide range of tensile and compressive strains. The usage of such AZO-CNF films allowed the development of flexible, transparent and naturally degradable TENGs. In general, the series of development of nanocellulose and its derivatives for TENG applications provided a promising material platform for developing high-performance mechanical energy harvesting devices from renewable and cost-effective natural energy resources.

11.2 Triboelectric Nanogenerator and Mechanical Energy Harvesting

With the rapid development of the Internet of Things, harvesting ambient mechanical energy is becoming particularly attractive for powering the increasing numbers of small and distributed electronics due to its abundant, renewable, and the most accessible nature [27]. Among many efforts on mechanical energy harvesting, nanogenerator was invented on the basis of the piezoelectric effect. Delicate piezoelectric nanostructures as the functional building blocks provide the nanogenerator with the advantages of high sensitivity, potential superior efficiency and long lifetime [29]. Over the past decade, lots of efforts have been made to develop high-performance piezoelectric nanogenerators for harvest and utilization of various forms of mechanical energy, such as physical deflection [30, 31], acoustic waves [32, 33], air and fluid flow [34–36], and body movements [37–39]. In 2012, triboelectric nanogenerator (TENG), an emerging mechanical energy harvesting technology, was invented based on the principles of traditional electrostatic energy harvesters by Wang's group [40]. Due to the simple structure and easy scalable/integratable nature of TENG, its area and volume output power density were rapidly improved to 500 W/m^2 and 15 MW/m^3 respectively within a few years. [41] An instantaneous energy conversion efficiency of 70.6% and a total energy conversion efficiency of 85% were achieved [42, 43]. Benefitting from its high output and energy conversion efficiency, various small commercial electronics were successfully powered by TENGs, such as smart phones [44, 45], light bulbs [41, 45], alarm systems [46, 47], cardiac pacemakers [48], large numbers of sensors [49–54], and light-emitting diodes [55–60]. Generally, TENGs have four basic working modes according to its configuration: vertical contact-separation mode, contact-sliding mode, single-electrode mode, and freestanding triboelectric layer mode. Multiple working modes make TENGs an ideal choice for harvesting various forms of mechanical energy, such as water waves [61–63], air flow [64, 65], water droplets [56, 66], vehicle [58, 67] and human movements [43, 68, 69]. These two advantages make TENGs a promising way for the development of practical self-powered systems as well as the large-scale generation of electricity. Some recent reviews also extensively summarized the development and broad applications of TENGs [27, 29, 70–74].

The operation of TENGs is based on the coupling effect of triboelectrification and electrostatic induction (Fig. 11.1). Triboelectrification is an ancient phenomenon that a material will be electrically charged when it physically contacts with a different one, which was discovered 2600 years ago. When two materials with different triboelectric polarity contacts each other, electrons, holes or charged ions/molecules will transfer from one to the other so that the surface potential difference between them can be balanced. This charge transfer during the contact electrification process drives the conversion of mechanical energy to electric energy in TENGs. The common used positive materials in TENGs include polyamides [75], metals [41], indium tin oxide (ITO) [76], and zinc oxide [77], and negative materials are fluorinated ethylene propylene (FEP) [11, 12], polytetrafluoroethylene (PTFE) [75] and polyvinylidene fluoride

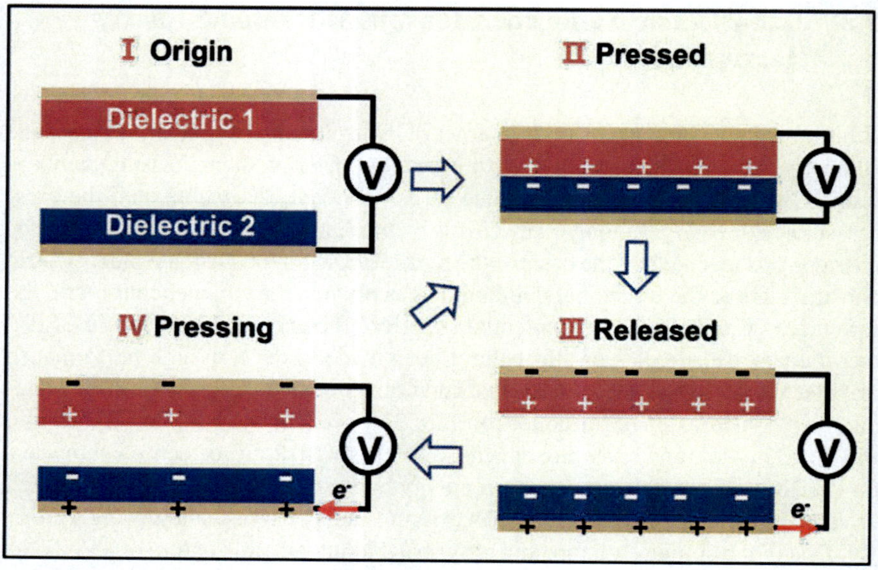

Fig. 11.1 Schematic diagram illustrating the operating mechanism of TENG

(PVDF) [41], polydimethylsiloxane (PDMS) [76–79], and polyethylene terephthalate (PET) [78–81]. Triboelectric materials in a TENG are usually attached to conductive electrodes to form a capacitor like structure, in which electrostatic charges are induced by triboelectric charges. When the capacitance of the TENG is changed by an external force, free charges flow between the two conductive electrodes and thereby outputting electric power. Therefore, increasing the surface charge density on the triboelectric materials is considered as the most fundament strategy for enhancing the output performance of TENGs. Generally, the surface charge density mainly depends on the number of charge carrying locations and their ability of attracting or repelling electrons. The two factors can be effectively regulated by morphological and chemical modification of the triboelectric polymers. By enlarging the surface roughness factor, the amount of charge carrying sites can be simply increased. For example, morphologies like nanowires, [82–84] nanoparticles [57, 85] or other nanoscale patterns [78, 86–91] are often chosen for raising the surface area of triboelectric materials and thereby achieving more charge carrying sites. As for the tendency of these charge carrying sites, it can be regulated through changing their chemical properties. Cellulose-based thin films, offer both morphological and chemical merits owing to their nanoscale fibrous surface features and high-density surface functional groups, respectively. Therefore, investigations and applications of cellulose-based materials for TENGs bring unique merits for improving the device performance as well as the eco-friendliness.

11.3 Nanocellulose Based Triboelectric Nanogenerator Development

Because of containing abundant oxygen atoms, natural cellulose shows a tendency to lose electrons and thus turns into positively charged [92] Therefore, natural cellulose can be used to fabricate TENGs by assembling with lots of triboelectric negative materials, such as FEP, PTFE, and PET. The first cellulose-based TENG was developed using nanostructured CNF thin films with a good flexible and transparent property. In this TENG, CNF acted as the triboelectric positive material and FEP was selected as the triboelectric negative material. The device worked in vertical contact-separation mode.

11.3.1 Fabrication of TENG Using CNF Thin Films

The nanocellulose thin film as a triboelectric active layer was obtained by filtering a water-diluted CNF hydrogel under an air pressure of 0.55 MPa. After pouring the slurry into a filtration chamber with a PTFE membrane, the water molecule is passed through the membrane, and finally leaves behind a CNF filter cake. After peeling off from the PTFE membrane, the CNF cake was dried at room temperature by putting it between layers of waxy coated paper, filter paper, and caul plates. After natural drying for several hours, the solid content of the film reached 80–90%, and it was placed in a 60 °C oven for further drying to obtain the CNF film.

During the thin film fabrication, when water was removed from the CNF suspension, the adjacent individual nanofibrils were pulled together by the strong hydrogen bonds between the hydroxyl groups on the surfaces of the nanofibrils, and a dense and flexible film can thus be obtained. Due to the fine nanofibril diameter, the film is also featured by optical transparency. Scanning electronic microscopy (SEM) image (inset of Fig. 11.2a) showed that the film was quite flat and compact. Atomic force microscopy (AFM) topography scan of the film surface showed obvious fibrous features in the size scale of ~200 nm (Fig. 11.2b). The maximum height difference was about 300 nm, demonstrating a surface roughness of a few hundreds of nanometers.

To assemble a TENG unit, an indium tin oxide/polyethylene terephthalate (ITO/PET) was used as the substrate. Then, an as-prepared CNF film was attached to ITO/PET film as the top electrode (Fig. 11.2c). As for the bottom electrode, another polymer film with the same size was attached on another ITO/PET substrate. This polymer film can be $TiCl_4$ treated CNF, pristine CNF, or FEP. Here, to obtain $TiCl_4$ treated CNF, $TiCl_4$ vapor was pulsed on CNF substrate for 30 cycles in an ALD system. The two electrodes were 1 mm separated using spacers and other compensation films, and connected to the external circuit through copper tapes.

In the CNF-based TENG configuration, FEP was chosen as the counter triboelectric material because it is one of the most tribo-negative polymers due to the

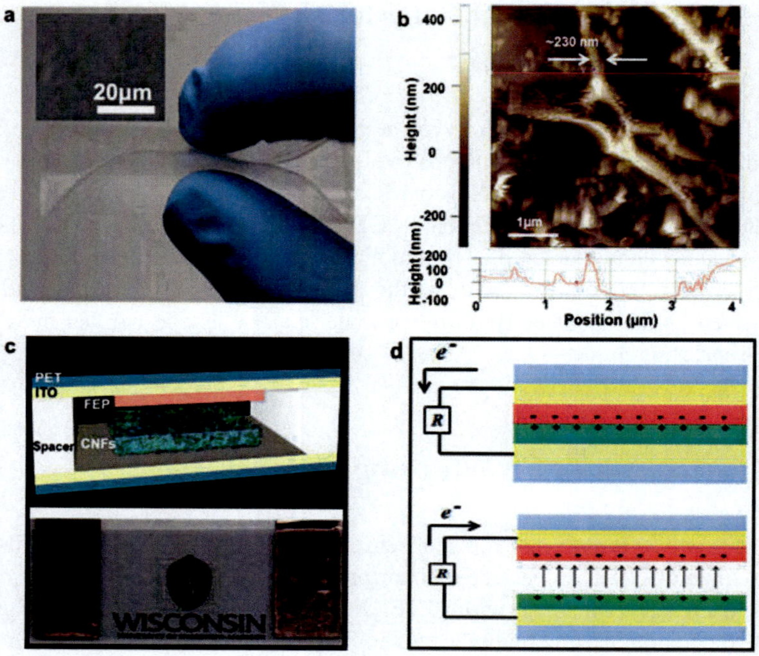

Fig. 11.2 CNF film and CNF based TENG. **a** Flexible CNF film. Inset is the SEM image of the CNF film. **b** AFM topography scan of CNF film. **c** Schematic diagram and a photo of the TENG testing sample which uses CNF film as an active layer. **d** Schematic diagram illustrating the operating principle of the TENG. Reproduced with permission [93]. Copyright 2016, Elsevier

electron-withdrawing ability of the fluorine element. For comparison, $TiCl_4$-treated CNF film and pristine CNF film were also used as the other surface, respectively. As schematically shown in Fig. 11.2d, CNF and FEP have a tendency in donating and withdrawing electrons, respectively. As a result of the huge difference in electron affinity, electrons are transferred from CNF to FEP upon the contact of CNF film and FEP film. Then, the CNF and FEP films accumulated a lot of positive charges and negative charges respectively, offering TENG with primary electrical energy. When the two triboelectric surfaces were separated, the stored electrical energy generated a potential difference and induced current flow in the external circuit, which was recorded as output current pulse and output voltage pulse. Furthermore, when putting two identical CNF films without any modification or functionalization together, there are no redistribution of surface charge with negligible electric outputs. The CNF-$TiCl_4$ treated CNF pair was expected to act in a similar way as CNF-FEP pair, since the chlorine element in the $TiCl_4$ treated CNF film can possibly increase the electron affinity thus triboelectrically negativity of the material with respect to the pristine CNF film.

11.3.2 Electric Output of CNF-Based TENG

The electric output performance of the TENG was evaluated by providing a periodical external force upon the top electrode, which was pushed to rub against the CNF film at the bottom with a fixed frequency. Meanwhile, the open-circuit voltage (V_{oc}) and the short-circuit current (I_{sc}) between the top and bottom electrodes were measured, respectively. Figure 11.3a, b are the experimental results of I_{sc} and V_{oc} of the TENGs with the three counter electrode configurations. An individual current peak was inserted in Fig. 11.3a, which clearly shows the profile of the $I_{sc} - t$ curve of the CNF-FEP pair in the triboelectricity process. In the measurement process, the electrodes attached with the CNF and FEP films were connected to the positive and negative electrodes of the current meter respectively. Thus, a positive output peak represents the current flowing from the pristine CNF to FEP, and vice versa. When both surfaces were brought into contact, there will be a positive current pulse generated at first, which suggests that electrons are transferred from CNF electrode to FEP electrode. And then, once the two surfaces were separated, the local charges were disequilibrium again and led electrons to flow back, generating a negative current pulse. The average peak values of the I_{sc} ~ 7 μA and V_{oc} ~ 5 V were achieved for the CNF-FEP TENG. These values are comparable to the performance of TENGs composed of two typical triboelectric materials with different electron affinities such as PDMS-ITO [78] and teflon-metal [94]. For the untreated CNF pair, 0.1 μA of I_{sc} and 0.1 V of V_{oc} were detected. Although the CNF pair should not exhibit any electrical output theoretically, the small output value may be due to subtle changes in terms of size, thickness, and surface roughness of CNF film. The device composed of CNF-$TiCl_4$ treated CNF pair showed 0.4 μA of I_{sc} and 4 V of V_{oc}. These values are a little lower than the CNF-FEP pair, but still high enough to be regarded as good triboelectric performance. The $TiCl_4$ treatment provided a simple pathway to tune the electron affinity of CNF.

Figure 11.3c shows a surface area-related V_{oc} variation. Here, five TENGs were made of CNF/FEP films with different active surface areas (A) between 1 cm^2 to 40 cm^2. In order to better understand the V_{oc} variation toward the surface area, the average peak value of V_{oc} was calculated and plotted in Fig. 11.3d as a function of the active surface area using solid circles. Because the ratio between initial surface charge Q_0 and active surface area A remained constant for the TENG, V_{oc} should be a constant with the change of A according to the equation $V = \frac{Q_0}{C_0} = \frac{Q_0 d}{A\varepsilon}$. This voltage-surface area relation was most likely the result of TENG's discharge behavior in a resistor-capacitor (RC) circuit, in which the oscilloscope's resistance is not infinitely large. As C is positively correlated with A, the TENG with a larger area could have a greater time constant ($\tau = RC$). Consequently, when giving all TENGs with a certain initial V_{oc}, the $V \sim t$ correlation curves can be extrapolated. TENG with larger A decays slowly as shown in Fig. 11.3e. Given the limited loading resistance and data sampling rate (1.6×10^{-5} s) of the oscilloscope, even for the same initial V_{oc} for all TENGs, their A-related voltage values could be well recorded at any moment t ($t > 0$). In Fig. 11.3d, the curve with blue blank circle was the voltage

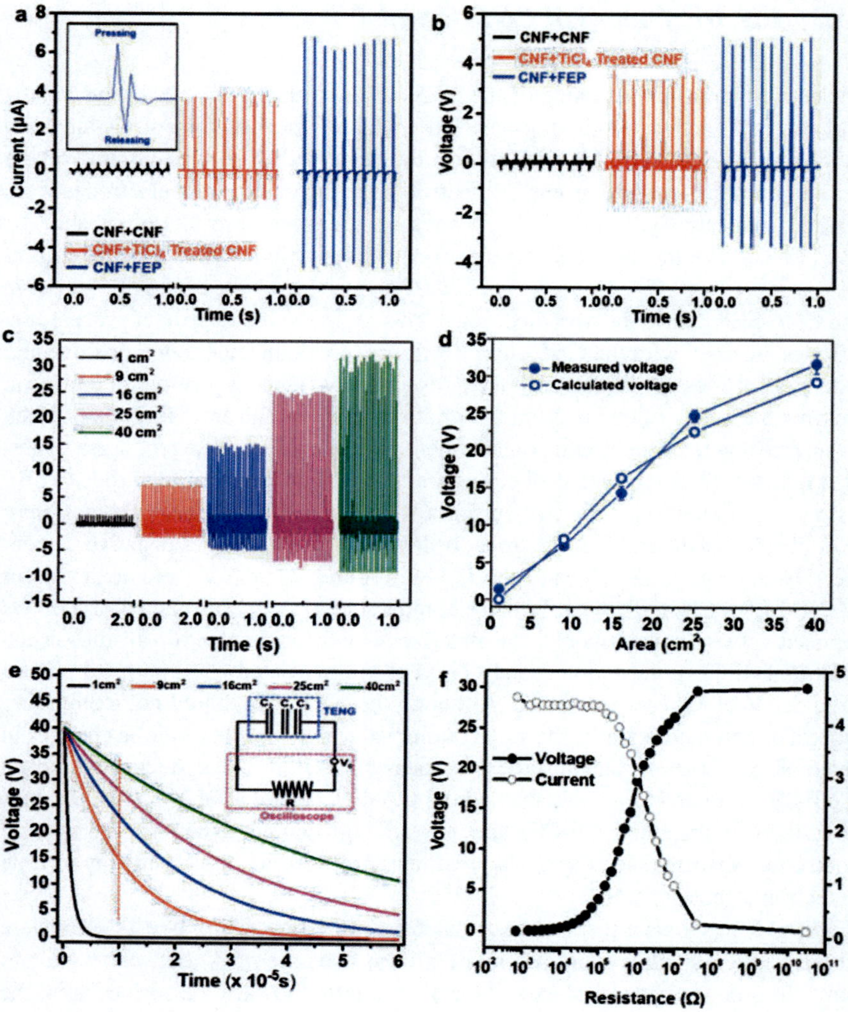

Fig. 11.3 Performance of the CNF based TENGs. **a** Short-circuit current (I_{sc}) and **b** open-circuit voltage (V_{oc}) output of three TENGs. All three TENGs used CNF film as an active layer, the other layer is CNF film, $TiCl_4$ treated CNF film, and FEP film, respectively. **c** Open-circuit voltage of the CNF-FEP TENG depending on the active area of CNF film surface, **d** Measured voltage output (dots) and calculated voltage output (circles) as a function of the active area of CNF film. **e** Calculated *RC* discharging curves for TENGs with a different active area of CNF film. **f** Open-circuit voltage and short-circuit current depending on the resistance. The inset in (**a**) is the enlarged short-circuit current output created by one pressing and releasing cycle. Reproduced with permission [93]. Copyright 2016, Elsevier

value calculated at $t = 1.6 \times 10^{-5}$ s (shown by the red dotted line in Fig. 11.3e), which has a good matching trend with the experimental value. The matching test of external loading resistance was done to evaluate the power output of the CNF-based TENG. In this experiment, the thickness and surface areas of the CNF film in the used TENG were 70 μm and 40 cm^2, respectively. Figure 11.3f exhibits the peak value of V_{oc} and I_{sc} at different external load resistance. In common with other TENGs, when increasing the load resistance, the V_{oc} increased monotonically, but the I_{sc} was reduced due to the Ohmic loss. Corresponding output power calculated from the product of V_{oc} and I_{sc} yields an optimal instantaneous power output of 0.56 mW at ~1 MΩ external load.

The CNF TENG was further used to charge a capacitor to investigate its possibility of serving as a direct power source [95]. The test was performed through directly connecting the CNF TENG with the capacitor, and no rectification unit was necessary. While the TENG converted energy in mechanical motions to electric energy, the voltage on the capacitor was measured every 5 s until the capacitor reached its saturation voltage. Figure 11.4a shows the charging behavior of a 10 μF capacitor. As expected for a capacitor that is charging when connected to a power source, a higher frequency yields a higher saturation voltage, and the lowest capacitance capacitor saturates faster than the others. 33 μF and 68 μF capacitors were also charged and demonstrated the same trends as the 10 μF capacitors. Moreover, the electrical energy stored in the capacitor was calculated with the equation $E = CU^2/2$, and it was derived from $P = CU^2/2t$, where P is the power needed to charge a capacitor of capacitance C, until it reaches a saturation voltage U on a time t. The result of this calculation was plotted in Fig. 11.4b. It can be found that in the tested range, the output power monotonically increased from 0.03 to 5.3 μW when increasing the frequency. The inset of Fig. 11.4b shows 155 LEDs lightened by a CNF TENG with a size of 40 cm^2.

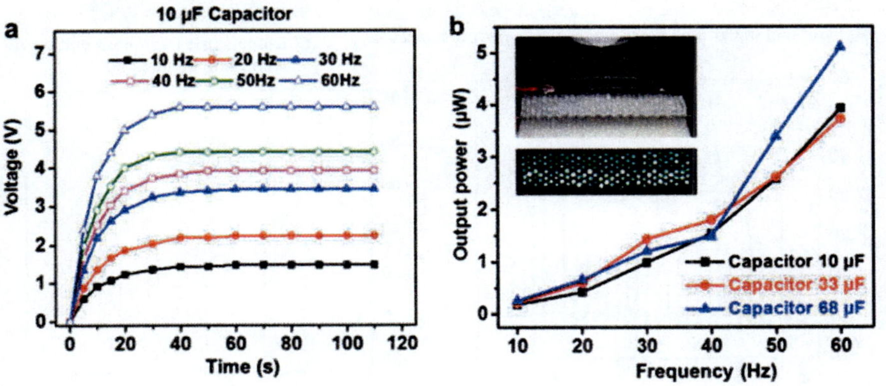

Fig. 11.4 Capacitor charging by the CNF TENG. **a** Voltage on a 10 μF capacitor as a function of charging time. **b** Energy stored in each capacitor dependent on the frequency of the mechanical impact applied on the TENG. Inset of (**b**) shows the TENG powering 155 LEDs that are connected in series. Reproduced with permission [93]. Copyright 2016, Elsevier

11.3.3 Development of Energy Harvesting Fiber Boards

The CNF-TENG unit was used to assemble a triboelectric fiberboard for harvesting energy in walking. Figure 11.5a demonstrated how the triboelectric fiberboard was fabricated. The wooden panel was made from fibers recovered from cardboard, and formed by vacuum sucking followed by cold pressing. There were no chemicals added throughout the manufacturing process. An 8×8 cm^2 CNF TENG was embedded without further treatment in the wet fiber mat during the vacuum sucking process. The final fiberboard product has a diameter of ~20 cm. To demonstrate its capability as a power supply, an experimenter with average weight constantly stepped on and off the TENG-integrated fiberboard and the corresponding electric output signals were real-time recorded. From Fig. 11.5b, c, we can get that the output peak value of V_{oc} was about ~10–30 V and the output peak value of I_{sc} was in a range of 10–90 μA, respectively, which can be attributed to the random stepping

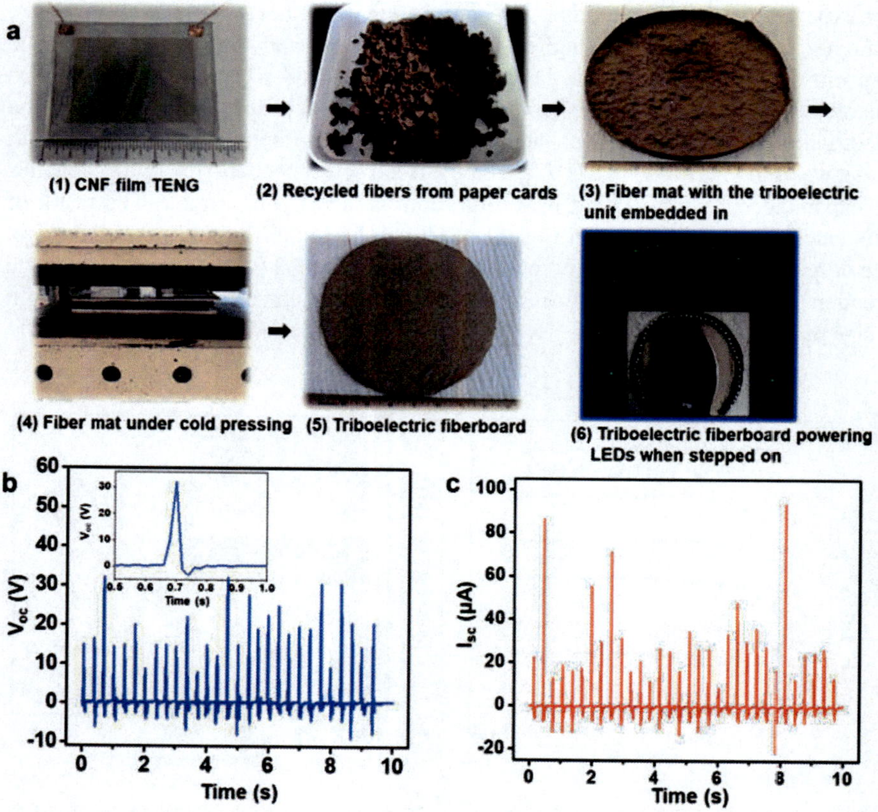

Fig. 11.5 Fabrication and performance of triboelectric fiberboard. **a** Fabrication process of the triboelectric fiberboard. **b** V_{oc} and **c** I_{sc} of the triboelectric fiberboard. Inset of (**b**) is the V_{oc} during one cycle of impact application. Reproduced with permission [93]. Copyright 2016, Elsevier

force and velocity. More practically, 35 green LEDs connected in series could be easily lighted up by the electrical energy harvested from one single step motion (panel 6 in Fig. 11.5a). These results showed the great potential of this design for harvesting the energy in walking to power LED lights, sensors, etc.

To estimate the energy conversion efficiency of the TENG system, charge output efficiency and mechanical-to-electrical energy conversion efficiency were then calculated. For charge output efficiency calculation, the voltage peak in Fig. 11.5b inset was used. For this TENG, the voltage peak could be given in the stepping cycle, so the maximum instantaneous output power P_{max}, the output energy E, and the transferred charge amount Q could be calculated as follows:

$$P_{\max} = \frac{V^2}{R} = \frac{32^2}{1 \times 10^6}\left(\frac{V^2}{\Omega}\right) = 1.024\,\mathrm{mW}$$

$$\mathrm{E} = \int_{0.66}^{0.72} P dt = \int_{0.66}^{0.72} \frac{V^2(t)}{R} dt = \frac{1}{R}\int_{0.66}^{0.72} V^2 dt = 22.6592 \times 10^{-6}\,\mathrm{J} = 0.022\,\mathrm{mJ}$$

$$\mathrm{Q} = \frac{1}{R}\int_{0.66}^{0.72} V(t) dt = 8.4 \times 10^{-7}\,A.s = 8.4 \times 10^{-7}\,\mathrm{C}$$

For the TENG fiberboard, the output energy in one cycle equals to the difference between the energy of the initial-state and that of the final-state. Considering the resistor, namely, the oscilloscope consumed some energy, the output energy became the calculated energy E. The initial-state TENG corresponds to one capacitor, while the final-state TENG was equivalent to three capacitors.

Energy stored on a capacitor can be given following the formula $E_{capacitor} = \frac{Q^2 x}{2A\varepsilon}$. Therefore, the output energy of the TENG fiberboard can be calculated as follows:

$$E = E_{output} = \frac{Q_0^2 x_1}{2A_{eff}\varepsilon_1} - \frac{(Q_0 - Q_2)^2 x_1}{2A_{eff}\varepsilon_1} - \frac{Q_2^2 x_2}{2A_{eff}\varepsilon_2} - \frac{Q_2^2 x_3}{2A_{eff}\varepsilon_3}, \text{ where}$$

$$\varepsilon_1 = 1 \times 8.8541878176 \times 10^{-12}\left(\frac{F}{m}\right),$$

$$x_1 = 80 \times 10^{-6}\,\mathrm{m}$$

$$\varepsilon_2 = 3 \times 8.8541878176 \times 10^{-12}\left(\frac{F}{m}\right),$$

$$x_2 = 80 \times 10^{-6}\,\mathrm{m},$$

$$\varepsilon_3 = 2.1 \times 8.8541878176 \times 10^{-12}\left(\frac{F}{m}\right),$$

$$x_3 = 110 \times 10^{-6}\,\mathrm{m}$$

$$\text{For } E_{output} = 0.022 \times 10^{-3}\,\text{J},\ Q_2 = Q = 8.4 \times 10^{-7}\,\text{C},$$
$$\text{we obtain: } Q_0 = 8.53552 \times 10^{-7}\,\text{C},$$

which was the quantity of the electric charges at the initial-state generated on the triboelectric surface of the TENG fiberboard when contact occurs. Therefore, for this TENG fiberboard, the charge transfer fraction was:

$$\eta_1 = \frac{Q_2}{Q_0} = \frac{8.4 \times 10^{-7}\,\text{C}}{8.54 \times 10^{-7}\,\text{C}} = 98.36\%$$

For mechanical-to-electrical energy conversion efficiency calculation, in one output voltage peak, the input energy from the mechanical motion can be estimated as:

$$W_1 = F \times S = m \times a \times x_1 = 10 \times 0.333 \times 80 \times 10^{-6}\,\text{J} = 0.266\,\text{mJ}$$

where m is an estimated effective weight, a is the one-step acceleration, and x_1 is the active layer spacing in the TENG. Here, m can be obtained using a weight scale, and a can be calculated from the fact that a still foot 0.06 m above the surface of triboelectric fiberboard reached the surface in 0.6 s.

Above result has given that the output energy E was 0.022 mJ. Then, the ratio of output electrical energy to input mechanical energy is:

$$\eta_2 = \frac{E}{W} = \frac{0.022\,\text{mJ}}{0.266\,\text{mJ}} = 8.3\%$$

These calculations showed that, the triboelectric fiberboard could drive nearly 98% of the triboelectric charge to the external circuit, which reveals a very high charge output efficiency. The conversion efficiency calculated as the ratio between the output electrical energy and the input mechanical energy was about ~8.3%, which is relatively low, and might be attributed to most input mechanical energy stored as mechanical deformation of the rigid fiberboard.

In summary, the abundant, renewable, and biodegradable natural CNF shows a great potential application in the development of TENG as a new mechanical energy converting system with good environmental friendliness. The CNF-FEP TENG's electrical output is comparable to that of the TENGs based on typical synthetic polymers. Then, using a chemical-free cold pressing method, CNF-based TENG was integrated with the fiberboard which comes from the recycled cardboard fibers. This fiberboard was able to maintain good integrity under mechanical impact and is highly recyclable. This CNF-based TENG fiberboard generated an output voltage of 30 V and an output current of 90 μA when it was driven by an experimenter's stepping. The converted electric energy could light up 35 green LEDs. Some calculations have been given, and it showed that the TENG fiberboard has a charge transfer ratio of about 98%. This development implied huge potential toward manufacturing environmentally sustainable triboelectric wood/fiber boards or floorings on a large-scale. Combining CNF and wooden fibers' eco-friendliness nature and unique mechanical energy harvesting ability via triboelectric effect, and the wide modification possibilities of CNF, it can foresee that CNF and other natural wood-extracted materials possess the excellent ability of harvesting environmental mechanical energy, which will play a vital role in fabricating and developing industrial-level flooring, packaging and supporting infrastructures.

11.4 Chemical Modification and All-Cellulose Triboelectric Nanogenerators

By using CNF in pair with FEP, a flexible, transparent TENG was developed [93]. While this initial research introduced a new promising material candidate for the development of TENG using recyclable and natural materials, this work still involved synthetic polymer to match with CNF for effectively separating the triboelectric charges. The capability of generating triboelectric charges of the CNF was not strong due to its weak polarization property, which made their output performance quite low compared with those synthetic polymers. Considering that there are abundant hydroxyl groups in cellulose molecule, different functionalities were introduced to obtain CNF with different tribo-polarities via a chemical approach. Tribo-positive and tribo-negative CNF were prepared by attaching methyl group and nitro group, respectively. Both of them demonstrated significantly enhanced Figure of Merit (FOM). The chemical modification enabled all-CNF-based TENGs with comparable outputs to FEP-based TENGs, in which FEP is one of the most negative triboelectric polymers.

11.4.1 Functionalization of CNFs

Before preparation of CNF film, methyl functionalized-CNF film, and nitro functionalized-CNF film, CNF hydrogel was firstly synthesized [93, 96]. Then, by filtering the CNF hydrogel and drying it under certain pressure, the CNF film could be got. For the synthesis of methyl-CNF and nitro-CNF films, the CNF hydrogel was freeze-dried at first and then disassembled into fibers for subsequent modification with functional groups. By being treated with nitration acid mixture including HNO_3, H_2SO_4 and water, the CNF fiber was turned into nitro-CNF fiber. And through reaction with dimethyl sulfate, mercerized CNF fiber was turned into methyl-CNF fiber.

The chemical structure of these CNF samples were characterized through FTIR analysis (Fig. 11.6). Every chemical structure was marked at the corresponding position in the FTIR curves. The black curve represents the pristine CNF, in which adsorptions of C-O-C stretching within the pyranose ring skeletal (at 1050 cm^{-1}), C-H stretching (2900 cm^{-1}), and O-H stretching (3500 cm^{-1}) were observed [97, 98]. For nitro-CNF, the hydroxyl groups were replaced by the nitrate ester groups

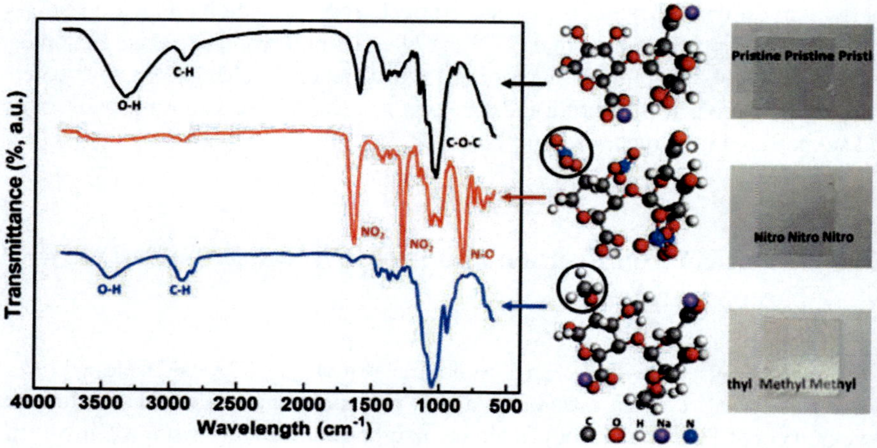

Fig. 11.6 FTIR spectrum and molecular structure of pristine CNF (top row), nitro-CNF (middle row) and methyl-CNF (bottom row). Photos show the transparent films fabricated from these three cellulosic materials. Reproduced with permission [96]. Copyright 2017, Wiley

after nitration treatment [99]. The red curve represents the nitro-CNF and compared with the pristine CNF, three new intense peaks were clearly observed at 1638, 1275 and 828 cm^{-1}, which respectively correspond to the asymmetric and symmetric stretching of the NO_2 group and stretching of the N-O bonds. Besides, it can be found two bands at 692, and 747 cm^{-1} in the red curve, which can be attributed to the scissoring δNO_2 and the wagging $\gamma w NO_2$ vibrations, respectively [100]. In addition, the hydroxyl group's adsorption intensity in nitro-CNF was obviously decreased when compared with the pristine CNF. These analysis results proved that lots of hydroxyls in pristine CNF fiber have been successfully replaced by nitro groups. For the methyl-CNF fiber, the hydroxyls could be converted into methoxides after the methylation process [101–103]. The blue curve in Fig. 11.6 is the FTIR spectrum of the as-synthesized methyl-CNF. Compared with the pristine CNF, the same peaks appeared in methyl-CNF, however, the peak intensities were significantly changed after methylation. Because the methylation treatment could not affect the pyranose ring skeletal in cellulose molecule and C-O-C stretching within the pyranose ring skeletal is at 1500 cm^{-1}, the intensity of one peak from different samples can be marked and compared by normalizing at 1500 cm^{-1}. The results of the methyl-CNF showed that the absorption intensity of the O-H stretching peak was obviously reduced. At the same time, the intensity of C-H stretching was increased. Therefore, the above results clear proofs the effective introduction of methyl groups to CNF molecules. After the modification process was completed, the two kinds of functionalized CNF fibers were further processed into the morphology of the thin film. By varying the hydrogel/solution amount, the thickness of the films could be remained in 100–150 μm. The optical photos on the right side of Fig. 11.6 were the as-prepared films from the pristine CNF and the functionalized CNF, which were all fairly flat, flexible, and transparent.

11.4.2 Quantification of Figure of Merit (FOM) and Triboelectric Output

Surface charge density (σ) is a critical factor that greatly influences the output performance of TENGs. For TENGs, the expression of FOM has proved that σ is the only material-related parameter [104], and their output performance could be greatly improved by the optimization of σ because the FOM is determined by the squared value of normalized surface triboelectric charge density. Thus, to evaluate the potential as active triboelectric materials, the σ from three-types of CNF-based TENGs was characterized. In the measurement, a home-made setup was used as the inset of Fig. 11.7a shown, in which Ga-In eutectic liquid metal was selected as the opposite triboelectrification layer material. By periodically contacting and separating the two triboelectrification layers in a pre-set way, triboelectric charges are transferred between the CNF film and the liquid metal electrode. Here, the pristine CNF, nitro-CNF, methyl-CNF films and FEP film were paired with Ga-In to assemble TENG, respectively. Under the contact-separation working mode, their output current was recorded and shown in Fig. 11.7a–d. Their current peak values were -15.9×10^{-3}, 144.5×10^{-3}, 102.6×10^{-3} and -74.7×10^{-3} μA, respectively. Table 11.1 shows the calculated charge density (-13.3, 120.9, 85.8, -62.5 $\mu C\ m^{-2}$, respectively). Here, the aim of using FEP film was as a standard to evaluate the output performance of the CNF-based TENGs. The charge density of FEP obtained in this work is very close to the value in the reference reported (133.24 $\mu C\ m^{-2}$ for FEP paired with Galinstan) [104], which confirms the validity of our experiment. The negative signal in CNF and methyl-CNF-based TENGs suggests that electrons migrated from the surface of CNF/methyl-CNF to the surface of Ga-In during the contact-separation process, i.e., the triboelectric surface potentials of CNF and methyl-CNF were relatively positive. In the same way, the surface potentials of FEP and nitro-CNF were relatively negative. These results are consistent with the nature of these materials. CNF is slightly tribo-positive because of the rich existence of oxygen atoms in cellulose and a small quantity of carboxylic groups introduced during the process of cellulose pulp' nan fibrillation. Methyl-CNF is also tribo-positive due to the methyl groups with the electron-donating property. Nitro-CNF is tribo-negative because it carries nitro groups with a strong electron-withdrawing property.

Comparing the output performance of TENGs composed of untreated and chemical functionalized CNF, it can be found that methylation and nitration largely increased the surface potential of CNF films by a factor of ~4–6. In addition, functionalizing the slightly positive CNF could obtain both positive and negative tribopolarities, which suggests that CNF's polarity can be effectively engineered by rationally controlling the introduced functional groups. Taking the charge density of the FEP as the benchmark, the charge density of the CNF-based films was normalized and recorded as the relative triboelectric charge densities (σN), which were found to be -0.110, 0.710 and -0.517 for CNF, nitro-CNF and methyl-CNF, respectively. The square of σN could give the FOMs of methyl-NC (0.28) and nitro-NC (0.51), which are comparable to and even better than other synthetic polymers commonly

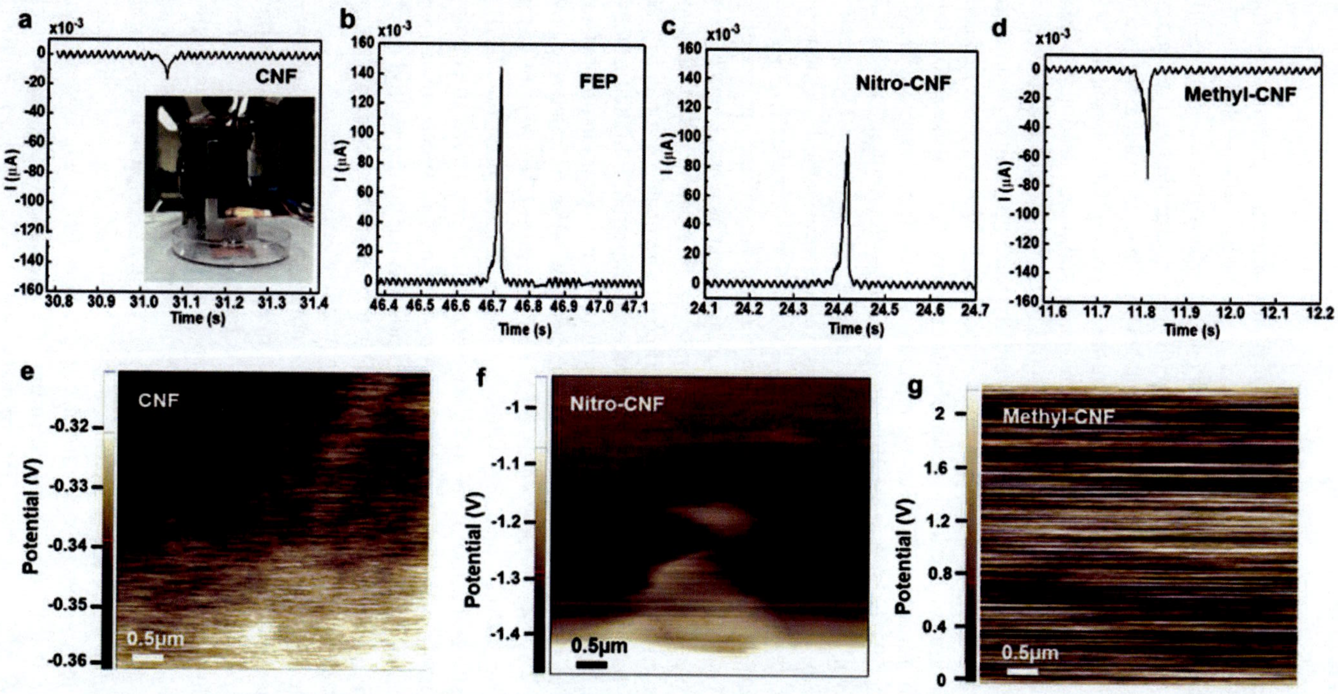

Fig. 11.7 **a–d** Current flow measured during one contact-separation event when electric contact was made between Ga-In eutectic liquid metal and the triboelectric film of CNF (**a**), FEP (**b**), nitro-CNF, and methyl-CNF. Inset in (**a**) is the setup that can move vertically for making the contacts. **e–g** The SKPM surface potential mapping of pristine CNF (**e**), nitro-CNF (**f**), and methyl-CNF (**g**). Reproduced with permission [96]. Copyright 2017, Wiley

Table 11.1 Triboelectric properties of pristine and functionalized CNF films in comparison to a FEP film and corresponding TENG output performance when paired with Cu

Material	Surface charge density [$\mu C/m^2$]	Normalized triboelectric charge density (σN)	Figure of merit (FOM)	Average current output from TENG when paired with Cu [V]	Average voltage output from TENG when paired with Cu [μA]
FEP	120.9	1	1	6.9	7.8
CNF	−13.3	−0.11	0.012	0.9	0.8
Nitro-CNF	85.8	0.71	0.504	4.9	5.0
Methyl-CNF	−62.5	0.52	0.267	3.7	3.9

used in TENGs, such as Kapton (a Dupont polyimide) (0.36), PVDF (0.20) and polyethylene (0.18). [104]

The scanning kelvin probe microscopy (SKPM) was used to characterize the CNF-based film's surface potential via non-contact mode. The SKPM result of the pristine CNF in Fig. 11.7e showed its average surface potential was about −330 mV. Figure 11.7f showed the surface potential of the nitro-CNF reached about −1.3 V, which is more negative due to its strong triboelectric negative polarity. Figure 11.7g demonstrated that the surface potential of the methyl-CNF was about 1.1 V, which can be attributed to its high triboelectric positive polarity. The great change of the surface potential further proved that the electronic property of natural CNF could be effectively regulated after nitration and methylation.

To evaluate the enhanced performance of TENGs by CNF functionalization, four different TENGs were fabricated using the three CNF-based films and the FEP film as one triboelectric layer, respectively, and the Cu electrode as the other triboelectric layer. All of them worked in vertical contact-separation mode to test their triboelectric outputs. The output current and voltage were recorded and Table 11.1 has given their average current and voltage values. Compared with pristine CNF, Cu shows a relatively slight triboelectric negativity, so the pristine CNF-Cu pair generated a quite small average current and voltage of ~0.8 μA and ~0.8 V, respectively. For the nitro-CNF-Cu pair, the average current and voltage reached about 5.0 μA/4.9 V. And for the methyl-CNF-Cu pair, the average current and voltage reached about 3.9 μA/3.7 V. Both the output of methyl-CNF-Cu and nitro-CNF-Cu pairs were obviously enhanced and close to the highest output of the FEP-Cu pair, whose average current and voltage were about 7.8 μA/6.9 V. These results fully evidenced that the triboelectric performance of nature CNF films could be effectively improved via functionalizing by chemical groups.

Based on the remarkable triboelectric polarity tuning property, the TENG can be designed with both the two triboelectric layers using CNF-based materials with different chemical functionalization. Here, TENGs were fabricated with different pairs of pristine and functionalized CNF in vertical contact mode and their output performance were measured under the same conditions. For a pair of pristine CNF films, its output was nearly negligible (Fig. 11.8a). When using the pristine CNF film

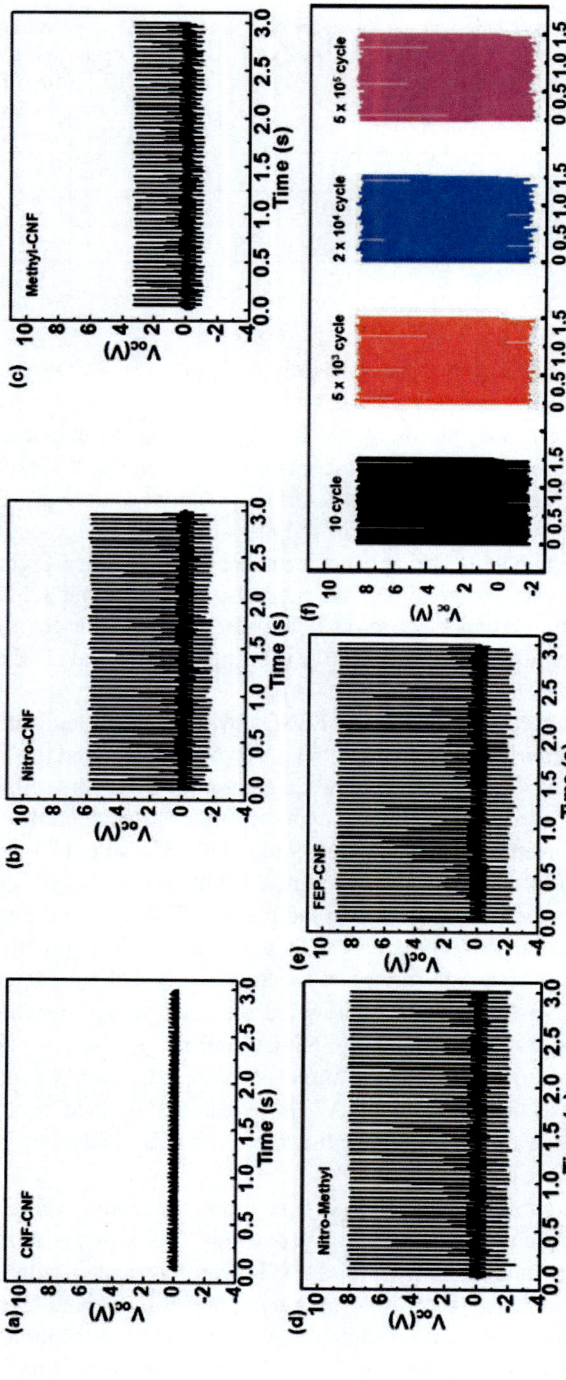

Fig. 11.8 Output voltage of TENGs made from various film pairs of **a** CNF/CNF, **b** nitro-CNF/CNF, **c** methyl-CNF/CNF, **d** nitro-CNF/methyl-CNF, and **e** FEP/CNF. **f** Long-term voltage output measured from nitro-CNF/methyl-CNF TENG. Reproduced with permission [96]. Copyright 2017, Wiley

as one triboelectric layer, and methyl-CNF film and nitro-CNF film as another triboelectric layer, respectively, both the output of the two kinds TENGs were enhanced with a high output of 5.9 V for the nitro-CNF film (Fig. 11.8b) and 3.3 V for methyl-CNF (Fig. 11.8c). Since the triboelectric polarization of the nitro- and methyl-functionalization was changed toward the opposite directions, so the average output voltage was raised to 8 V when using the two functionalized CNF films as the triboelectric pair materials (Fig. 11.8d). It is worth to note that the output voltage of the TENG using the methyl-CNF-nitro-CNF triboelectric pair was close to that of the TENG using FEP-CNF triboelectric pair (9 V, Fig. 11.8e). Because this polarization tuning by functionalization was done on a level of molecule, the enhanced output of the treated CNF films was quite stable. Long-term output performance of the nitro-CNF-methyl-CNF TENG was tested and showed in Fig. 11.8f, from which it can be found that the output signal without decay after continuously working for 5×10^5 cycles. While FEP is recognized as one of the most tribo-negative materials, the comparable output from the functionalized CNF pair fully proved the great potential of biodegradable and renewable CNF based films to replace commercial synthetic polymers with comparable triboelectric performance.

In summary, introducing nitro and methyl groups on CNF by chemical methods made these two functionalized CNFs possess opposite high tribo-polarity. The surface charge density of pristine CNF was about $-13.3\ \mu C\ m^{-2}$, but after chemical treatment, the surface charge density of the nitro-CNF and methyl-CNF reached to $-85.8\ \mu C\ m^{-2}$ and $62.5\ \mu C\ m^{-2}$, respectively, which is substantially enhanced. Their charge density was close to 71% and 52% of that for the most triboelectric negative material FEP, while pristine CNF could only approach 11% of that for FEP. The quantified FOM of the nitro-CNF and methyl-CNF were up to 0.51 and 0.28, respectively. Moreover, different TENGs were fabricated through combining the functionalized CNF thin film with Cu electrode and among each other. Triboelectric output performance was measured and both the output of nitro-CNF and methyl-CNF films-based TENG outperformed that of pristine CNF film-based TENG. When the TENG was fabricated using nitro-CNF paired with methyl-CNF, the output voltage was further increased to a value very close to that of TENG using the EFP-pristine CNF pair. These results further expanded the technology platform for the development of TENG devices with high-performance using cost-effective renewable and natural resources.

11.5 CNF-Based Flexible, Transparent and Conductive Films for Degradable TENG

Since cellulose itself is not conductive, previous developments all used a metal electrode for charge collection. In order to make CNF-based TENG transparent, flexible and eventually biocompatible and degradable, new types of charge collectors need to be implemented. Among several material candidates for transparent flexible electrodes, aluminum-doped zinc oxide (AZO) is particularly promising. It has the potential to reach very low resistivity on the scale of 10^{-4} Ω·cm. It is based on non-toxic elements that are cheap and abundant on earth and could be eventually used for biomedical applications. Based on this rational, a flexible and nature degradable transparent CNF film was developed with a layer of AZO thin film coated in the equipment of atomic layer deposition (ALD). A TENG fabricated using this CNF film demonstrated excellent transparency, flexibility, as well as high electric output. The entire TENG device could also be fully dissolved in warm water showing the potential to achieve self-degradability.

11.5.1 Fabrication of AZO-CNF Film for TENG

CNF film was prepared using the as-processed CNF hydrogel in the same way as introduced in Sect. 11.3. To grow AZO, as-prepared CNF film was loaded in an ALD reactor chamber at a temperature of 150 °C and under a constant nitrogen gas flow of 40 sccm. Dimethyl zinc ($Zn(CH_3)_2$) (DEZn) and trimethylaluminum ($Al(CH_3)_3$) (TMA) were chosen as the precursors. One ZnO cycle was sequentially composed of 0.5 s DEZn pulsing, 30 s purging with N_2, 0.5 s pulsing of H_2O, and 30 s purging with N_2. One cycle of Al_2O_3 was 0.5 s pulsing of TMA, 30 s purging with N_2, 0.5 s pulsing with H_2O and 30 s purging with N_2. The cycle ratio for deposition of DEZn/TMA (X:Y) was changed from 20:1 to 5:1 to obtain AZO films with different doping concentrations of Al element. Total deposition cycle for each sample consisted of a fixed ZnO cycle (1500) plus a varying Al_2O_3 cycle (0, 75, 100, 150, 300) to achieve different deposition cycle ratios (20:1, 15:1, 10:1, 5:1). TENGs (vertical contact mode) were then fabricated by assembling an AZO-CNF film against a tribo-polarity-tuned CNF. The tribo-polarity-tuned CNF was achieved by $TiCl_4$ infiltration (Sect. 11.4). The infiltration of $TiCl_4$ was implemented in the ALD chamber with 10 cycles of $TiCl_4$ vapor exposure in 5-s and N_2 purge in 60-s.

11.5.2 Property Characterization of AZO-CNF Films

High transparency CNF film in the thickness of 50 μm was employed as the supporting substrate for AZO deposition (top of Fig. 11.9a). A layer AZO of 285 nm in thickness was deposited in ALD at low-temperature, showing very good film uniformity and conformity. After coating AZO at 150 °C for nearly 27 h, the CNF film maintained a good mechanical integrity and a high optical transparency (bottom of Fig. 11.9a). Low-magnification SEM imaging demonstrated that the as-deposited surface was fairly smooth (inset of Fig. 11.9b). For the film of AZO with Al content of 3.7 at. %, worm-like grains in the length of 50–150 nm and the width of 16–20 nm presented and evenly distributed over the whole surface (Fig. 11.9b). As increasing the Al content, the grain sizes gradually decreased to ~60–80 nm. AFM topology scanning was performed to further study the surface morphologies of the coated

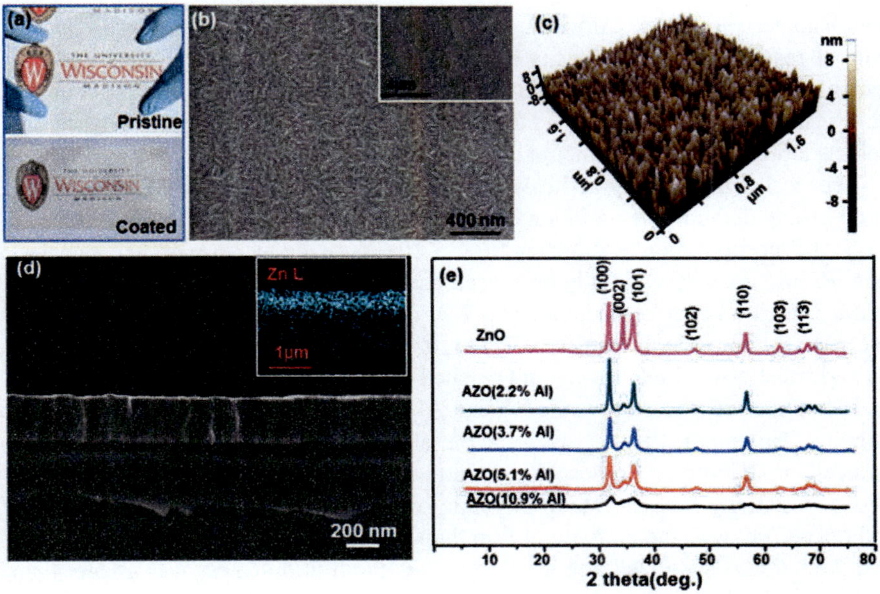

Fig. 11.9 Morphology and crystalline structure of AZO-CNF films. **a** Photograph of an as-fabricated CNF film from the CNF gel (top) and an AZO-coated CNF film (bottom). **b** SEM image of the AZO surface with 3.7 at. % Al. Inset is a low-magnification SEM image. **c** AFM image of the AZO surface with 3.7 at. % Al. **d** Cross-sectional image of the AZO-CNF film. Inset is the corresponding EDS mapping of the Zn elements. **e** XRD spectra of AZO-CNF films with different Al concentrations. Reproduced with permission [105]. Copyright 2017, Springer Nature

surface. The AZO film exhibited a markedly uniform grain size distribution and a very small root mean roughness (RMS) of ~2 nm (Fig. 11.9c). This indicates that the introduction of Al into the ZnO film could reduce the average size of the grains and improve surface smoothness, which could help improve the optical transparency of films.

The cross-sectional SEM image showed that the AZO layer had good adhesion on CNF film and no obvious cracks or peeling phenomena were observed at the interface (Fig. 11.9d). The typical structure of these grains was columnar. The grain boundary became less distinguishable at high Al concentrations (>5.1%), and Zn signal was clearly detected inside the CNF film. This could be contributed to the mesoporous structure of CNF film that allowed Zn precursor's infiltration. When ZnO grows a certain distance into the CNF film, the binding between the AZO layer and the CNF film could be enhanced, thus improving the mechanical integrity.

It has been reported that ZnO with doped Al could lead to the substitution of Al^{3+} ions for Zn^{2+} ions, or lead to the interstitial Al defects in the ZnO lattice [106]. In order to study the Al dopant type in the ALD-prepared AZO films, XRD scanning was performed and the resultant XRD spectra of different Al concentrations were presented in Fig. 11.9e. ZnO of hexagonal wurtzite phase was detected in all deposited samples, where the (100), (002), (101), (102), (110), (103) and (113) were the characteristic peaks. Comparing the spectra of films with varying Al concentrations revealed that the intensity of these characteristic peaks decreased with the increase of Al concentration, but no peak shifts or appearance of new peaks could be observed. This analysis confirmed that the presence of Al in the form of substitutive defects reduced the crystallinity of AZO film [107].

The electrical resistivity of the AZO-coated CNF films was measured by the four-point probe method. The variation curve of the sheet resistance (*Rs*) and resistivity with Al concentration was shown in Fig. 11.10a. With the increase of Al concentration, *Rs* and resistivity initially decreased and dropped to the lowest value of 96 Ω and 2.7 × 10^{-3} Ω.cm, respectively at 3.7 at.% Al. This suggested that Al atoms are successfully doped in the lattice of ZnO [108]. As the Al content further increased, *Rs* and resistivity raised sharply and reached the highest value of 6942.6 Ω and 2.4 × 10^{-1} Ω.cm, respectively at the highest Al concentration (11%) within the study window. This might be the reason for the clustering of AlO_x compounds and the reduced crystallinity at higher Al concentrations [109, 110]. Effects of film thickness on resistivity were also studied by fixing the Al concentration at 3.7%. As shown in Fig. 11.10b, with the increase of film thickness, *Rs* and resistivity first decreased and then almost stabilized at a low level when the film thickness was greater than 170 nm. This observation agreed well with the variation of thickness-dependent *Rs* for ITO films as reported elsewhere, where the relationship was attributed to the Fuchs-Sondheimer theory of electron scattering [111–114].

AZO-CNF films with different Al concentrations between 0% and 10.9% were prepared and their light transmittance was measured within the wavelength range of 200–800 nm. As shown in Fig. 11.10c, a high transmittance of >84% was observed for all samples in the visible region from 400 to 760 nm, indicating these samples are acceptable for TCOs applications [115]. The fluctuation of the absorption curve was a result of the uniform thickness, which induced optical interference. For these AZO-CNF films, the optical transmittance at 550 nm versus the sheet resistance *Rs* was presented in Fig. 11.10d. For comparison purposes, the value of commonly used transparent conductive substrates, e.g., graphene on PET and ITO substrates, were also plotted in the figure [116, 117]. The 3.7 at.% Al doped AZO film demonstrated the best *Rs* and transparency combination, whose testing values were identical to the values reported for graphene and ITO.

As these AZO-CNF films were designed for flexible TCO applications, the influences of strain on film conductivity were studied. The 3.76 at. % Al doped AZO-CNF film was wrapped around a serial of cylindrical glass molds with definite radius to obtain different kinds of strains. Compressive strain was introduced by wrapping with the AZO side down, and the tensile strain was introduced by wrapping with the AZO side up, as indicated in the inset of Fig. 11.11a, b. *Rs* was measured under these strain conditions and strain-related *Rs* changes were shown in Fig. 11.11a (compressive strain) and Fig. 11.6b (tensile strain). Under compressive strain, *Rs* exhibited a small variation within the studied bending radius in the range of 5–45 mm. Considering that the thickness of AZO-CNF film is 50 μm, the corresponding compressive strain range was calculated to be −0.057% to −0.457%. Under tensile strain, *Rs* first showed the same small variation for bending radii over 10 mm, which corresponds

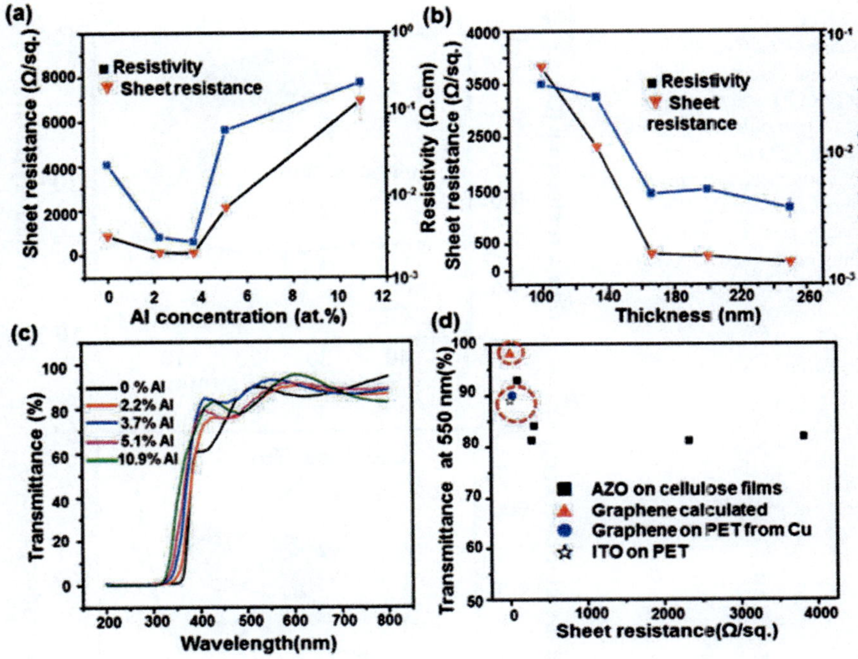

Fig. 11.10 The electrical and optical properties of AZO-CNF films. **a** The sheet resistance and resistivity measured as a function of Al concentration. **b** The sheet resistance and resistivity measured as a functional of AZO film thickness. All the AZO films had an Al concentration of 3.7 at.%. **c** Transmittance of the AZO-CNF films with different Al concentrations. **d** Transmittance (at 550 nm) and sheet resistance of AZO-CNF in comparison with other typical flexible transparent electrode candidates. Reproduced with permission [105]. Copyright 2017, Springer Nature

to tensile strains of less than 0.196% (Fig. 11.11b). When the bending radius reached 5 mm (0.45% tensile strain), *Rs* increased abruptly. This indicated the integrity of AZO film degraded. The low tensile strain tolerance is a typical behavior of ceramic materials, in which the quality of the film depends on the propagation of a major defect. However, when radius was in the range >10 mm, the film was in a stable region, at which the AZO-CNF film's *Rs* was approaching to that of these flexible and transparent ITO based films [118]. Therefore, this AZO-CNF film holds potential as a new kind of transparent and flexible substrate.

11.5.3 Degradable TENG Development

As the AZO-CNF film is comprised of green CNF material and water soluble AZO, it was expected to show a good ability to degrade and recycle. Figure 11.12a showed a piece of AZO-CNF film immersed in 70 °C city water under stirring. After 60 min, the AZO-CNF film was completely dissolved in the water and no solid piece could

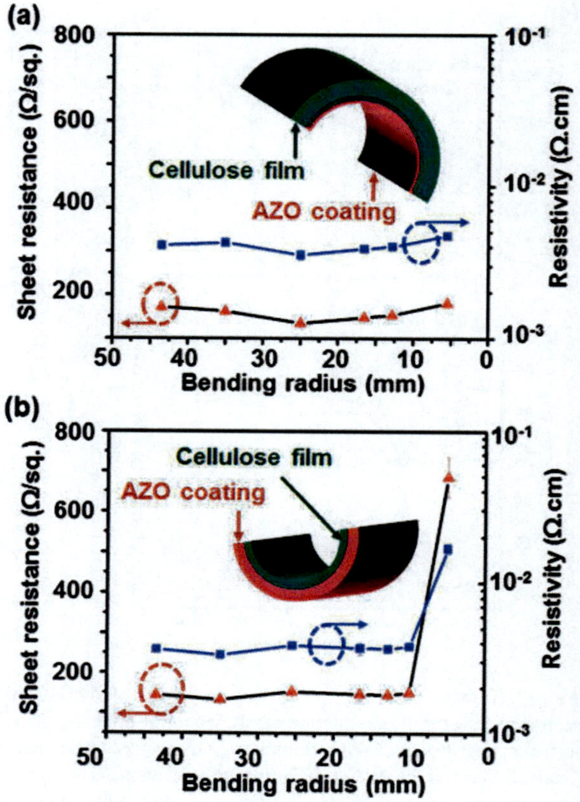

Fig. 11.11 The strain tolerance of AZO-CNF films at different bending radius **a** Change of Rs and resistivity as a function of bending radii when the AZO film was under compressive strains. **b** Change of *Rs* and resistivity of coating as a function of bending radii when the AZO film was under tensile strains. Reproduced with permission [105]. Copyright 2017, Springer Nature

be observed. And a clear CNF gel solution was obtained except for a few Zn and Al elements from the coated AZO film. With further concentrating and filtering, solid-state CNF film could be obtained from the CNF gel solution, demonstrating an excellent recyclable capacity.

A TENG was fabricated using the AZO-CNF film to harvest mechanical energy. Figures 11.12b, c were the schematic configuration and photograph of the contact-separation mode TENG, respectively. A piece of as-prepared AZO-CNF film was assembled against a piece of $TiCl_4$ treated AZO-CNF film (to improve the electron affinity of the CNF film). Both films had the AZO side facing outward. AZO acted as the back electrode, and the charge generation occurred upon the contact between the pristine CNF of the top AZO-CNF film and the $TiCl_4$ treated CNF of the bottom AZO-CNF film. Since both the positive and negative triboelectric films were the AZO-CNF film, so the TENG showed very good flexibility and transparency as shown in Fig. 11.12c. It should also be pointed out that, the entire device could be completely degraded in water because no external metal electrodes were used. Figure 11.12d–f showed the output performance of the TENG under a periodical 10 Hz vertical compression-and-release force. The maximum voltage output, current output, and charge transfer reached ~7 V, ~0.7 μA, and ~77 $\mu C/m^2$, respectively,

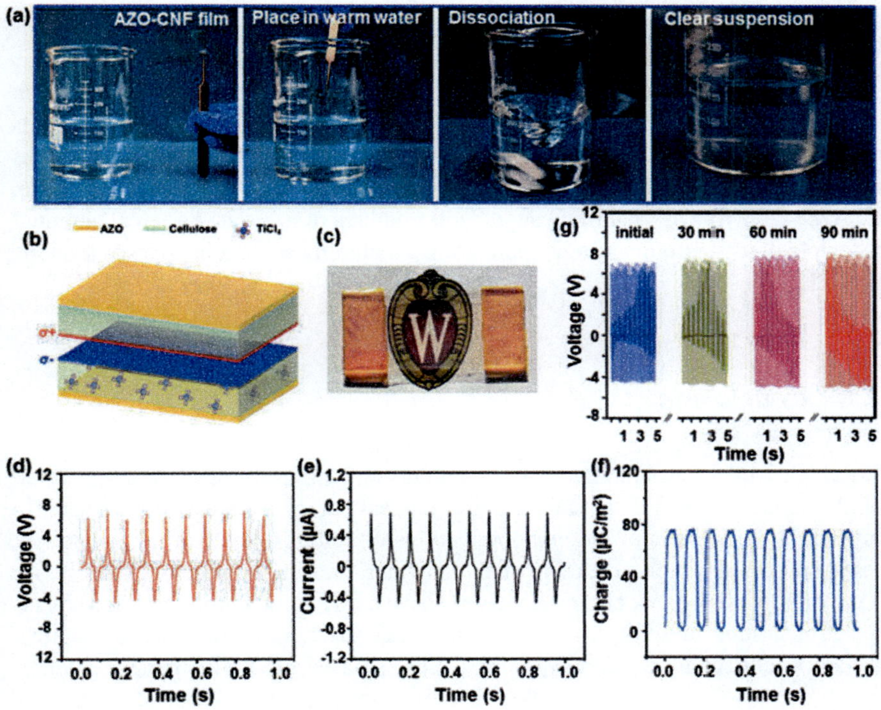

Fig. 11.12 Degradation and TENG application of AZO-CNF films. **a** A series of photos showing the process of dissolving an AZO-CNF paper in city water at 70 °C. The AZO-CNF film was completely dissolved after 60-min stirring. **b** Schematic image of a TENG made from a pair of AZO-CNF films. **c** A photograph of the transparent all-CNF TENG device. **d–f** The voltage (**d**), current (**e**) and charge (**f**) outputs of the TENG device were measured at a 10 Hz frequency. **g** The long-term stability of AZO-CNF-based TENG. It showed a stable output after a 90-min operation. Reproduced with permission [105]. Copyright 2017, Springer Nature

which were comparable to previously reported CNF-based TENGs (Sects. 11.3 and 11.4). To test the durability of this TENG, 90-min of continuous compression force was applied at a 2 Hz frequency and the output voltage was measured every 30 min (Fig. 11.12g). The output was stable over the entire study period, showing the AZO coating had very good durability even under constant strong force application.

In summary, coating an AZO thin layer by low-temperature ALD technique could lead to a new type of TCO film with good flexibility and transparency. The AZO thin layer was tightly adhered on the CNF surface because the mesoporous structure inside the CNF film was favorable for infiltrating AZO precursor. The AZO coating had a high ability against strains from the tensile and compressive force. The 3.7 at. % Al doped AZO film the best *Rs* and transparency combination, whose testing values were identical to graphene and ITO. The 3.76 at. % Al doped AZO-CNF demonstrated the best *Rs* and transparency combination character, whose testing values were identical to the values reported for commonly used flexible and transparent conductive substrates, such as PET substrates coated with ITO or graphene. Besides, it was demonstrated that this flexible TOC film could be naturally degradable. TENGs with good flexibility and transparency were fabricated by only using the AZO-CNF films with the assistance of $TiCl_4$ treatment. The output performance of these TENGs was comparable to those fabricated from other common synthetic polymers. Therefore, this AZO-CNF film as a naturally degradable conductive substrate has great potential in flexible optoelectronic system applications, including solar cells, optical sensors, and LEDs.

11.6 Conclusion

This chapter overviews the latest developments and applications of nanostructured cellulose materials in TENG to harvest mechanical energy from the environment. In comparison with the conventional applications of cellulose in bio-fuel generation, these applications emphasize more on the polymeric material nature of cellulose. As a kind of polymers, nanocellulose materials have many unique properties and advantages, integrating unique properties of chemical, structural, dielectric, and optical that are rarely found in other polymer systems. Based on above mentioned characteristics, the usage of nanocellulose materials in modern devices and systems as a functional component may lead to a new direction in the innovative application of materials. Although functional cellulose nanomaterial just emerged in the most recent few years, it has shown unprecedented possibility in many applications. It can be predicted that nanocellulose material will perform an increasingly important role in the functional soft material family. Of course, there are also challenges, and a lot of efforts are needed to overcome them.

Firstly, the synthesis of cellulose nanomaterials with controllable polymerization degree. This is the basic and necessary technique for us to reliably understand the properties and structural relationships of cellulose nanomaterials, and further predict their functionality.

Secondly, understanding how cellulose molecules are arranged into different phases at the molecular-level, i.e., the organization and interaction of the molecular chains in short- and long-term in an orderly manner. Although many crystalline forms of cellulose have been discovered, how to control them to regulate their macroscopic properties has not been fully unveiled. Understanding of above issue will fundamentally reveal the unprecedented potential to utilize cellulose nanomaterials as vital functional components for many modern electronic and energy systems.

Thirdly, new methods for functionalization, modification, and assembly of cellulose nanomaterials with different length scales. At present, we lack the related capability, which is a key step in leading the basic and conceptual research of cellulose nanomaterials to practical applications.

In conclusion, the research on functional cellulose nanomaterials is still in its infancy. Taking the presented TENG development as an example, functional cellulose nanomaterials exhibit great material science value and potential applications in energy-related fields. With the help of advanced techniques in synthesis, processing and characterization, and the cooperation from different scientific disciplines, it can be expected that a powerful material platform with multifunctionality may be soon achieved by cellulose nanomaterials that nature has provided us to take our materials technology into a new perspective.

Acknowledgements XW thanks the support of the National Science Foundation DMR-1709025. The authors also thank Dr. Yanhao Yu, Dr. Zhiyong Cai and Dr. Bing Yang for their contribution to the work reviewed in this chapter. The work reviewed in this chapter is partially supported by Wisconsin Alumni Research Foundation (WARF) and USDA Forest Product Laboratory (14-JV-11111124-029).

References

1. B.P. Dieter Klemm, T. Hein, Comprehensive cellulose chemistry 1, in *Fundamentals And Analytical Methods* (Wiley-VCH, Weinheim, 1998)
2. H.A. Krässig, *Cellulose: Structure, Accessibility and Reactivity* (Gordon and Breach Science Publishers, 1993)
3. D. Klemm, B. Heublein, H.P. Fink, A. Bohn, Cellulose: fascinating biopolymer and sustainable raw material. Angew. Chem. Int. Ed. **44**, 3358–3393 (2005)
4. R.P. Chandra, R. Bura, W. Mabee, d.A. Berlin, X. Pan, J. Saddler, *Substrate Pretreatment: The Key to Effective Enzymatic Hydrolysis of Lignocellulosics?* Biofuels (Springer, 2007), pp. 67–93
5. D. Klemm, D. Schumann, U. Udhardt, S. Marsch, Bacterial synthesized cellulose—artificial blood vessels for microsurgery. Prog. Polym. Sci. **26**, 1561–1603 (2001)
6. A. Bodin, S. Concaro, M. Brittberg, P. Gatenholm, Bacterial cellulose as a potential meniscus implant. J. Tissue Eng. Regen. Med. **1**, 406–408 (2007)
7. I. Degreef, S. Tejpar, L. De Smet, Improved postoperative outcome of segmental fasciectomy in Dupuytren disease by insertion of an absorbable cellulose implant. J. Plast. Surg. Hand Surg. **45**, 157–164 (2011)
8. R.A. Gross, B. Kalra, Biodegradable polymers for the environment. Science **297**, 803–807 (2002)
9. K.G. Satyanarayana, G.G. Arizaga, F. Wypych, Biodegradable composites based on lignocellulosic fibers—an overview. Prog. Polym. Sci. **34**, 982–1021 (2009)
10. Y.H. Jung, T.-H. Chang, H. Zhang, C. Yao, Q. Zheng, V.W. Yang, H. Mi, M. Kim, S.J. Cho, D.-W. Park, High-performance green flexible electronics based on biodegradable cellulose nanofibril paper. Nat. Commun. **6** (2015)
11. A. Sani, Y. Dahman, Improvements in the production of bacterial synthesized biocellulose nanofibres using different culture methods. J. Chem. Technol. Biotechnol. **85**, 151–164 (2010)
12. I. Usov, G. Nyström, J. Adamcik, S. Handschin, C. Schütz, A. Fall, L. Bergström, R. Mezzenga, Understanding nanocellulose chirality and structure-properties relationship at the single fibril level. Nat. Commun. **6** (2015)
13. Y. Habibi, L.A. Lucia, O.J. Rojas, Cellulose nanocrystals: chemistry, self-assembly, and applications. Chem. Rev. **110**, 3479–3500 (2010)
14. K.E. Shopsowitz, W.Y. Hamad, M.J. MacLachlan, Flexible and iridescent chiral nematic mesoporous organosilica films. J. Am. Chem. Soc. **134**, 867–870 (2012)

15. J. Araki, M. Wada, S. Kuga, T. Okano, Flow properties of microcrystalline cellulose suspension prepared by acid treatment of native cellulose. Colloids Surf., A **142**, 75–82 (1998)
16. T. Saito, M. Hirota, N. Tamura, S. Kimura, H. Fukuzumi, L. Heux, A. Isogai, Individualization of nano-sized plant cellulose fibrils by direct surface carboxylation using TEMPO catalyst under neutral conditions. Biomacromol **10**, 1992–1996 (2009)
17. R.J. Moon, A. Martini, J. Nairn, J. Simonsen, J. Youngblood, Cellulose nanomaterials review: structure, properties and nanocomposites. Chem. Soc. Rev. **40**, 3941–3994 (2011)
18. D. Klemm, F. Kramer, S. Moritz, T. Lindström, M. Ankerfors, D. Gray, A. Dorris, Nanocelluloses: a new family of nature-based materials. Angew. Chem. Int. Ed. **50**, 5438–5466 (2011)
19. I. Siró, D. Plackett, Microfibrillated cellulose and new nanocomposite materials: a review. Cellulose **17**, 459–494 (2010)
20. S. Eichhorn, A. Dufresne, M. Aranguren, N. Marcovich, J. Capadona, S. Rowan, C. Weder, W. Thielemans, M. Roman, S. Renneckar, Review: current international research into cellulose nanofibres and nanocomposites. J. Mater. Sci. **45**, 1–33 (2010)
21. K. Saalwächter, W. Burchard, P. Klüfers, G. Kettenbach, P. Mayer, D. Klemm, S. Dugarmaa, Cellulose solutions in water containing metal complexes. Macromolecules **33**, 4094–4107 (2000)
22. B. Seger, W. Burchard, Structure of cellulose in cuoxam, in *Macromolecular Symposia*, pp. 291–310 (1994)
23. B.E. Dale, G.T. Tsao, A microcalorimetric study of complex formation between alkaline sodium tartrate and iron (III). J. Polym. Sci. Polym. Chem. Ed. **18**, 3163–3175 (1980)
24. M. Hattori, T. Koga, Y. Shimaya, M. Saito, Aqueous calcium thiocyanate solution as a cellulose solvent. Structure and interactions with cellulose. Polym. J. **30**, 43–48 (1998)
25. B.P. Dieter Klemm, T. Hein, Comprehensive cellulose chemistry 2, *in Functionlization of Cellulose* (Weinheim, 1998)
26. S. Fischer, H. Leipner, K. Thümmler, E. Brendler, J. Peters, Inorganic molten salts as solvents for cellulose. Cellulose **10**, 227–236 (2003)
27. Z.L. Wang, Triboelectric nanogenerators as new energy technology for self-powered systems and as active mechanical and chemical sensors. ACS Nano **7**, 9533–9557 (2013)
28. S. Wang, L. Lin, Z. Wang, Triboelectric nanogenerators as self-powered active sensors. Nano Energy 436–462 (2015)
29. Z.L. Wang, J. Chen, L. Lin, Progress in triboelectric nanogenerators as a new energy technology and self-powered sensors. Energy Environ. Sci. **8**, 2250–2282 (2015)
30. Y. Mao, P. Zhao, G. McConohy, H. Yang, Y. Tong, X. Wang, Sponge-like piezoelectric polymer films for scalable and integratable nanogenerators and self-powered electronic systems. Adv. Energy Mater. **4** (2014)
31. K.I. Park, M. Lee, Y. Liu, S. Moon, G.T. Hwang, G. Zhu, J.E. Kim, S.O. Kim, K. Kim do, Z.L. Wang, K.J. Lee, Flexible nanocomposite generator made of BaTiO(3) nanoparticles and graphitic carbons. Adv. Mater. **24**, 2999–3004, 2937 (2012)
32. X. Wang, J. Song, J. Liu, Z.L. Wang, Direct-current nanogenerator driven by ultrasonic waves. Science **316**, 102–105 (2007)
33. S.N. Cha, J.S. Seo, S.M. Kim, H.J. Kim, Y.J. Park, S.W. Kim, J.M. Kim, Sound-driven piezoelectric nanowire-based nanogenerators. Adv. Mater. **22**, 4726–4730 (2010)
34. X. Wang, Piezoelectric nanogenerators—harvesting ambient mechanical energy at the nanometer scale. Nano Energy **1**, 13–24 (2012)
35. J.-H. Lee, K.Y. Lee, B. Kumar, N.T. Tien, N.-E. Lee, S.-W. Kim, Highly sensitive stretchable transparent piezoelectric nanogenerators. Energy Environ. Sci. **6**, 169–175 (2013)
36. X.D. Wang, J. Liu, J.H. Song, Z.L. Wang, Integrated nanogenerators in biofluid. Nano Lett. **7**, 2475–2479 (2007)
37. Y. Qin, X. Wang, Z.L. Wang, Microfibre-nanowire hybrid structure for energy scavenging. Nature **451**, 809–813 (2008)

38. C. Sun, J. Shi, D.J. Bayerl, X. Wang, PVDF microbelts for harvesting energy from respiration. Energy Environ. Sci. **4**, 4508–4512 (2011)
39. M. Lee, C.Y. Chen, S. Wang, S.N. Cha, Y.J. Park, J.M. Kim, L.J. Chou, Z.L. Wang, A hybrid piezoelectric structure for wearable nanogenerators. Adv. Mater. **24**, 1759–1764 (2012)
40. F.-R. Fan, Z.-Q. Tian, Z. Lin Wang, Flexible triboelectric generator. Nano Energy **1**, 328–334 (2012)
41. G. Zhu, Y.S. Zhou, P. Bai, X.S. Meng, Q. Jing, J. Chen, Z.L. Wang, A shape-adaptive thin-film-based approach for 50% high-efficiency energy generation through micro-grating sliding electrification. Adv. Mater. **26**, 3788–3796 (2014)
42. W. Tang, T. Jiang, F.R. Fan, A.F. Yu, C. Zhang, X. Cao, Z.L. Wang, Liquid-metal electrode for high-performance triboelectric nanogenerator at an instantaneous energy conversion efficiency of 70.6%. Adv. Funct. Mater. **25**, 3718–3725 (2015)
43. Y. Xie, S. Wang, S. Niu, L. Lin, Q. Jing, J. Yang, Z. Wu, Z.L. Wang, Grating-structured freestanding triboelectric-layer nanogenerator for harvesting mechanical energy at 85% total conversion efficiency. Adv. Mater. **26**, 6599–6607 (2014)
44. S. Wang, L. Lin, Z.L. Wang, Nanoscale triboelectric-effect-enabled energy conversion for sustainably powering portable electronics. Nano Lett. **12**, 6339–6346 (2012)
45. G. Zhu, J. Chen, T. Zhang, Q. Jing, Z.L. Wang, Radial-arrayed rotary electrification for high performance triboelectric generator. Nat. Commun. **5**, 3426 (2014)
46. G. Zhu, W.Q. Yang, T. Zhang, Q. Jing, J. Chen, Y.S. Zhou, P. Bai, Z.L. Wang, Self-powered, ultrasensitive, flexible tactile sensors based on contact electrification. Nano Lett. **14**, 3208–3213 (2014)
47. Z. Wen, J. Chen, M.-H. Yeh, H. Guo, Z. Li, X. Fan, T. Zhang, L. Zhu, Z.L. Wang, Blow-driven triboelectric nanogenerator as an active alcohol breath analyzer. Nano Energy **16**, 38–46 (2015)
48. Q. Zheng, B. Shi, F. Fan, X. Wang, L. Yan, W. Yuan, S. Wang, H. Liu, Z. Li, Z.L. Wang, In vivo powering of pacemaker by breathing-driven implanted triboelectric nanogenerator. Adv. Mater. **26**, 5851–5856 (2014)
49. J. Yang, J. Chen, Y. Su, Q. Jing, Z. Li, F. Yi, X. Wen, Z. Wang, Z.L. Wang, Eardrum-inspired active sensors for self-powered cardiovascular system characterization and throat-attached anti-interference voice recognition. Adv. Mater. **27**, 1316–1326 (2015)
50. R. Zhang, S. Wang, M.H. Yeh, C. Pan, L. Lin, R. Yu, Y. Zhang, L. Zheng, Z. Jiao, Z.L. Wang, A streaming potential/current-based microfluidic direct current generator for self-powered nanosystems. Adv. Mater. (2015)
51. P.K. Yang, L. Lin, F. Yi, X. Li, K.C. Pradel, Y. Zi, C.I. Wu, J.H. He, Y. Zhang, Z.L. Wang, A flexible, stretchable and shape-adaptive approach for versatile energy conversion and self-powered biomedical monitoring. Adv. Mater. **27**, 3817–3824 (2015)
52. J. Chen, G. Zhu, J. Yang, Q. Jing, P. Bai, W. Yang, X. Qi, Y. Su, Z.L. Wang, Personalized keystroke dynamics for self-powered human–machine interfacing. ACS Nano **9**, 105–116 (2015)
53. X. Wang, S. Wang, Y. Yang, Z.L. Wang, Hybridized electromagnetic-triboelectric nanogenerator for scavenging air-flow energy to sustainably power temperature sensors. ACS Nano **9**, 4553–4562 (2015)
54. X. Fan, J. Chen, J. Yang, P. Bai, Z. Li, Z.L. Wang, Ultrathin, rollable, paper-based triboelectric nanogenerator for acoustic energy harvesting and self-powered sound recording. ACS Nano **9**, 4236–4243 (2015)
55. M. Kanik, M.G. Say, B. Daglar, A.F. Yavuz, M.H. Dolas, M.M. El-Ashry, M. Bayindir, A motion- and sound-activated, 3D-printed, chalcogenide-based triboelectric nanogenerator. Adv. Mater. **27**, 2367–2376 (2015)
56. Z.H. Lin, G. Cheng, S. Lee, K.C. Pradel, Z.L. Wang, Harvesting water drop energy by a sequential contact-electrification and electrostatic-induction process. Adv. Mater. **26**, 4690–4696 (2014)
57. J. Chun, J.W. Kim, W.-S. Jung, C.-Y. Kang, S.-W. Kim, Z.L. Wang, J.M. Baik, Mesoporous pores impregnated with Au nanoparticles as effective dielectrics for enhancing triboelectric nanogenerator performance in harsh environments. Energy Environ. Sci. **8**, 3006–3012 (2015)

58. Y. Mao, D. Geng, E. Liang, X. Wang, Single-electrode triboelectric nanogenerator for scavenging friction energy from rolling tires. Nano Energy **15**, 227–234 (2015)
59. J.M. Wu, C.K. Chang, Y.T. Chang, High-output current density of the triboelectric nanogenerator made from recycling rice husks. Nano Energy **19**, 39–47 (2016)
60. Y. Yang, G. Zhu, H. Zhang, J. Chen, X. Zhong, Z.H. Lin, Y. Su, P. Bai, X. Wen, Z.L. Wang, Triboelectric nanogenerator for harvesting wind energy and as self-powered wind vector sensor system. ACS Nano **7**, 9461–9468 (2013)
61. X. Wang, S. Niu, Y. Yin, F. Yi, Z. You, Z.L. Wang, Triboelectric nanogenerator based on fully enclosed rolling spherical structure for harvesting low-frequency water wave energy. *Advanced* Energy Mater. (2015)
62. J. Chen, J. Yang, Z. Li, X. Fan, Y. Zi, Q. Jing, H. Guo, Z. Wen, K.C. Pradel, S. Niu, Z.L. Wang, Networks of triboelectric nanogenerators for harvesting water wave energy: a potential approach toward blue energy. ACS Nano **9**, 3324–3331 (2015)
63. G. Zhu, Y. Su, P. Bai, J. Chen, Q. Jing, W. Yang, Z.L. Wang, Harvesting water wave energy by asymmetric screening of electrostatic charges on a nanostructured hydrophobic thin-film surface. ACS Nano **8**, 6031–6037 (2014)
64. S. Wang, X. Mu, X. Wang, A.Y. Gu, Z.L. Wang, Y. Yang, Elasto-aerodynamics-driven triboelectric nanogenerator for scavenging air-flow energy. ACS Nano **9**, 9554–9563 (2015)
65. Y. Xie, S. Wang, L. Lin, Q. Jing, Z.H. Lin, S. Niu, Z. Wu, Z.L. Wang, Rotary triboelectric nanogenerator based on a hybridized mechanism for harvesting wind energy. ACS Nano **7**, 7119–7125 (2013)
66. H.R. Zhu, W. Tang, C.Z. Gao, Y. Han, T. Li, X. Cao, Z.L. Wang, Self-powered metal surface anti-corrosion protection using energy harvested from rain drops and wind. Nano Energy **14**, 193–200 (2015)
67. H. Zhang, Y. Yang, X. Zhong, Y. Su, Y. Zhou, C. Hu, Z.L. Wang, Single-electrode-based rotating triboelectric nanogenerator for harvesting energy from tires. ACS Nano **8**, 680–689 (2014)
68. S. Wang, Y. Xie, S. Niu, L. Lin, Z.L. Wang, Freestanding triboelectric-layer-based nanogenerators for harvesting energy from a moving object or human motion in contact and non-contact modes. Adv. Mater. **26**, 2818–2824 (2014)
69. G. Zhu, P. Bai, J. Chen, Z. Lin Wang, Power-generating shoe insole based on triboelectric nanogenerators for self-powered consumer electronics. Nano Energy **2**, 688–692 (2013)
70. Z.L. Wang, Triboelectric nanogenerators as new energy technology and self-powered sensors—principles, problems and perspectives. Faraday Discuss. **176**, 447–458 (2014)
71. G. Zhu, B. Peng, J. Chen, Q. Jing, Z. Lin Wang, Triboelectric nanogenerators as a new energy technology: from fundamentals, devices, to applications. Nano Energy **14**, 126–138 (2015)
72. S. Wang, L. Lin, Z.L. Wang, Triboelectric nanogenerators as self-powered active sensors. Nano Energy **11**, 436–462 (2015)
73. X.-S. Zhang, M.-D. Han, B. Meng, H.-X. Zhang, High performance triboelectric nanogenerators based on large-scale mass-fabrication technologies. Nano Energy **11**, 304–322 (2015)
74. R. Hinchet, W. Seung, S.W. Kim, Recent progress on flexible triboelectric nanogenerators for selfpowered electronics. Chemsuschem **8**, 2327–2344 (2015)
75. Y. Yang, H. Zhang, R. Liu, X. Wen, T.C. Hou, Z.L. Wang, Fully enclosed triboelectric nanogenerators for applications in water and harsh environments. Adv. Energy Mater. **3**, 1563–1568 (2013)
76. F.R. Fan, J. Luo, W. Tang, C. Li, C. Zhang, Z. Tian, Z.L. Wang, Highly transparent and flexible triboelectric nanogenerators: performance improvements and fundamental mechanisms. J. Mater. Chem. A **2**, 13219–13225 (2014)
77. Y.H. Ko, G. Nagaraju, S.H. Lee, J.S. Yu, PDMS-based triboelectric and transparent nanogenerators with ZnO nanorod arrays. ACS Appl. Mater. Interfaces. **6**, 6631–6637 (2014)
78. F.-R. Fan, L. Lin, G. Zhu, W. Wu, R. Zhang, Z.L. Wang, Transparent triboelectric nanogenerators and self-powered pressure sensors based on micropatterned plastic films. Nano Lett. **12**, 3109–3114 (2012)

79. B. Meng, W. Tang, Z.-H. Too, X. Zhang, M. Han, W. Liu, H. Zhang, A transparent single-friction-surface triboelectric generator and self-powered touch sensor. Energy Environ. Sci. **6**, 3235–3240 (2013)
80. K.Y. Lee, M.K. Gupta, S.-W. Kim, Transparent flexible stretchable piezoelectric and triboelectric nanogenerators for powering portable electronics. Nano Energy (2014)
81. L.M. Zhang, F. Xue, W.M. Du, C.B. Han, C. Zhang, Z.L. Wang, Transparent paper-based triboelectric nanogenerator as a page mark and anti-theft sensor. Nano Res. **7**, 1215–1223 (2014)
82. L. Zheng, Z.-H. Lin, G. Cheng, W. Wu, X. Wen, S. Lee, Z.L. Wang, Silicon-based hybrid cell for harvesting solar energy and raindrop electrostatic energy. Nano Energy **9**, 291–300 (2014)
83. L. Lin, Y. Xie, S. Niu, S. Wang, P.K. Yang, Z.L. Wang, Robust triboelectric nanogenerator based on rolling electrification and electrostatic induction at an instantaneous energy conversion efficiency of approximately 55%. ACS Nano **9**, 922–930 (2015)
84. W. Seung, M.K. Gupta, K.Y. Lee, K.S. Shin, J.H. Lee, T.Y. Kim, S. Kim, J. Lin, J.H. Kim, S.W. Kim, Nanopatterned textile-based wearable triboelectric nanogenerator. ACS Nano **9**, 3501–3509 (2015)
85. G. Zhu, Z.H. Lin, Q. Jing, P. Bai, C. Pan, Y. Yang, Y. Zhou, Z.L. Wang, Toward large-scale energy harvesting by a nanoparticle-enhanced triboelectric nanogenerator. Nano Lett. **13**, 847–853 (2013)
86. B. Dudem, Y.H. Ko, J.W. Leem, S.H. Lee, J.S. Yu, Highly Transparent and flexible triboelectric nanogenerators with subwavelength-architectured polydimethylsiloxane by a nanoporous anodic aluminum oxide template. ACS Appl. Mater. Interfaces **7**, 20520–20529 (2015)
87. D. Choi, D. Yoo, D.S. Kim, One-step fabrication of transparent and flexible nanotopographical-triboelectric nanogenerators via thermal nanoimprinting of thermoplastic fluoropolymers. Adv. Mater. **27**, 7386–7394 (2015)
88. X.S. Zhang, M.D. Han, R.X. Wang, F.Y. Zhu, Z.H. Li, W. Wang, H.X. Zhang, Frequency-multiplication high-output triboelectric nanogenerator for sustainably powering biomedical microsystems. Nano Lett. **13**, 1168–1172 (2013)
89. C.K. Jeong, K.M. Baek, S. Niu, T.W. Nam, Y.H. Hur, D.Y. Park, G.T. Hwang, M. Byun, Z.L. Wang, Y.S. Jung, K.J. Lee, Topographically-designed triboelectric nanogenerator via block copolymer self-assembly. Nano Lett. **14**, 7031–7038 (2014)
90. D. Kim, S.-B. Jeon, J.Y. Kim, M.-L. Seol, S.O. Kim, Y.-K. Choi, High-performance nanopattern triboelectric generator by block copolymer lithography. Nano Energy **12**, 331–338 (2015)
91. K.Y. Lee, J. Chun, J.H. Lee, K.N. Kim, N.R. Kang, J.Y. Kim, M.H. Kim, K.S. Shin, M.K. Gupta, J.M. Baik, S.W. Kim, Hydrophobic sponge structure-based triboelectric nanogenerator. Adv. Mater. **26**, 5037–5042 (2014)
92. A. Diaz, R. Felix-Navarro, A semi-quantitative tribo-electric series for polymeric materials: the influence of chemical structure and properties. J. Electrostat. **62**, 277–290 (2004)
93. C. Yao, A. Hernandez, Y. Yu, Z. Cai, X. Wang, Triboelectric nanogenerators and power-boards from cellulose nanofibrils and recycled materials. Nano Energy **30**, 103–108 (2016)
94. S. Wang, Y. Xie, S. Niu, L. Lin, C. Liu, Y.S. Zhou, Z.L. Wang, Maximum surface charge density for triboelectric nanogenerators achieved by ionized-air injection: methodology and theoretical understanding. Adv. Mater. **26**, 6720–6728 (2014)
95. Y. Mao, P. Zhao, G. McConohy, H. Yang, Y. Tong, X. Wang, Sponge-like piezoelectric polymer films for scalable and integratable nanogenerators and self-powered electronic systems. Adv. Energy Mater. **4** (2014)
96. C. Yao, X. Yin, Y. Yu, Z.Cai, X.D. Wang, Chemically functionalized natural cellulose materials for effective triboelectric nanogenerator development. Adv. Funct. Mater. **1700794** (2017)
97. T. Kondo, C. Sawatari, A Fourier transform infra-red spectroscopic analysis of the character of hydrogen bonds in amorphous cellulose. Polymer **37**, 393–399 (1996)
98. H. Yang, R. Yan, H. Chen, D.H. Lee, C. Zheng, Characteristics of hemicellulose, cellulose and lignin pyrolysis. Fuel **86**, 1781–1788 (2007)
99. B.C. Beard, Appl. Surf. Sci. **45**, 221–227 (1990)

100. V. Kovalenko, R. Mukhamadeeva, L. Maklakova, N. Gustova, Interpretation of the IR spectrum and structure of cellulose nitrate. J. Struct. Chem. **34**, 540–547 (1994)
101. R.L. Oliveira, J.G. Vieira, H.S. Barud, R. Assunção, G.R. Filho, S.J. Ribeiro, Y. Messadeqq, Synthesis and characterization of methylcellulose produced from bacterial cellulose under heterogeneous condition. J. Braz. Chem. Soc. **26**, 1861–1870 (2015)
102. P.L. Nasatto, F. Pignon, J.L. Silveira, M.E.R. Duarte, M.D. Noseda, M. Rinaudo, Methylcellulose, a cellulose derivative with original physical properties and extended applications. Polymers **7**, 777–803 (2015)
103. R.G. Viera, G. Rodrigues Filho, R.M. de Assunção, C.d.S. Meireles, J.G. Vieira, G.S. de Oliveira, Synthesis and characterization of methylcellulose from sugar cane bagasse cellulose. Carbohydr. Polym. **67**, 182–189 (2007)
104. Y. Zi, S. Niu, J. Wang, Z. Wen, W. Tang, Z.L. Wang, Standards and figure-of-merits for quantifying the performance of triboelectric nanogenerators. Nat. Commun. **6** (2015)
105. B. Yang, C. Yao, Y. Yu, Z. Li, X.D. Wang, Nature flexible, and transparent conductive substrates from green and earth-abundant materials. Sci. Rep. **7**, 4936 (2017)
106. Y. Geng, L. Guo, S.-S. Xu, Q.-Q. Sun, S.-J. Ding, H.-L. Lu, D.W. Zhang, Influence of Al doping on the properties of ZnO thin films grown by atomic layer deposition. J. Phys. Chem. C **115**, 12317–12321 (2011)
107. J. Lu, Z. Ye, Y. Zeng, L. Zhu, L. Wang, J. Yuan, B. Zhao, Q. Liang, Structural, optical, and electrical properties of (Zn, Al) O films over a wide range of compositions. J. Appl. Phys. **100**, 073714 (2006)
108. Q. Hou, F. Meng, J. Sun, Electrical and optical properties of Al-doped ZnO and ZnAl 2 O 4 films prepared by atomic layer deposition. Nanoscale Res. Lett. **8**, 144 (2013)
109. C.H. Ahn, H. Kim, H.K. Cho, Deposition of Al doped ZnO layers with various electrical types by atomic layer deposition. Thin Solid Films **519**, 747–750 (2010)
110. P. Banerjee, W.-J. Lee, K.-R. Bae, S.B. Lee, G.W. Rubloff, Structural, electrical, and optical properties of atomic layer deposition Al-doped ZnO films. J. Appl. Phys. **108**, 043504 (2010)
111. M. Benoy, E. Mohammed, M. Suresh Babu, P. Binu, B. Pradeep, Thickness dependence of the properties of indium tin oxide (ITO) FILMS prepared by activated reactive evaporation. Braz. J. Phys. **39**, 629–632 (2009)
112. Y.H. Kim, C. Sachse, M.L. Machala, C. May, L. Müller-Meskamp, K. Leo, Highly conductive PEDOT: PSS electrode with optimized solvent and thermal post-treatment for ITO-free organic solar cells. Adv. Func. Mater. **21**, 1076–1081 (2011)
113. E.H. Sondheimer, The mean free path of electrons in metals. Adv. Phys. **1**, 1–42 (1952)
114. A. Mayadas, M. Shatzkes, Electrical-resistivity model for polycrystalline films: the case of arbitrary reflection at external surfaces. Phys. Rev. B **1**, 1382 (1970)
115. J. Tauc, The optical properties of solids. Opt. Prop. Solids (1966)
116. F. Wong, M. Fung, S. Tong, C. Lee, S. Lee, Flexible organic light-emitting device based on magnetron sputtered indium-tin-oxide on plastic substrate. Thin Solid Films **466**, 225–230 (2004)
117. S. Bae, H. Kim, Y. Lee, X. Xu, J.-S. Park, Y. Zheng, J. Balakrishnan, T. Lei, H.R. Kim, Y.I. Song, Roll-to-roll production of 30-inch graphene films for transparent electrodes. Nat. Nanotechnol. **5**, 574 (2010)
118. J. Yun, Y.H. Park, T.-S. Bae, S. Lee, G.-H. Lee, Fabrication of a completely transparent and highly flexible ITO nanoparticle electrode at room temperature. ACS Appl. Mater. Interfaces **5**, 164–172 (2012)

Chapter 12
Biomedical Applications of Nanocellulose

Muhammad Wajid Ullah, Manjila Adhikari, Omar Mohammad Atta, Umer Farooq, Mazhar Ul-Islam, Ajmal Shahzad, Sehrish Manan, and Guang Yang

Abstract Nanocellulose having at least one dimension equal or less than 100 nm is the most abundant polymer on Earth. It is mainly obtained from the plants as well as produced by some microorganisms, algae, and animals, and synthetized in vitro by the cell-free enzyme systems. Over the past few decades, nanocellulose is receiving tremendous attention for its applications in different areas owing to its easy availability, renewability, and unique structural and chemical features, and inert and biocompatible nature. Taking the advantages of such features, the current biomaterials science aims at developing nanocellulose-based functional biomaterials with healing and regenerative properties, with the aim to meet the human demand, especially in tissue engineering and regenerative medicines. This chapter provides an overview of the synthesis of different types of nanocellulose, with special focus given to the development of nanocellulose-based functional biomaterials for various biomedical applications, including the synthetic organs, delivery systems, tissue regeneration materials, biosensors, and the biological scaffolds.

M. W. Ullah · M. Adhikari · O. M. Atta · A. Shahzad · S. Manan (✉) · G. Yang (✉)
Biofuels Institute, School of the Environment and Safety Engineering, Jiangsu University, Zhenjiang 212013, China
e-mail: sehrish_manan@ujs.edu.cn

G. Yang
e-mail: gyang-hust@hust.edu.cn; yang_sunny@yahoo.com

M. W. Ullah · S. Manan
Department of Biomedical Engineering, College of Life Science and Technology, Huazhong University of Science and Technology, Wuhan 430074, PR China

U. Farooq
Advanced Biomaterials & Tissue Engineering Centre, Huazhong University of Science and Technology, Wuhan 430074, PR China

M. Ul-Islam
Department of Chemical Engineering, College of Engineering, Dhofar University, Salalah 211, Oman

L. Hu et al. (eds.), *Emerging Nanotechnologies in Nanocellulose*, NanoScience and Technology, https://doi.org/10.1007/978-3-031-14043-3_12

12.1 Introduction

Cellulose with fiber diameter in the nanoscale range is known as nanocellulose. It is mainly obtained from plants, produced by some microbial cells, and synthesized enzymatically by the cell-free systems. Regardless of source, the chemical structure of nanocellulose is same comprised of asymmetric chain of repeated glucose units where one terminus contains the reducing aldehyde (–CHO) group while the other end is non-reducing in character [1, 2]. Based on its dimension, function, and synthesis method, nanocellulose is categorized into three main types, including nanocrystalline cellulose (NCC) or cellulose nanocrystals (CNCs), nanofibrillated cellulose (NFC) or cellulose nanofibers (CNF), and bacterial cellulose (BC) or bacterial nanocellulose (BNC). Figure 12.1 illustrates a size-range comparison of different living and synthetic nano-sized materials with nanocellulose.

As a renewable and abundant natural polymer having been obtained from various sources and possessing varying morphological, physiological, and biological properties, the nanocellulose and nanocellulose-based composites find applications in different fields [3], including the biomedical [4–7], energy [8], pharmaceutics [9], infrastructure [10], additive manufacturing [11–13], paper [14], textile [15], environment [16], food [17–20], textile [21], and several others. This trend is supported by the increasing number of publications in last 10 years (Fig. 12.2a) in different research areas (Fig. 12.2b). Over the past couple of decades, different types of nanocellulose, especially BNC, have been extensively explored for biomedical applications such as artificial heart valves [22], blood vessels [23], bone and cartilage tissues [24, 25], drug delivery systems [26–29], regenerative medicines [30], wound healing [31–33], cosmetics [4], and 3D/4D printed scaffolds [34] owing to their unique features such as reticulate three-dimensional (3D) network [35], high mechanical strength, liquid and gas absorbing ability, hydrophilicity, biocompatibility, biodegradability, non-toxicity, transparency, high aspect ratio and surface area, moldabilitty, and unique

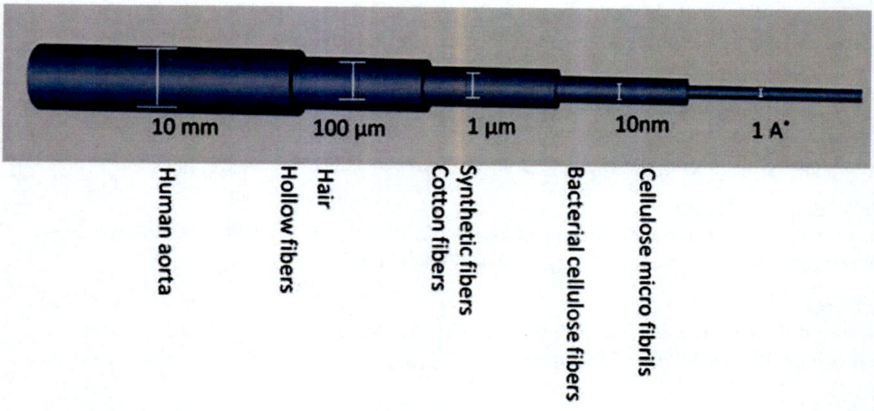

Fig. 12.1 Illustration of size-range comparison of nanocellulose with different natural and synthetic nanomaterials

surface chemistry [15, 36, 37]. The intrinsic properties of nanocellulose are modulated in several ways. For instance, plant-derived nanocellulose with varying degree of polymerization, mechanical strength, and structural morphologies is obtained through different treatment methods, i.e., by varying the reagents (acid and alkali) and mechanical treatment methods [38]. Similarly, the intrinsic properties of microbial-derived cellulose are tuned by changing the growth conditions, cultivation methods, strain type, and through the supplementation of additives [39–41]. The structural, chemical, mechanical, thermal, and biological features of all types of nanocellulose are also tuned by making their hybrids with other materials, such as natural and synthetic polymers [42–45], metals and ions [46–51], peptides [52, 53], ceramics [54, 55], clays [56, 57], plant extracts [58, 59], and others [36, 46, 60–63]. The surface chemistry as well as morphological features of nanocellulose could be modulated at macro-, micro-, and nano-scales [64]. All types of nanocellulose possess exciting features both in dry (i.e., membrane, powder, and aerogel) and wet forms (i.e., hydrogels) [65]; however, compared to other forms, BNC is anisotropic in nature and forms the web-shaped arrangement of fibrils, which is attributed to the motile nature of the BNC-producing microorganisms [66]. The anisotropic structure of BNC could be transformed into isotropic alignment by controlling the movement of microbial cells through various techniques such as microfluidic channels [67], honeycomb-patterned microgrooves [68], and by applying electric or magnetic fields [66, 69], as well as using different templates such as chitin [70] and poly (dimethyl siloxane) [71]. The improved alignment of BNC fibers accounts for the enhanced structural, mechanical, and biological features. For instance, the aligned fibers better respond to the applied pressure and support the growth of cells, and thus could find applications in the development of patterned scaffolds. The potential of BNC for various biomedical applications is supported by the high number of publications in the biomedical-related fields (Fig. 12.2c).

To date, different top-down and bottom-up approaches have been developed for fabrication of nanocellulose-based biomaterials for their applications in engineering of different tissue constructs and scaffolds. In the top-down approach, scaffolds are first fabricated followed by seeding of cells which then synthesize the extracellular matrix (ECM). In contrast, the cells are mixed with the scaffold materials followed by fabrication of microstructured tissues in the bottom-up approaches. The cells grow and form the large tissue constructs. Nanocellulose has been extensively utilized in both such approaches for the development of biological scaffolds. The unique nanofibrous structure of nanocellulose makes it a suitable material for mimicking the biological functionalities for wide range biomedical applications [50, 72]. The potential of nanocellulose has been explored for the growth of different kinds of cells and tissues, such as neurons to bone cells and from cardiovascular devices to ocular implants.

This chapter provides an overview of different types of nanocellulose with emphasis on their types and biomedical applications. Different types of nanocellulose such as CNC, CNF, and BNC are described in terms of their synthesis, structure, properties, and applications. A large portion of chapter is devoted to various biomedical applications of different types of nanocellulose and their composites.

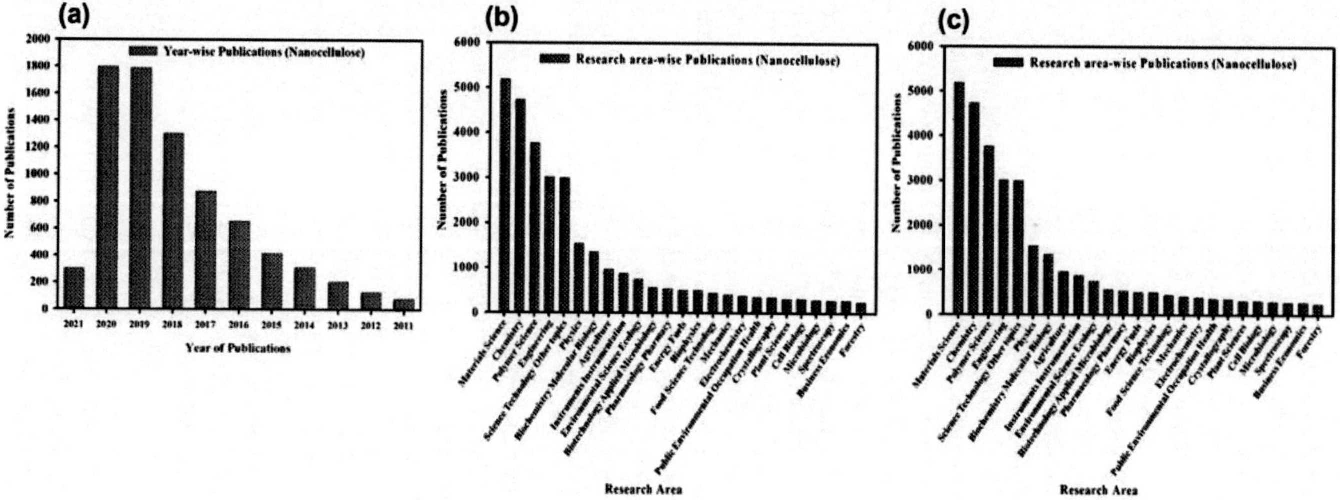

Fig. 12.2 Number of publications based on nanocellulose with respect to **a** year of publication and **b** area of research (*Source* Clarivate analytics-web of science; search key word (title): Nanocellulose). **c** Number of publications of bacterial cellulose in biomedical-related areas (*Source* Clarivate analytics-web of science; search key word (title): Bacterial cellulose). Data excessed on March 19, 2021

Finally, it describes various challenges associated with the existing synthesis methods and potential of nanocellulose-based functional materials for advanced biomedical applications.

12.2 Nanocellulose Types

12.2.1 Nanocrystalline Cellulose

Nanocrystalline cellulose (NCC) or cellulose nanocrystals (CNCs) are synthesized through acid hydrolysis of amorphous region of native cellulose in a two-step process: pre-treatment of raw materials and acid hydrolysis. The acid only hydrolyzes the amorphous region without disturbing the crystalline regions of native cellulose [1]. These are also called as cellulose nanowhiskers (CNWs) due to their appearance of a whisker. CNCs appear as the rod-like materials under transmittance electron macroscopic (TEM). The length and diameter of CNCs are dependent on the type of starting raw material, type of acid used, and the hydrolysis conditions such as temperature. A study reported the synthesis of CNCs with diameter 200 to 300 nm and length 10–30 nm through hydrolysis of cotton fibers at 65 °C using sulfuric acid. In contrast, the hydrolysis of ascidian under the same experimental conditions produced cellulose crystals at micron scale [73], thus confirming the importance of start material for the synthesis of CNCs of desired dimensions. A demonstration of acid hydrolysis, surface chemistry, and physical properties of CNCs is presented in Fig. 12.3.

CNCs have gained significant interest in different areas due to their superior physiochemical properties and low manufacturing cost as compared to the glass fiber and carbon nanotubes [74]. Also these are non-toxic, biocompatible, biodegradable, and possess excellent physical properties, such as high surface area, mechanical strength, and large number of free hydroxyl groups on their surface, which provide excellent opportunities for surface modification and applications in different areas (Fig. 12.3). Its applications scope is further expanded when combined with other materials. The microscopic (physical and surface chemistry) and macroscopic (rheology and stability) properties of CNC greatly affect its usage in wastewater treatment, electronics, biomedical, industrial, and also in energy sectors [75–77]. The surface of CNCs is modified for various biomedical applications such as antiviral and antibacterial agent as well as scaffolds for biomarkers or detectors, gene vectors, drug delivery carriers, tissue engineering, and biocatalyst scaffolds [78]. The addition of silver nanoparticles to the surface of CNCs turned them into an effective antimicrobial agent [79]. CNCs have been commonly reviewed for their tissue engineering application owing to their ability to support cell adhesion and growth as well as enhancing the features of polymer matrices [80]. CNCs are commonly used for protein immobilization as these form a stable membrane that can improve the protein activity in the surrounding environment and facilitates its removal from the media [81].

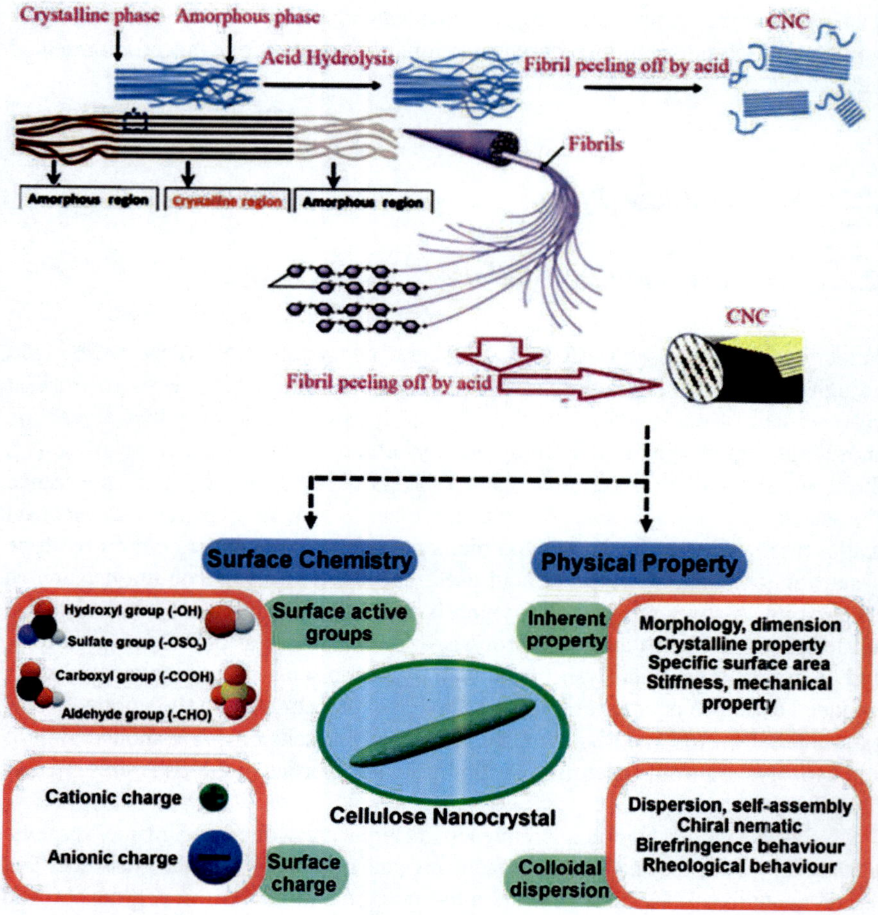

Fig. 12.3 Illustration of synthesis of cellulose nanocrystals (CNC) through acid hydrolysis and their physical and chemical properties

12.2.2 Nanofibrillated Cellulose

Nanofibrillated cellulose (NFC) or cellulose nanofibers (CNFs) are synthesized in a much simpler way than CNCs, and are generally produced through physico-mechanical and chemical treatment [82]. For example, CNFs are produced physico-mechanically via grinding, homogenization, ultrasonication, and chemically through 2,2,6,6-tetramethylpiperidine-1-oxy (TEMPO)-oxidation under magnetic stirring, which result in formation of individual CNFs [83, 84]. Under certain circumstances, both physical and chemical methods are employed together for the synthesis of CNFs. For example, carboxymethylation together with high-pressure homogenization lead to formation of uniformly distributed CNFs. CNFs are generally prepared from wood as well as other materials such as cotton, tunicate, and agricultural residues (e.g., canola straw and sugar cane bagasse) [85, 86]. The physical properties such as diameter of CNFs depend on the synthesis method and the raw material. Due to high aspect ratio, CNFs are preferred materials as enhancer for the polymer enhancement. A cellulose fiber is typically comprised of three major nanostructures including nanocrystal, primary fibril, and nanofibril cellulose [87]. Figure 12.4 represents the different types of CNFs obtained from different raw materials. As a chemical synthesis method, TEMPO oxidation is one of the most effective nanocellulose pretreatment processes with minimum energy consumption [88]. The TEMPO-mediated cellulose oxidation translates the C6 primary cellulose hydroxyl groups into C6 water carboxylates in the pH range 10–11. This method is often used for separation of nanofibrils from the pulp of wood and linters of cotton [89]. The TEMPO-oxidized CNFs are seen as nano-size biomaterial because of their amorphous and crystalline areas as compared to the simple CNFs [90].

CNFs have been focused by researchers owing to their flexibility and strength, which can be used for the development of hydrogels and films compared to the native cellulose [91]. Furthermore, CNFs have good dispersion ability in water without the need for harsh organic solvents or severe pH changes and are renewable materials [92]. Due to their non-toxic nature, CNFs are exciting materials for the development of eco-friendly nanocomposites and functional materials [93]. Importantly, a

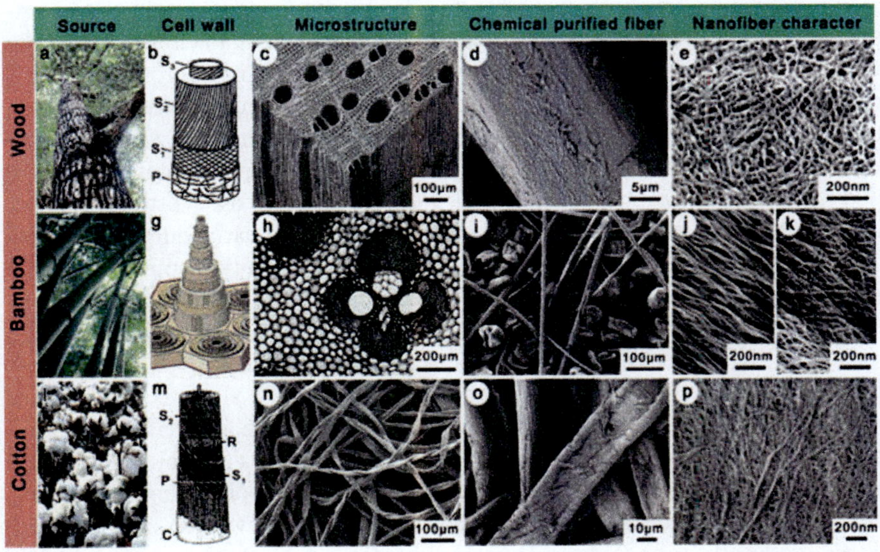

Fig. 12.4 Utilization of cell wall of wood, bamboo, and cotton for preparation of cellulose nanofibers. **a, f, l** digital photographs, **b, g, m** typical cell wall models of the selected plants, **c, h, n** microstructures of raw materials, **d, i, o** TEM micrographs of chemically purified cellulose after removal of the matrix, and **e, j, k, p** representative images of cellulose nanofibers obtained from the selected plants. The grouped figure reproduced from [97]

hydrogel composite of CNF with Ag nanoparticles was able to regulate the evaporation of water and showed excellent antimicrobial activity and non-toxic characteristics, thus showing its potential in wound healing [94]. Also, CNFs have a large axial modulus (~150 GPa) in its cellulose nanostructure [95], and thus can compete with the commercial fibers like polyamide, carbon, and glass [95, 96].

12.2.3 Bacterial Cellulose/Nanocellulose

Bacterial nanocellulose (BNC), also known as bacterial cellulose (BC), microbial cellulose (MC), or bio-cellulose, is a type of nanocellulose produced by the microbial cells. BNC is known to be produced by acetic acid bacteria [98] and some algae and fungi by utilizing specific media containing different carbon sources [99, 100]. Compared to the nanocellulose obtained from different biomasses (plant sources) [101], the microbial cells produce the purest form of cellulose in the form of β-1,4-glucan chains which are protruded out as protofibrils through the terminal complexes (TCs) present in the outer cellular membrane. The fibrils crystallize and form the ribbon-shaped microfibrils which then form pellicles at the top of liquid medium. The pellicles are comprised of bundles, which then form ribbon [102–104] (Fig. 12.5a).

Currently, BNC research is mainly focused on its low-cost production, identification of novel microbial strains and modifying the known BC-producing strains through genetic and metabolic engineering strategies, its improved structural features, and development of its functional materials for applications in different fields. For low-cost production of BNC, the potential of different wastes such as sugarcane molasses [105], agricultural wastes [106], industrial wastes [107, 108], food wastes [109], and beer wastes [110], as well as low-cost substrates such as fruit juices [111] has been explored as carbon sources. Similarly, different fermentation strategies, developing engineered strains, and co-culturing approached have been developed for enhancing the yield and productivity of BNC [39]. The addition of various additives such as carboxymethyl cellulose (CMC) [112], organic acids [113], lignosulfonate [114], sugar-linked glucuronic acid-based oligosaccharides [115], agar [116], xanthan [117], sodium alginate [118], alcohols [119], acetan [120], dextran [117], and yeast extracts [121] to the culture medium has significantly improved the structural, physico-chemical, mechanical, thermal, and biological properties of BNC. The yield, productivity, structural morphology, physiological features, molar mass and molar mass distribution, and production of unwanted metabolites by the BC-producers can be controlled by selecting appropriate substrates, culture conditions, suitable additives, and active bacterial strain [51, 122]. Moreover, the addition of nanoparticles, polymers, clay materials, and other active agents to its matrix by various strategies, including in situ, ex situ, and regeneration methods [37, 51, 123] not only enhance its existing properties but also bless it with "added properties" such as antimicrobial and catalytic activities, conductive and magnetic features, and sensing ability for various applications in tissue engineering, regenerative medicines, biosensing, drug delivery, and others [50, 72, 124].

12.2.4 Bio-cellulose

Park and co-workers reported the development of a cell-free enzyme system through bead beating of *Gluconacetobacter hansenii* PJK and utilized the cellular lysate as a source of enzymes and cofactors for synthesis of cellulose in a cell-free environment. The in vitro synthesized cellulose was termed as "bio-cellulose" that was chemically same as BNC, nevertheless possessed enhanced thermal, mechanical, and morphological features [102, 125, 126]. The concept of cell-free synthesis of biocommodities is not a new approach and was first reported by Buchner for the production of bioethanol [127]. The cell-free enzyme system represents an efficient, controllable, feasible, and cost-effective approach for the synthesis of a variety of bioproducts [128, 129].

The cell-free system follows the same chemical route of cellulose synthesis as by the microbial cells for BNC production; however, both systems differ in three main aspects. (1) unlike microbial cell system where β-1,4-glucan chains are produced inside the microbial cells, the enzymes synthesize these in the culture medium (Fig. 12.5b) which then self-aggregate and form the micro/macrofibrils (Fig. 12.5c) and pellicles (Fig. 12.5d). The pellicles then move to the top of the medium due to the density gradient and form a cellulose hydrogel (Fig. 12.5e). (2) Due to the absence of external barrier in the form of a cell wall or a membrane in the cell-free enzyme systems, the complex energy-dependent extrusion of nanocellulose fibers is bypassed [102]. (3) Unlike the microbial cell system, the older nanofibrils are present at top surface while the newly synthesized nanofibrils are added at the bottom surface, indicating the independency of this process for oxygen supply [102]. A schematic illustration of cellulose synthesizing machinery in microbial and cell-free system is shown in Fig. 12.6. Compared to the microbial cell system, a cell-free system is advantageous in that it offers high yield (>58%), anaerobic synthesis, simple chemical composition of the production medium, energy-deficient process, and supporting batches operation [102, 125]. The nanocellulose produced by cell-free system is different from BNC in terms of fiber density, diameter, and compactness, as well as possesses stronger hydrogen bonding which in turn contribute to high water and other liquids' absorption capacity and controlled release. Together, all these features account for better mechanical and thermal properties [103].

12.3 Biomedical Applications

Nanocellulose offers promising applications in biomedical field owing to its versatile fiber morphology, high water and other liquid absorption capacity, controlled release of liquids and solids, 3D ultrafine structure, higher mechanical and thermal properties, and biocompatibility [130]. The concept of combining natural polymer with synthetic ones not only improves the biocompatibility of nanocellulose but also blesses it with

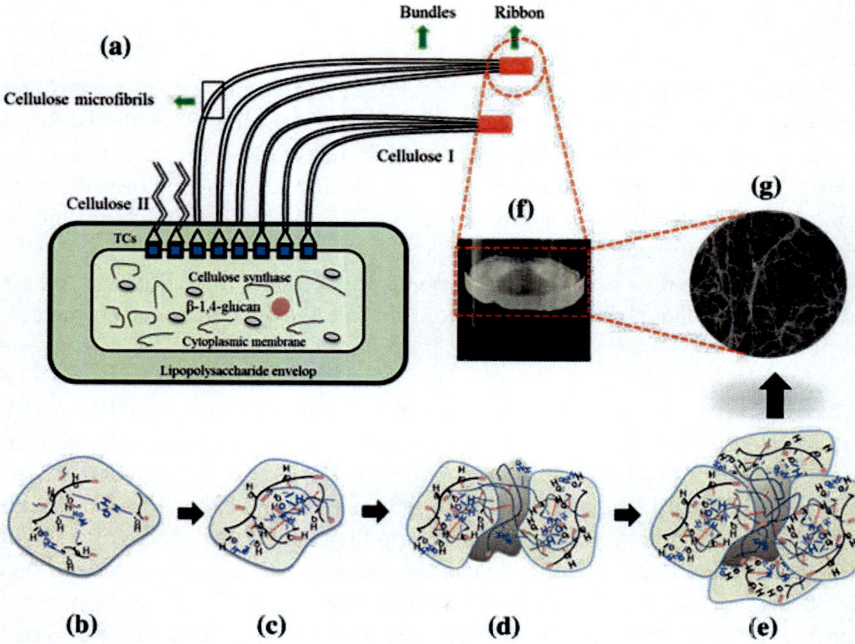

Fig. 12.5 **a** In vivo synthesis and extrusion of β-1,4-glucan chains, **b–e** in vitro organization of cellulose fibrils into high-order structures, **f** formation of cellulose hydrogel at the air-medium interface, and **g** SEM micrograph of fibrous structure of BNC. Figure adapted from [22, 102]

additional properties [131]. Composite of nanocellulose with wide range of material has been prepared for various biomedical applications. Nanocellulose has been widely explored for its applications in biomedical field such as in the development of wound dressing and tissue regenerating materials, drug carrier, artificial gut and blood vessels, scaffolds for bone regeneration and nasal septal proliferation, enzyme immobilization, and biosensors [22, 63, 132, 133]. Nanocellulose could potentially replace the synthetic materials for various biomedical applications because it is biocompatible with better mechanical properties and environment-friendly nature. The size of synthetic and natural materials is a key determinant for choosing a suitable material for desired biomedical applications; however, unique requirements for specific cells allow to modify nanocellulose with desired properties for wide range of biomedical applications [134]. A wide variety of biomedical applications of nanocellulose are summarized in Figs. 12.7, 12.8, 12.9, 12.10 and 12.11 and Table 12.1.

12.3.1 Cardiovascular Devices

Artery embolism is one of the life-threating cardiovascular disease that causes blockage of blood flow in artery. Likewise, breakdown of vein due to thickening of

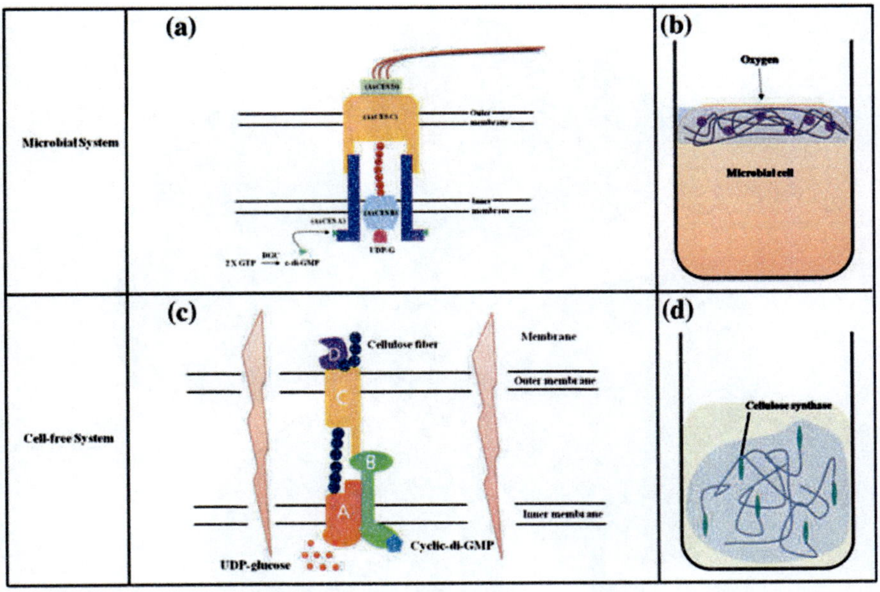

Fig. 12.6 Schematic illustration of **a**, **c** cellulose synthesis machinery in microbial and cell-free enzyme systems and **b**, **d** cellulose production and accumulation in the culture medium. Figure reproduced from [102]

arterial wall is another major cardiac problem. These problems could be potentially treated by developing artificial blood vessels [135]. Nanocellulose is hemocompatible and immunocompatible which makes it suitable for artificial organ implants. Studies have reported the development of various cardiovascular devices such as blood vessels, heart valves, aorta, prosthesis, and others by using different forms of nanocellulose as the scaffolds [22, 136]. An earlier study reported the development of PVA/CNF nanocomposite that demonstrated high mechanical strength comparable to that of the cardiovascular tissues such as heart valve leaflets and aorta [137]. In a similar study, a cellulose nanocomposite with poly (vinyl alcohol) (PVA) is highly experimented for its usefulness as an aortic heart valve prosthesis [138]. In another study, a tube-like structure was developed from BNC that supported the adhesion and proliferation of different types of cells such as human umbilical vein endothelial cells (HUVECs), smooth muscle cells (SMCs), and fibroblasts, and thus could be used as the artificial blood vessel for replacing the damaged and blocked veins [135]. The addition of fibrin protein to the nanocellulose/fibrin nanocomposite greatly improved the cell proliferation. The developed nanocomposite served as collagen-elastin inducer, and thus can find potential applications in vascular graft surgery [139]. Similarly, the potential of nanocellulose-based composite with peptides was investigated for the development of artificial blood vessels [23]. Similarly, composites of peptides and proteins with CMC and nanocellulose have been fabricated for the development of artificial blood vessel. Heparin, a naturally occurring glycosaminoglycan

Fig. 12.7 Illustration of biomedical applications of nanocellulose

having anticoagulant properties, was composited with BNC for its application as cardiovascular tissue engineering scaffold [140]. These and few other cardiovascular applications of nanocellulose and nanocellulose-based functional biomaterials are summarized in Fig. 12.8. Few cellulose-based products like BASYC®, Securian®, and SyntheCel® are already commercialized as cardiovascular implants [141].

12.3.2 Skin Tissues Regeneration

The use of nanocellulose in skin tissue regeneration is not new. The similarities in size of fiber diameter (32.08 ± 10.85 nm) and pores among the fibers (254.16 ± 76.65 nm) with the size of human fibroblast (10–15 nm) along with the non-toxic and biocompatible nature make BNC a suitable material for its application in developing tissue engineering scaffolds for treating damaged skin caused by injuries, burns, or microbial infections. The nanofibrous structure of cellulose provides the attachment sites for the growth of fibroblasts. In case of BNC, the upper surface is more dense and compact than lower surface, and thus it facilitates the cell attachment and migration into the lower highly porous matrix [141]. The porosity of nanocellulose can be further improved through post-synthesis treatment such as heating and freeze-drying

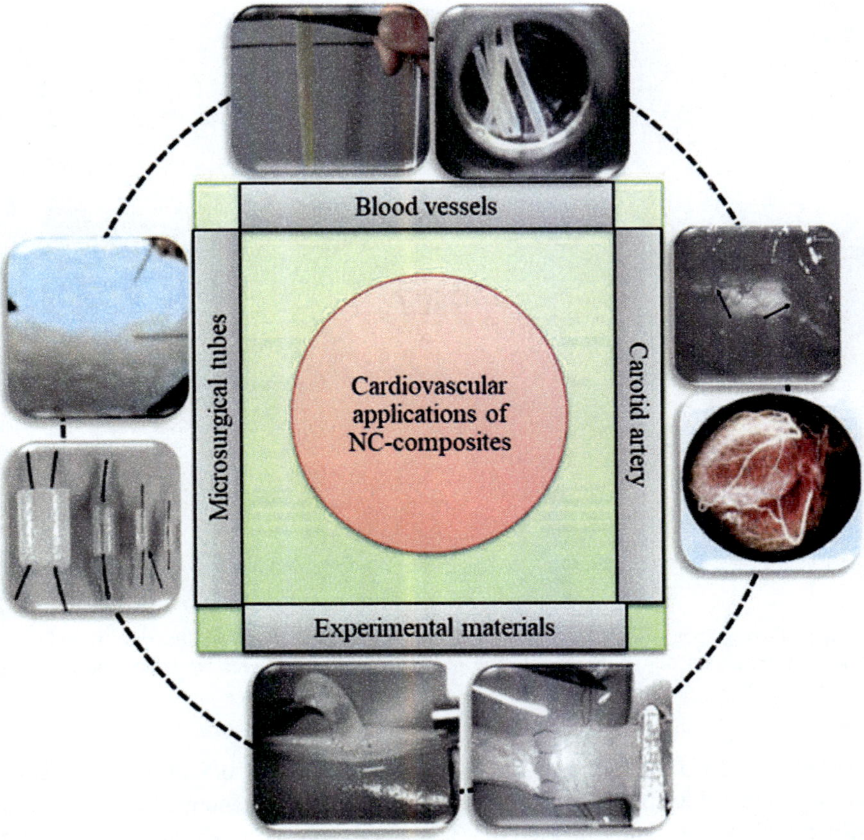

Fig. 12.8 Illustration of various cardiovascular applications of nanocellulose and nanocellulose-based functional biomaterials

which result in macroporous structure, without interrupting its fibrous morphology [142, 143]. Further, to meet the requirements for the growth of different cells, nanocellulose can be easily modified such as through the reinforcement of peptide, proteins, and other active molecules, which promote cell adhesion and proliferation. Following this approach, BNC was modified with collagen and fibronectin and cells-derived adhesive proteins to induce the fibroblast adhesion and proliferation, which make BNC suitable for soft tissue engineering applications [144]. BNC also provides a smooth surface for the adhesion, proliferation, and NIH-3T3 cells as well as their infiltration into the scaffolds, thus making it a suitable material as dressing for large area skin transplantation and development of artificial skin graft [145]. Antimicrobial activity is also essential for skin tissue regeneration. A study reported the development of ion crosslinked nanocellulose hydrogels which showed better antibacterial activity than non-crosslinked hydrogel [146]. In addition to fibrous BNC, the regenerated BNC is also receiving immense consideration for biomedical applications.

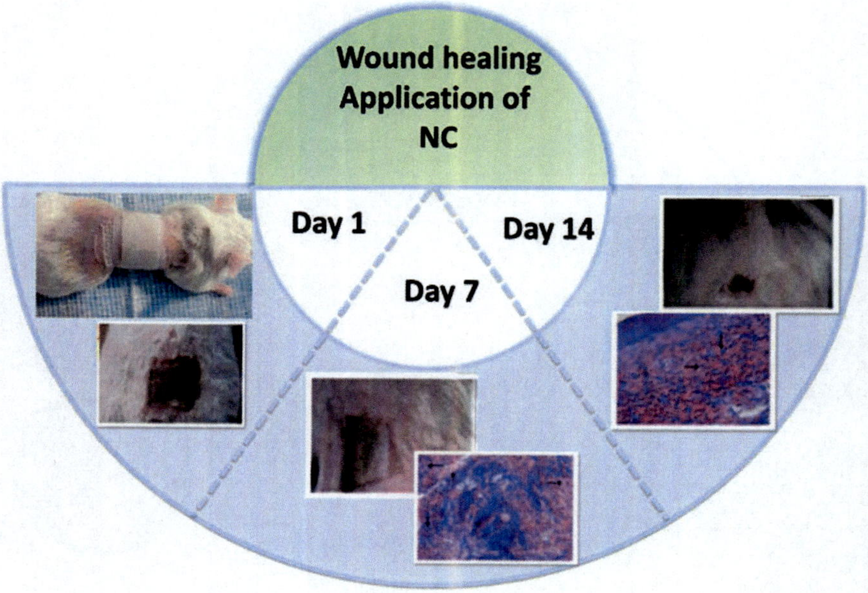

Fig. 12.9 Photographs of BNC-based wound dressings applied on full-thickness skin lesion induced in mice after day 1, 7, and 14. The photographs show progressive wound healing and closure

Regenerated BNC has porous geometry because of changing of cellulose I to cellulose II polymorphic structure [51]. In vivo evaluation of regenerated BNC showed proliferation of animal fibroblasts (NIH-3T3), indicating its suitability for skin tissue regeneration [147]. Besides, other types of nanocellulose are also used in the development of skin tissues. For example, Ghafari et al. developed a bilayer CNF/PVA aerogel scaffold through a one-step lyophilization approach. The developed scaffold possessed interconnected pores with two distinct pore sizes in each layer corresponding to the two layers of dermis and epidermis of the skin tissue. The developed CNF/PVA scaffold supported the adhesion and proliferation of fibroblasts [148].

12.3.3 Bone and Cartilage

Bone is the strongest and stiff tissue made up of organic materials collagen and non-collagenous proteins and an inorganic material hydroxyapatite [149–152], while cartilage is soft and elastic connective tissue. Cartilage has limited regeneration ability, thus it cannot be regenerated as other tissues of body [140]. The potential of several polymeric materials has been investigated to address the bone- and cartilage-defect-related problems. The similarity of structural morphology and diameter of

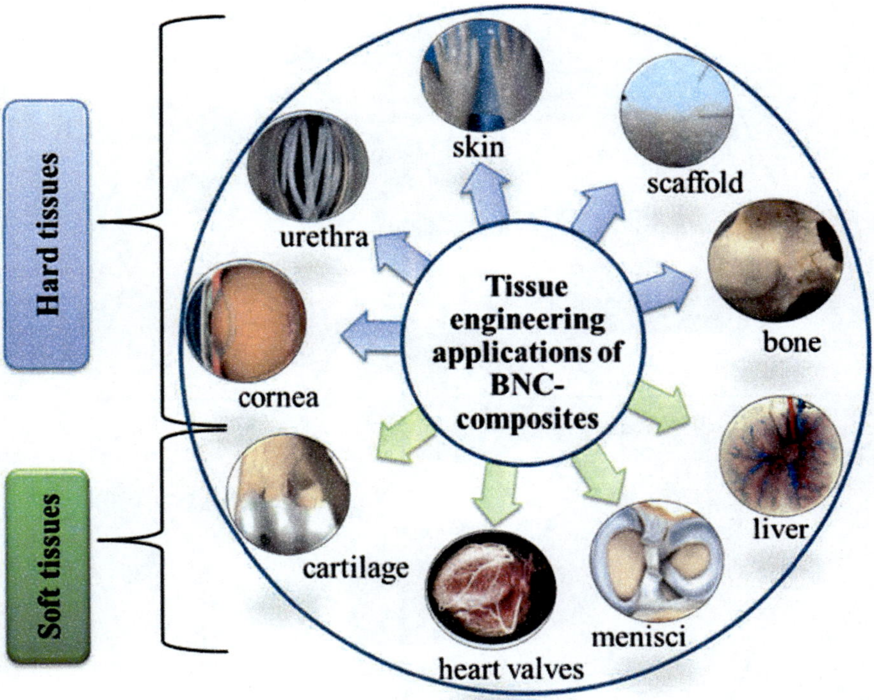

Fig. 12.10 Various biomedical applications of BNC-based composite materials. Figure reproduced from [161]

BNC fibrils with the collagen fibers (100 nm diameter) make them a suitable material for the development of soft tissues, for instance, cartilage. Also, BNC assists the growth of chondrocytes on its surface, further confirming its suitability for engineering of cartilage tissues [153]. Studies have reported the growth of nasoseptal chondrocytes on nanocellulose and alginate scaffolds, and thus these could be used as the auricular cartilage [24, 154]. BNC has also been reported to have potential to be used in replacement and repair of bone and cartilage tissues [24]. 3D printed human chondrocytes along with cellulose and alginate as bioink showed better cell proliferation properties [30]. Similarly, electrospun cellulose/hydroxyapatite nanocomposite nanofibers (ECHNN) with high mechanical properties supported the attachment and proliferation of human dental follicles, indicating its suitability for bone tissue engineering [155]. Small fibril network of BNC led to good adhesion, growth, and proliferation of human-adipose-derived stem cell (HASCs), which were successfully differentiated into an osteogenic lineage leading to regeneration of new bone in a 3D BNC scaffold [156]. A composite of nanocellulose, hydroxyapatite, and cellulose nanocrystals (CNC) was reported to be useful in bone tissue engineering [157]. A study reported the development of aligned electrospun cellulose/CNCs

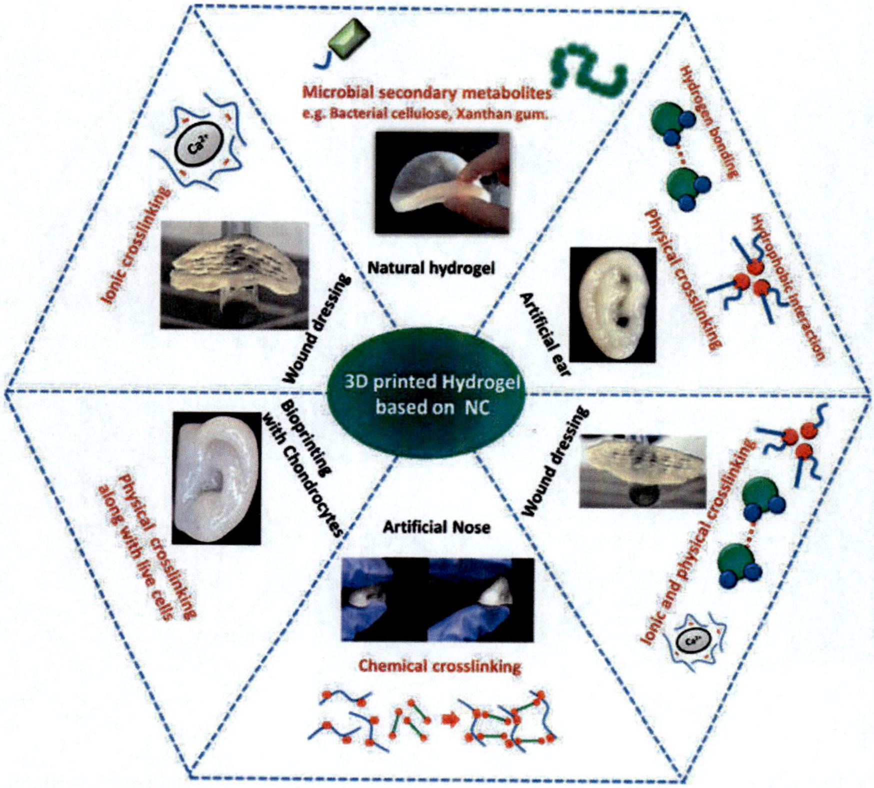

Fig. 12.11 3D printed and crosslinked hydrogels of nanocellulose for various biomedical applications

nanocomposite nanofibers, which supported the growth and osteogenic differentiation of human mesenchymal stem cells in the presence of bone morphogenic protein BMP-2. A better alignment of cells was observed in vitro and improved assembly of collagen proteins better favored cortical bone formation in vivo for aligned scaffold than non-aligned scaffold [158]. In a similar recent study, TEMPO-oxidized CNF/gelatin scaffolds with tuned degradation rate and high mechanical strength were developed through crosslinking with dehydrothermal treatment (DHT) or by a combination of hexamethylenediamine, genipin, and DHT. The scaffolds supported the adhesion, proliferation, spreading, and osteogenic differentiation of human bone marrow mesenchymal stem cells [159]. Among all kind of nanocellulose, BNC is less affected by the different sterilization methods [160] (Table 12.1).

Table 12.1 Enhanced properties of nanocellulose with diverse reinforcement materials for various biomedical applications

Cellulose types	Reinforcement materials	Enhanced properties	Applications	References
BNC	Chitosan	Biocompatible, antimicrobial	Tissue engineering, wound dressing, burn wound dressing	[217–219]
	Chitosan	Anticoagulant	Hemocompatible biomaterial	[220]
	Chitosan and diamond nanoparticles	Enhanced electric modulus, biocompatible	Wound dressings	[222]
	Xyloglucan	Enhanced mechanical properties, biocompatible	Characterize and design biomaterial	[203]
	–	Stripe ordered BC	Scaffolds for nerve, skeleton and hamstring	[156]
	–	Biocompatible	Large area skin transplantation	[145]
	–	Biocompatible and suitable mechanical properties	Artificial kidney and liver	[223]
	–	Bilayer, mechanical stability, porous	Neo-cartilage formation	[24]
	–	Aligned fiber	Neural cell proliferation	[156]
	–	pH-dependent release rate, 3D structure	Drug delivery	[224]
	–	Tubular structure, biocompatible	Blood vessels	[23]
	–	High cellulose content	Wound dressing material	[225]
	–	Tubular and biocompatible	Artificial blood vessels	[135]
	–	Multilayer, biocompatible	Skin tissue material	[226]
	–	Biocompatible, transparent, and suitable mechanical properties	Artificial cornea	[181]
	–	Transparent, high WHC	Ocular bandage	[227]

(continued)

Table 12.1 (continued)

Cellulose types	Reinforcement materials	Enhanced properties	Applications	References
	RGO	Biocompatible, conductive, hydrophilic	Bioelectronics, tissue engineering	[228]
	ZnO	Biocompatible, antimicrobial	Wound dressing	[229]
	GO	High tensile strength and biocompatible	Tissue engineering	[230]
	GO	Less crystalline	Neural cell proliferation	[164]
	GO	Highly porous, spherical, biocompatible	Wound healing	[167]
	PAni	pH sensitive and electro-conductive	Drug delivery	[26]
	Fibrin	Biocompatible	Vascular graft	[139]
	Peptide	Biocompatible	Blood vessels	[23]
	NaCl	Porous	Tissue engineering	[147]
	CNC, protein	Biocompatible	Bone tissue engineering	[158]
	MTT	Antimicrobial	Wound healing	[172]
	Vaccarin	Biocompatible	Wound dressing	[171]
	PVA	Suitable mechanical properties and anisotropic behaviour	Heart valve	[138]
	PVA	Transparent, UV absorbent, high mechanical strength, and thermal stability	Artificial cornea	[45]
	Hydroxyapatite, CNC	Improved thermal properties, biocompatible	Bone tissue engineering	[157]
	PMETAC	UV-A and UV-B blocker, viscoelastic, fungicidal, biocompatible	Drug delivery	[194]
CNF	Nanofibrous structure	Porous	3D cell culture	[204]
	–	–	Treatment of *C. albicans* infections	[176]

(continued)

Table 12.1 (continued)

Cellulose types	Reinforcement materials	Enhanced properties	Applications	References
	Chitosan and PVA	Antibacterial	Wound healing	[31]
	Alginate	3D structure, biocompatible	Cartilage regeneration	[30]
	Hydroxyapatite	Extraordinary mechanical properties, biocompatible	Bone tissue engineering	[155]
CNC	CNF, alginate	Controlled pore size, biocompatible	Tissue engineering	[160]
	Ag	Antimicrobial	Antimicrobial agent	[231]
	CMC, starch, and pectin	pH resistance	Probiotics delivery	[195]
	GO	Proximity sensing ability	Optoelectronic sensing devices	[198]
	GO	Flexible, transparent, conductive	Biofluid separation	[214]

12.3.4 Neuronal Tissue Regeneration

Neurons are the longest cells in human body; however, these have limited ability of repair in case of some damage. In contrast, the neuronal stem cells can self-renew and differentiate into all types of neurons in the presence of specific biological and topological clues. The differentiation of neural stem progenitor cell depends upon the stiffness of the scaffold. A study investigated the effect of surface stiffness of scaffolds on the differentiation ability of neural stem cell and concluded that the oligodendrocyte differentiation was better in stiff surface rather than the astrocyte. However, oligodendrocyte differentiation and myelination was observed in scaffold having surface stiffness less than 1 kPa [162]. Likewise, the fiber diameter is also important for the proliferation and differentiation of neural cells. For example, in the case of laminin-coated electrospun poly (ether sulfone) fiber, a decrease in fiber diameter resulted in improved cell proliferation and low aggregation [163]. Nanocarbon-hybridized cellulose hydrogels were prepared by in situ modification of BNC by using amphiphilic comb-like polymer (APCLP) as stabilizer. The developed BNC-based composite showed low crystallinity than the native cellulose which enhanced the neuronal bilayer formation. Similarly, a GO-BC composite demonstrated the formation of distinguished multilayers of neural networks with vertical axonal outgrowth showing the neuronal cell proliferation in vivo [164]. In another study, a conductive bioink comprised of CNF and CNTs was formulated and printed into 3D scaffolds under optimized conditions of pH-dependent surface charges of CNFs and CNTs.

The 3D CNF/CNT scaffolds supported the adhesion, proliferation, and differentiation of neurons in the 3D printed conductive guidelines (3.8×10^{-1} S/cm) of less than 1 mm diameter [165]. Besides the conventional approach of developing tissue scaffolds for repairing of neuronal tissue damage, the development of 3D printed scaffolds could be better choice for treating neural tissue damages such as spinal cord injury. A recent review discussed the potential development of neuron-based bioinks and their printing into neural tissues for spinal cord injury repair [166].

12.3.5 *Wound Healing*

Nanocellulose possesses high WHC and slow WRR, making it a suitable material for its application in wound dressing. The 3D fibrous structure of cellulose nanofibers due to its hydrophilic nature maintains a moist environment condition around the wound [49]. In case of burn wound healing, it may help to absorb the wounds exudates making nanocellulose a better burn wound-healing material. At the site of chronic wounds, the human neutrophil elastase enzyme (HNE) is predominantly secreted by human body which is responsible for breakdown of collagen and growth factors.

In order to monitor the HNE release, the nanocellulose-peptide biosensor was developed to monitor the healing of chronic wounds [139]. Hollow microspheres of BNC produced using microfluidic technique have also been used for wound healing [167]. However, as pure nanocellulose itself lacks antimicrobial properties, its composites with other reinforcement material, such as polymers and nanoparticles, have been prepared to bless it with antimicrobial properties [46, 47, 51, 141, 168]. For example, nanocellulose loaded with polyhexanide and povidone-iodine showed excellent antibacterial activity. Similarly, the use of nanoparticles such as Ag, Au, ZnO as well as antibiotics and antibacterial peptides as antimicrobial agents are not new; however, these are still used to bless nanocellulose with antimicrobial activities [60, 158]. Among them, Ag nanoparticles have both antibacterial and antifungal properties; however, its high concentration is toxic to human cells [169]. In addition to nanoparticles, composites of other materials such as chitosan [170], vaccarin [171], and montmorillonite [172], and chitosan and poly (ethylene oxide) [173] with nanocellulose have also shown considerable wound-healing ability. In a study, antibacterial and antifungal electrospun nanofibers based on CNC, chitosan, poly(ethylene oxide), and acacia plant-based extract were developed. The presence of CNC in the composite nanofibers blessed them with improved morphological, physical, and thermal properties. The developed nanofibers not only showed biocompatible properties against non-cancer cells but also demonstrated sustained release of acacia extract for up to 24 h. Thus, the developed CNC-based nanofibers system could be used as the localized drug delivery system for wound care applications [174].

The ability of nanocellulose to form translucent films with good liquid absorption capacity makes it an ideal choice for its application in the development of wound dressing materials where a considerable mechanical strength, moist environment, and an appropriate management of absorption of exudates are required. In a study, a nanocomposite of CNF with poly(ethylene glycol) was developed that showed high tensile strength, elongation, and elastic modulus in wet conditions, i.e., in water and phosphate buffer saline, and high water absorption capacity, thus demonstrating its potential to be used as a wound dressing material [175]. Similarly, the crosslinking ability of nanocellulose makes it an ideal platform for the development of functional hydrogels [65]. The BNC crosslinked with poly ([2-(methacryloyloxy) ethyl] trimethylammonium chloride) (PMETAC) showed antifungal properties against *Candida albicans*, which is also an important property for wound-healing materials [176]. Several commercial products based on nanocellulose are available for wound dressing, such as Bioprocess™, Biofill^R, Dermafill™, CELMAT® Wound/Eye/Face, NexFill®, and Suprasorb® Xb + PHMB [177]. Similarly, NFC-based hydrogels were developed through Ca^{2+} crosslinking. The developed hydrogels demonstrated fibrous network and solid-like behavior. The biocompatibility analysis showed the adhesion and proliferation of dermal fibroblasts on the hydrogels during incubation and even after removal of the scaffold. Moreover, the scaffolds showed non-toxic behavior when evaluated for cytokine secretion and generation of reactive oxygen species. These features demonstrate the potential applications of the crosslinked nanocellulose hydrogels as the wound dressing materials [178]. In another study, glutaraldehyde crosslinked NFC threads supported the adhesion, proliferation, and migration of human adipose mesenchymal stem cells for their application in wound dressing [179]. Currently, the 3D printing of nanocellulose with fibroblast cells is considered a safe approach for the development of biocompatible wound-healing materials as it allows to maintain the desired shape and size of material with better fibroblast cell proliferation properties [144, 180].

12.3.6 Eye Implants

Cornea is the most important structure in eye. Eyelids, conjunctiva, and tears gland play vital roles to protect cornea and maintain its transparency. There is no blood supply in cornea of eye, but very rich optical nerve fibers are present in eyes making it very sensitive to external stimuli. In condition for corneal disease, currently there is no synthetic cornea replacement available. To address this issue, extensive efforts have been made by using nanocellulose as base material to develop synthetic cornea replacement. An ideal cornea substitute should be mechanically strong, transparent, biocompatible, permeable to oxygen, and supportive for epithelization [45]. Growth of human corneal stromal cell on nanocellulose supported its proliferation, making them suitable as scaffold for engineering of cornea tissue [181]. A composite of nanocellulose with PVA was reported to have suitable optical and mechanical properties for cornea replacement [45]. In addition to corneal transplantation, the potential of nanocellulose was also investigated for ophthalmic drug delivery systems and found to be a better candidate for sustainable drug release and ocular bandage [182].

12.3.7 Drug Delivery Systems

Targeted specific drug delivery approach is gaining much attention in pharmaceutical field. Various polymeric systems including nanocellulose-based systems have been extensively investigated as potential drug carrier vehicles [183]. The hydroxyl (OH) group on nanocellulose surface makes it fit to bind with the hydrophilic drug and targeting ligands.

Due to its non-toxic and biocompatible nature, nanocellulose-based drug oral, transdermal, and local delivery systems could be developed [184]. A study reported the development of CNF/alginate beads for the oral delivery of metformin hydrochloride, where CNF contributed to improved swelling behavior and mechanical strength to the beads. The developed system demonstrated sustained drug release for extended period [185]. In another study, a CNF/gelatinized corn starch/urea–formaldehyde nanocomposite was developed and loaded with dimethyl phthalate. In the nanocomposite, the network structure formed by the CNF and starch prevented the burst release of the drug [186]. In a more recent study, a CNC/PCL composite demonstrated a controlled release of tetracycline, where the concentration of CNC was inversely proportional to the rate of drug release indicating the direct interaction of cellulose crystals with the drug [187]. Similarly, different forms of nanocellulose could be used in the development of transdermal drug delivery system. For example, a study reported the development of CNF/chitosan nanocarrier film where CNFs supported the reinforcement of chitosan and contributed to the controlled transdermal delivery of ketorolac tromethamine [188]. In contrast to controlled release in the oral and transdermal drug delivery systems, a nanocellulose-based system for local drug delivery could also be developed for rapid drug release. In a study, Laurén et al. developed mucoadhesive films comprised of CNF and anionic CNF complemented with mucin, pectin, and chitosan as the functional adhesive activators. The developed films showed a rapid release of 60–80% metronidazole only in 80 min for the treatment of periodontitis [189].

Different physiological forms of nanocellulose such as microspheres, hydrogels, membrane, and sandwich structure are used as the drug carriers, gene delivery vehicles, and substrates for enzymes and protein immobilization [26, 190]. It is also used for in vitro release of antibiotics, fibroblast growth factors, epidermal growth factors, and vitamin B12 [191]. For the delivery of hydrophobic drug, nanocellulose is modified by treating with cationic surfactant like cetyl trimethyl ammonium bromide (CTAB). For passive targeted delivery, the nanocellulose is used in the form of nanoparticles and injected into the blood circulatory system which can avoid phagocytosis by the reticuloendothelial system. In contrast, for active targeting, the drug along with targeting ligand or antibody is conjugated and delivered to the targeted site using specific carrier [192]. Even in diverse pH range of digestive system, BNC has proved to be non-toxic, thus confirming its suitability for oral drug delivery [139]. The 3D ultrafine network of BNC ensures a sustainable and controlled release of drug [26, 193]. NC matrix prepared from regenerated cellulose by electrospinning techniques has also shown to be promising transdermal drug release system [194].

A study has also reported that the fridge drying of nanocellulose does not affect the drug release profile, indicating that nanocellulose-based hydrogels could be stored even in dry condition [146]. Owing to the porous nature of BNC, the impregnation of conductive polyaniline (PAni) resulted in the fabrication of pH-electroactive-controlled drug release system [26]. BNC-pec-CMC and BNC-pec-starch system was developed for oral prebiotics delivery. BNC provides protective surface and matrices for controlled release in harass condition of gastrointestinal tract [195]. A study also reported the adhesion of mucous cells on nanocellulose, where cell adhesion was greatly dependent on the type and concentration of nanocellulose and the site of scaffold implantation in the gastrointestinal tract (GIT) [196].

12.3.8 Biosensors

Biosensors are devices capable of detecting biological analyte such as a biomolecule, a biological structure, or a microorganism, when combined with the conductive materials serving as a transducer. The unique and tunable surface chemistry, high surface area and aspect ratio, flexibility, and excellent mechanical properties make nanocellulose a suitable material for the development of different types of biosensors. Nanocellulose in combination with other polymers like PAni and conductive nanomaterials such as carbon nanotubes (CNTs) and graphene confers conductive properties to it. Nanocomposites of BC-PAni/SWCNT and BC-polyethyleneimine (PEI)/graphene oxide membrane synthesized by oxidative polymerization showed improved conductive properties, and thus can be used as electrochemical biosensors [63, 197]. CNCs-g-GO with the treatment of isophorene diisocyanate (IPDI) has the ability to detect human finger interface [198]. Fluorescent labeling of nanocellulose could be applied in biomedical fields such as bioimaging, bio-probe development, and fluorescence bioassay [139]. In a study, an electrochemical biosensor comprised of CNC, polyaniline, and ionic liquid was developed on the modified screen printed electrode. By using the cholesterol oxidase as the biorecognition element, the developed sensor was reproducible and demonstrated operational repeatability, low limit of detection, and high sensitivity for monitoring cholesterol level [199]. In another study, Farooq et al. developed a BNC-based electrochemical biosensor by using the immobilized phages as the biorecognition elements for the selective detection of *Staphylococcus aureus*. Here, BNC was used as the matrix that offered a high surface area for the reinforcement of carbon nanotubes while its surface modification through polyethyleneimine facilitated the oriented phage immobilization to develop a BNC/CNTs-PEI-phage bio-interface. The developed biosensor showed high-density immobilization of phage particles, low limit of detection and stability over extended period, and high specificity for selective detection of *S. aureus* in a mixture of live and dead bacterial cells [200]. In a study, BNC membranes decorated with gold nanoparticles were used for the development of glucose sensor for measuring the blood glucose level [201].

12.3.9 3D/4D Printed Scaffolds and Hydrogels

The 3D/4D printed scaffolds provide structural support to the growth of cells and development of tissues, and thus could potentially replace the conventional organ transplantation. Nanocellulose is a potential candidate for the development of bioink for printing of scaffolds owing to its unique biological, mechanical, and rheological properties. Besides, its in vitro and in vivo biocompatibility and similar structure to natural ECM in terms of porosity and interconnected fibrous network enables it to support different cellular functions. In addition, nanocellulose supports crosslinking and thickening effect in aqueous dispersion, and thus forms hydrogels as the potential bioink formulations for tissue engineering applications [202]. Nanocellulose hydrogels prepared by using xyloglucan as a crosslinker showed adhesive properties are suitable for tissue engineering applications [203]. Nanocellulose aerogels of different swelling ratio and porosity were prepared by varying the charge density. This unique strategy of fabricating scaffolds also provides better cell proliferation matrices for 3D cell culture [204].

The 3D constructs of nanocellulose have shown promising ability for the development of different tissue substitutes such as fat, cartilage, tendons, and others [205]. In a study, 2 wt.% CNF with gelatin methacrylamide (GelMA) were printed to develop mechanically strong and elastic human nose [34]. In another study, the 2 wt. % CNF-based bioink prepared by blending with lignosulfonate showed excellent shear thinning performance, high yield stress, and viscoelastic behavior [206]. As the OH group on nanocellulose allows its reaction with a variety of functional groups, the incorporation of antibacterial components such as functionalized peptide ε-polylysine [207] could lead to the production of functional 3D hydrogels for wound-healing applications. In a recent study, a comparative study of preparing ink from CNFs, TEMPO-oxidized CNFs, and acetylated CNFs showed that the later produced dimensionally stable 3D scaffolds. The scaffolds remained stable upon drying and rewetting, and supported the adhesion, proliferation, and infiltration of cardiac myoblast cells up to 21 days without causing any toxicity, and thus could be used for tissue engineering applications [208]. Various 3D printed and crosslinked hydrogels of nanocellulose for various biomedical applications are summarized in Fig. 12.11.

12.3.10 Others

In addition to the above-mentioned specific applications, nanocellulose and its composites find several other applications, like skeletal muscle [209], cell encapsulation [210, 211], hygiene products [212], antimicrobials [124], and chemophototothermal synergistic cancer therapy [213]. For example, nanocellulose impregnated with Ag nanoparticles was used as air filtering and antimicrobial face mask where a composite of nanocellulose with PVA was used as an antivirus mask [193]. Similarly, nanocellulose crystals impregnated with 2D graphene showed desirable mechanical strength, flexibility, optical transparency in the visible range, and electrically conductive properties, for application in biofluid separation [214]. In another study, artificial tongue tissues were developed by using BNC nanopaper as the substrate [215]. In another study, Khan et al. developed a highly conducting and biocompatible biology-device interface comprised of BNC, gold nanoparticles, and poly(3,4-ethylenedioxythiophene) polystyrene sulfonate [216].

12.4 Conclusions and Prospects of Using Nanocellulose-Based Composites for Biomedical Applications

As biomaterials, nanocellulose and its derivatives have been extensively used in biomedical field due to its unique features including high crystallinity, mechanical strength, and ultrafine 3D fibrillar network. The unique features of various nanocellulose types allow them to be selected for various applications. For instance, the pure nature, unique structural morphology, physiological properties, and non-toxic nature make BNC an ideal candidate for biomedical applications. Although nanocellulose is non-toxic, its limited biocompatibility due to absence of active sites for cell adhesion as well as limited in vivo biodegradability and highly hydrophilic nature limit its direct use as biomaterial for biomedical applications. In addition, it lacks innate antimicrobial activity and possesses limited optical transparency, which further limit its application in wound dressing and tissue regeneration applications. These issues have been addressed to large extent through surface modification, such as by making its composites with biological active materials such as biopolymers, nanomaterials, and RGD peptides. Considering the importance of size of nanofibers rendering unique features and functionality, various strategies have been developed for controlling the size of nanofibers and nanocrystals. For example, rotary drum bioreactor is one technique used to produce nanocellulose fibers of high mechanical strength and uniform morphological structure [232]. A computational analysis of the surface data could be used to describe the longitudinal axis of cellulose microfibrils from which one can calculate the nanoscale curvature of single fibers.

Various nanocellulose-based functional materials have been successfully developed for applications in biomedical field such as engineering of hard and soft tissues. However, nanocellulose alone cannot fulfill the requirements for the adhesion and growth of all types of cells in the human body, and thus its modification with other reinforcement materials is one of the techniques used to synthesize biomaterials with advanced features and activities.

Although extensive research has been done about biomedical applications of nanocellulose over the past few decades, there still lack the commercialization of nanocellulose-based products in the market; nevertheless, considering the potential of nanocellulose, it is assumed that continuous research and effort will lead nanocellulose becoming a milestone material in the biomedical field. It is also anticipated that nanocellulose could emerge as a cost-effective commercial source of biomaterials in near future, and could potentially replace the commercial inorganic products available in market [134].

Acknowledgements This work was supported by the National Natural Science Foundation of China (21774039, 51973076); BRICS STI Framework Programme 3rd call 2019 (2018YFE0123700); China Postdoctoral Science Foundation (2016M602291); and the Fundamental Research Funds for Central Universities, Open Research Fund of State Key Laboratory of Polymer Physics and Chemistry, and Changchun Institute of Applied Chemistry, Chinese Academy of Sciences.

References

1. Y. Habibi, L.A. Lucia, O.J. Rojas, Cellulose Nanocrystals: Chemistry, Self-Assembly, and Applications. Chem. Rev. **110**, 3479–3500 (2010). https://doi.org/10.1021/cr900339w
2. M.S. Islam, L. Chen, J. Sisler, K.C. Tam, Cellulose nanocrystal (CNC)-inorganic hybrid systems: Synthesis, properties and applications. J. Mater. Chem. B. (2018). https://doi.org/10.1039/c7tb03016a
3. M.W. Ullah, M. Ul-Islam, F. Wahid, G. Yang, Editorial: Nanocellulose: A Multipurpose Advanced Functional Material, Volume II, Front. Bioeng. Biotechnol. 10 (2022). https://doi.org/10.3389/FBIOE.2022.931256.
4. L. Fu, J. Zhang, G. Yang, Present status and applications of bacterial cellulose-based materials for skin tissue repair. Carbohydr. Polym. **92**, 1432–1442 (2013). https://doi.org/10.1016/j.carbpol.2012.10.071
5. M. Ul-Islam, F. Subhan, S.U. Islam, S. Khan, N. Shah, S. Manan, M.W. Ullah, G. Yang, Development of three-dimensional bacterial cellulose/chitosan scaffolds: Analysis of cell-scaffold interaction for potential application in the diagnosis of ovarian cancer. Int. J. Biol. Macromol. **137**, 1050–1059 (2019). https://doi.org/10.1016/j.ijbiomac.2019.07.050
6. F.V. Ferreira, C.G. Otoni, K.J. De France, H.S. Barud, L.M.F. Lona, E.D. Cranston, O.J. Rojas, Porous nanocellulose gels and foams: Breakthrough status in the development of scaffolds for tissue engineering. Mater. Today. **37**, 126–141 (2020). https://doi.org/10.1016/j.mattod.2020.03.003
7. A. Fatima, S. Yasir, M. Ul-Islam, T. Kamal, M.W. Ahmad, Y. Abbas, S. Manan, M.W. Ullah, G. Yang, Ex situ development and characterization of green antibacterial bacterial cellulose-based composites for potential biomedical applications. Adv. Compos. Hybrid Mater. **5**, 307–321 (2022). https://doi.org/10.1007/S42114-021-00369-Z/FIGURES/7
8. S. Li, D. Huang, B. Zhang, X. Xu, M. Wang, G. Yang, Y. Shen, Flexible Supercapacitors Based on Bacterial Cellulose Paper Electrodes. Adv. Energy Mater. **4**, 1301655 (2014). https://doi.org/10.1002/aenm.201301655
9. M. Badshah, H. Ullah, A.R. Khan, S. Khan, J.K. Park, T. Khan, Surface modification and evaluation of bacterial cellulose for drug delivery. Int. J. Biol. Macromol. **113**, 526–533 (2018). https://doi.org/10.1016/j.ijbiomac.2018.02.135
10. M. Nasir, R. Hashim, O. Sulaiman, M. Asim, Nanocellulose, in: Cellul. Nanofibre Compos., Elsevier, 2017: pp. 261–276. https://doi.org/10.1016/B978-0-08-100957-4.00011-5.
11. W. Aljohani, M.W. Ullah, X. Zhang, G. Yang, Bioprinting and its applications in tissue engineering and regenerative medicine. Int. J. Biol. Macromol. **107**, 261–275 (2018). https://doi.org/10.1016/j.ijbiomac.2017.08.171
12. R.R. McCarthy, M.W. Ullah, E. Pei, G. Yang, Antimicrobial Inks: The Anti-Infective Applications of Bioprinted Bacterial Polysaccharides, Trends Biotechnol. 37 (2019) 1153–1155. https://doi.org/10.1016/j.tibtech.2019.05.004.

13. R.R. McCarthy, M.W. Ullah, P. Booth, E. Pei, G. Yang, The use of bacterial polysaccharides in bioprinting. Biotechnol. Adv. **37**, 107448 (2019). https://doi.org/10.1016/j.biotechadv.2019.107448
14. J. George, S. S N, Cellulose nanocrystals: synthesis, functional properties, and applications, Nanotechnol. Sci. Appl. (2015) 45. https://doi.org/10.2147/NSA.S64386.
15. R.J. Moon, A. Martini, J. Nairn, J. Simonsen, J. Youngblood, Cellulose nanomaterials review: Structure, properties and nanocomposites. Chem. Soc. Rev. **40**, 3941–3994 (2011). https://doi.org/10.1039/c0cs00108b
16. A. Shoukat, F. Wahid, T. Khan, M. Siddique, S. Nasreen, G. Yang, M.W. Ullah, R. Khan, Titanium oxide-bacterial cellulose bioadsorbent for the removal of lead ions from aqueous solution. Int. J. Biol. Macromol. **129**, 965–971 (2019). https://doi.org/10.1016/j.ijbiomac.2019.02.032
17. Z. Shi, Y. Zhang, G.O. Phillips, G. Yang, Utilization of bacterial cellulose in food. Food Hydrocoll. **35**, 539–545 (2014). https://doi.org/10.1016/j.foodhyd.2013.07.012
18. O.M. Atta, S. Manan, M. Ul-Islam, A.A.Q. Ahmed, M.W. Ullah, G. Yang, Development and characterization of plant oil-incorporated carboxymethyl cellulose/bacterial cellulose/glycerol-based antimicrobial edible films for food packaging applications. Adv. Compos. Hybrid Mater. **2022**, 1–18 (2022). https://doi.org/10.1007/S42114-021-00408-9
19. O.M. Atta, S. Manan, A. Shahzad, M. Ul-Islam, M.W. Ullah, G. Yang, Biobased materials for active food packaging: A review, Food Hydrocoll. (2021) 107419. https://doi.org/10.1016/J.FOODHYD.2021.107419.
20. O.M. Atta, S. Manan, A.A.Q. Ahmed, M.F. Awad, M. Ul-Islam, F. Subhan, M.W. Ullah, G. Yang, Development and Characterization of YeasAtta, O. M., Manan, S., Ahmed, A. A. Q., Awad, M. F., Ul-Islam, M., Subhan, F., Ullah, M. W., & Yang, G. (2021). Development and Characterization of Yeast-Incorporated Antimicrobial Cellulose Biofilms for Edible Foo, Polym. 2021, Vol. 13, Page 2310. 13 (2021) 2310. https://doi.org/10.3390/POLYM13142310.
21. S.M. Salah, Application of Nano-Cellulose in Textile, J. Text. Sci. Eng. 03 (2013). https://doi.org/10.4172/2165-8064.1000142.
22. M. Ul-Islam, S. Khan, M.W. Ullah, J.K. Park, Bacterial cellulose composites: Synthetic strategies and multiple applications in bio-medical and electro-conductive fields. Biotechnol. J. **10**, 1847–1861 (2015). https://doi.org/10.1002/biot.201500106
23. S.P. Lin, I. Loira Calvar, J.M. Catchmark, J.R. Liu, A. Demirci, K.C. Cheng, Biosynthesis, production and applications of bacterial cellulose, Cellulose. 20 (2013) 2191–2219. https://doi.org/10.1007/s10570-013-9994-3.
24. H. Martínez Ávila, E.M. Feldmann, M.M. Pleumeekers, L. Nimeskern, W. Kuo, W.C. de Jong, S. Schwarz, R. Müller, J. Hendriks, N. Rotter, G.J.V.M. van Osch, K.S. Stok, P. Gatenholm, Novel bilayer bacterial nanocellulose scaffold supports neocartilage formation invitro and invivo, Biomaterials. 44 (2015) 122–133. https://doi.org/10.1016/j.biomaterials.2014.12.025.
25. S. Khan, R. Siddique, D. Huanfei, M.A. Shereen, G. Nabi, Q. Bai, S. Manan, M. Xue, M.W. Ullah, H. Bowen, Perspective Applications and Associated Challenges of Using Nanocellulose in Treating Bone-Related Diseases. Front. Bioeng. Biotechnol. **9**, 350 (2021). https://doi.org/10.3389/fbioe.2021.616555
26. S. Li, A. Jasim, W. Zhao, L. Fu, M.W. Ullah, Z. Shi, G. Yang, Fabrication of pH-electroactive Bacterial Cellulose/Polyaniline Hydrogel for the Development of a Controlled Drug Release System, ES Mater. Manuf. (2018) 41–49. https://doi.org/10.30919/esmm5f120.
27. J. Supramaniam, R. Adnan, N.H. Mohd Kaus, R. Bushra, Magnetic nanocellulose alginate hydrogel beads as potential drug delivery system, Int. J. Biol. Macromol. 118 (2018) 640–648. https://doi.org/10.1016/j.ijbiomac.2018.06.043.
28. T.M. Sampath Udeni Gunathilake, Y.C. Ching, C.H. Chuah, H.A. Illias, K.Y. Ching, R. Singh, L. Nai-Shang, Influence of a nonionic surfactant on curcumin delivery of nanocellulose reinforced chitosan hydrogel, Int. J. Biol. Macromol. 118 (2018) 1055–1064. https://doi.org/10.1016/j.ijbiomac.2018.06.147.
29. M. Ul-Islam, S. Ul-Islam, S. Yasir, A. Fatima, M.W. Ahmed, Y.S. Lee, S. Manan, M.W. Ullah, Potential Applications of Bacterial Cellulose in Environmental and Pharmaceutical Sectors.

Curr. Pharm. Des. **26**, 5793–5806 (2020). https://doi.org/10.2174/13816128266620100816 5241
30. K. Markstedt, A. Mantas, I. Tournier, H. Martínez Ávila, D. Hägg, P. Gatenholm, 3D bioprinting human chondrocytes with nanocellulose-alginate bioink for cartilage tissue engineering applications, Biomacromolecules. 16 (2015) 1489–1496. https://doi.org/10.1021/acs.biomac.5b00188.
31. R. Poonguzhali, S.K. Basha, V.S. Kumari, International Journal of Biological Macromolecules Novel asymmetric chitosan / PVP / nanocellulose wound dressing : In vitro and in vivo evaluation. Int. J. Biol. Macromol. **112**, 1300–1309 (2018). https://doi.org/10.1016/j.ijbiomac.2018.02.073
32. L. Wang, L. Mao, F. Qi, X. Li, M. Wajid Ullah, M. Zhao, Z. Shi, G. Yang, Synergistic effect of highly aligned bacterial cellulose/gelatin membranes and electrical stimulation on directional cell migration for accelerated wound healing, Chem. Eng. J. 424 (2021) 130563. https://doi.org/10.1016/j.cej.2021.130563.
33. L. Mao, L. Wang, M. Zhang, M.W. Ullah, L. Liu, W. Zhao, Y. Li, A.A.Q. Ahmed, H. Cheng, Z. Shi, G. Yang, In Situ Synthesized Selenium Nanoparticles-Decorated Bacterial Cellulose/Gelatin Hydrogel with Enhanced Antibacterial, Antioxidant, and Anti-Inflammatory Capabilities for Facilitating Skin Wound Healing. Adv. Healthc. Mater. **10**, 2100402 (2021). https://doi.org/10.1002/adhm.202100402
34. B. Mbituyimana, L. Mao, S. Hu, M.W. Ullah, K. Chen, L. Fu, W. Zhao, Z. Shi, G. Yang, Bacterial cellulose/glycolic acid/glycerol composite membrane as a system to deliver glycolic acid for anti-aging treatment. J. Bioresour. Bioprod. (2021). https://doi.org/10.1016/j.jobab.2021.02.003
35. S. Shin, S. Park, M. Park, E. Jeong, K. Na, H.J. Youn, J. Hyun, Cellulose nanofibers for the enhancement of printability of low viscosity gelatin derivatives, BioResources. 12 (2017) 2941–2954. https://doi.org/10.15376/biores.12.2.2941-2954.
36. S. Khan, M. Ul-Islam, M.W. Ullah, Y. Zhu, K.B. Narayanan, S.S. Han, J.K. Park, Fabrication strategies and biomedical applications of three-dimensional bacterial cellulose-based scaffolds: A review. Int. J. Biol. Macromol. **209**, 9–30 (2022). https://doi.org/10.1016/J.IJBIOMAC.2022.03.191
37. W. Czaja, A. Krystynowicz, S. Bielecki, R.M. Brown, Microbial cellulose - The natural power to heal wounds. Biomaterials **27**, 145–151 (2006). https://doi.org/10.1016/j.biomaterials.2005.07.035
38. N. Shah, M. Ul-Islam, W.A. Khattak, J.K. Park, Overview of bacterial cellulose composites: A multipurpose advanced material. Carbohydr. Polym. **98**, 1585–1598 (2013). https://doi.org/10.1016/j.carbpol.2013.08.018
39. J.D.P. de Amorim, K.C. de Souza, C.R. Duarte, I. da Silva Duarte, F. de Assis Sales Ribeiro, G.S. Silva, P.M.A. de Farias, A. Stingl, A.F.S. Costa, G.M. Vinhas, L.A. Sarubbo, Plant and bacterial nanocellulose: production, properties and applications in medicine, food, cosmetics, electronics and engineering. A review, Environ. Chem. Lett. (2020). https://doi.org/10.1007/s10311-020-00989-9.
40. M. Ul-Islam, M.W. Ullah, S. Khan, N. Shah, J.K. Park, Strategies for cost-effective and enhanced production of bacterial cellulose. Int. J. Biol. Macromol. **102**, 1166–1173 (2017). https://doi.org/10.1016/j.ijbiomac.2017.04.110
41. M. Ul-Islam, M.W. Ullah, S. Khan, J.K. Park, Production of bacterial cellulose from alternative cheap and waste resources: A step for cost reduction with positive environmental aspects. Korean J. Chem. Eng. **37**, 925–937 (2020). https://doi.org/10.1007/s11814-020-0524-3
42. M. Ul-Islam, M.W. Ullah, T. Khan, J.K. Park, Bacterial cellulose: Trends in synthesis, characterization, and applications, in: G.O. Phillips, P.A. Williams (Eds.), Handb. Hydrocoll., 3rd ed., Elsevier, 2021: pp. 923–974. https://doi.org/10.1016/B978-0-12-820104-6.00010-3.
43. S. Khan, M. Ul-Islam, M. Ikram, M.W. Ullah, M. Israr, F. Subhan, Y. Kim, J.H. Jang, S. Yoon, J.K. Park, Three-dimensionally microporous and highly biocompatible bacterial cellulose–gelatin composite scaffolds for tissue engineering applications. RSC Adv. **6**, 110840–110849 (2016). https://doi.org/10.1039/C6RA18847H

44. H.S. Barud, S.J.L. Ribeiro, C.L.P. Carone, R. Ligabue, S. Einloft, P.V.S. Queiroz, A.P.B. Borges, V.D. Jahno, Optically transparent membrane based on bacterial cellulose/polycaprolactone. Polímeros. **23**, 135–142 (2013). https://doi.org/10.1590/S0104-14282013005000018
45. J. Wang, C. Gao, Y. Zhang, Y. Wan, Preparation and in vitro characterization of BC/PVA hydrogel composite for its potential use as artificial cornea biomaterial. Mater. Sci. Eng. C. **30**, 214–218 (2010). https://doi.org/10.1016/j.msec.2009.10.006
46. M.W. Ullah, M. Ul-Islam, S. Khan, Y. Kim, J.H. Jang, J.K. Park, In situ synthesis of a bio-cellulose/titanium dioxide nanocomposite by using a cell-free system. RSC Adv. **6**, 22424–22435 (2016). https://doi.org/10.1039/c5ra26704h
47. S. Khan, M. Ul-Islam, W.A. Khattak, M.W. Ullah, J.K. Park, Bacterial cellulose-titanium dioxide nanocomposites: nanostructural characteristics, antibacterial mechanism, and biocompatibility. Cellulose (2015). https://doi.org/10.1007/s10570-014-0528-4
48. A. Khalid, H. Ullah, M. Ul-Islam, R. Khan, S. Khan, F. Ahmad, T. Khan, F. Wahid, Bacterial cellulose-TiO 2 nanocomposites promote healing and tissue regeneration in burn mice model. RSC Adv. (2017). https://doi.org/10.1039/c7ra06699f
49. Z. Di, Z. Shi, M.W. Ullah, S. Li, G. Yang, A transparent wound dressing based on bacterial cellulose whisker and poly(2-hydroxyethyl methacrylate). Int. J. Biol. Macromol. **105**, 638–644 (2017). https://doi.org/10.1016/j.ijbiomac.2017.07.075
50. W. He, X. Huang, Y. Zheng, Y. Sun, Y. Xie, Y. Wang, L. Yue, In situ synthesis of bacterial cellulose/copper nanoparticles composite membranes with long-term antibacterial property. J. Biomater. Sci. Polym. Ed. (2018). https://doi.org/10.1080/09205063.2018.1528518
51. M. Ul-Islam, W.A. Khattak, M.W. Ullah, S. Khan, J.K. Park, Synthesis of regenerated bacterial cellulose-zinc oxide nanocomposite films for biomedical applications. Cellulose **21**, 433–447 (2014). https://doi.org/10.1007/s10570-013-0109-y
52. H. Bäckdahl, M. Esguerra, D. Delbro, B. Risberg, P. Gatenholm, Engineering microporosity in bacterial cellulose scaffolds. J. Tissue Eng. Regen. Med. **2**, 320–330 (2008). https://doi.org/10.1002/term.97
53. M. Rouabhia, J. Asselin, N. Tazi, Y. Messaddeq, D. Levinson, Z. Zhang, Production of Biocompatible and Antimicrobial Bacterial Cellulose Polymers Functionalized by RGDC Grafting Groups and Gentamicin. ACS Appl. Mater. Interfaces. **6**, 1439–1446 (2014). https://doi.org/10.1021/am4027983
54. N. Tazi, Z. Zhang, Y. Messaddeq, L. Almeida-Lopes, L.M. Zanardi, D. Levinson, M. Rouabhia, Hydroxyapatite bioactivated bacterial cellulose promotes osteoblast growth and the formation of bone nodules. AMB Express (2012). https://doi.org/10.1186/2191-0855-2-61
55. J. Wang, Y. Wan, J. Han, X. Lei, T. Yan, C. Gao, Nanocomposite prepared by immobilising gelatin and hydroxyapatite on bacterial cellulose nanofibres. Micro Nano Lett. **6**, 133 (2011). https://doi.org/10.1049/mnl.2010.0209
56. M. Ul-Islam, T. Khan, W.A. Khattak, J.K. Park, Bacterial cellulose-MMTs nanoreinforced composite films: Novel wound dressing material with antibacterial properties. Cellulose **20**, 589–596 (2013). https://doi.org/10.1007/s10570-012-9849-3
57. W. Sajjad, T. Khan, M. Ul-islam, R. Khan, Z. Hussain, A. Khalid, Development of modi fi ed montmorillonite-bacterial cellulose nanocomposites as a novel substitute for burn skin and tissue regeneration, Carbohydr. Polym. 206 (2019) 548–556. https://doi.org/10.1016/j.carbpol.2018.11.023.
58. M. Ul-Islam, F. Ahmad, A. Fatima, N. Shah, S. Yasir, M.W. Ahmad, S. Manan, M.W. Ullah, Ex situ Synthesis and Characterization of High Strength Multipurpose Bacterial Cellulose-Aloe vera Hydrogels, Front. Bioeng. Biotechnol. 9 (2021). https://doi.org/10.3389/fbioe.2021.601988.
59. A. Fatima, S. Yasir, M.S. Khan, S. Manan, M.W. Ullah, M. Ul-Islam, Plant extract-loaded bacterial cellulose composite membrane for potential biomedical applications. J. Bioresour. Bioprod. **6**, 26–32 (2021). https://doi.org/10.1016/j.jobab.2020.11.002

60. N. Lin, A. Dufresne, Nanocellulose in biomedicine: Current status and future prospect. Eur. Polym. J. **59**, 302–325 (2014). https://doi.org/10.1016/j.eurpolymj.2014.07.025
61. W. Aljohani, W. Li, M.W. Ullah, X. Zhang, G. Yang, Application of Sodium Alginate Hydrogel. IOSR J. Biotechnol. Biochem. **03**, 19–31 (2017). https://doi.org/10.9790/264X-03031931
62. W. Sajjad, F. He, M.W. Ullah, M. Ikram, S.M. Shah, R. Khan, T. Khan, A. Khalid, G. Yang, F. Wahid, Fabrication of Bacterial Cellulose-Curcumin Nanocomposite as a Novel Dressing for Partial Thickness Skin Burn. Front. Bioeng. Biotechnol. **8**, 553037 (2020). https://doi.org/10.3389/fbioe.2020.553037
63. A. Jasim, M.W. Ullah, Z. Shi, X. Lin, G. Yang, Fabrication of bacterial cellulose/polyaniline/single-walled carbon nanotubes membrane for potential application as biosensor. Carbohydr. Polym. **163**, 62–69 (2017). https://doi.org/10.1016/j.carbpol.2017.01.056
64. E. Niinivaara, E.D. Cranston, Bottom-up assembly of nanocellulose structures. Carbohydr. Polym. **247**, 116664 (2020). https://doi.org/10.1016/j.carbpol.2020.116664
65. Z. Shi, X. Gao, M.W. Ullah, S. Li, Q. Wang, G. Yang, Electroconductive natural polymer-based hydrogels, Biomaterials. 111 (2016). https://doi.org/10.1016/j.biomaterials.2016.09.020.
66. Z. Shi, X. Shi, M.W. Ullah, S. Li, V.V. Revin, G. Yang, Fabrication of nanocomposites and hybrid materials using microbial biotemplates. Adv. Compos. Hybrid Mater. (2017). https://doi.org/10.1007/s42114-017-0018-x
67. G. Wang, X.F. Chen, X.D. Shi, L.J. Yu, B.F. Liu, G. Yang, Bio-Fabrication of Patterned Cellulose Nano-Fibers, Multi-Functional. Mater. Struct. **47**, 1359–1362 (2008). https://doi.org/10.4028/www.scientific.net/AMR.47-50.1359
68. Y. Uraki, J. Nemoto, H. Otsuka, Y. Tamai, J. Sugiyama, T. Kishimoto, M. Ubukata, H. Yabu, M. Tanaka, M. Shimomura, Honeycomb-like architecture produced by living bacteria, Gluconacetobacter xylinus. Carbohydr. Polym. **69**, 1–6 (2007). https://doi.org/10.1016/j.carbpol.2006.08.021
69. L. Wang, S. Hu, M.W. Ullah, X. Li, Z. Shi, G. Yang, Enhanced cell proliferation by electrical stimulation based on electroactive regenerated bacterial cellulose hydrogels, Carbohydr. Polym. 249 (2020) 116829. https://doi.org/10.1016/j.carbpol.2020.116829.
70. T. Kondo, W. Kasai, M. Nojiri, Y. Hishikawa, E. Togawa, D. Romanovicz, R.M. Brown, Regulated patterns of bacterial movements based on their secreted cellulose nanofibers interacting interfacially with ordered chitin templates. J. Biosci. Bioeng. (2012). https://doi.org/10.1016/j.jbiosc.2012.02.020
71. A. Putra, A. Kakugo, H. Furukawa, J.P. Gong, Y. Osada, T. Uemura, M. Yamamoto, Production of Bacterial Cellulose with Well Oriented Fibril on PDMS Substrate. Polym. J. (2008). https://doi.org/10.1016/0029-554X(81)91167-8
72. C. Sharma, N.K. Bhardwaj, Bacterial nanocellulose: Present status, biomedical applications and future perspectives. Mater. Sci. Eng. C. (2019). https://doi.org/10.1016/j.msec.2019.109963
73. A. Pakzad, J. Simonsen, R.S. Yassar, Gradient of nanomechanical properties in the interphase of cellulose nanocrystal composites. Compos. Sci. Technol. (2012). https://doi.org/10.1016/j.compscitech.2011.11.020
74. M. Mariano, N. El Kissi, A. Dufresne, Cellulose nanocrystals and related nanocomposites: Review of some properties and challenges. J. Polym. Sci. Part B Polym. Phys. (2014). https://doi.org/10.1002/polb.23490
75. S.K. Arya, A. Singh, R. Naidoo, P. Wu, M.T. McDermott, S. Evoy, Chemically immobilized T4-bacteriophage for specific Escherichia coli detection using surface plasmon resonance. Analyst. **136**, 486–492 (2011). https://doi.org/10.1039/C0AN00697A
76. N. Grishkewich, N. Mohammed, J. Tang, K.C. Tam, Recent advances in the application of cellulose nanocrystals. Curr. Opin. Colloid Interface Sci. (2017). https://doi.org/10.1016/j.cocis.2017.01.005
77. M. Ul-Islam, M.W. Ullah, S. Khan, T. Kamal, S. Ul-Islam, N. Shah, J.K. Park, Recent advancement in cellulose based nanocomposite for addressing environmental challenges. Recent Pat. Nanotechnol. **10**, 169–180 (2016). https://doi.org/10.2174/1872210510666160429144916

78. C. Endes, S. Mueller, C. Kinnear, D. Vanhecke, E.J. Foster, A. Petri-Fink, C. Weder, M.J.D. Clift, B. Rothen-Rutishauser, Fate of Cellulose Nanocrystal Aerosols Deposited on the Lung Cell Surface In Vitro. Biomacromol (2015). https://doi.org/10.1021/acs.biomac.5b00055
79. S. Wang, J. Sun, Y. Jia, L. Yang, N. Wang, Y. Xianyu, W. Chen, X. Li, R. Cha, X. Jiang, Nanocrystalline Cellulose-Assisted Generation of Silver Nanoparticles for Nonenzymatic Glucose Detection and Antibacterial Agent. Biomacromol (2016). https://doi.org/10.1021/acs.biomac.6b00642
80. R.M.A. Domingues, M.E. Gomes, R.L. Reis, The potential of cellulose nanocrystals in tissue engineering strategies. Biomacromol (2014). https://doi.org/10.1021/bm500524s
81. V. Incani, C. Danumah, Y. Boluk, Nanocomposites of nanocrystalline cellulose for enzyme immobilization. Cellulose (2013). https://doi.org/10.1007/s10570-012-9805-2
82. J. Liu, R. Korpinen, K.S. Mikkonen, S. Willför, C. Xu, Nanofibrillated cellulose originated from birch sawdust after sequential extractions: A promising polymeric material from waste to films. Cellulose (2014). https://doi.org/10.1007/s10570-014-0321-4
83. W. Chen, Q. Li, Y. Wang, X. Yi, J. Zeng, H. Yu, Y. Liu, J. Li, Comparative study of aerogels obtained from differently prepared nanocellulose fibers. Chemsuschem (2014). https://doi.org/10.1002/cssc.201300950
84. A. Dufresne, Nanocellulose: A new ageless bionanomaterial. Mater. Today. **16**, 220–227 (2013). https://doi.org/10.1016/j.mattod.2013.06.004
85. H. Yousefi, M. Faezipour, T. Nishino, A. Shakeri, G. Ebrahimi, All-cellulose composite and nanocomposite made from partially dissolved micro-and nanofibers of canola straw. Polym. J. (2011). https://doi.org/10.1038/pj.2011.31
86. L. Wågberg, G. Decher, M. Norgren, T. Lindström, M. Ankerfors, K. Axnäs, The build-up of polyelectrolyte multilayers of microfibrillated cellulose and cationic polyelectrolytes. Langmuir (2008). https://doi.org/10.1021/la702481v
87. E. Afra, S. Mohammadnejad, A. Saraeyan, Cellulose nanofibils as coating material and its effects on paper properties. Prog. Org. Coatings. (2016). https://doi.org/10.1016/j.porgcoat.2016.09.018
88. A. Isogai, Wood nanocelluloses: Fundamentals and applications as new bio-based nanomaterials. J. Wood Sci. (2013). https://doi.org/10.1007/s10086-013-1365-z
89. A. Isogai, T. Saito, H. Fukuzumi, TEMPO-oxidized cellulose nanofibers. Nanoscale (2011). https://doi.org/10.1039/c0nr00583e
90. L. Dai, B. Wang, Z. Long, L. Chen, D. Zhang, S. Guo, Properties of hydroxypropyl guar/TEMPO-oxidized cellulose nanofibrils composite films. Cellulose (2015). https://doi.org/10.1007/s10570-015-0691-2
91. J. Song, C. Chen, S. Zhu, M. Zhu, J. Dai, U. Ray, Y. Li, Y. Kuang, Y. Li, N. Quispe, Y. Yao, A. Gong, U.H. Leiste, H.A. Bruck, J.Y. Zhu, A. Vellore, H. Li, M.L. Minus, Z. Jia, A. Martini, T. Li, L. Hu, Processing bulk natural wood into a high-performance structural material. Nature (2018). https://doi.org/10.1038/nature25476
92. O. Nechyporchuk, M.N. Belgacem, J. Bras, Production of cellulose nanofibrils: A review of recent advances. Ind. Crops Prod. (2016). https://doi.org/10.1016/j.indcrop.2016.02.016
93. I. Siró, D. Plackett, Microfibrillated cellulose and new nanocomposite materials: A review. Cellulose (2010). https://doi.org/10.1007/s10570-010-9405-y
94. R. Liu, L. Dai, C. Si, Z. Zeng, Antibacterial and hemostatic hydrogel via nanocomposite from cellulose nanofibers. Carbohydr. Polym. (2018). https://doi.org/10.1016/j.carbpol.2018.04.085
95. H. Sehaqui, N. Ezekiel Mushi, S. Morimune, M. Salajkova, T. Nishino, L.A. Berglund, Cellulose nanofiber orientation in nanopaper and nanocomposites by cold drawing, ACS Appl. Mater. Interfaces. (2012). https://doi.org/10.1021/am2016766.
96. W. Gindl-Altmutter, S. Veigel, M. Obersriebnig, C. Tippelreither, J. Keckes, High-modulus oriented cellulose nanopaper. ACS Symp. Ser. (2012). https://doi.org/10.1021/bk-2012-1107.ch001
97. W. Chen, H. Yu, S.Y. Lee, T. Wei, J. Li, Z. Fan, Nanocellulose: A promising nanomaterial for advanced electrochemical energy storage. Chem. Soc. Rev. (2018). https://doi.org/10.1039/c7cs00790f

98. C. Seo, H.W. Lee, A. Suresh, J.W. Yang, J.K. Jung, Y.C. Kim, Improvement of fermentative production of exopolysaccharides from Aureobasidium pullulans under various conditions. Korean J. Chem. Eng. **31**, 1433–1437 (2014). https://doi.org/10.1007/s11814-014-0064-9
99. H. Ahrem, D. Pretzel, M. Endres, D. Conrad, J. Courseau, H. Müller, R. Jaeger, C. Kaps, D.O. Klemm, R.W. Kinne, Laser-structured bacterial nanocellulose hydrogels support ingrowth and differentiation of chondrocytes and show potential as cartilage implants. Acta Biomater. **10**, 1341–1353 (2014). https://doi.org/10.1016/j.actbio.2013.12.004
100. K. Kubiak, M. Kurzawa, M. Jedrzejczak-Krzepkowska, K. Ludwicka, M. Krawczyk, A. Migdalski, M.M. Kacprzak, D. Loska, A. Krystynowicz, S. Bielecki, Complete genome sequence of Gluconacetobacter xylinus E25 strain-Valuable and effective producer of bacterial nanocellulose. J. Biotechnol. (2014). https://doi.org/10.1016/j.jbiotec.2014.02.006
101. M. Ul-Islam, S. Khan, M.W. Ullah, J.K. Park, Comparative study of plant and bacterial cellulose pellicles regenerated from dissolved states. Int. J. Biol. Macromol. **137**, 247–252 (2019). https://doi.org/10.1016/j.ijbiomac.2019.06.232
102. Y. Kim, M.W. Ullah, M. Ul-Islam, S. Khan, J.H. Jang, J.K. Park, Self-assembly of bio-cellulose nanofibrils through intermediate phase in a cell-free enzyme system. Biochem. Eng. J. **142**, 135–144 (2019). https://doi.org/10.1016/j.bej.2018.11.017
103. M.W. Ullah, M. Ul-Islam, S. Khan, Y. Kim, J.K. Park, Structural and physico-mechanical characterization of bio-cellulose produced by a cell-free system. Carbohydr. Polym. **136**, 908–916 (2016). https://doi.org/10.1016/j.carbpol.2015.10.010
104. S. Manan, M.W. Ullah, M. Ul-Islam, Z. Shi, M. Gauthier, G. Yang, Bacterial cellulose: Molecular regulation of biosynthesis, supramolecular assembly, and tailored structural and functional properties. Prog. Mater. Sci. **129**, 100972 (2022). https://doi.org/10.1016/J.PMATSCI.2022.100972
105. W.A. Khattak, T. Khan, M. Ul-Islam, M.W. Ullah, S. Khan, F. Wahid, J.K. Park, Production, characterization and biological features of bacterial cellulose from scum obtained during preparation of sugarcane jaggery (gur). J. Food Sci. Technol. **52**, 8343–8349 (2015). https://doi.org/10.1007/s13197-015-1936-7
106. Z. Cheng, R. Yang, X. Liu, X. Liu, H. Chen, Green synthesis of bacterial cellulose via acetic acid pre-hydrolysis liquor of agricultural corn stalk used as carbon source. Bioresour. Technol. **234**, 8–14 (2017). https://doi.org/10.1016/j.biortech.2017.02.131
107. W.A. Khattak, T. Khan, M. Ul-Islam, F. Wahid, J.K. Park, Production, Characterization and Physico-mechanical Properties of Bacterial Cellulose from Industrial Wastes. J. Polym. Environ. **23**, 45–53 (2015). https://doi.org/10.1007/s10924-014-0663-x
108. J. Zhou, Y. Chen, Y. Zhang, S. Sun, M.W. Ullah, W. Xu, Biotransformation of nylon-6,6 hydrolysate to bacterial cellulose. Green Chem. **23**, 7805–7815 (2021). https://doi.org/10.1039/D1GC02129J
109. V. Revin, E. Liyaskina, M. Nazarkina, A. Bogatyreva, M. Shchankin, Cost-effective production of bacterial cellulose using acidic food industry by-products, Brazilian. J. Microbiol. **49**, 151–159 (2018). https://doi.org/10.1016/j.bjm.2017.12.012
110. S. Bian, M. He, J. Sui, H. Cai, Y. Sun, J. Liang, Y. Fan, Colloids and Surfaces B : Biointerfaces The self-crosslinking smart hyaluronic acid hydrogels as injectable three-dimensional scaffolds for cells culture. Colloids Surfaces B Biointerfaces. **140**, 392–402 (2016). https://doi.org/10.1016/j.colsurfb.2016.01.008
111. I. Algar, S.C.M. Fernandes, G. Mondragon, C. Castro, C. Garcia-Astrain, N. Gabilondo, A. Retegi, A. Eceiza, Pineapple agroindustrial residues for the production of high value bacterial cellulose with different morphologies, J. Appl. Polym. Sci. 132 (2015). https://doi.org/10.1002/app.41237.
112. K.C. Cheng, J.M. Catchmark, A. Demirci, Effects of CMC addition on bacterial cellulose production in a biofilm reactor and its paper sheets analysis. Biomacromol **12**, 730–736 (2011). https://doi.org/10.1021/bm101363t
113. H. Lu, Q. Jia, L. Chen, L. Zhang, Effect of Organic Acids on Bacterial Cellulose Produced by Acetobacter xylinum, J. Microbiol. Biotechnol. 5 (2015) 30–37. https://www.rroij.com/open-access/effect-of-organic-acids-on-bacterial-cellulose-produced-byacetobacter-xylinum-.php?aid=72595.

114. S. Keshk, K. Sameshima, Influence of lignosulfonate on crystal structure and productivity of bacterial cellulose in a static culture. Enzyme Microb. Technol. **40**, 4–8 (2006). https://doi.org/10.1016/j.enzmictec.2006.07.037
115. J.H. Ha, N. Shah, M. Ul-Islam, T. Khan, J.K. Park, Bacterial cellulose production from a single sugar alpha-linked glucuronic acid-based oligosaccharide. Process Biochem. **46**, 1717–1723 (2011). https://doi.org/10.1016/j.procbio.2011.05.024
116. S. Bae, Y. Sugano, M. Shoda, Improvement of Bacterial Cellulose Production by Addition of Agar in a Jar Fermentor. J. Biosci. Bioeng. **97**, 33–38 (2004). https://doi.org/10.1016/S1389-1723(04)70162-0
117. Y. Chao, M. Mitarai, Y. Sugano, M. Shoda, Effect of addition of water-soluble polysaccharides on bacterial cellulose production in a 50-L airlift reactor. Biotechnol. Prog. **17**, 781–785 (2001). https://doi.org/10.1021/bp010046b
118. D. Shan, G. Cheng, D. Zhu, H. Xue, S. Cosnier, S. Ding, Direct electrochemistry of hemoglobin in poly(acrylonitrile-co-acrylic acid) and its catalysis to H2O2, Sensors Actuators. B Chem. (2009). https://doi.org/10.1016/j.snb.2008.11.029
119. Z. Lu, Y. Zhang, Y. Chi, N. Xu, W. Yao, B. Sun, Effects of alcohols on bacterial cellulose production by Acetobacter xylinum 186. World J. Microbiol. Biotechnol. **27**, 2281–2285 (2011). https://doi.org/10.1007/s11274-011-0692-8
120. T. Ishida, Y. Sugano, T. Nakai, M. Shoda, Effects of Acetan on Production of Bacterial Cellulose by *Acetobacter xylinum*. Biosci. Biotechnol. Biochem. (2002). https://doi.org/10.1271/bbb.66.1677
121. M. Matsuoka, T. Tsuchida, K. Matsushita, O. Adachi, F. Yoshinaga, A Synthetic Medium for Bacterial Cellulose Production by *Acetobacter xylinum* subsp. *sucrofermentans*, Biosci. Biotechnol. Biochem. (1996). https://doi.org/10.1271/bbb.60.575.
122. A.R.P. Figueiredo, A.G.P.R. Figueiredo, N.H.C.S. Silva, A. Barros-Timmons, A. Almeida, A.J.D. Silvestre, C.S.R. Freire, Antimicrobial bacterial cellulose nanocomposites prepared by in situ polymerization of 2-aminoethyl methacrylate. Carbohydr. Polym. **123**, 443–453 (2015). https://doi.org/10.1016/j.carbpol.2015.01.063
123. W.A. Khattak, M. Ul-Islam, M.W. Ullah, S. Khan, J.K. Park, Endogenous hydrolyzing enzymes: Isolation, characterization, and applications in biological processes, 2015. https://doi.org/10.1007/978-3-319-16298-0_55.
124. M.A. Taher, K.A. Zahan, M.A. Rajaie, L.C. Ring, S.A. Rashid, N.S. Mohd Nor Hamin, T.W. Nee, T.W. Yenn, Nanocellulose as drug delivery system for honey as antimicrobial wound dressing, Mater. Today Proc. 31 (2020) 14–17. https://doi.org/10.1016/j.matpr.2020.01.076.
125. M.W. Ullah, M. Ul-Islam, S. Khan, Y. Kim, J.K. Park, Innovative production of bio-cellulose using a cell-free system derived from a single cell line. Carbohydr. Polym. **132**, 286–294 (2015). https://doi.org/10.1016/j.carbpol.2015.06.037
126. M.W. Ullah, M. Ul Islam, S. Khan, N. Shah, J.K. Park, Recent advancements in bioreactions of cellular and cell-free systems: A study of bacterial cellulose as a model, Korean J. Chem. Eng. 34 (2017) 1591–1599. https://doi.org/10.1007/s11814-017-0121-2.
127. E. Buchner, Alkoholische Gärung ohne Hefezellen. Berichte Der Dtsch. Chem. Gesellschaft. **30**, 1110–1113 (1897). https://doi.org/10.1002/cber.189703001215
128. M.W. Ullah, W.A. Khattak, M. Ul-Islam, S. Khan, J.K. Park, Metabolic engineering of synthetic cell-free systems: Strategies and applications, Biochem. Eng. J. 105 (2016). https://doi.org/10.1016/j.bej.2015.10.023.
129. W.A. Khattak, M.W. Ullah, M. Ul-Islam, S. Khan, M. Kim, Y. Kim, J.K. Park, Developmental strategies and regulation of cell-free enzyme system for ethanol production: a molecular prospective. Appl. Microbiol. Biotechnol. **98**, 9561–9578 (2014). https://doi.org/10.1007/s00253-014-6154-0
130. B. Deepa, C.J. Chirayil, L.A. Pothan, S. Thomas, Chapter 4 - Lignocellulose-Based Nanoparticles and Nanocomposites: Preparation, Properties, and Applications, in: H. Ariffin, S.M. Sapuan, M.A. Hassan (Eds.), Lignocellul. Futur. Bioeconomy, Elsevier, 2019: pp. 41–69. https://doi.org/10.1016/B978-0-12-816354-2.00004-9.

131. G.P. Chen, T. Ushida, T. Tateishi, Scaffold design for tissue engineering. Macromol. Biosci. **2**, 67–77 (2002). https://doi.org/10.1002/1616-5195(20020201)2:2%3c67::aid-mabi67%3e3.3.co;2-6
132. W.K. Czaja, D.J. Young, M. Kawecki, R.M. Brown, The future prospects of microbial cellulose in biomedical applications. Biomacromol **8**, 1–12 (2007). https://doi.org/10.1021/bm060620d
133. N.H.C.S. Silva, P. Garrido-Pascual, C. Moreirinha, A. Almeida, T. Palomares, A. Alonso-Varona, C. Vilela, C.S.R. Freire, Multifunctional nanofibrous patches composed of nanocellulose and lysozyme nanofibers for cutaneous wound healing. Int. J. Biol. Macromol. **165**, 1198–1210 (2020). https://doi.org/10.1016/j.ijbiomac.2020.09.249
134. R. Curvello, V.S. Raghuwanshi, G. Garnier, Engineering nanocellulose hydrogels for biomedical applications, Adv. Colloid Interface Sci. 267 (2019) 47–61. https://doi.org/10.1016/j.cis.2019.03.002.
135. S. Zang, R. Zhang, H. Chen, Y. Lu, J. Zhou, X. Chang, G. Qiu, Z. Wu, G. Yang, Investigation on artificial blood vessels prepared from bacterial cellulose. Mater. Sci. Eng. C. **46**, 111–117 (2015). https://doi.org/10.1016/j.msec.2014.10.023
136. A. Subhedar, S. Bhadauria, S. Ahankari, H. Kargarzadeh, Nanocellulose in biomedical and biosensing applications: A review. Int. J. Biol. Macromol. **166**, 587–600 (2021). https://doi.org/10.1016/j.ijbiomac.2020.10.217
137. L.E. Millon, W.K. Wan, The polyvinyl alcohol-bacterial cellulose system as a new nanocomposite for biomedical applications, J. Biomed. Mater. Res. - Part B Appl. Biomater. (2006). https://doi.org/10.1002/jbm.b.30535.
138. H. Mohammadi, Nanocomposite biomaterial mimicking aortic heart valve leaflet mechanical behaviour, in: Proc. Inst. Mech. Eng. Part H J. Eng. Med., 2011: pp. 718–722. https://doi.org/10.1177/0954411911399826.
139. A. Karimian, H. Parsian, M. Majidinia, M. Rahimi, S.M. Mir, H. Samadi Kafil, V. Shafiei-Irannejad, M. Kheyrollah, H. Ostadi, B. Yousefi, Nanocrystalline cellulose: Preparation, physicochemical properties, and applications in drug delivery systems, Int. J. Biol. Macromol. 133 (2019) 850–859. https://doi.org/10.1016/j.ijbiomac.2019.04.117.
140. H.G. de Oliveira Barud, R.R. da Silva, H. da Silva Barud, A. Tercjak, J. Gutierrez, W.R. Lustri, O.B. de Oliveira, S.J.L. Ribeiro, A multipurpose natural and renewable polymer in medical applications: Bacterial cellulose, Carbohydr. Polym. 153 (2016) 406–420. https://doi.org/10.1016/j.carbpol.2016.07.059.
141. K. Ludwicka, M. Jedrzejczak-Krzepkowska, K. Kubiak, M. Kolodziejczyk, T. Pankiewicz, S. Bielecki, Chapter 9 - Medical and Cosmetic Applications of Bacterial NanoCellulose, in: M. Gama, F. Dourado, S. Bielecki (Eds.), Bact. Nanocellulose, Elsevier, Amsterdam, 2016: pp. 145–165. https://doi.org/10.1016/B978-0-444-63458-0.00009-3.
142. C. Chang, L. Zhang, J. Zhou, L. Zhang, J.F. Kennedy, Structure and properties of hydrogels prepared from cellulose in NaOH/urea aqueous solutions. Carbohydr. Polym. **82**, 122–127 (2010). https://doi.org/10.1016/j.carbpol.2010.04.033
143. S. Khan, M. Ul-Islam, M. Ikram, S.U. Islam, M.W. Ullah, M. Israr, J.H. Jang, S. Yoon, J.K. Park, Preparation and structural characterization of surface modified microporous bacterial cellulose scaffolds: A potential material for skin regeneration applications in vitro and in vivo. Int. J. Biol. Macromol. **117**, 1200–1210 (2018). https://doi.org/10.1016/j.ijbiomac.2018.06.044
144. M. Osorio, I. Ortiz, P. Gañán, T. Naranjo, R. Zuluaga, T.G. van Kooten, C. Castro, Novel surface modification of three-dimensional bacterial nanocellulose with cell-derived adhesion proteins for soft tissue engineering. Mater. Sci. Eng. C. **100**, 697–705 (2019). https://doi.org/10.1016/j.msec.2019.03.045
145. L. Fu, P. Zhou, S. Zhang, G. Yang, Evaluation of bacterial nanocellulose-based uniform wound dressing for large area skin transplantation. Mater. Sci. Eng. C. (2013). https://doi.org/10.1016/j.msec.2013.03.026
146. H. Du, W. Liu, M. Zhang, C. Si, X. Zhang, B. Li, Cellulose nanocrystals and cellulose nanofibrils based hydrogels for biomedical applications. Carbohydr. Polym. **209**, 130–144 (2019). https://doi.org/10.1016/j.carbpol.2019.01.020

147. S. Khan, M. Ul-Islam, M.W. Ullah, M. Ikram, F. Subhan, Y. Kim, J.H. Jang, S. Yoon, J.K. Park, Engineered regenerated bacterial cellulose scaffolds for application in vitro tissue regeneration. RSC Adv. **5**, 84565–84573 (2015). https://doi.org/10.1039/C5RA16985B
148. S. Khan, M. Ul-Islam, M.W. Ullah, M. Ikram, F. Subhan, Y. Kim, J.H. Jang, S. Yoon, J.K. Park, Engineered regenerated bacterial cellulose scaffolds for application in in vitro tissue regeneration. RSC Adv. **5**, 84565–84573 (2015). https://doi.org/10.1039/C5RA16985B
149. R. Ghafari, M. Jonoobi, L.M. Amirabad, K. Oksman, A.R. Taheri, Fabrication and characterization of novel bilayer scaffold from nanocellulose based aerogel for skin tissue engineering applications. Int. J. Biol. Macromol. **136**, 796–803 (2019). https://doi.org/10.1016/j.ijbiomac.2019.06.104
150. I. Ullah, W. Li, S. Lei, Y. Zhang, W. Zhang, U. Farooq, S. Ullah, M.W. Ullah, X. Zhang, Simultaneous co-substitution of Sr2+/Fe3+in hydroxyapatite nanoparticles for potential biomedical applications. Ceram. Int. **44**, 21338–21348 (2018). https://doi.org/10.1016/j.ceramint.2018.08.187
151. I. Ullah, A. Gloria, W. Zhang, M.W. Ullah, B. Wu, W. Li, M. Domingos, X. Zhang, Synthesis and Characterization of Sintered Sr/Fe-Modified Hydroxyapatite Bioceramics for Bone Tissue Engineering Applications, ACS Biomater. Sci. Eng. (2019) acsbiomaterials.9b01666. https://doi.org/10.1021/acsbiomaterials.9b01666.
152. I. Ullah, W. Zhang, L. Yang, M.W. Ullah, O.M. Atta, S. Khan, B. Wu, T. Wu, X. Zhang, Impact of structural features of Sr/Fe co-doped HAp on the osteoblast proliferation and osteogenic differentiation for its application as a bone substitute. Mater. Sci. Eng. C. **110**, 110633 (2020). https://doi.org/10.1016/j.msec.2020.110633
153. S. Khan, M.W. Ullah, R. Siddique, Y. Liu, I. Ullah, M. Xue, G. Yang, H. Hou, Catechins-modified selenium-doped hydroxyapatite nanomaterials for improved osteosarcoma therapy through generation of reactive oxygen species, Front. Oncol. 9 (2019). https://doi.org/10.3389/fonc.2019.00499.
154. A. Svensson, E. Nicklasson, T. Harrah, B. Panilaitis, D.L. Kaplan, M. Brittberg, P. Gatenholm, Bacterial cellulose as a potential scaffold for tissue engineering of cartilage. Biomaterials (2005). https://doi.org/10.1016/j.biomaterials.2004.02.049
155. L. Han, X. Lu, K. Liu, K. Wang, L. Fang, L.T. Weng, H. Zhang, Y. Tang, F. Ren, C. Zhao, G. Sun, R. Liang, Z. Li, Mussel-Inspired Adhesive and Tough Hydrogel Based on Nanoclay Confined Dopamine Polymerization. ACS Nano **11**, 2561–2574 (2017). https://doi.org/10.1021/acsnano.6b05318
156. C. Ao, Y. Niu, X. Zhang, X. He, W. Zhang, C. Lu, Fabrication and characterization of electrospun cellulose/nano-hydroxyapatite nanofibers for bone tissue engineering. Int. J. Biol. Macromol. **97**, 568–573 (2017). https://doi.org/10.1016/j.ijbiomac.2016.12.091
157. S. Zang, Z. Sun, K. Liu, G. Wang, R. Zhang, B. Liu, G. Yang, Ordered manufactured bacterial cellulose as biomaterial of tissue engineering. Mater. Lett. (2014). https://doi.org/10.1016/j.matlet.2014.04.183
158. T. Niamsap, N.T. Lam, P. Sukyai, Production of hydroxyapatite-bacterial nanocellulose scaffold with assist of cellulose nanocrystals, Carbohydr. Polym. 205 (2019) 159–166. https://doi.org/10.1016/j.carbpol.2018.10.034.
159. X. Zhang, C. Wang, M. Liao, L. Dai, Y. Tang, H. Zhang, P. Coates, F. Sefat, L. Zheng, J. Song, Z. Zheng, D. Zhao, M. Yang, W. Zhang, P. Ji, Aligned electrospun cellulose scaffolds coated with rhBMP-2 for both in vitro and in vivo bone tissue engineering, Carbohydr. Polym. 213 (2019) 27–38. https://doi.org/10.1016/j.carbpol.2019.02.038.
160. I.E. Carlström, A. Rashad, E. Campodoni, M. Sandri, K. Syverud, A.I. Bolstad, K. Mustafa, Cross-linked gelatin-nanocellulose scaffolds for bone tissue engineering. Mater. Lett. **264**, 127326 (2020). https://doi.org/10.1016/j.matlet.2020.127326
161. A. Al-Sabah, S.E.A. Burnell, I.N. Simoes, Z. Jessop, N. Badiei, E. Blain, I.S. Whitaker, Structural and mechanical characterization of crosslinked and sterilised nanocellulose-based hydrogels for cartilage tissue engineering. Carbohydr. Polym. **212**, 242–251 (2019). https://doi.org/10.1016/j.carbpol.2019.02.057

162. M.U. Islam, S. Khan, W.A. Khattak, J.K. Park, Synthesis, Chemistry, and Medical Application of Bacterial Cellulose Nanocomposite, in: Eco-Friendly Polym. Nanocomposites, 2015: pp. 399–337. https://doi.org/10.1007/978-81-322-2473-0.
163. N.D. Leipzig, M.S. Shoichet, The effect of substrate stiffness on adult neural stem cell behavior. Biomaterials **30**, 6867–6878 (2009). https://doi.org/10.1016/j.biomaterials.2009.09.002
164. G.T. Christopherson, H. Song, H.Q. Mao, The influence of fiber diameter of electrospun substrates on neural stem cell differentiation and proliferation. Biomaterials **30**, 556–564 (2009). https://doi.org/10.1016/j.biomaterials.2008.10.004
165. D. Kim, S. Park, I. Jo, S.M. Kim, D.H. Kang, S.P. Cho, J.B. Park, B.H. Hong, M.H. Yoon, Multiscale Modulation of Nanocrystalline Cellulose Hydrogel via Nanocarbon Hybridization for 3D Neuronal Bilayer Formation. Small **13**, 1–10 (2017). https://doi.org/10.1002/smll.201700331
166. V. Kuzmenko, E. Karabulut, E. Pernevik, P. Enoksson, P. Gatenholm, Tailor-made conductive inks from cellulose nanofibrils for 3D printing of neural guidelines. Carbohydr. Polym. (2018). https://doi.org/10.1016/j.carbpol.2018.01.097
167. T. Bedir, S. Ulag, C.B. Ustundag, O. Gunduz, 3D bioprinting applications in neural tissue engineering for spinal cord injury repair. Mater. Sci. Eng. C. **110**, 110741 (2020). https://doi.org/10.1016/j.msec.2020.110741
168. J. Yu, T.R. Huang, Z.H. Lim, R. Luo, R.R. Pasula, L. De Liao, S. Lim, C.H. Chen, Production of Hollow Bacterial Cellulose Microspheres Using Microfluidics to Form an Injectable Porous Scaffold for Wound Healing. Adv. Healthc. Mater. **5**, 2983–2992 (2016). https://doi.org/10.1002/adhm.201600898
169. Ullah MW, Manan S, Kiprono SJ, Ul-Islam M, Yang G, Synthesis, structural, and properties of bacterial cellulose, in: Nanocelluose From Fundam. to Adv. Mater., Wiley-VCH Verlag GmbH & Co. KGaA., 2019: pp. 81–113.
170. X. Zhang, H. Sun, S. Tan, J. Gao, Y. Fu, Z. Liu, Hydrothermal synthesis of Ag nanoparticles on the nanocellulose and their antibacterial study. Inorg. Chem. Commun. **100**, 44–50 (2019). https://doi.org/10.1016/j.inoche.2018.12.012
171. P. Zhang, L. Chen, Q. Zhang, F.F. Hong, Using in situ dynamic cultures to rapidly biofabricate fabric-reinforced composites of chitosan/bacterial nanocellulose for antibacterial wound dressings. Front. Microbiol. **7**, 1–15 (2016). https://doi.org/10.3389/fmicb.2016.00260
172. Y. Qiu, L. Qiu, J. Cui, Q. Wei, Bacterial cellulose and bacterial cellulose-vaccarin membranes for wound healing. Mater. Sci. Eng. C. **59**, 303–309 (2016). https://doi.org/10.1016/j.msec.2015.10.016
173. M. Ul-Islam, T. Khan, W.A. Khattak, J.K. Park, Bacterial cellulose-MMTs nanoreinforced composite films: Novel wound dressing material with antibacterial properties. Cellulose (2013). https://doi.org/10.1007/s10570-012-9849-3
174. D.A. Gopakumar, S. Thomas, Y. Grohens, Nanocelluloses as Innovative Polymers for Membrane Applications. Multifunct. Polym. Nanocomposites Based Cellul. Reinf. (2016). https://doi.org/10.1016/B978-0-323-44248-0.00008-0
175. A.S. Ribeiro, S.M. Costa, D.P. Ferreira, R.C. Calhelha, L. Barros, D. Stojković, M. Soković, I.C.F.R. Ferreira, R. Fangueiro, Chitosan/nanocellulose electrospun fibers with enhanced antibacterial and antifungal activity for wound dressing applications. React. Funct. Polym. **159**, 104808 (2021). https://doi.org/10.1016/j.reactfunctpolym.2020.104808
176. F. Sun, H.R. Nordli, B. Pukstad, E. Kristofer Gamstedt, G. Chinga-Carrasco, Mechanical characteristics of nanocellulose-PEG bionanocomposite wound dressings in wet conditions, J. Mech. Behav. Biomed. Mater. 69 (2017) 377–384. https://doi.org/10.1016/j.jmbbm.2017.01.049.
177. C. Vilela, H. Oliveira, A. Almeida, A.J.D. Silvestre, C.S.R. Freire, Nanocellulose-based antifungal nanocomposites against the polymorphic fungus Candida albicans, Carbohydr. Polym. 217 (2019) 207–216. https://doi.org/10.1016/j.carbpol.2019.04.046.

178. K. Ludwicka, M. Jedrzejczak-Krzepkowska, K. Kubiak, M. Kolodziejczyk, T. Pankiewicz, S. Bielecki, Medical and Cosmetic Applications of Bacterial NanoCellulose. Bact. Nanocellulose From Biotechnol. to Bio-Economy (2016). https://doi.org/10.1016/B978-0-444-63458-0.00009-3
179. A. Basu, J. Lindh, E. Ålander, M. Strømme, N. Ferraz, On the use of ion-crosslinked nanocellulose hydrogels for wound healing solutions: Physicochemical properties and application-oriented biocompatibility studies. Carbohydr. Polym. **174**, 299–308 (2017). https://doi.org/10.1016/j.carbpol.2017.06.073
180. H. Mertaniemi, C. Escobedo-Lucea, A. Sanz-Garcia, C. Gandía, A. Mäkitie, J. Partanen, O. Ikkala, M. Yliperttula, Human stem cell decorated nanocellulose threads for biomedical applications. Biomaterials **82**, 208–220 (2016). https://doi.org/10.1016/j.biomaterials.2015.12.020
181. G. Chinga-Carrasco, N.V. Ehman, D. Filgueira, J. Johansson, M.E. Vallejos, F.E. Felissia, J. Håkansson, M.C. Area, Bagasse—A major agro-industrial residue as potential resource for nanocellulose inks for 3D printing of wound dressing devices. Addit. Manuf. **28**, 267–274 (2019). https://doi.org/10.1016/j.addma.2019.05.014
182. H. Jia, Y. Jia, J. Wang, Y. Hu, Y. Zhang, S. Jia, Potentiality of bacterial cellulose as the scaffold of tissue engineering of cornea, in: Proc. 2009 2nd Int. Conf. Biomed. Eng. Informatics, BMEI 2009, 2009. https://doi.org/10.1109/BMEI.2009.5305657.
183. J.T. Orasugh, G. Sarkar, N.R. Saha, B. Das, A. Bhattacharyya, S. Das, R. Mishra, I. Roy, A. Chattoapadhyay, S.K. Ghosh, D. Chattopadhyay, Effect of cellulose nanocrystals on the performance of drug loaded in situ gelling thermo-responsive ophthalmic formulations. Int. J. Biol. Macromol. **124**, 235–245 (2019). https://doi.org/10.1016/j.ijbiomac.2018.11.217
184. G. Tiwari, R. Tiwari, B. Sriwastawa, L. Bhati, S. Pandey, P. Pandey, S.K. Bannerjee, Drug delivery systems: An updated review., Int. J. Pharm. Investig. (2012). https://doi.org/10.4103/2230-973X.96920.
185. N. Raghav, M.R. Sharma, J.F. Kennedy, Nanocellulose: A mini-review on types and use in drug delivery systems. Carbohydr. Polym. Technol. Appl. **2**, 100031 (2021). https://doi.org/10.1016/j.carpta.2020.100031
186. T. Guo, Y. Pei, K. Tang, X. He, J. Huang, F. Wang, Mechanical and drug release properties of alginate beads reinforced with cellulose. J. Appl. Polym. Sci. (2017). https://doi.org/10.1002/app.44495
187. M.D. Patil, V.D. Patil, A.A. Sapre, T.S. Ambone, A.T. Torris, P.G. Shukla, K. Shanmuganathan, Tuning Controlled Release Behavior of Starch Granules Using Nanofibrillated Cellulose Derived from Waste Sugarcane Bagasse. ACS Sustain. Chem. Eng. (2018). https://doi.org/10.1021/acssuschemeng.8b01545
188. A. Hivechi, S.H. Bahrami, R.A. Siegel, Drug release and biodegradability of electrospun cellulose nanocrystal reinforced polycaprolactone. Mater. Sci. Eng. C. (2019). https://doi.org/10.1016/j.msec.2018.10.037
189. G. Sarkar, J.T. Orasugh, N.R. Saha, I. Roy, A. Bhattacharyya, A.K. Chattopadhyay, D. Rana, D. Chattopadhyay, Cellulose nanofibrils/chitosan based transdermal drug delivery vehicle for controlled release of ketorolac tromethamine. New J. Chem. (2017). https://doi.org/10.1039/c7nj02539d
190. P. Laurén, H. Paukkonen, T. Lipiäinen, Y. Dong, T. Oksanen, H. Räikkönen, H. Ehlers, P. Laaksonen, M. Yliperttula, T. Laaksonen, Pectin and Mucin Enhance the Bioadhesion of Drug Loaded Nanofibrillated Cellulose Films. Pharm. Res. **35**, 145 (2018). https://doi.org/10.1007/s11095-018-2428-z
191. A. Sharma, M. Thakur, M. Bhattacharya, T. Mandal, S. Goswami, Commercial application of cellulose nano-composites – A review, Biotechnol. Reports. 21 (2019) e00316. https://doi.org/10.1016/j.btre.2019.e00316.
192. M. Ramos, A. Valdés, M.C. Garrigós, Chapter 6 - Multifunctional Applications of Nanocellulose-Based Nanocomposites, in: D. Puglia, E. Fortunati, J.M. Kenny (Eds.), Multifunct. Polym. Nanocomposites Based Cellul. Reinf., William Andrew Publishing, 2016: pp. 177–204. https://doi.org/10.1016/B978-0-323-44248-0.00006-7.

193. T.H. Tan, H.V. Lee, W.A.Y. Dabdawb, S.B.B.O.A.A. Hamid, Chapter 5 - A review of nanocellulose in the drug-delivery system, in: A.-M. Holban, A.M. Grumezescu (Eds.), Mater. Biomed. Eng., Elsevier, 2019: pp. 131–164. https://doi.org/10.1016/B978-0-12-816913-1.00005-2.
194. J.M. Rajwade, K.M. Paknikar, J.V. Kumbhar, Applications of bacterial cellulose and its composites in biomedicine. Appl. Microbiol. Biotechnol. (2015). https://doi.org/10.1007/s00253-015-6426-3
195. Y. Liu, A. Nguyen, A. Allen, J. Zoldan, Y. Huang, J.Y. Chen, Regenerated cellulose micro-nano fiber matrices for transdermal drug release. Mater. Sci. Eng. C. **74**, 485–492 (2017). https://doi.org/10.1016/j.msec.2016.12.048
196. A.C. Khorasani, S.A. Shojaosadati, Starch- and carboxymethylcellulose-coated bacterial nanocellulose-pectin bionanocomposite as novel protective prebiotic matrices. Food Hydrocoll. **63**, 273–285 (2017). https://doi.org/10.1016/j.foodhyd.2016.09.002
197. Y.J. Lin, J.A. Shatkin, F. Kong, Evaluating mucoadhesion properties of three types of nanocellulose in the gastrointestinal tract in vitro and ex vivo. Carbohydr. Polym. **210**, 157–166 (2019). https://doi.org/10.1016/j.carbpol.2019.01.029
198. X.-Y. Yang, C. Huang, H.-J. Guo, L. Xiong, J. Luo, B. Wang, X.-Q. Lin, X.-F. Chen, X.-D. Chen, Bacterial cellulose production from the litchi extract by Gluconacetobacter xylinus. Prep. Biochem. Biotechnol. **46**, 39–43 (2016). https://doi.org/10.1080/10826068.2014.958163
199. K.K. Sadasivuni, A. Kafy, L. Zhai, H.U. Ko, S. Mun, J. Kim, Transparent and flexible cellulose nanocrystal/reduced graphene oxide film for proximity sensing. Small **11**, 994–1002 (2015). https://doi.org/10.1002/smll.201402109
200. M.M. Abdi, R.L. Razalli, P.M. Tahir, N. Chaibakhsh, M. Hassani, M. Mir, Optimized fabrication of newly cholesterol biosensor based on nanocellulose. Int. J. Biol. Macromol. (2019). https://doi.org/10.1016/j.ijbiomac.2019.01.001
201. U. Farooq, M.W. Ullah, Q. Yang, A. Aziz, J. Xu, L. Zhou, S. Wang, High-density phage particles immobilization in surface-modified bacterial cellulose for ultra-sensitive and selective electrochemical detection of Staphylococcus aureus. Biosens. Bioelectron. **157**, 112163 (2020). https://doi.org/10.1016/j.bios.2020.112163
202. W. Wang, H.Y. Li, D.W. Zhang, J. Jiang, Y.R. Cui, S. Qiu, Y.L. Zhou, X.X. Zhang, Fabrication of Bienzymatic Glucose Biosensor Based on Novel Gold Nanoparticles-Bacteria Cellulose Nanofibers Nanocomposite. Electroanalysis (2010). https://doi.org/10.1002/elan.201000235
203. C. Yadav, A. Saini, W. Zhang, X. You, I. Chauhan, P. Mohanty, X. Li, Plant-based nanocellulose: A review of routine and recent preparation methods with current progress in its applications as rheology modifier and 3D bioprinting. Int. J. Biol. Macromol. **166**, 1586–1616 (2021). https://doi.org/10.1016/j.ijbiomac.2020.11.038
204. M.R. Bonilla, P. Lopez-Sanchez, M.J. Gidley, J.R. Stokes, Micromechanical model of biphasic biomaterials with internal adhesion: Application to nanocellulose hydrogel composites. Acta Biomater. **29**, 149–160 (2016). https://doi.org/10.1016/j.actbio.2015.10.032
205. J. Liu, F. Cheng, H. Grénman, S. Spoljaric, J. Seppälä, J.E. Eriksson, S. Willför, C. Xu, Development of nanocellulose scaffolds with tunable structures to support 3D cell culture. Carbohydr. Polym. **148**, 259–271 (2016). https://doi.org/10.1016/j.carbpol.2016.04.064
206. S. Sultan, G. Siqueira, T. Zimmermann, A.P. Mathew, 3D printing of nano-cellulosic biomaterials for medical applications. Curr. Opin. Biomed. Eng. **2**, 29–34 (2017). https://doi.org/10.1016/j.cobme.2017.06.002
207. Y. Shao, D. Chaussy, P. Grosseau, D. Beneventi, Use of Microfibrillated Cellulose/Lignosulfonate Blends as Carbon Precursors: Impact of Hydrogel Rheology on 3D Printing. Ind. Eng. Chem. Res. **54**, 10575–10582 (2015). https://doi.org/10.1021/acs.iecr.5b02763
208. O. Fourmann, M.K. Hausmann, A. Neels, M. Schubert, G. Nyström, T. Zimmermann, G. Siqueira, 3D printing of shape-morphing and antibacterial anisotropic nanocellulose hydrogels. Carbohydr. Polym. **259**, 117716 (2021). https://doi.org/10.1016/j.carbpol.2021.117716

209. R. Ajdary, S. Huan, N. Zanjanizadeh Ezazi, W. Xiang, R. Grande, H.A. Santos, O.J. Rojas, Acetylated Nanocellulose for Single-Component Bioinks and Cell Proliferation on 3D-Printed Scaffolds, Biomacromolecules. 20 (2019) 2770–2778. https://doi.org/10.1021/acs.biomac.9b00527.
210. J.M. Dugan, R.F. Collins, J.E. Gough, S.J. Eichhorn, Oriented surfaces of adsorbed cellulose nanowhiskers promote skeletal muscle myogenesis. Acta Biomater. **9**, 4707–4715 (2013). https://doi.org/10.1016/j.actbio.2012.08.050
211. M. Park, D. Lee, J. Hyun, Nanocellulose-alginate hydrogel for cell encapsulation. Carbohydr. Polym. **116**, 223–228 (2015). https://doi.org/10.1016/j.carbpol.2014.07.059
212. L. Ning, C. You, Y. Zhang, X. Li, F. Wang, Synthesis and biological evaluation of surface-modified nanocellulose hydrogel loaded with paclitaxel. Life Sci. **241**, 117137 (2020). https://doi.org/10.1016/j.lfs.2019.117137
213. J. Henschen, P.A. Larsson, J. Illergård, M. Ek, L. Wågberg, Bacterial adhesion to polyvinylamine-modified nanocellulose films. Colloids Surfaces B Biointerfaces. **151**, 224–231 (2017). https://doi.org/10.1016/j.colsurfb.2016.12.018
214. A. T. S., C.S. V., S. F., J.P. Thomas, Effect of dual stimuli responsive dextran/nanocellulose polyelectrolyte complexes for chemophotothermal synergistic cancer therapy, Int. J. Biol. Macromol. (2019). https://doi.org/10.1016/j.ijbiomac.2019.05.218.
215. R. Xiong, K. Hu, A.M. Grant, R. Ma, W. Xu, C. Lu, X. Zhang, V.V. Tsukruk, Ultrarobust Transparent Cellulose Nanocrystal-Graphene Membranes with High Electrical Conductivity. Adv. Mater. **28**, 1501–1509 (2016). https://doi.org/10.1002/adma.201504438
216. S. Abbasi-Moayed, H. Golmohammadi, M.R. Hormozi-Nezhad, A nanopaper-based artificial tongue: A ratiometric fluorescent sensor array on bacterial nanocellulose for chemical discrimination applications. Nanoscale (2018). https://doi.org/10.1039/c7nr05801b
217. S. Khan, M. Ul-Islam, M.W. Ullah, M. Israr, J.H. Jang, J.K. Park, Nano-gold assisted highly conducting and biocompatible bacterial cellulose-PEDOT:PSS films for biology-device interface applications. Int. J. Biol. Macromol. **107**, 865–873 (2018). https://doi.org/10.1016/j.ijbiomac.2017.09.064
218. Y. Jia, X. Wang, M. Huo, X. Zhai, F. Li, C. Zhong, Preparation and characterization of a novel bacterial cellulose/chitosan bio-hydrogel. Nanomater. Nanotechnol. **7**, 184798041770717 (2017). https://doi.org/10.1177/1847980417707172
219. F. Wahid, X.H. Hu, L.Q. Chu, S.R. Jia, Y.Y. Xie, C. Zhong, Development of bacterial cellulose/chitosan based semi-interpenetrating hydrogels with improved mechanical and antibacterial properties. Int. J. Biol. Macromol. (2019). https://doi.org/10.1016/j.ijbiomac.2018.10.105
220. P. Zhang, L. Chen, Q. Zhang, L.J. Jönsson, F.F. Hong, Using in situ nanocellulose-coating technology based on dynamic bacterial cultures for upgrading conventional biomedical materials and reinforcing nanocellulose hydrogels. Biotechnol. Prog. **32**, 1077–1084 (2016). https://doi.org/10.1002/btpr.2280
221. W. Song, Q. Zeng, X. Yin, L. Zhu, T. Gong, C. Pan, Preparation and anticoagulant properties of heparin-like electrospun membranes from carboxymethyl chitosan and bacterial cellulose sulfate. Int. J. Biol. Macromol. **120**, 1396–1405 (2018). https://doi.org/10.1016/j.ijbiomac.2018.09.133
222. F. Ostadhossein, N. Mahmoudi, G. Morales-Cid, E. Tamjid, F.J. Navas-Martos, B. Soriano-Cuadrado, J.M.L. Paniza, A. Simchi, Development of chitosan/bacterial cellulose composite films containing nanodiamonds as a potential flexible platform for wound dressing. Materials (Basel). (2015). https://doi.org/10.3390/ma8095309
223. D.O.S. Recouvreux, C.R. Rambo, F.V. Berti, C.A. Carminatti, R.V. Antônio, L.M. Porto, Novel three-dimensional cocoon-like hydrogels for soft tissue regeneration. Mater. Sci. Eng. C. **31**, 151–157 (2011). https://doi.org/10.1016/j.msec.2010.08.004
224. L. Huang, X. Chen, T.X. Nguyen, H. Tang, L. Zhang, G. Yang, Nano-cellulose 3D-networks as controlled-release drug carriers. J. Mater. Chem. B. (2013). https://doi.org/10.1039/c3tb20149j

225. J.D. Fontana, A.M. De Souza, C.K. Fontana, I.L. Torriani, J.C. Moreschi, B.J. Gallotti, S.J. De Souza, G.P. Narcisco, J.A. Bichara, L.F.X. Farah, Acetobacter cellulose pellicle as a temporary skin substitute. Appl. Biochem. Biotechnol. **24–25**, 253–264 (1990). https://doi.org/10.1007/BF02920250
226. L. Fu, Y. Zhang, C. Li, Z. Wu, Q. Zhuo, X. Huang, G. Qiu, P. Zhou, G. Yang, Skin tissue repair materials from bacterial cellulose by a multilayer fermentation method. J. Mater. Chem. **22**, 12349–12357 (2012). https://doi.org/10.1039/c2jm00134a
227. M. Patchan, J.L. Graham, Z. Xia, J.P. Maranchi, R. Mccally, O. Schein, J.H. Elisseeff, M.M. Trexler, Synthesis and properties of regenerated cellulose-based hydrogels with high strength and transparency for potential use as an ocular bandage. Mater. Sci. Eng. C. **33**, 3069–3076 (2013). https://doi.org/10.1016/j.msec.2013.03.037
228. H.-W. Kang, S.J. Lee, I.K. Ko, C. Kengla, J.J. Yoo, A. Atala, A 3D bioprinting system to produce human-scale tissue constructs with structural integrity. Nat. Biotechnol. **34**, 312–319 (2016). https://doi.org/10.1038/nbt.3413
229. N. Janpetch, N. Saito, R. Rujiravanit, Fabrication of bacterial cellulose-ZnO composite via solution plasma process for antibacterial applications. Carbohydr. Polym. **148**, 335–344 (2016). https://doi.org/10.1016/j.carbpol.2016.04.066
230. H. Si, H. Luo, G. Xiong, Z. Yang, S.R. Raman, R. Guo, Y. Wan, One-step in situ biosynthesis of graphene oxide-bacterial cellulose nanocomposite hydrogels. Macromol. Rapid Commun. **35**, 1706–1711 (2014). https://doi.org/10.1002/marc.201400239
231. F. Luzi, D. Puglia, L. Torre, 10 - Natural fiber biodegradable composites and nanocomposites: a biomedical application, in: D. Verma, E. Fortunati, S. Jain, X. Zhang (Eds.), Biomass, Biopolym. Mater. Bioenergy, Woodhead Publishing, pp. 179–201 (2019). https://doi.org/10.1016/B978-0-08-102426-3.00010-2.
232. G. Chen, L. Chen, W. Wang, F.F. Hong, M. Zhu, Manufacture of a novel anisotropic bacterial nanocellulose hydrogel membrane by using a rotary drum bioreactor. Carbohydr. Polym. **211**, 281–288 (2019). https://doi.org/10.1016/j.carbpol.2019.01.072

Index

L. Hu et al. (eds.), *Emerging Nanotechnologies in Nanocellulose*, NanoScience and Technology, https://doi.org/10.1007/978-3-031-14043-3

G

H

I

K

L